The Evolution of the Sensitive Soul

The Evolution of the Sensitive Soul

Learning and the Origins of Consciousness

Simona Ginsburg and Eva Jablonka

with illustrations by Anna Zeligowski

The MIT Press
Cambridge, Massachusetts
London, England

This book was set in Stone by Westchester Publishing Services. Printed and bound in the United States of America.

Library of Congress Cataloging-in-Publication Data.

Names: Ginsburg, Simona, author. | Jablonka, Eva, author.
Title: The evolution of the sensitive soul : learning and the origins of
   consciousness / Simona Ginsburg and Eva Jablonka.
Description: Cambridge, MA : MIT Press, 2019. | Includes bibliographical
   references and index.
Identifiers: LCCN 2018019048 | ISBN 9780262039307 (hardcover : alk. paper)
Subjects: LCSH: Consciousness. | Consciousness--Physiological aspects. |
   Evolution (Biology)
Classification: LCC BF311 .G5325 2019 | DDC 153--dc23
LC record available at https://lccn.loc.gov/2018019048

10 9 8 7 6 5 4 3 2 1

To the next generations—
Dafna and Aviad
Iddo, Eliana, and Amalia
Marco and Giulia

Contents

Preface

There is nothing more intimate and cherished, and nothing more elusive, than subjective experiencing. In this book we use an evolutionary approach to explore the biological basis of such experiencing or, as it is usually called, "consciousness." We argue that consciousness emerged in the context of the evolution of learning, and we maintain that by figuring out how the evolutionary transition to subjectively experienced living occurred, we can gain an insight into the nature of this mode of being. We are therefore interested in consciousness-as-we-know-it—in animal consciousness, the only type of consciousness that we know exists, rather than in hypothetical machine consciousness. This is reflected in the title of our book: the "sensitive soul" is the apt term used by Aristotle to describe the ability of animals to subjectively experience percepts and feelings.

We build our evolutionary account on what we have learned from studies of neurobiology, cognitive science, animal learning, philosophy of mind, and evolutionary biology. Some of our interpretations and conclusions are likely to be proven wrong, but we think that the framework we use—embedding subjective experiencing in the biological, evolutionary processes of neural animals rather than in medium-free, multirealizable computations—will remain useful.

Our approach focuses on the *evolutionary transition to minimal consciousness* and is based on both the teleological framework developed by Aristotle twenty-four hundred years ago and on our current understanding of biological evolution, the twenty-first-century developmentally informed theory that has its origins in the nineteenth-century work of Jean-Baptiste Lamarck, Charles Darwin, and their followers. It is a theory that is constantly updated as new developments occur in molecular and developmental biology, paleontology, ecology, and other biological disciplines. Evolutionary theory is, for us, the most general framework for understanding the biological world. It is a conceptual bottleneck through which any theory of life and mind must pass. If a biological (or psychological, or sociological) theory fails to pass through this bottleneck, it is likely there is something seriously wrong with it.

Because evolution is so central to biological investigations, it is natural to assume that it has been incorporated into the framework of consciousness studies, both as a yardstick for measuring the validity of new theories and as a source of insights. But in fact, until very recently there has been a strange lacuna in the field. Although most scientists and philosophers who write about consciousness are now convinced that it is a biological process that is a product of evolution, its evolutionary origins are rarely central to their discussions. Indeed, it was not until the first decade of the twenty-first century, after more than one hundred years of academic silence about the evolution of consciousness, that serious attempts to understand it began to emerge again. It seems that the skepticism about the possibility of conducting a scientific study of consciousness that prevailed until the 1990s, coupled with conceptual difficulties, were important reasons for this neglect.

For an evolutionary account of the origins of experiencing, biologists must agree on what the first type of subjectively experiencing animal was like. This means that one has to try and characterize minimal consciousness or, alternatively, find some good criteria (or markers) that indicate that it is in place. This is a difficult task because it is not clear how to identify consciousness in animals very different from us. Nevertheless, such difficulties have not stopped evolutionary biologists from trying to solve comparable problems. Inquiries into the origin of life presented similar conceptual and theoretical difficulties, but scientists did not shy away from the problem, and an evolutionary approach proved to be extremely fruitful. Because questions about the origin of life, the origin of minimal consciousness, and the origin of human abstract values are all concerned with the emergence of new types of goal-directed systems and confront similar conceptual and methodological challenges, we modeled our approach to the evolution of minimal consciousness on the best-developed research program of the three, research on the origin of life.

The benefits of the evolutionary, origin-focused approach are obvious: if we can locate when and how in evolutionary history the transition from an organism that lacked consciousness to one with minimal consciousness occurred, it becomes possible to explore the processes and organizational principles involved without being misled by later derived dissociations and integrations that mask the fundamental properties of subjective experiencing. We approach this task by characterizing the essential features and dynamics of consciousness and singling out a diagnostic, tractable, biological capacity that was necessary for its presence—the evolutionary transition marker of consciousness. Studying the evolutionary transition to conscious animals by tracing the evolutionary history of this diagnostic capacity allows us to identify and reconstruct (i.e., reverse engineer) the type of system of which it is part. Our focus is therefore on *minimal animal consciousness*—human reflective consciousness is not the subject of this book, although we believe (and argue) that our approach has important implications for its study.

As we have discovered, trying to understand minimal consciousness is a mammoth undertaking. When we started our work, we were blissfully ignorant of its real dimensions. This was a good thing, for had we been aware of the magnitude of the task, we would probably not have proceeded. As we became immersed in the project, we realized that we (and our prospective readers) needed a lot of background knowledge in order to have a foundation on which to build an evolutionary account. The result is this rather fat book. The size of the book led us to divide it into two parts, with part I (chapters 1–5) providing the historical, biological, and conceptual foundations on which we build, and part II (chapters 6–10) developing the evolutionary arguments. We know that long science books are not fashionable, and we are aware that some people may not be interested in all of the background chapters. For example, people who are wary of history can skip chapter 2, those who are allergic to philosophy can skim over chapter 4, and nonbiologists may hum through the neurobiological and biochemical details. We do hope, however, that we have managed to convey some of the excitement and humility that we felt as we researched and wrote our book—and that Anna Zeligowski's illustrations will engage our readers' aesthetic sensibilities, deepening the sense of inquiry and wonder that the subject elicits.

Acknowledgments

This book took a decade to write. It could not have been written without the help of colleagues, students, and friends. Their kushiot (the Hebrew and Aramaic word combining hard question and challenge) helped clarify our ideas and develop the themes we conceived.

Our home institutions, the Cohn Institute at Tel Aviv University and the Open University of Israel, supported and facilitated our research. Two courses taught at Tel Aviv University made us realize which of our ideas had not been clear to ourselves. We thank our students who kept us on our toes and forced us to sharpen our thoughts. We were enlightened by the comments and queries raised after many lectures delivered at various conferences and workshops in Israel and abroad. In particular, we are indebted to colleagues who articulated some of our suggestions in ways that were superior to or more subtle than our own formulations. Special thanks are due to Patricia Smith Churchland and Denis Noble for their encouragement and kind words at critical stages of our work.

Many people read and commented on chapters of the manuscript at different stages, and we have incorporated some of their suggestions and have hopefully corrected most of the flaws they found. We are especially indebted to the following colleagues for their detailed input: Yemima Ben-Menahem (chapter 4); Mazviita Chirimuuta (chapters 1, 2, and 4); Pietro Corsi (chapter 2); Leon Deouell (chapter 3); Daniel Dor (chapter 10); Ran Feldesh (chapters 2, 7, and 8); Snait Gissis (chapter 2); Amir Horowitz (chapter 4); Fred Keijzer (chapter 6); Rafael Malach (chapter 8); Jessica Riskin (chapters 1 and 2); Nicholas Shea (chapters 7 and 8); Vincent Torley (chapters 3 and 5); Mary Jane West-Eberhard (chapters 1 and 9); Wayne Wu (chapters 1 and 4).We are especially grateful to Margaret Yekutieli, who read all ten chapters of the book and made many valuable suggestions, improving both content and style. John Donohue and Wendy Lawrence of Westchester Publishing Services did an excellent job editing the manuscript and detecting errors, omissions, and other faults, and we thank them for setting them right. We are also greatly indebted to Alessandro Altamura, who helped us with the graphic aspects of the figures, for his skill, aesthetic sense, and infinite patience.

Marion Lamb and Zohar Bronfman made the greatest contributions to our work, and we thank them from the bottom of our hearts. If any parts of this book are elegantly written and well argued, Marion is responsible. She not only read the manuscript several times, checked every word we wrote (including each endnote, figure legend, and reference), and corrected many of them, but she also made us rethink, clarify, and sophisticate the structure of some of our arguments; abandon some of them; and present the material in a way that is accessible to nonspecialists. Zohar Bronfman, who was our coauthor on several papers written while we were composing this text, helped formulate the notion of unlimited, open-ended learning; introduced us to several cognitive-science concepts; and was a driving force behind the UAL toy model presented in detail in chapter 8. This brilliant young man (who earned two PhD degrees in parallel, one in the cognitive sciences and one in the philosophy of biology under E. J.'s guidance), is also an unusually modest, cheerful, and generous person, and it was a great pleasure to work with him. And of course, we wish to thank Anna Zeligowski, our partner on the book, for her artwork. We think that her pictures speak for themselves, providing a necessary dimension of experiencing to the understanding of the sensitive soul.

Many of the discussions that led to this book were carried out while strolling among the stalls of Mahane Yehuda in Jerusalem and eating at Azura; we wish to thank the cooks and waiters of that market restaurant for nourishing us. Finally, we are grateful for another down-to-earth influence, provided by our families, who kept us from forcing them to eat Aristotle for breakfast and Dennett for dinner.

Figure Credits

1.2. Yeshayahu Leibowitz. Portrait by Bracha L. Ettinger ©.   14

1.3. The autopoietic system. Figure 1 (Views on agency), p. 15 in E. Di Paolo, "Extended Life," *Topoi* 28 (2009): 9–21. Reproduced by permission of Springer.   21

2.1. Jean-Baptiste Pierre Antoine de Monet, Chevalier de Lamarck. Reproduced by kind permission of Wellcome Library, London.   46

2.4. Charles Darwin and his son. Reproduced by kind permission of Cambridge University Library. Classmark: Dar 225:129.   64

3.6a. Neural selectionist schemes. Loosely based on figure 65, p. 303 in J.-P. Changeux, *L'Homme neuronal* (Paris: Fayard, 1983).   121

3.6b. Neural selectionist schemes. Figure 5, p. 40 in G. M. Edelman, *Wider than the Sky: The Phenomenal Gift of Consciousness* (New Haven, CT: Yale University Press, 2004). Reproduced by permission of Yale University Press.   121

3.7. Reentrant pathways leading to primary consciousness. Figure 1, p. 5522 in G. M. Edelman, "Naturalizing Consciousness: A Theoretical Framework," *Proceedings of the National Academy of Sciences USA* 100 (2003): 5520–5524. Reproduced by permission of The National Academy of Sciences, U.S.A.   125

3.11. The global neural workspace (GNW) model. Figure 1, p. 14530 in S. Dehaene, M. Kerszberg, and J.-P. Changeux, "A Neuronal Model of a Global Workspace in Effortful Cognitive Tasks," *Proceedings of the National Academy of Sciences USA* 95(24) (1998): 14529–14534. Adapted by permission of The National Academy of Sciences, U.S.A.   131

4.5. The "reality space" model. Figure 5, p. 73 in B. Merker, "Consciousness without a Cerebral Cortex: A Challenge for Neuroscience and Medicine," *Behavioral and Brain Sciences* 30(1) (2007): 63–81. Adapted by permission of Cambridge University Press.   178

5.5. Lateral views of the brains of some extant vertebrates. Figure 1, p. 744 in R. G. Northcutt, "Understanding Vertebrate Brain Evolution," *Integrative and Comparative Biology* 42 (2002): 743–756. Reproduced by permission of Oxford University Press on behalf of The Society for Integrative and Comparative Biology.   210

5.9. Panskepp's evolutionary view of the emotional organization of the brain. Figure 3, p. 6 in J. Panksepp, "Cross-Species Affective Neuroscience: Decoding of the Primal Affective Experiences of Humans and Related Animals," *PLoS One* 6(9) (2011): e21236. Reproduced by permission (unrestricted open access).   217

II.1. An imaginary exploration-stabilization system. Figure 11.8, p. 451 in E. Jablonka and M. J. Lamb, *Evolution in Four Dimensions*, 2nd ed. (Cambridge, MA: MIT Press, 2014). Reproduced by permission of The MIT Press.   246

6.4b. Two scenarios of evolving neurons. Loosely based on Figure 2a, p. 918 in G. Jékely, "Origin and Early Evolution of Neural Circuits for the Control of Ciliary Locomotion," *Proceedings of the Royal Society B* 278 (2011): 914–922.   264

7.1. The flight simulator constructed to study conditioning. Adapted (from photographs) by permission of B. Brembs.   299

7.2. Instrumental conditioning of pigeons. Figure 4.1, p. 47 in S. F. Walker, *Learning and Reinforcement* (Essential Psychology Series) (London: Methuen, 1976). Adapted by permission of Routledge.   305

7.5 a–c. Sensitization in *Aplysia*. Figure 1, p. 2 in M. Mayford, S. A. Siegelbaum, and E. R. Kandel, "Synapses and Memory Storage," *Cold Spring Harbor Perspectives in Biology* 4 (2012): a005751. Modified by permission of Cold Spring Harbor Laboratory Press.   310

7.5d. Classical conditioning in *Aplysia*. Figure 2, p. 4 in R. D. Hawkins and J. H. Byrne, "Associative Learning in Invertebrates," *Cold Spring Harbor Perspectives in Biology* 7(5) (2015): a021709. Modified by permission of Cold Spring Harbor Laboratory Press.   310

7.7. Cell memory and cell heredity through self-sustaining loops. Figure B.3, p. 418 in S. B. Gissis and E. Jablonka, eds., *Transformations of Lamarckism: From Subtle Fluids to Molecular Biology* (Cambridge, MA: MIT Press, 2011). Reproduced by permission of The MIT Press.   317

7.8. Cell memory and cell heredity through three-dimensional templating. Figure B.4, p. 420 in S. B. Gissis and E. Jablonka, eds.,

Transformations of Lamarckism: From Subtle Fluids to Molecular Biology (Cambridge, MA: MIT Press, 2011). Reproduced by permission of The MIT Press. 318

7.9. Cell memory and cell heredity through the maintenance of DNA methylation patterns. Figure B.1, p. 415 in S. B. Gissis and E. Jablonka, eds., *Transformations of Lamarckism: From Subtle Fluids to Molecular Biology* (Cambridge, MA: MIT Press, 2011). Reproduced by permission of The MIT Press. 319

7.10. Three mechanisms of cell memory and cell heredity mediated by small regulatory RNAs. Figure B.2, p. 417 in S. B. Gissis and E. Jablonka, eds., *Transformations of Lamarckism: From Subtle Fluids to Molecular Biology* (Cambridge, MA: MIT Press, 2011). Reproduced by permission of The MIT Press. 320

7.13. Phylogenetic relationships of major animal phyla. Figure 1, p. R877 in M. J. Telford, G. E. Budd, and H. Philippe, "Phylogenomic Insights into Animal Evolution," *Current Biology* 25(19) (2015): R876–R887. Reproduced by permission of Elsevier. 330

8.1 a, c. Interactions in the primate ventral visual system. Figure 1, p. 2 in M. Manassi, B. Sayim, and M. H. Herzog, "When Crowding of Crowding Leads to Uncrowding," *Journal of Vision* 13(13) (2013): 10. Adapted by permission of The Association for Research in Vision and Ophthalmology. 358

8.6. A comparison between some midbrain and higher brain structures. Figure 3, p. 4 in T. Mueller, "What Is the Thalamus in Zebrafish?," *Frontiers in Neuroscience* 6 (2012): 64. Adapted by permission of Thomas Mueller. 387

8.7. A section through the insect brain. Reproduced by permission of www.cronodon.com. 388

8.8. The octopus brain. Figure 23, p. 103 in J. Z. Young, *A Model of the Brain* (Oxford: Clarendon Press, 1964). Reproduced by permission of Oxford University Press. 389

9.2. Phylogenetic relationships among major animal phyla. Figure 1, p. R877 in M. J. Telford, G. E. Budd, and H. Philippe, "Phylogenomic Insights into Animal Evolution," *Current Biology* 25(19) (2015): R876–R887. Reproduced by permission of Elsevier. 408

10.1. Comparison between the structures of the ocelloid and the vertebrate eye. Figure S1 (A and F) in S. Hayakawa, Y. Takaku, J. S. Hwang,

Introduction to Part I Rationale and Foundations

When the answer cannot be put into words, neither can the question be put into words. The riddle does not exist. If a question can be framed at all, it is also possible to answer it.

—Wittgenstein 1922, 6.5

In this book we try to answer the question: How did minimal animal consciousness originate during animal evolution? We argue that this is an answerable question if one can uncover a capacity that is a good marker of the evolutionary transition from preconscious to conscious animals. This can be, we maintain, an Archimedean point to explore the biological nature of consciousness, of sentience. We present our evolutionary transition–oriented account of consciousness in part II of the book. Part I provides background information about the explanatory framework we use, the history of the evolutionary approach to mentality, and current neurobiological and philosophical approaches to consciousness. Part I ends with a chapter that may be viewed as a bridge to part II, as it deals with present ideas, including our own, about the distribution of consciousness in the animal world. The background chapters, particularly 2 (history) and 4 (philosophy), are somewhat idiosyncratic: since ours is an evolutionary transition–oriented perspective, we highlight those facets of history and philosophy that have an evolutionary orientation or that make evolutionary sense. We do not, therefore, engage with philosophers who hold dualistic positions with regard to the mind-body problem or with those who do not regard consciousness as the product of biological evolution.

But how can one study the biological nature and evolution of minimal consciousness? Consciousness is a goal-directed process, and the study of subjective experiencing, which is the hallmark of animals with "sensitive souls," has conceptual challenges similar to those presented by other teleological systems. These include striving to go on living through self-maintaining activities (having a "nutritive soul" such as that of a plant) and being rational—striving for abstract values such as justice, beauty, and truth—something that a human with a "rational soul" possesses. These similarities, which we pursue in chapter 1, suggest that the study of the evolutionary origins of the different teleological systems may be amenable to similar theoretical analyses and that insights gained through the study of one system, such as the successful origin-of-life research program, may be a useful guide to the study of the less tractable origin-of-consciousness problem. We discuss the assumptions and implications of our evolutionary-transition approach and situate them within a broad evolutionary framework.

Chapter 2 charts the historical origins of the evolutionary approaches to the mind. This outline is presented because a historical approach not only provides a deep cultural context for any subject but also because of the insights gained by early evolutionary psychologists. We focus on Jean-Baptiste Lamarck, Herbert Spencer, and Charles Darwin, and then explore the foundations on which modern consciousness studies are based, going back to William James's 1890 *Principles of Psychology* and highlighting his selectionist approach. The chapter ends by discussing the fall and rise of consciousness studies in the twentieth century.

Another necessary foundation for our evolutionary analysis is the neuroscience of consciousness. Although we cannot do justice to the many studies in this area, chapter 3 summarizes what scientists regard as the major neurobiological characteristics and dynamics of consciousness and presents the common ground they share. And although this is not intended to be a book about the philosophy of mind, chapter 4 considers the insights and problems identified by modern philosophers of mind who have adopted a naturalistic stance. It is these philosophers who point most clearly to the kind of questions that an evolutionary theory of minimal consciousness must answer. Mental causation, the biological construction of consciousness and "self," and the function of consciousness are questions addressed in this chapter, although our answers sometime diverge from traditional opinions. We suggest, for example, that consciousness has telos rather than function and is best considered as a way of being that is constructed by multiple cognitive functional capacities.

Part I ends with chapter 5, where we summarize the different views of biologists and psychologists regarding the distribution of minimal consciousness in the animal world. These views are highly divergent: some argue that only linguistically endowed humans, with their highly evolved neocortex, are subjectively experiencing, while others maintain that a cortex is not a necessary condition for consciousness and that all or most vertebrates and some invertebrates are subjectively experiencing. The chapter closes with our own proposal that the evolutionary-transition marker for consciousness is unlimited (open-ended) associative learning (UAL). This, we argue, was the phylogenetically earliest manifestation and driver of the evolution of sustainable minimal consciousness. UAL refers to an organism's ability to attach motivational value to a compound, multifeatured stimulus and a new action pattern and to use it as the basis for future learning. We argue that UAL is a good transition marker because the features that neurobiologists and philosophers regard as essential for consciousness are also required for UAL. If UAL is accepted as a transition marker, one can identify this capacity in different taxa and provide an account of the distribution of consciousness in the animal world—a major issue with important biological and ethical implications. This chapter therefore both ends the first part of the book and begins part II, where we give an evolutionary analysis of the initiating, realizing, and stabilizing biological conditions for the evolution of UAL and animal consciousness and their further elaborations during the evolutionary history of animals.

1 Goal-Directed Systems: An Evolutionary Approach to Life and Consciousness

How can consciousness be studied? Our point of departure is Aristotle's "soul," the organizational dynamics of living beings, and its different manifestations in different types of organisms: the "nutritive and reproductive soul," which involves self-maintenance and reproduction and is present in all living things; the "sensitive soul," which is equated with the living organization of sentient, subjectively experiencing beings; and the "rational soul," which is special to reasoning humans. Our main interest is in the sensitive soul and we ask: Is it possible for scientists to study the sensitive souls of bees, of dogs, and of humans? More generally, we inquire whether it is possible to relate teleological and mechanistic causations or whether there is an unbridgeable explanatory gap between them. We start with the life gap—the gap between inanimate matter and animate beings with nutritive souls—and the study of the origin of life, a research program with a long and successful history. Although we cannot yet construct living organisms from inanimate matter, the evolutionary transition to a living organization is no longer seen as a mystery. We then ask whether the investigations of the life gap can illuminate the qualia gap—the enigma of how living matter gives rise to subjective experiencing, to sensitive souls. Adopting Daniel Dennett's evolutionary hierarchy of goal-directed systems, which parallels Aristotle's teleological hierarchy of souls, we suggest that an evolutionary, transition-oriented approach not only may lead to biological insights but also may settle some thorny philosophical problems.

Life and consciousness seem to be the very core of what it means to be a sentient biological being. It is therefore not surprising that both life and consciousness are notoriously difficult to define and analyze and that they have long frustrated the philosophers and biologists who have attempted to account for them in naturalistic terms. The Cartesian view that the living body is a material machine, whereas the mind is nonmaterial, deeply influenced Western thought from the seventeenth century onward and gave rise to the infamous mind-body problem. However, by the dawn of the twentieth century, life and consciousness—body and mind—seemed to many people to be intimately related. This was not just because consciousness or subjective experiencing (we shall be using these terms interchangeably) could be understood as a product of the evolution of living organisms but also because life and consciousness were both seen as ongoing, self-organizing processes. This communality is probably why Henri Bergson, for example, equated the two notions, suggesting that life is creative becoming and charged with consciousness:

> The evolution of life, from its early origins up to man, presents to us the image of a current of consciousness flowing against matter, determined to force for itself a subterranean passage, making tentative attempts to the right and to the left, pushing more or less ahead, for the most part encountering rock and breaking itself against it, yet in one direction at least succeeding in piercing its way through into the light. That direction is the line of evolution which ends in man. (Bergson 1920, pp. 27–28)

There is beauty in this "current" metaphor, as in so much of Bergson's prose, but although the continuity between life and consciousness is self-evident because all known conscious beings are alive, we do not endorse Bergson's position that consciousness and life are identical. It is not surprising that this nebulous view did not lead to a scientific approach to the subject. On the contrary, it reinforced the generally shared feeling at that time that the nature of life and consciousness would remain forever elusive, forever inaccessible to scientific inquiry.[1] This impasse started to be overcome later in the twentieth century, however, and today the nature of life, though recognized as a very difficult problem, is no longer seen as scientifically impenetrable. With consciousness this is not yet the case, but more and more biologists, psychologists, and philosophers believe that the increasing understanding of the nervous system, the insights into the biology of cognition and affect, the progress in computational biology and in brain imaging, and the advances in the naturalistic philosophy of mind all point in the same hopeful direction.

We have already indicated that alongside the term "consciousness," which is widely used and therefore unavoidable, we will be using the term "subjective experiencing." "Subjective experiencing" is a self-explanatory, intuitive term encompassing the paradigmatic processes that we identify with conscious experiences: it refers to what happens to us and in us when we have not eaten for a few days, when we trip over a rock and sprain our ankle, when we taste a ripe banana, when we watch the starry night sky in the desert, when we hold and smell a baby, when our beloved mother dies, when we are attacked at night, when we have a nightmare, or when we solve a difficult mathematical problem. It is also what disappears when we fall into dreamless sleep or into a coma. People often refer to humans and animals who subjectively experience as "sentient," and that is how we use the words "sentient" and "sentience" here.

In addition to the intuitive appeal of "subjective experiencing," we like the term because "experiencing" is a verbal noun, so the dynamic nature of the processes involved is explicit. We realize that this casual characterization of subjective experiencing may irritate some of our readers, and we can only repeat the argument of Patricia Churchland, who maintained that starting a discussion of consciousness by defining it is not necessary because "we use the same strategy here as we use in the early stages of any science: delineate the paradigmatic cases, and then bootstrap our way up from there."[2] However, for those who want more, we offer a provisional characterization (not a definition), which we will expand on later. We suggest that consciousness is not a property or a capacity of a system such as having sight, nor is it a processes such as metabolism. We see subjective experiencing *as a mode of being* that involves activities that generate temporally persistent, dynamic, integrated, and embodied neurophysiological states that ascribe values to complex stimuli emanating from the external world, from the body, and from bodily actions. Although perceptual consciousness (e.g., seeing a red poppy) and affective consciousness (e.g., feeling pain or fear) can be distinguished, and it seems that the first (perception) can occur in the absence of the second (feeling), they are a unified aspect of experience, something that is evident when the evolutionary history of consciousness is addressed. Inevitably, this characterization is, at this point, rather opaque and clumsy, and certainly it is lacking in poetry, but we hope that as we proceed some flesh will be put on its dry bones.

Our term "subjective experiencing" is thus equivalent to both sentience and consciousness, but consciousness researchers have qualified the latter term in many different ways, some of which are overlapping and often confusing.[3] It is important to stress here that human consciousness, which laypeople usually associate with the term "consciousness" and which we discuss from an evolutionary perspective in the last chapter, is not the main topic of this book. Our book is about the origins and evolution of sentience, of *minimal animal consciousness*—the ability to have basic subjective experiences—rather than the ability to reflect about those subjective experiences, which seems to be the peculiar gift and curse of humans.

Interpretive problems plague not only the notion of consciousness but also the related concepts of awareness, mind, soul, self, mentality, and cognition. "Awareness" usually refers to a state of wakeful attention and precludes the subjective, nonreflective experiencing that occurs when we learn implicitly or when our thoughts are just roaming, whereas "self-awareness" is similar to self-consciousness but commonly has more affective connotations, as in shyness. "Mind" and "cognition" are usually, though inconsistently, used in a very broad way. Cognition, in the broad sense, refers to any information processing that involves interactions between sensors and effectors; it is used not only when referring to all types of neural processing in animals but also for describing processes involving flexible sensor-effector interactions and signal transduction networks in nonneural organisms like bacteria, paramecia, fungi, and plants.[4] The commonly used terms "mind" and "mentality" usually refer to intellectual faculties and to thinking but are sometimes used more generally—for example, as in "the mind-body problem." "Spirit" is used for a nonmaterial, mental, psychological "something" that is separated from the body, while "self" refers to a subjectively felt distinction between the subject and the world.

The term "soul" is also ambiguous, referring, in most of the monotheistic theological texts that followed the rise of Christianity, to something that is usually separated from the body; something that is responsible for morality and that remains after death. This usage, however, was not universally shared in the ancient world. In Genesis 1:25, God is said to have created the animals. All animals, beginning with the creatures swarming in the seas and ending with man, are what the Hebrew biblical text calls a "living soul" (נפש חיה) or an "animal soul" (חיה is both "animal" and "living" in Hebrew), rendered in the King James translation as "the moving creature that hath life." Significantly, "living soul" is not an attribute attached to plants, which were created much earlier, on the third day; plants grow and reproduce after their own kind but are not said to be living souls, so a clear distinction between plants and animals is made, with only the latter being ensouled. Some pre-Socratic philosophers were more liberal, granting both life and soul to plants and even to magnets. Nevertheless, both the ancient Hebrews and the pre-Socratic philosophers seem to have regarded the soul as an intrinsic part of the entity in question, in contradistinction to Plato and his school, who attributed an autonomous existence to the soul and regarded it as a separable entity. This latter notion had a profound influence on theological and philosophical reflections in the Western world.

We have used the problematic term "soul" in the title of this book. However, our usage will not follow the Platonic or theological traditions in the Western world. We use the term as a tribute to Aristotle, the greatest-ever philosopher of living things and the founder of the life sciences, and to his great treatise *De Anima* (*On the Soul*). In *De Anima* Aristotle carved the living world at its teleological joints. He characterized the soul as the principle of life, the cause and source of the living body: "It [the soul] is the source of movement, it is the end, it is the essence of the whole living body."[5] The Aristotelian soul is the dynamic embodied form (organization) that makes an entity teleological in the intrinsic sense—having internal goals that are not externally designed for it but that are dynamically constructed by it.

Aristotle distinguished three nested levels of organizational goal-directedness—three levels of soul: the nutritive and reproductive, the sensitive, and the rational. The first, most basic nutritive soul characterizes life and is involved in self-maintenance. The telos (a term derived from the Greek word meaning aim, purpose, and even duty) of this soul is inherent in those aspects of organization that dynamically maintain the organism's existence through ontogenetic (individual) time and often through historical (lineage) time. All living things—plants as well as animals—have this basic nutritive and reproductive soul. The second level, the sensitive soul, is built on the first one. This is the living organization of the perceiving/feeling animal, which has goals set by its perceptions and feelings that usually, although not always, lead to goal-directed movement.[6] Aristotle believed that all animals have this second level of soul, although he recognized important differences among species in the sensory modalities available to them and in their behavioral sophistication. Wanting, which is a subjective feeling, is found in "sensitive" animals, and its telos is manifest through their striving to satisfy the needs they feel. The third level of soul, which in life on Earth is based on the first and second ones, is rationality. Rationality is an additional hierarchical level of living organization that is specific to humans. Symbolic values such as goodness, truth, and beauty are teloi of the rational soul, and their existence depends on a high level of abstraction and representation, which in the Gods, and in the Gods alone, has no material basis.[7] Figure 1.1 depicts the three levels of the Aristotelian soul.

As we see it, subjective experiencing or consciousness refers to the sensitive soul—to the soul of animals. We realize that our decision to restrict consciousness to animals may not satisfy everyone, and we shall no doubt offend and irritate some people by what they regard as the arbitrary exclusion or inclusion of some natural entities in the category of "conscious" beings.[8] Our own position, like Aristotle's, is that nonanimal living organisms are not conscious. One reason for denying many living entities sentience and consciousness is that for the distinctions between the conscious, unconscious, and nonconscious states to make sense, consciousness cannot be defined too broadly. Cells (including neurons) in an unconscious human under deep anesthesia are fully alive, but in spite of their living and dynamic state,

Figure 1.1
Aristotle's three soul levels: the plant's nutritive and reproductive soul; the elephant's sensitive—motor and sensory—soul, which includes the nutritive and the reproductive; and the human's rational soul, built on the nutritive and the sensitive, which enables it to evaluate scientific theorems and Japanese poems.

it is the person who can be said to be unconscious; her cells are neither conscious nor unconscious—they are nonconscious. Similarly, living cells taken from an animal such as a human may be grown in culture, differentiate into neurons, and form interesting neural networks. However, to say that neural networks are conscious and have subjective experiences would imply that they can be unconscious, and this seems to us to be a completely vacuous statement. Since we are interested in living beings who can *lose consciousness*, we cannot attribute sentience or consciousness to a motile bacterium or a ripe tomato.[9] The distinction between conscious and unconscious makes no sense with bacteria and tomatoes. We therefore agree with Aristotle that plants have splendid nutritive souls, but they *do not have* (losable) sensitive souls like those of humans and cats.

But are complex adaptive behaviors and nervous systems, the hallmarks of animals, sufficient for rendering them conscious? It is clear that complex adaptive responses can occur without subjective experiencing. We can build robots that exhibit adaptive behaviors, but these robots are not deemed sentient because they do not satisfy the list of characteristics considered necessary and sufficient for minimal consciousness.[10] Similarly, cells form complex networks in petri dishes, and they can also learn. Organized neural systems, such as severed spinal cords, can exhibit learning too, but as has been found with some unfortunate victims of terrible accidents, spinal learning is not associated with subjective experiencing, so such a network cannot confer consciousness.[11] Adaptive behaviors and learning, even when involving neurons, are therefore not sufficient criteria for identifying consciousness. If we do not want to render the distinction between neutrally instantiated conscious and unconscious states unintelligible, we have to qualify the kind of nervous system and the kind of neural dynamics that generate subjective experiencing.

Although they are not sufficient, there are reasons for thinking that in the biological world, a *nervous system* and a *brain* are necessary for subjective experiencing to occur. First, since it is the whole organism that experiences rather than only a part of an organism, subjective experiencing must involve a *systemic reaction*. The response to a stimulus must be integrated with the overall state of the organism in a way that preserves the specificity of the response in terms of its location, modality, or strength yet leads to whole-organism subjective experiencing and a particular coordinated action, which depend on multiple reciprocal interactions. Second, in a multicellular body, different types of stimuli, in different locations and acting on different sense organs, must be able to elicit integrated yet specific subjective experiencing—something that requires both elaborate connectivity and a common "language" for communication. Third, as integration occurs, the stimulus and the integrated state must persist in order for feeling to occur, and this seems to require both spatial convergence and temporal synchronization in a center of communication. The rapid transmission of signals by electrical impulses, neural connectivity, temporal binding, and a brain in which signals are combined and where rapid and persistent feedback loops can occur fulfill all of these requirements. Since bacteria and tomatoes do not have such central transmission-integration systems, most people would agree they seem to be unendowed with the phenomenal consciousness of animals—with felt needs and perceptions.

For all these and other reasons that we shall come to in later chapters, we follow Aristotle and maintain that only animals (and, as we argue later, not *all* animals) have a sensitive soul, the second level of living organization in Aristotle's soul hierarchy. Of course, organisms endowed with rationality (Aristotle's third level of soul) also have sensitive souls, but as we suggest in chapter 10, their sensitive souls are radically different from those of nonrational animals because the evolution of rationality involved profound changes in subjective experiencing.

Although we adopt the Aristotelian hierarchy and find Aristotle's intrinsically teleological stance and his emphasis on the functional unity of the organism inspiring and useful,[12] our evolutionary approach to consciousness is, in a way, very non-Aristotelian because the historical-evolutionary perspective came into use only two hundred years ago. Aristotle's nonevolutionary approach does not tell us much about the temporal changes and gradations between the different soul levels. Is there continuity between the nutritive (plant) soul and the sensitive (animal) soul, or is the difference between them qualitative? Are there gray areas that are particularly difficult to categorize? Aristotle's position here is not entirely clear: most of his writings suggest that he saw the differences as qualitative.

From an evolutionary point of view, understanding the transitions that resulted in the three Aristotelian goal-directed systems is enormously challenging. The first problem, understanding *the transition to the first living system*, to the nutritive soul, is still not fully solved, although great strides have been made in this domain. Very little is known about the second, understanding *the transition to subjective experiencing*, the evolutionary origin of the sensitive soul. The third, understanding *the transition to rationalizing, symbolizing animals*, to the rational (human) soul, is one of the hottest topics in present-day evolutionary-cognitive biology, and progress is being made. All of these goal-directed systems are the products of chemical and biological evolution, and there is an evolutionary continuity between them. Studying the transitions that led to their emergence may therefore provide valuable insights into the dynamic organization of the systems and also tell us something about the way in which they are related.

But can the evolutionary approach really tell us what life, consciousness, and rationality are, or will these aspects of being evade biological explanation? Since all are inherently dynamic systems driven by goals, and goals presuppose a criterion that enables evaluation, are they amenable to conventional scientific investigation? Can we find only the correlates of life, consciousness, and rationality but never attain a full explanation of such systems in biological terms? Is there an unbridgeable explanatory gap? Or more than one? The claim for such a gap was forcibly made by Yeshayahu Leibowitz, an Israeli philosopher, biologist, and theologian who, from a point of view diametrically opposed to our own, made us recognize the special difficulties and challenges inherent in understanding goal-directed systems.

The Leibowitz Challenge: The Kantian Epistemological Gap

For it is quite certain that in terms of merely mechanical principles of nature we cannot even adequately become familiar with, much less explain, organized beings and how they are internally possible. So certain is this idea that we may boldly state that it is absurd for human beings even to attempt it, or to hope that perhaps some day another Newton might arise who would explain to us, in terms of natural laws unordered by any intention, how even a mere blade of grass is produced.

—Kant 1790/1987, pp. 282–283

Modern science is based on the concept of cause that Aristotle called the mechanical cause, and thus it [modern science] creates a rift between science and the theory of values. No deep philosophical reflection is needed to clearly recognize that the concept of goal is connected to that of value, which is the meaning that we attribute to things.

—Leibowitz 1985, p. 27

Yeshayahu Leibowitz (figure 1.2) is little known outside the Israeli political and cultural scene, but he was probably the most outstanding and controversial Israeli intellectual during the second half of the twentieth century. His intellectual authority was based on his acerbic eloquence and vast knowledge of matters both secular and religious: he was a physician, had a doctorate in chemistry, was an eminent scholar of Judaism and an interpreter of

Figure 1.2
Yeshayahu Leibowitz (1903–1994). Portrait by Bracha L. Ettinger ©.

Maimonides, and was also an ordained rabbi belonging to the rational-
istic Orthodox Jewish tradition. An outspoken critic of Israeli politics, a
staunch opponent of retaining any of the territories seized during the Six-
Day War, a supporter of conscientious objection to military service in the
Occupied Territories and Lebanon, and an advocate of the separation of
religion and state, Leibowitz was seen by many Israelis as the modern incar-
nation of a fierce biblical prophet. He also looked like one—tall, thin, and
stooping, with a high forehead, bright eyes, and a slight Eastern European
accent accentuating his formidable Hebrew, he was universally admired
and feared. Professor Leibowitz, as he was referred to (he was a professor
of biochemistry at the Hebrew University), was also the first philosopher
of biology in Israel. The three issues on which he focused his interest in
this field included the nature of life, the relationship between genetics and
embryological development, and the mind-body (or as he called it, "the
psycho-physical") problem. Goal-directedness was the common denominator of all three, and this was what passionately interested him. Leibowitz, a
great admirer of Immanuel Kant, followed Kant's argument that there is an
epistemological, explanatory gap between the mechanismic (mechanism-
based) and teleological descriptions of natural and psychological processes.

He fully endorsed Kant's view that the teleological nature of the processes has to be *assumed* to permit analysis: it is not itself amenable to scientific (mechanism-based) study. The origin of life and the goal-directed processes of embryological development cannot be fully described in terms of molecular biology and genetics, he argued. He vehemently denied that the mind—willing, feeling, and thinking, which are all purposeful processes—can ever be understood in terms of neural mechanisms, however sophisticated.

Like all aspiring Israeli intellectuals, we had to face up to Leibowitz's Kantian challenges. Although we agreed with his political views and admired his civic courage, we strongly objected to his epistemological dualism. However, a person's gut feelings and vague scientific optimism will not do when confronting Leibowitz: a flood of angry and learned arguments will crush the poor offender. We had to confront head-on the challenge of explaining goal-oriented systems in some kind of mechanismic terms. We had to explain how a goal, a term that implies some kind of evaluation, can be accounted for within the framework of science, which has no room for values.

Leibowitz's challenge, like Kant's, was a general claim about the incommensurability between mechanismic and teleological explanations, and although Leibowitz is now dead, the challenge is not: we are still engaged in lively discussions with his argumentative ghost. His unifying focus on value has encouraged us to adopt a comparable unifying approach: since orientation toward a goal is a hallmark of a living system such as a food-seeking bacterium, or an embryological process that seems to "strive" toward a steady state, or a subjectively experiencing state of being such as that of a thirsty cat or a moralizing prophet, we believe that what scientists reveal by studying one goal-directed system may be worth exploring in others. The question is how best to study these types of system. What is the best approach for studying life, embryological development, or subjective experiencing?

We find an evolutionary approach focusing on the transition from a preteleological system to a teleological one particularly attractive and promising for four related reasons. First, if we can identify the evolutionary transitions to unicellular organisms, to multicellular organisms, to embryologically developing organisms, and to conscious, subjectively experiencing organisms, and describe them in terms of the changes in the systems' organization, it can help to characterize the mechanisms and dynamics that make them goal-directed. Second, the goal-directed system that appears immediately after a transition does not carry the baggage of later evolved structures and processes and will therefore enable us to recognize the most fundamental features of the organization that allow the system to work. Third, since there is an evolutionary continuity between the different goal-directed systems, the transitional, fuzzy gray areas that defy definition may uncover the evolutionary scaffolds (structures, processes), which have been subsequently discarded or greatly modified, on which the posttransition system was constructed. Fourth, there may be common principles that all these goal-directed systems share, so understanding a better-researched transition (such as the transition to life) may provide tools for understanding others that are less well understood.

Another motivation (although not exactly a reason) for taking an evolutionary approach is that it allows some cautious optimism: the evolutionary origins of the first and third souls—that of living beings (the nutritive soul) and that of rational, symbolizing animals (the rational soul)—have been intensely studied and are increasingly better understood. In the last twenty years, the evolution of the features that demarcate humans from all other animals, in particular the ability to think and communicate through symbolic language (an ability that seems to be the basis for rationality), has been a focus of much research. Although it is recognized as one of the most challenging subjects in evolutionary biology, with very few exceptions no secular scientist or philosopher regards it as a mystery requiring more than a profound and creative understanding of evolutionary biology. This is also true for the origin of the first living systems, a problem that in the nineteenth and early twentieth centuries was seen as the greatest challenge for evolutionary theory. What biologists now know is not how to give a succinct definition of life, nor how to reconstruct a fully blown living system from inanimate matter: we still cannot do that, although there are several promising approaches and scenarios. What we have learned is how to think about the problem and how to study it. Research into the origin of life has transformed the notion of life from a mystery into a very tough scientific problem that awaits full experimental fleshing out.[13]

Our hope is that an evolutionary, transition-oriented approach to consciousness will help dispel the fog surrounding the subject and show that this evolutionary transition, like the transition to life and the transition to rationality, led to new modes of being. As we see it, the big and elusive questions about *what* consciousness is, *how*, mechanistically, it is instantiated, and *why* it has evolved are all related. We cannot fully answer the "what" and "how" questions before we begin to answer the "why" question, nor can we answer the "why" question without some clues about the "what" and the "how." The answers are interlaced, so progress in answering one question sheds light on the others. We have to build our explanatory framework the way an orb spider builds a web, with ever-growing spirals of explanation based on evolutionary reasoning. But before we start weaving our web, we need to take a closer look at the insights that have been gained from the well-studied teleological transition to life to see how they can serve us when we approach the transition to consciousness.

The Life Gap: From a Mystery to a Scientific Problem

> Life must be considered a process. This process takes place in a special system of reversible reactions so that the system is thermodynamically unstable and its state is sustained by spontaneous uptake of nutrients and energy.
> —Gánti 1971/1987, p. 31

In the second quarter of the twentieth century, Alexander Oparin in the Soviet Union and J. B. S. Haldane in the United Kingdom suggested scenarios for the origin of life. Their ideas were anchored in two related convictions and three empirical findings. The two convictions had been shared by all evolutionary biologists since the beginning of the nineteenth century, when Jean-Baptiste Lamarck first expressed them. They were, first, that life is to be understood as a certain type of dynamic organization; and second, that no vitalist principles and assumptions are needed: the dynamic organization that we call life is constituted by coupled chemical reactions, the components of which can be identified in nonliving complex chemical systems. The three empirical findings were that cell metabolism can be understood in biochemical terms, that heredity can be understood in terms of self-replication, and that the kind of geochemical conditions and reactions that may have prevailed on ancient Earth and could have generated the first living entities can be reconstructed. The scenarios and the attempts to simulate them in the lab (and later in silico) made the nature of life an in-principle, answerable (albeit hugely difficult) scientific question.[14]

Lists

For a satisfactory naturalistic explanation of the origin of life,[15] there has to be agreement on what the basic properties of a living system are. If we find a system that has these properties, even if it is very different from the systems with which we are familiar (as the first living beings on Earth are likely to have been), then it is living.

Lists of the characteristics of life have been drafted since the beginning of the nineteenth century, when Lamarck, the first biologist to give an account of living beings from an evolutionary perspective, enumerated their properties:

> All these [living beings] possess individuality, either simple or compound; have a shape peculiar to their species; are born at the moment life begins to exist in them or when they are separated from the body whence they spring; are permanently or temporarily animated by a special force which stimulates their vital movements; are only preserved through nutrition which more or less restores their losses of substances; grow for a limited period by internal development; form for themselves the compound substances of which they are made; reproduce and multiply so as to carry on the species like themselves; lastly, all reach a period when the state of their organization no longer permits of the maintenance of life within them. (Lamarck 1809/1914, p. 195)

Individuality, metabolism, growth, reproduction, some form of heredity, and death were the characteristics of life, according to Lamarck. The special force of life that he listed as one of his characteristics was a *physical* force that resulted from the flux of subtle fluids in the self-organizing material body, the subtle fluids being electricity and heat fluxes. Lamarck was a committed and sophisticated materialist and, as we shall describe later, abhorred any kind of nonphysical incursion into the study of life (and also of mind).

There are other catalogs of life characteristics that are similar to Lamarck's list. They share many features, and this consensus has been important for investigations into the origin of life.[16] In table 1.1 we present several representative twentieth-century lists that emphasize the biochemical-metabolic, genetic-molecular, and evolutionary aspects of living systems. Although the motivations and emphases of these compilations are different, they are clearly related and describe the basic processes of living systems as biologists perceive them. Other compilations have a slightly different focus, many stressing the more abstract and holistic properties of living systems, such as emergence and self-organization. Margaret Boden, who analyzes the relations between life and mind from the perspectives of artificial intelligence (AI) and artificial life (AL), claims that in almost all lists of properties, self-organization, autonomy, emergence, development, adaptation, responsiveness, evolution, reproduction, growth, and metabolism loom large.[17]

Lists reflect history-laden scientific ideas, but they can be used as pointers and guides for research. Of course, they are only the first step in a very long journey. To convince the skeptic that a naturalistic explanation of life is feasible, additional questions must be answered. What are the organizational principles and the dynamics of a system that *generates* the above

Table 1.1
Representative twentieth-century lists of characteristics of minimal living systems.

Author	Characteristics	Emphasis
Gánti **1971/1987**	(1) Inherent unity; (2) metabolism; (3) inherent stability; (4) information-carrying subsystem; (5) program control; (6) growth and multiplication; (7) hereditary system enabling open-ended evolution; (8) mortality	Self-organization, evolution
Orgel 1973	(1) Functionally complex organization; (2) subject to natural selection; (3) replication of genetic material; (4) information for specifying the living system stored in stable chemical molecules	Information, evolution
Maturana and Varela 1980	(1) Individuality (closure); (2) self-production; (3) responsiveness; (4) regulation and selectivity	Self-organization
Mayr 1982	(1) Complexity and organization; (2) chemical uniqueness (living organisms are composed of large polymers); (3) quality (some relations between aspects of the living world can only be described qualitatively); (4) uniqueness and variability; (5) possession of a genetic program; (6) historical nature; (7) subject to natural selection; (8) indeterminacy (biological systems have emergent properties)	Evolution
de Duve 1991	(1) Manufacturing its own constituents; (2) extracting energy and converting it to work for the system; (3) catalyzing the system's reactions; (4) having information systems, enabling reproduction; (5) closure (individuality); (6) regulation; (7) multiplication	Metabolism

characteristics? How did such a system emerge during evolutionary history? We look at possible answers to these questions in the next two sections.

Organizational Principles

> The soul is the first grade of actuality of a natural body having life potentially in it. The body so described is a body which is organized.
> —Aristotle 1984d, 412a, 27–28

In 1971, Humberto Maturana and Francisco Varela introduced a new concept into discussions about the nature of life.[18] This concept, "autopoiesis," was originally formulated to describe the dynamic organization of a machine representing a minimal living system (the cell is the paradigmatic example):

> An autopoietic machine is a machine organized (defined as a unity) as a network of processes of production (transformation and destruction) of components which: (i) through their interactions and transformations continuously regenerate and realize the network of processes (relations) that produced them; and (ii) constitute it (the machine) as a concrete unity in space in which they (the components) exist by specifying the topological domain of its realization as such a network. (Maturana and Varela 1980, p. 78)

Autopoiesis has been a useful concept, aiding theorizing about both simple and extended manifestations of life. Maturana and Varela's focus was on the dynamic organization of an individual entity and its spatial and temporal persistence. One of the major motivations for their approach was to describe life in cognitive terms. A cognitive description of a very simple system requires that cognition is defined very broadly, as indeed it was: it was seen as the ability of the autopoietic entity to regulate its relations with the environment. The system achieves this through active sensor-effector relationships.[19] Figure 1.3 schematically describes a simple preautopoietic system (*A*), a basic autopoietic system (*B*), and an adaptive autopoietic system (*C*) that shows agency—the ability to act in the world in a goal-directed manner. The sensory components of the constitutive systems in figure 1.3*C* are coupled to effectors through feedback loops and enable the system to adaptively regulate its responses to a changing environment and to noise from within.[20]

In the same year (1971) that Maturana and Varela introduced the autopoiesis concept, the organic chemist Tibor Gánti published a book titled *The Principle of Life*, in which he developed a more concrete, chemical model of minimal life. Although less famous than Varela and Maturana's autopoietic model, which is deliberately very abstract and intended to capture the logical (dynamic, formal) structure of a living system, the chemical, cyclical-stoichiometric perspective of Gánti's minimal protocell model captures both the formal dynamics of living autopoietic processes and fleshes out their mechanical and chemical (material) facets. Although it is still idealized, it is extremely useful as a guide for theoretical and empirical approaches to the origin of life.[21]

Gánti started by enumerating the basic criteria for life (table 1.1, row 1). He regarded his first five criteria (individuality, metabolism, stability, a subsystem that carries information about the system as whole,[22] and regulation) as "absolute," by which he meant that these properties have to be

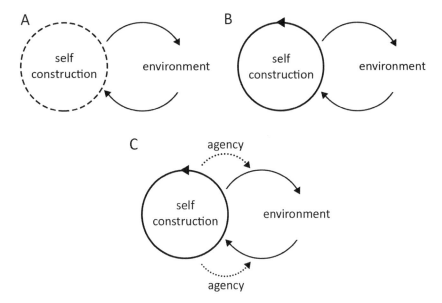

Figure 1.3
The autopoietic system: *A*, a preautopoietic system coupled to the environment, constituted by it and altering it; *B*, a self-sustaining and self-constructing system exhibiting elementary autopoiesis; *C*, a more advanced autopoietic system in which elements of the system adaptively control the way the system responds and constructs the environment. Modified by permission of Springer.

found in every living being and are jointly necessary and sufficient for life. His additional three criteria, which he called "potential" (the ability to grow and multiply, the capacity to exhibit hereditary variation and evolutionary change, and irreversible disintegration) are necessary for the ongoing, *long-term persistence* of the living state. He then constructed the simplest dynamic theoretical-chemical toy model that satisfied these criteria. He called his toy model the "chemoton" (figure 1.4). His chemoton is a chemical system made up of three indissolubly coupled autocatalytic subsystems that form a stable, functional entity. The links between the subsystems mean that they grow and reproduce in a regulated and coordinated manner. The "engine" of the chemoton is the autocatalytic metabolic cycle that transforms nutrients into the substances needed in the other two subsystems (the membrane and the information polymer) as well as for the cycle's own reproduction. Growth of the chemoton leads to the growth of the membrane, which, when it reaches a critical size, becomes

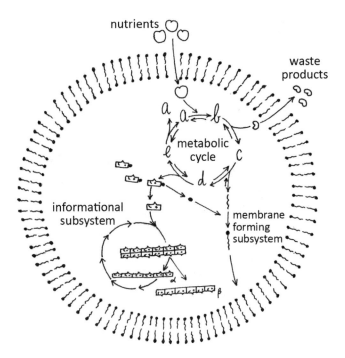

Figure 1.4

Gánti's chemoton, made up of three tightly coupled subsystems. At its core is the autocatalytic metabolic cycle, where molecule *a* combines with a substance formed spontaneously in the external environment and is transformed to *b*; *b* releases a waste product and forms *c*, which dissociates into *d* and a component of the membrane; *d* dissociates into *e* and a precursor of the polymer unit; and *e* dissociates to form two *a* molecules. Since the metabolic cycle forms more cycles like itself (because *a* →2*a*), it can be said to grow. One of the by-products of the metabolic cycle forms the hydrophobic part (*squiggle*) of the membrane unit, which combines with a hydrophilic component (*black dot*) to form the mature unit of the lipid membrane. The mature units spontaneously self-organize to form the membrane subsystem, which forms a boundary between the chemoton and its environment. In the third subsystem, the linear double-strand polymer reproduces by the template-based addition of units to the single "strands" of the initially double-strand structure, thus forming two double-strand polymers. Note that the membrane system and the linear polymer are functionally linked because the growth of the membrane depends on a by-product formed as the linear molecule polymerizes. This template-based polymerization occurs only when the units that are components of the polymer reach a critical concentration. When this happens, the original double-stranded polymer separates into two single structures (α and β), and the polymer components are consumed as they join to form new double structures (not shown). The length of the polymer therefore controls the number of turns of the metabolic cycle that are needed to produce the units necessary for its own reproduction and for the production of the by-product needed for forming membrane units. *Source*: Based on Gánti 2003, figure 1.1, p. 4, and the modification of this figure by Jablonka and Lamb 2006.

spatially deformed, leading to fission. In this way, the chemoton as a whole reproduces.

Gánti's chemoton does indeed satisfy all his "absolute" and "potential" life criteria. The chemoton metabolizes, it is dynamically stable, there is simple regulation of its general size and rate of growth, it has a subsystem carrying information about the system as a whole, its system properties are emergent (nonadditive), and its destruction is irreversible (it can therefore "die"). It grows and multiplies, and displays heredity. Moreover, chemotons can heritably vary: different chemotons can have different chemical metabolic cycles, leading to different rates of production of components; they can vary in the chemical composition of their membrane, thus affecting the influx of materials into the chemoton, the timing of fission, and the effect of the external microenvironment (through the selective elimination of some waste products) with which the chemoton interacts; and they can have different lengths of linear polymers that affect the rate of chemoton fission. Selection among chemotons is therefore inevitable. It is important to note that the most significant source of hereditary variation is the length of the linear polymer because this is the only subsystem in which variation does not involve a change in the chemical nature of the components, only a change in organization. Variation in the length of the polymer would allow some hereditary variants in growth rates, whereas variations in the chemical cycle and membrane components would be more constrained by the special functions of these subsystems. Gánti showed that the constraints that limit heritable variability in the model shown in figure 1.4 are overcome if the linear polymer is made up of *two* types of units rather than one, and it is its sequence, rather than its length, that controls reproduction. In principle, with two polymer units, the variations among chemotons with modest polymer length can be great enough to allow open-ended evolution.[23]

We have described the chemoton in some detail because it enables us to envisage how, once a list of characteristics for life has been drawn up, they can be instantiated by a concrete model of chemical (protocellular) organization. The factors and processes that constitute the chemoton can be thought of as the building blocks of this minimal (biological) living system. If we formulate it in Aristotelian terms, chemoton dynamics can explain life in terms of the Aristotelian material, mechanical, and formal causes: the factors involved are the building blocks of the chemoton (the material cause), and their dynamic relations are both the mechanical and the formal cause of the chemoton. Moreover, as Aristotle knew, in living beings the formal cause is also the intrinsic teleological cause.[24] In the next section we describe, very sketchily, some of the many experiments and additional theoretical models that show how chemical evolution could have led to the emergence of living, highly complex, and intrinsically teleological, chemoton-like entities on ancient Earth.

Scenarios and Simulations for the Origin of Life

> It is useless to speak of the origin of entelechial suspension in nature; useless, *i.e., to discuss the origin of life*. It is absolutely impossible for us to say anything definite on this subject.
> —Driesch 1914, p. 205

> Given the suitable initial conditions, the emergence of life is highly probable and governed by the laws of chemistry and physics.
> —de Duve 2002, p. 55

Between the impossibility and the inevitability of the emergence of life on Earth through physical-chemical processes, there is also the possibility that life is indeed probable, but not *very* probable, and that contingent conditions (some luck) played a role in its emergence.[25] Whatever the position adopted on this issue, there is a fundamental difference between the view of vitalists such as Driesch and that of modern investigators of life's origins. That life emerged on Earth (or on some other planet and was brought here by meteors) is not doubted. That God or an almighty alien intelligence is not a player in this history is an epistemological commitment of most biologists. So, once we know the properties that living systems must display and the kind of organizational principles involved in such systems, we can ask under what circumstances they came into being.

In a famous, oft-quoted speculation made in 1871 in a letter to his friend Hooker, Charles Darwin suggested that living systems may have emerged in a "warm little pond, with all sorts of ammonia and phosphoric salts, light, heat, electricity, and so on, present, so that a protein-compound was chemically formed ready to undergo still more complex changes."[26] The "warm soup" idea was developed about fifty years later, in the 1920s, by J. B. S. Haldane in the United Kingdom and Alexander Oparin in the Soviet Union. Although Haldane stressed the genetic aspects (self-reproduction) and Oparin stressed the metabolic aspects of the first living entities, both shared the basic assumption of a warm soup beginning.[27] Oparin proposed that in such conditions simple organic monomers, which form and assemble spontaneously, are held together by hydrophobic forces and form tiny spherical droplets. These droplets grow by accretion and inevitably fragment ("reproduce"). Since different droplets differ in reproducible properties that affect their growth and fragmentation, evolution by chemical selection can occur. The droplets, claimed Oparin, are the forerunners of the first cellular organisms. Haldane's starting point was different. He began with self-replicating, virus-like elements, which formed spontaneously in the hot soup. These viral-like systems evolved through chemical-natural selection into more complex entities, leading eventually to a cell-like organism.

In the mid-twentieth century, the crystallographer John Desmond Bernal took up the challenge of understanding the transition from chemical to biological entities. He suggested that first we must have some idea of the geochemical conditions on the ancient Earth and then study—and try to simulate—three stages: (1) the emergence of complex monomers that can form the building blocks of biological beings (e.g., amino acids, pyrimidines); (2) the emergence from those monomers of more complex and more stable polymers and systems of interaction; and (3) the emergence from these latter systems of the first bona fide biological organisms. These three stages constitute what he described as the "first act," which led to the simplest, still-fragile life forms. The first act was followed by a "second act," in which the first living forms were stabilized, and by a "third act," which led to present-day complex organisms.[28]

Speculations like Bernal's led to great and heated debates among origin-of-life researchers. What kind of atmosphere did the ancient world have? Where did life emerge? At what temperature? What were the first monomers? What were the first polymers and self-sustaining systems? How did the jump to a protocell-like system occur? Soon, experiments began to provide answers. Here we can mention only a few of the landmarks in the history of this research, but they should be enough to give a sense of how the question of the origin of life moved into the realms of science.

The first demonstration that the basic monomers that characterize life can be formed in ancient Earth conditions was the famous Miller-Urey experiment. In 1952, Stanley Miller, a graduate student, and Harold Urey, his supervisor, took a mixture of water vapor and gases (methane, ammonia, and hydrogen), which they believed simulated the reducing conditions of the primeval atmosphere, and passed electric sparks ("lightning") through it. Organic monomers such as amino acids, which according to the warm pond scenario are some of the types of precursor molecules required for life, formed readily in these conditions. A few years later, in 1961, the Catalonian biochemist Joan Oró showed that the nucleic acid base adenine, as well as many amino acids, could be formed by heating an aqueous solution of ammonia and hydrogen cyanide. Although there has been no consensus about the conditions that prevailed on ancient Earth, these experiments established the feasibility of the first of Bernal's stages, and they were followed by others showing that monomers can be formed under many different conditions. The second stage, the generation of self-sustaining complexes and cycles of reaction, is much more tricky, but there has been progress here too. For example, in line with some of Oparin's ideas, Sydney Fox showed in the late 1950s that amino acids spontaneously link to generate small peptides that form tiny closed spherical blobs that have many of the basic characteristics of life, demonstrating something like growth and a very crude form of reproduction through fragmentation.[29]

A more recent and very ingenious hypothesis, which combined Bernal's first two stages (the production of monomers and then polymers and self-sustaining systems of interactions), has been worked out in great chemical detail by the German chemist and patent lawyer Günter Wächterhäuser. It is rather different from the warm pond scenarios. Wächterhäuser suggested that life began in volcanic vents in the deep sea, in conditions of high pressure and high temperature, where pressurized hot water with dissolved volcanic gases (like carbon monoxide, ammonia, and hydrogen sulfide) flowed over catalytic solid surfaces (e.g., iron and nickel sulfides). As a result, organic compounds containing carbon were formed and bound to the catalytic surface, and the process became autocatalytic. Once such a primitive autocatalytic metabolism was established, it began to produce ever more complex organic compounds, ever more complex pathways, and ever more complex catalytic centers. The ancient system can be thought of as a scaffold for the formation of more complex systems that eventually came to have nucleic acids and the other familiar constituents of present-day biological systems. The scaffold was then discarded, although a few traces of its prior existence can still be discerned.

Many other scenarios stretch and challenge the imagination, and it is quite possible that several of them describe processes and products that occurred on ancient Earth. We still do not know which system or systems (they may have become combined) led to protocellular, chemoton-like life, and solving this problem is one of the biggest hurdles in origin-of-life research. Some hypotheses postulate that nucleic acids (RNA or something similar) came first ("genes first"); others suggest that biochemical reactions and pathways involving proteins came first ("metabolism first"); yet others stress the priority of a containing, enclosing (membrane) system.[30] Today, hybrid models that combine these three aspects seem most promising. Whatever their priorities and models, chemists and physicists today appreciate the complexity of a reaction system that can implement living organization. The first living systems, as well as their precursors, were all chemically complex.[31]

The Transition Marker

One of the questions about the origin of life that has remained unanswered is: At what point in the evolutionary process do we decide that a system is living? Is a self-replicating, 10-base-pair-long RNA molecule living? Is the simple chemoton shown in figure 1.7 or even its precursors (maybe something like the autopoietic system shown in figure 1.5) really living, or do we need a more complex system? Intuitively, it seems that a small RNA molecule, a very simple autopoietic system, or even a protochemoton are not really alive—they are somehow too simple. But this may just be a prejudice, an unjustified intuition. Even if most people have the feeling that life has to be more complex than a self-replicating, 10-base-pair-long RNA molecule or an elementary autopoietic system, these gut feelings require a scientific articulation. Is there a marker, a capacity or part that will allow us to reconstruct the whole system from it, a threshold beyond which we can agree that a system is alive?

Maturana and Varela presented powerful arguments in favor of an autopoietic organization as the manifestation of life. However, they did not suggest a marker and did not consider the conditions that would enable an autopoietic system to exhibit long-term persistence. A criterion that can be used to mark forms of life that can persist over time (we call such a criterion a "transition marker") was suggested by Gánti, and Maynard Smith and Szathmáry developed the idea.[32] They highlighted one characteristic of the system, heredity, and distinguished between limited and unlimited heredity. Systems that can have only very few hereditary variants are *limited heredity systems*, and they reside in the gray area between the nonliving and living stages. They are on the evolutionary route to fully fledged life if they evolve further and the number of their hereditary functional variations becomes great enough to be practically unlimited. Without open-ended heredity and the open-ended evolvability that comes with it (that also generates new niches of which the new variants are part), the lineage would soon go extinct. For Maynard Smith and Szathmáry, it is *the transition to unlimited heredity* that identifies sustainable living entities. According to this view, rather than a single and inevitably highly contentious line between life and nonlife, there is a gradual transition. A gray zone, rather than a transition point, marks the road to sustainable life.

A New (Living) Way of Being: Goals, Functions, and
Functional Information

The transition from a chemical to a living system involved more than new and more sophisticated chemical structures, mechanisms, and dynamics. Life is not just wonderfully complicated chemistry but a drastically new way of being. With life, mere chemical processes and mechanisms became organized into systems to which a goal (self-maintenance) can be ascribed, and the parts and processes of such systems can be said to have functions.

Function is something that only parts or processes in goal-directed systems can have. Living beings, which reconstruct themselves and their parts, are paradigmatic goal-directed systems; so are systems designed by living creatures, such as human artifacts or termites' nests, and so are chemical systems on the verge of living, like the system described in figure 1.5. Biological function (also known as teleofunction) is defined as the role that a part, a process, or a mechanism plays within an encompassing system—a role that contributes to the goal-directed behavior of that system.[33] As we have already noted, the most basic goal-directed behavior of living organisms is self-maintenance (survival) and, in the long-term, reproduction.[34] *Functional information* is any difference that makes a systematic, causal difference to the goal-directed behavior of an encompassing system and in the case of simple living forms, to the system's self-sustaining dynamics.[35] Chemical processes that do not organize into self-maintaining entities do not have functional information since they are not parts of a goal-directed system. Function is not a new high-level chemical process or trait. In Aristotelian terms, it is a facet of the teleological cause "that for the sake of which" things exist. Functions and functional information are the very essence of living organisms and are irreducible to descriptions in terms of chemistry.[36]

Before the realization that matter is inherently active, before the recognition that life evolved, and before the early twentieth-century advances in the understanding of biochemical cycles, the dynamic goal-directed organization that is the hallmark of living organisms was seen as a deep mystery, even by biologically well-informed philosophers and naturalists. Kant could not envisage how a self-organizing living being (what he called "a product of nature") could be constructed in such a way that everything in it has a goal and yet is also, reciprocally, a (mechanism-based) means.[37] Now, however, as a result of the experimenting, theorizing, and philosophizing about the origin of life, the nature of living entities and their origins have lost their aura of unreachable mystery. The Kantian gap has been bridged by our better understanding of the dynamic nature of matter and our ideas about how certain types of autopoietic dynamics can instantiate life (chemoton dynamics is one example); we can theoretically simulate autopoietic systems and figure out how new functions can arise through a process of natural selection. We have moved beyond the rather narrow notions of matter and mechanism that Kant had assumed. In fact, as Wittgenstein pointed out, the metaphysical problem of life has vanished: "The solution of the problem of life is seen in the vanishing of the problem."[38] The question we now ask is whether the approach to the evolutionary transition to life can serve as a model for understanding the transition to consciousness.

Back to Consciousness: The Qualia Gap

> How it is that any thing so remarkable as a state of consciousness comes about as the result of irritating nervous tissue, is just as unaccountable as the appearance of the Djin when Aladdin rubbed his lamp.
> —Huxley and Youmans 1868, p. 178

We believe that if scientists are able to understand what the transition to a conscious system entails, characterize this new way of being, provide a model describing the kind of biological dynamics that instantiates it, and define a transition marker, then subjective experiencing will become as well explained as the state of being alive and, as was the case with the latter, the mystery will slowly vanish. The case of the transition to life shows that the Kantian explanatory gap between mechanism-based and teleological descriptions of a living system can in principle be bridged, and there is almost universal agreement among scientists and philosophers that the problem is accessible, and its solution does not require Higher Intervention. It is not a coincidence that creationists are extremely worried about this kind of research.

Such a universally accepted dissolution of mystery has not yet happened with the problem of consciousness. Although the approaches of most philosophers and neuroscientists are firmly grounded in biology, some eminent philosophers still doubt the possibility of explaining consciousness using the traditional tools of biological investigation. The problem was well captured by the words of Huxley and Youmans quoted above, as well as by Thomas Nagel, who more than a century later fleshed it out in his article "What Is It Like to Be a Bat?" In this famous article, Nagel considers the subjective experiencing of species that sense the world in ways different from our own, such as that of a bat navigating using sonar; he points to our inability to understand the bat's subjective experiences through the detailed, third-person scientific knowledge of the bat's neural mechanisms.[39] We call this second Kantian explanatory gap the "qualia gap"—qualia being the contents of subjective experiences such as seeing red and feeling pain.

David Chalmers has called the bridging of this explanatory gap the "hard problem" of consciousness,[40] a notion that has become very popular in the consciousness literature. He sees no way to bridge the gap between subjective experiencing and its mechanismic-functional description other than to accept a drastic solution that requires an addition to the laws of physics: he suggests that consciousness needs to be explained in terms of a new physical primitive (like mass or electric charge), which would render a theory of consciousness more similar to elegant physics than to "messy biology."[41] This approach is taken even further by Thomas Nagel, who in a 2012 book applied this kind of approach not just to consciousness but also to the origin of life, claiming that the failure to scientifically account for subjective experiencing reflects back on the whole theory of evolution. The evolution of life, of complex adaptations like the DNA translation machinery, of subjective experiencing, and of human reason and human values cannot, he claims, be explained by existing materialist evolutionary theory. This supposed failure suggests to Nagel that basic physics is in dire need of foundational enrichment. To fully explain the appearance of life, mentality, and human reason, an intrinsic mental aspect that pervades all matter is required, in conjunction with a teleological law of nature.[42]

Here, it seems, Chalmers and Nagel disagree. Unlike Nagel, Chalmers regards the case of subjective experiencing as unique in its irreducibility. It is, he maintains, very different from the problem of living, which he agrees is amenable to standard materialistic evolutionary explanations. In response to Chalmers, Daniel Dennett argues that the problem of subjective experiencing *is* analogous to the old problem of living and that Chalmers's arguments for the irreducibility of subjective experiencing are similar to the failed arguments of the early twentieth-century vitalists. An imaginary vitalist, says Dennett, would tell the materialist that the solution of "easy" problems such as the mechanisms of reproduction, development, and growth does not help us understand the mysterious, holistic essence of life, which is the hard problem.[43]

Chalmers rejects this analogy, claiming the two types of problems belong to different explanatory categories:

> When it comes to the problem of life, for example, it is just obvious that what needs explaining is structure and function: How does a living system self-organize?

How does it adapt to its environment? How does it reproduce? Even the vitalists recognized this central point: their driving question was always "How could a mere physical system perform these complex functions?", not "Why are these functions accompanied by life?" It is no accident that Dennett's version of a vitalist is "imaginary". There is no distinct "hard problem" of life, and there never was one, even for vitalists. (Chalmers 1997, p. 6)

Chalmers's argument is wrong because he includes in his description of a living system the ability to adapt and have functions, as if these were self-evident and nonproblematic. As we have already indicated, they are indeed unproblematic for us today, accustomed as we have become to the hard-won understanding of self-organization and adaptation through natural selection. But the possibility of the emergence from mere chemistry of a system that can have functions was a huge problem for intellectual giants like Kant.[44] Although the processes supporting consciousness have functions, Chalmers is wrong in assuming that subjective experiencing is a new high-level biological, functional trait. Just as functioning is not a new high-level chemical process or trait but rather a facet of a new (reproductive) telos, so subjective experiencing is a facet of a new telos: ontogenetic values and felt needs, which open up a whole new realm of functions. We therefore need to work out how subjective experiencing can emerge from certain biological functions, structures, and mechanisms. For this (and Chalmers is right here) we need a theory or a theoretical framework. But we think that the theory required will be more like Gánti's model coupled with the theory of natural selection than a new theory of fundamental physics, and there is no need to add an extra "primitive" to the explanatory framework. In the second part of this book, we develop the argument that an expanded evolution-selection theory, which links the notion of variation to that of developmental and evolutionary stabilization, is a suitable theoretical framework for understanding the great teleological transitions to the sensitive and rational souls.

Three Explanatory Gaps

In addition to the second qualia gap, the transition to the rational soul—to human reflective culture and language-based conscious existence—is another fundamental teleological transition. Although evolutionary biologists have some good ideas about how the transition to human reflective and symbolic-cultural life occurred, they all agree that it is one of the most difficult questions in evolutionary biology. The existence of all three "soul" levels can therefore be said to present Kantian explanatory gaps because in all three cases something completely new, a new way of being, emerged. In the case of the transition to life, it was the emergence of function and functional information; in the case of the transition to subjective experiencing, it was the emergence of first-person experiences and subjective needs; and in the case of the transition to rationality, it was the emergence of symbolic concepts and symbolic values like truth, justice, beauty, and freedom. In all three cases, a new, open-ended realm of possibilities opened up: with life it was open-ended evolution; with subjective experiencing, as we argue in later chapters, it was open-ended associative learning; and with rationality it was open-ended imagination and reasoning.

Yes, we concede that a biological account of subjective experiencing is a special and challenging problem, but it is not unique: there are three such problems. We agree with Nagel that the three teleological transitions—to life, to subjective experiencing, to human values—do pose special philosophical and evolutionary challenges. But we take a diametrically opposite position to his regarding what these teleological transitions mean. As we see it, all are explicable within a sophisticated evolutionary framework, which once understood seeps down (or up?) and reformulates the philosophical problems. So our suggestion is exactly the reverse of that of Nagel, who argues that philosophical obstacles imply the fundamental explanatory insufficiency of evolutionary theory. We argue that once we can account for the teleological evolutionary transitions, many of the problems that were deemed insoluble dissolve. We think that Nagel's problem is the well-recognized one of the failure of evolutionary imagination. As Darwin confessed, the evolution of the eye made him shudder, but as he explained, this was the problem of a failure of his imagination rather than a failure of his evolutionary theory.[45] In fact, we believe that having three problems rather than one is very helpful because two of the "hard" problems—life and rationality—are actually beginning to yield to evolutionary investigations. Recognizing that there is continuity between living, subjective experiencing, and rationalizing and that each one is the product of a transition to a new teleological system can be informative. We can look for analogies between the first, second, and third transitions and see, in very general terms, what we can learn from them and whether they throw light on the nature of subjective experiencing.

The first and most obvious thing to recognize is that the only system in which consciousness has ever been found is a living system, so a good starting point is to investigate the kind of living organization that instantiates the essential properties of subjective experiencing. (At this point we are not interested in robots, although we shall discuss them later.) This is the approach taken by Evans Thompson, who combined an expanded autopoietic view of living with a phenomenological approach and emphasized the cognitive embodiment of biological systems, a view that he called "embodied dynamicism." This view redefines the mind-body problem as the *body-body* problem.[46] Feelings and thoughts cannot be attributed to a brain, however evolved: they can be attributed to an enbrained body, a living, active animal, which is a very different matter. A second facet of the continuity between living, subjectively experiencing, and rationalizing forms of life is the intriguing parallels between them (table 1.2). For example, just as life entails functions, so subjective experiencing entails qualia, and rationality entails symbolic concepts. Similarly, the teloi of minimal life are phylogenetic (survival and reproduction), those of consciousness are ontogenetic values (values that can be ascribed to newly learned complex stimuli and actions guiding open-ended learning), and those of rationality are symbolic values (symbolic categories ascribed to states and actions that guide human cultural behavior). Just as there is no life without the processes instantiating it having functions, so there is no consciousness without the instantiating processes having qualia, and no rationality without symbolic concepts. We come back to the table in the last chapter. Here we focus on the middle column, looking at subjective experiencing.

It is a remarkable fact that in spite of the self-evident and generally acknowledged usefulness of an evolutionary approach to all biological questions, including the origins of life and the origins of human reflective consciousness, attempts to investigate subjective experiencing using the traditional methods of evolutionary biology have been, until quite recently, surprisingly uncommon in both the biological and philosophical literature. Scientists studying subjective experiencing are committed to the theory of evolution, and some philosophers and several neurobiologists take it very seriously, but it was only during the first decade of the twenty-first century that detailed evolutionary scenarios began to be suggested.[47]

One of the reasons for the paucity of concrete, evolution-focused approaches seems to be the lack of agreement about how minimal subjective experiencing or minimal consciousness should be characterized. This makes it very difficult to decide which organisms have it (a problem known as "the distribution problem") and where and when in evolutionary history it first emerged. The problem is exacerbated by the nature of evolutionary changes, which usually are not sharp and clear. As we have already noted, when a new level of organization emerges, there is always a gray and fuzzy

Table 1.2
Suggested parallels between concepts used to describe living systems, experiencing animals, and rationalizing humans.

Living	Subjective experiencing	Symbolizing/rationalizing
Phylogenetic teloi: self-maintenance—survival and reproduction	Ontogenetic teloi: ascription of values to newly learned complex stimuli and actions	Symbolic teloi: symbolic values like freedom and justice
Function	Qualia	Symbolic concepts
Heredity (unlimited)	Memory/learning (unlimited)[a]	Transmission of adaptations involving symbolic representations (unlimited)[b]
Development	Recall	Social reconstruction[b]
Evolution (open-ended)[c]	Learning (unlimited); behavioral adaptation, open-ended[c]	History (open-ended)[c]

Notes:
[a] Our suggestion that unlimited memory and learning parallel unlimited heredity is a central theme in this book. We contend that only conscious living beings can learn in an unrestricted manner but *not* that all sentient beings (e.g., babies) have such learning ability.

[b] These notions can be seen as different facets of historical cultural change.

[c] The relationships between unlimited heredity; learning; symbolizing; and open-ended genetic, neural, and symbolic evolution are far from simple. We assume that at each level hereditary transmissible variations map onto functionally diverse, potentially novel phenotypes and lead to the construction of new selective environments. For a discussion of the relation between unlimited transmissible variations and open-ended evolution, see de Vladar, Santos, and Szathmáry 2017.

area where the classification of the system is uncertain; for most philosophers this presents a major problem, although biologists are much more tolerant of classificatory ambiguity—gray areas are inevitable, given evolutionary history. Nevertheless, once the transition has been identified, it is possible to recognize processes and properties in the pretransition stages, which, though not sufficient for the transition, are necessary for it to occur. Over evolutionary time, the necessary factors and processes accumulate, combine, and become sufficient. A new teleological system emerges.

Our strategy in this book is to employ the evolutionary, transition-oriented methodology that has proved so fruitful for the study of life to the study of the teleological system we call "consciousness." We therefore present and discuss the "lists" of characteristics that neuroscientists and philosophers of mind have associated with consciousness and the very preliminary dynamic models that neurobiologists have suggested. On the basis of theoretical and empirical considerations, we suggest a transition marker for consciousness: unlimited associative learning (UAL). UAL refers to an animal's ability to ascribe motivational value to a compound stimulus or action pattern and to use it as the basis for future learning. We show that the features that enable UAL are based on computational mechanisms and neural structures generally believed to underlie the ability to form mental representations and presuppose the list of criteria and the dynamic organization that scholars of consciousness suggest. We provide evidence that the groups that exhibit UAL are the same as those having the capacities in our list, even when learning-independent criteria are used. Following the evolutionary origins of UAL enables us to identify its building blocks and attempt to reconstruct the system of which it is part—a system that, we argue, instantiates minimal consciousness. Finally, we consider how understanding the evolution of UAL enables us to work out how consciousness has changed during evolutionary history.

Dennett's Hierarchy and Phylogenetic Distributions: Locating the Experiencing (EX) Factor

> I want to propose a framework in which we can place the various design options for brains, to see where their powers come from. It is an outrageously oversimplified structure, but idealization is the price we should often pay for synoptic insight. I call it the Tower of Generate-and-Test; as each new floor of the Tower gets constructed it empowers the organisms at that level to find better and better moves, and find them more efficiently.
> —Dennett 1995, p. 373

Dennett provided a general, evolution-inspired framework for describing different levels of goal-directedness that we find useful for investigating the evolution of subjective experiencing. He used the term "intentionality"—the ability to represent or to stand for things, properties, and states of affairs—and called his approach the "intentional stance." His evolutionary-selectionist framework can be seen as an extension of the Aristotelian teleological approach, and we make use of it throughout the book. However, our standpoint differs from Dennett's because whereas Dennett believes it is convenient to talk about living and sensitive creatures as if they had a telos, we think that these selection-based systems are intrinsically teleological.[48]

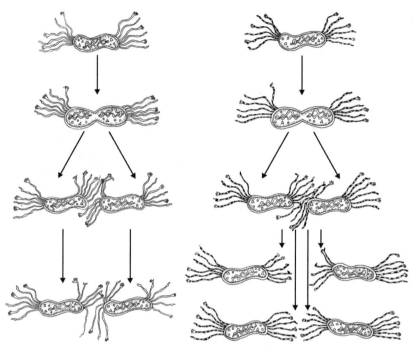

Figure 1.5
Darwinian organisms. These organisms, which multiply and transfer their characteristics to offspring, can "learn" through the mutation-selection mechanism. The lineage on the right is more adapted and has greater reproductive success than the lineage on the left.

Dennett calls the first type of goal-directed organisms, those that generate random variations and are tested by the external environment, "Darwinians organisms" (see figure 1.5). They are, of course, the products of the first great teleological transition—to life. They include organisms such as single-celled microorganisms, plants, fungi, and very simple animals like sponges, which Dennett thinks have limited flexibility (i.e., plasticity) and do not learn during their own lifetime. Although we think Dennett greatly underestimates their plasticity and we believe that all (or most) organisms, including the single-celled, exhibit some form of learning,[49] we agree with Dennett that Darwinian organisms do not experience (i.e., do not have phenomenal consciousness) because, as we claimed earlier, a central nervous system with a particular type of dynamics is necessary for consciousness.

Dennett's second type of organisms are the "Skinnerian organisms" (figure 1.6), which learn during their lifetimes (ontogenetically) through

Figure 1.6
A Skinnerian organism, represented by a nectar-drinking dragonfly. The insect learns from experience; it passes by the flower on the right and chooses the one on the left, which, as it knows from past experience, is richer in nectar.

trial and error. These organisms, which include animals such as worms, crabs, insects, fish, and frogs, are named after the behaviorist Burrhus Frederic Skinner, who saw ontogenetic learning as an extension of Darwinian natural selection (an idea that, as we shall show in the next chapters, was shared by prominent nineteenth-century evolutionists such as Herbert Spencer and William James). Dennett quotes Skinner's view on this: "Where inherited behavior leaves off, the inherited modifiability of the process of conditioning takes over."[50] One of the major arguments in this book is that subjective experiencing is an aspect of open-ended associative learning and that the evolution of open-ended associative learning led to the emergence of consciousness. Since some Skinnerian animals seem to have open-ended associative learning, they are, according to our view, subjectively experiencing organisms.

Dennett's "Popperian organisms" (figure 1.7) are those that can choose between imagined, recalled alternatives for action without having to try them out. They are named after Karl Popper, the twentieth-century philosopher of science who developed an evolutionary epistemological approach and famously explained that foresight is an internal selection principle that "permits our hypotheses to die in our stead." Selection based on recall and

Figure 1.7
Popperian organisms, represented by elephants. The elephant not only learns from experience but also plans ahead and, using its imagination, protects its offspring from the heat of the sun. (Although not actually observed, such a behavior is not beyond the intellectual capacities of elephants, which can prepare tools such as a "brush" made of branches to swat flies or to scratch themselves.)

imagination is required for this kind of feat, which may be characteristic not only of mammals and birds but also of other vertebrates (and maybe bright cephalopods, such as squid). From our point of view, these animals, which are all characterized by highly sophisticated learning and memory, have richer subjective experiencing than the subjectively experiencing Skinnerian organisms. With more evolution, a Popperian species can (though by no means must) evolve a rationalizing mode of being.

Dennett has a fourth category of organisms, the "Gregorian organisms" (figure 1.8), named after the psychologist Richard Gregory, who highlighted the role of the collectively designed human social environment in informing and extending the minds of humans. Although other animals have some precursors of a Gregorian mind, we agree with Dennett that a Gregorian organism has to have symbolic language. How then do Gregorian animals (humans) experience the world when compared with other (Popperian) animals? From our perspective, Gregorian organisms are symbolizing animals, the products of the third great transition, and we are interested in the evolutionary transformations that were necessary for their new type of goal-directedness to emerge. The evolution of symbolic representation and communication, of a new memory system, of the development of a "we" perspective, and of recursive mind reading are therefore of great importance. Moreover, as we see it, the ways in which symbolic language changed the subjective experiencing of humans, including their emotions, are of crucial

Figure 1.8
Gregorian organisms. The women in this orchestra create, by using symbols (musical notation) and artifacts, a new world of experience.

significance. Humans are not intellectual geniuses with the emotions of a chimpanzee. In profound ways we experience the world differently. The transition to having abstract values (justice, freedom, beauty, and others) that regulate social and individual behavior—the transition to rationality—occurred, as far as we know, only in the hominin lineage.

Dennett's generate-and-test hierarchy can be seen as an evolutionary (selectionist) interpretation of Aristotle's teleological hierarchy. Using Aristotelian terms, Darwinian organisms can be said to have a vegetative soul, while Skinnerian and Popperian organisms have a sensitive soul (although the Popperian ones are endowed with quite a bit of imagination), and the Gregorian organisms have a rational soul.

Although Dennett's hierarchy is useful as a general framework, evolutionary reasoning requires grounding in terms of phylogenies and scenarios. For example, if one claims that vertebrates are subjectively experiencing animals, the questions an evolutionist will ask are: All vertebrates? Only vertebrates? (Maybe some chordate ancestor was also endowed with a little bit of the EX factor? Maybe nonvertebrates, such as some arthropods or mollusks, are minimally conscious too?) Was it lost in some lineages? Did it evolve in parallel in some vertebrate and invertebrate lineages, and when did this happen? Can we work out the selection conditions (the scenarios) for each occurrence? What were the necessary (and jointly sufficient) molecular-neurological-cognitive-behavioral preconditions in the ancestral lineages, and how are these phylogenetically distributed? Every theory about consciousness suggests answers to these questions, and biologists have to produce the information needed. Even if one assumes that subjective experiencing is primitive, as Chalmers and Nagel think, or believes, like Lynn Margulis, that it is characteristic of all living organisms, the difference between a dog's and a tomato's levels of subjective experiencing is striking and requires a detailed and concrete biological explanation. Natural philosophers became fascinated by the evolutionary framing of this question as soon as theories of evolution emerged in the early nineteenth century. Their ideas paved the way, scaffolding and influencing subsequent research on the subject, and in spite of the many inevitable blind alleys they stumbled into, their work is full of intriguing and sometimes surprisingly relevant insights.

2   The Organization and Evolution of the Mind: From Lamarck to the Neuroscience of Consciousness

Beginning with the eighteenth-century associationists, we trace the origins and development of modern physiological and evolutionary ideas about mental life. The first evolutionary psychologist we discuss is Jean-Baptiste Lamarck, who at the beginning of the nineteenth century suggested that through adaptive self-organization mediated by use and disuse, apes were transformed into humans, and the nervous system changed from a system of communication in a humble polyp into the bicortical, sophisticated human brain. The second is Herbert Spencer, who argued that mental evolution proceeded from reflexes to habits and from habits to complex instincts, culminating in consciousness, whose different facets are feelings, reasoning, and willing. The third is Charles Darwin, who established the validity of the theory of descent with modifications and provided a powerful motor for adaptive evolution—natural selection. Although he carefully avoided the question of the origins of mentality, he used sexual selection to show that animals have mental traits, used the descent of man from the apes to show how mental faculties evolved in our lineage, and used the expression of emotion in man and the animals to show how emotions evolved through the processes of natural selection and use and disuse. The ideas of these thinkers deeply influenced Williams James, the father of modern consciousness studies, who applied the selection principle to the operation of the mind-brain, arguing that consciousness is adaptive scale tipping, which "loads the dice" of neural activities. For him, consciousness was a goal-directed, selection-based process that resolves perceptual ambiguities, guides actions, and enables inferences to be made. Although hugely influential, his view of psychology as the science of mind was soon supplanted by behaviorism, which dominated psychology for a large part of the twentieth century. Behaviorism redefined psychology as the study of controlling and manipulating observable behavior and, in its American version, banished mind and consciousness. The subsequent rise of the cognitive sciences paved the way for the return of James-inspired consciousness studies during the late twentieth century.

A full account of the development of evolutionary psychology in the nineteenth and twentieth centuries would require several book-length treatises. This is not just because the topic was of interest to so many scholars—it is also because those who had the most to say about it—Jean-Baptiste Lamarck, Herbert Spencer, Charles Darwin, and William James—were some of the most impressive and prolific natural philosophers of their time, and interesting lines of descent and dissent exist among them.[1] Lamarck laid down some of the theoretical evolutionary foundations on which Spencer developed his ideas, which then, together with Darwin's ideas on the evolution of human mentality and the expression of emotions, provided the scaffold for later ideas, such as those of William James. Although it is in the nature of scaffolds to be discarded after use, the intellectual structures erected by James and by his late twentieth-century followers form the basis of current theories of consciousness, including our own.

The common assumptions of all four thinkers sprang from physiological versions of the "associationist" theories developed in Britain and France during the late eighteenth century. These theories connected ideas to sensations and sensations to the physiology of the nervous system.[2] The notion that mental processes are made of (or as we will usually say, are constituted by) physiological processes in the nervous system facilitated the development of evolutionary ideas, and in the nineteenth century, it led to the inevitable conclusion that mental-physiological processes have evolved.

The Associationists

The view that mental activity is the result of the association of ideas is often traced back to Aristotle's reflections on recollection.[3] Like almost all of Aristotle's ideas, this proposal underwent many reformulations by theologians and philosophers, who broadened and qualified the basic claim that recollection occurs when one encounters features associated with the original experience.

In its eighteenth-century reincarnation, the explanation of recollection and other mental processes in terms of associations was based on several assumptions. The first, that complex thoughts and emotions are ultimately derived from sensations, was developed by John Locke. He believed that there are no innate ideas and that the mind is a blank slate, or, in his own words, "white paper," on which sensations are inscribed through the operation of innate, internal psychic principles.[4] The second assumption was that there are principles, like those enumerated by David Hume, that link sensation-derived ideas together: resemblance, spatial contiguity, cause and effect (which is for Hume contiguity in time), and contrast.[5] The third assumption was that this linkage is the result of the operation of some kind of force analogous to Newton's gravitational attraction. These associationist assumptions were not, at first, linked to materialistic or physiological notions; the elementary sensations that were the building blocks of complex ideas were seen as belonging to the realm of the mental, which was assumed to be different from that of the physical and the physiological. It was only with David Hartley (1705–1757) in England, and more explicitly and boldly with Pierre Jean George Cabanis (1757–1808) in France, that mental processes began to be understood in terms of the physiological activities of nerves.

David Hartley, an English physician and philosopher, was the first to develop a physiological account of the association of ideas.[6] He started with the assumption that the body's "component particles" are subject to the same "subtle laws" that govern all other material entities. Like all eighteenth-century scholars, Hartley was deeply influenced by Newton's theory of gravitation and sought general, Newtonian laws of human nature. But although Hartley's (1749) *Observations on Man, His Frame, His Duty, and His Expectations* was highly regarded by late eighteenth-century and early nineteenth-century English philosophers, it was in France that associationism became the basis of a secular science of man and mentality. This science was promoted in revolutionary and postrevolutionary France by the "ideologues"—Parisian intellectuals who were so called because they declared an ambition to construct a *science of ideas* that would establish a physiological basis for social and cultural reforms. As Destutt de Tracy, who used the term "ideology," boldly stated, ideology is a part of zoology.[7] The evolutionary theories of Lamarck, especially his theory of mental evolution, shared and extended the aspirations and assumptions of the ideologues.

Pierre Jean Georges Cabanis, a physician, philosopher, and social reformer, was one of the ideologue leaders. His principal work *On the Relations between the Physical and Moral Aspects of Man* (*Rapports du physique et du moral de l'homme*) was published in 1802. As the title indicates, he wanted to establish a bridge between the physical features of humankind and its highest mental-social aspects, and he did so through the development of what we may call "physiological psychology."[8] As a doctor and a social reformer, Cabanis was well aware of and fully appreciated research showing that faculties such as memory and reflection are associated with states of the brain, on the one hand, and with the physical and social environment, on the other. His conclusion was that "the moral is only the physique from a certain point of view."[9] He suggested that brain disturbances can bring about madness and frenzy, while hysteria and languor can result from disorders in the genital center of sensibility. Consciously following in the steps of the ancient Hippocratic doctors, he suggested that age, sex, temperament, illness, climate (a general term denoting the environment), and regimen (routines characterizing a lifestyle) shaped human physical and mental states. Moreover, the effects of these factors, when enduring and habitual, became hereditary. Therefore, the first stage in reforming human society was to alter the physical, social, educational, and working conditions of people's lives. By building new social institutions, such as a secular, universal school system and a reformed health system, new habits could be installed, and a fresh, physically and mentally balanced, healthy citizen could be formed.

For Cabanis, physical sensibility (or "sensitivity," a term often used interchangeably) was a property specific to living, neural organisms. Following the great Swiss anatomist and physiologist Albrecht von Haller, he claimed that it is sensibility that allows organisms to perceive and react to the impact of external objects, and it is sensibility that is the sole basis of sensations. However, Cabanis was not sure whether sensibility was a primary property of dynamic matter or a derived property stemming from gravitation. He was very clear, however, about the physiological nature of the sensations and thoughts that sensibility makes possible. The exquisitely complex and specialized brain, which all the physiological and medical studies had shown to be crucial for sensation and thought, was not fundamentally different, claimed Cabanis, from the stomach. In a much-quoted paragraph that was used by admirers as an inspiring manifesto of materialism, he wrote that the brain was

> a particular organ, especially destined to produce thought, just as the stomach and intestines digest, the liver filters bile, the parotid, maxillary and sublingual glands produce salivary juices. (Cabanis 1802/1891, p. 195)

With great intellectual energy and optimism, the ideologues, like the philosophers of the Enlightenment who inspired them, insisted on miracle-free reasoning based on experiments and observations, which would eventually explain the origin of living entities, the diversity of the species inhabiting the earth, human mentality, and social order. After all, if matter is inherently dynamic, why should it not metamorphose under the right conditions into a living organization? And if recurring behaviors can become habits and finally transmute into inborn behavioral tendencies and even full-blown instincts, as Diderot hinted and Cabanis claimed, what stands in the way of species transformations? The idea of a capacity for adaptation that can result in species transformation was quite prevalent among leading eighteenth-century intellectuals, and sensationalist theories played an important role in the rise of evolutionism, although, of course, sensationalism was just part of a suite of ideas and observations that affected the rise.[10] But only for one scholar, a biologist specializing in the taxonomy and systematics of plants and invertebrates, was biological evolution *the key* for explaining the diversity of the living world, the origins and variety of animal mentality, and human reason. Jean-Baptiste Lamarck, the first systematic evolutionist, not only *derived* evolution from sensationalism, the theory of habits, and spontaneous generations but also looked at things the other way around: understanding the evolutionary transition from nonminded to minded animals, and the subsequent evolutionary transformations of the nervous system, would, he believed, unlock the mind's secrets. In the third part of his famous (and infamous) *Philosophie zoologique*, Lamarck used his theory of evolution to develop new hypotheses about the nature of the mind and the ways of studying it.

The Evolutionary Psychology of Jean-Baptiste Lamarck

It is difficult to think of any other major biologist who has been as scandalously misrepresented, especially by fellow biologists, as Lamarck (figure 2.1).[11] Most nonhistorians who write about him have secondhand knowledge of his work and thought and depict him as the promoter of a naïve and erroneous theory of evolution through the inheritance of acquired characters, a theory that was finally dismissed during the twentieth century, when neo-Darwinism in its Modern Synthesis version finally triumphed.

Lamarck was a materialist in the sophisticated sense of late eighteenth-century materialism, which recognized the dynamic nature of matter.[12] In spite of references in his writings to the "Sublime Author of All Things" or "Supreme Power," references whose frequency waxed and waned with the political atmosphere in France, the Sublime Author was for him only a first cause, which for all scientific purposes could be safely ignored. He was fiercely opposed to the introduction of any transcendental principles into the study of nature. Moreover, it seems that he was hostile to religious observance, marrying his first wife only when she was on her deathbed and waiting until 1808 to baptize his children, when it became clear that not doing so might lead to trouble. He asserted that all biological phenomena, including the origin of life, the ability to feel, and the highest intellectual capacities of humans, could be explained by known physical and biological principles. He also claimed that humans evolved from apes and that religion was invented by the elites to subjugate the masses.[13] Such

Figure 2.1
Jean-Baptiste Pierre Antoine de Monet, Chevalier de Lamarck (1744–1828).
Source: Wellcome Library, London.

assertions, which he consistently repeated, match the penetrating characterization of Lamarck by the famous, temperamental nineteenth-century French literary critic Charles Augustin Sainte-Beuve (1804–1869), who as a very young man had attended elderly Lamarck's lectures in the Jardin des Plantes (box 2.1).

According to Lamarck, two basic principles accounted for the evolutionary changes that started with the tiny, fragile creatures produced by spontaneous generation and culminated (but did not end, since evolution is, at least in theory, open-ended) in humans. The first was the effect of the changing environmental conditions on physiological and hereditary processes. Lamarck assumed that living organization is very plastic. Indeed, to him, adaptive plasticity is inherent in the special type of self-organization that characterizes the living state. In young animals with malleable and soft tissues, adaptive plasticity is mediated through behavioral changes: changed environments lead to a need to cope with the change, and this leads to activities that alter the use of some organs, leading to new habits; when the conditions and habits persist for many generations, the new behavioral tendencies and their supporting organs become hereditarily stable. As a result, offspring develop more readily the characteristics that their ancestors had laboriously and imperfectly acquired, and the cumulative result of this process is the observed diversification and functional specialization of animals. The long neck of the giraffe is one of the many examples that Lamarck put forward to exemplify this process.[14] To this major principle, Lamarck added a supplementary process, hybridization. Once new races or species were formed, they sometimes hybridized, a process that further increased the diversity of living organisms and that can be seen in domestication.

The second basic principle behind evolutionary change explained the nature of adaptive plasticity. It was, Lamarck suggested, a result of the dynamics of the subtle fluids (heat, electricity) in the living entity that tended in time to increase complexity through activities involved in self-maintenance and growth. Self-maintenance was the result of a special kind of self-organization that occurred as the activities of the subtle fluids organized matter. This self-organization was the hallmark of life, and Lamarck elaborated on the process in the second part of *Philosophie zoologique*, which dealt with the origin of life. Becoming more complex was a physical inevitability, according to Lamarck, because the fluid movements were accelerated as the effects of their activities built up during the organism's growth and development. As we see it, Lamarck was expressing a very modern idea, which would today be described as a positive feedback reaction. He thought that the movements of fluids lead to the formation of better and deeper old and

Box 2.1

Charles Augustin Sainte-Beuve on Lamarck.

In his autobiographic novel, *Volupté*, Sainte-Beuve conveys his strong impressions of the lectures delivered by Lamarck, and of the man.

I attended frequently, several times per decade, the lectures in natural history of Mr de Lamarck, in the Jardin des Plantes; his teaching, of whose hypothetical paradoxes, and conflicts with other more positive and more advanced systems I was unaware, had a powerful attraction for me because of the serious and fundamental questions which it always raised and the passionate and almost sorrowful tone that was mingled with it. Mr de Lamarck was by then the last representative of this great school of physicists and observers who had reigned since the times of Thalès and Democritus until Buffon: he was profoundly opposed to the tiny chemists, experimentalists and analysts, as he called them. His hate, his philosophical hostility to the Flood, to Biblical Creation and anything reminiscent of Christian theology, was profound. His conceptions had much simplicity, bareness, and much sadness. He constructed the world with the least number of elements, the least number of upheavals and with as great a duration as possible. According to him, things arose of themselves, one by one, continuously, enduring for sufficiently long time, without instantaneous transformation for overcoming catastrophes, without disasters or commotions, without centers, growth-nodes or organs deliberately arranged to help them reduplicate. A long blind patience, this was his kind of world. The actual form of the earth, according to him, depended only on the slow deterioration of pluvial waters, on the daily rotations and the successive displacement of the seas; he admitted no great bowel movements in this Cymbeline [goddess of nature], nor the renewal of her earthly face by some temporary heavenly body. So too, within the organic order, the mysterious power of life was rendered by him as small and as basic as possible, assumed to develop by itself, order itself, construct itself little by little through time; the obtuse need, the only [source of] habit in diverse circumstances, eventually gave rise to organs, opposing the constant, destructive power of nature; For Mr de Lamarck separated life from nature. Nature, in his opinion, was stone and cinder, granite of the tomb, death! Life intervened there as an artful and strange accident, an extended battle, with more or less success or equilibrium here and there, but always finally conquered; cold immobility reigned before as well as after its occurrence. I loved these questions of origin and finality, this view of a dismal nature, these sketches of obscure energy. My reason was suspended, pushed to its limits, enjoying its own confusion. I was, of course, far from accepting these much too simple hypotheses, this uniform series of continuity that went against my exuberant feeling of creation and vigorous youth, but the boldness of a man of genius made me think. (Sainte-Beuve 1834/1986; translation by Eva Jablonka)

Lamarck's world was indeed devoid of grand cataclysms and mysterious forces: the origin of life was a recurring, chemically predictable accident; living beings evolved very slowly; individuals persisted for a very short time; and there were no miracles, no afterlives, no external telos. But it was, for Lamarck, a beautiful and rich world, a feeling that, it seems, was not apparent to the young and impatient Sainte-Beuve.

new paths that facilitate further movements, that allow growth, that promote further activities, and so on. The challenges of new conditions usually (not always!) add to rather than diminish the existing order. Together, the cumulative effects of these two related factors—complexification and adaptation to contingent conditions—generated over very long periods of time the diversity and increase in organizational complexity observed in the living world. The result is a branching tree, which started from several events of spontaneous generation (probably three—one leading to plants and two leading to animals) that became firmly established and gave rise to all plants and all animals. This progressive and branching process, presented in the first part of *Philosophie zoologique*, accounts both for the diversity and for the increase in complexity in the living world and leads to a natural system of classification.

The Physiological Evolution of the Mind

> In this fictitious entity [mind], which is not like anything else in nature, I see a mere invention for the purpose of resolving the difficulties that follow from inadequate knowledge of the laws of nature.
> —Lamarck 1809/1914, p. 286

Philosophie zoologique was part of a grand project: that of *Biologie*, the study of *all* living bodies—plants and animals—as well as their products—minerals. In the more modest project of *Philosophie zoologique*, the subject matter was animal diversity and functional biology explained through their physiology. The evolutionary origin and sophistication of animal life and mind were part of this scheme, and Lamarck devoted two-thirds of the book to the problems of the origin and evolution of life (part II) and mind (part III), dwelling with special pleasure, as he confessed, on the question of the origins and the evolution of mentality. His starting point, like that of Cabanis, was the assumption that the moral is an aspect of the physical.

Lamarck went further, however, than other ideologues, suggesting that only through an evolutionary analysis, starting from the simplest living beings, can one discover the basic principles underlying mental phenomena. Beginning with complex animals like man, on which Cabanis and many other philosophers focused, cannot lead very far, Lamarck argued, because the complexity of the organization of these already highly evolved animals makes their organization "the most difficult from which to infer the origin of so many phenomena."[15] It is only when we start with the simplest organisms that manifest mental life and follow the gradual evolution of mentality in animal lineages that we can discover what feelings and ideas actually are and how human mentality came about. To do this we must study the organs and processes that constitute and generate feelings: we must study the anatomy and physiology of the nervous system.

Anatomically, the nervous system envisaged by Lamarck consists of three major parts: (1) a pulpy medullary mass, which is a kind of center of communication, with many extensions and threads; (2) protective sheaths enclosing the central mass and the threads (nerves) emanating from it; and (3) a subtle fluid that he said is fundamentally electric current, which is the factor active in transmitting sensory inputs, eliciting motor outputs, and tracing activity marks on the soft medullary mass. Although not all animals have a nervous system, claimed Lamarck, these three major elements are present in all of those that do. However, nervous system organization differs in different taxa, and there are many variations and gradations between the different systems. That is why the nervous system provides the best suite of characters for the classification of neural animals, just as in plants the best organs for classification are the reproductive parts, the flowers.

Lamarck recognized three major stages in the evolution of the nervous system, with endless gradations in-between. The simplest system was a nervous system with a medullary mass or masses (ganglia) with nerves, which could control and coordinate motor actions but not feelings; this system, he suggested, seems to characterize creatures such as sea urchins and possibly sea anemones. In the next major stage, seen in most invertebrates and in simple vertebrates, a more centralized sensory system evolved. In these animals, one ganglion (clusters of nervous tissue), the head ganglion, a true brain (according to Lamarck, in vertebrates it was the brain stem), was responsible for the integration of sensory inputs that produced feelings, and these feelings could then guide and coordinate behaviors by communicating with the more ancient motor ganglia. During the third stage, the two cortical hemispheres on top of the primary brain, which enable feelings to be combined and reorganized into thoughts, emerged.

Lamarck was well aware that the ability to have feelings was generally assumed to be an immaterial faculty, and he therefore emphasized again and again that he regarded it as the effect of an entirely physical, and rather simple, process. He claimed that the faculty of receiving sensations from the sense organs and the brain constitutes a feeling, a physical sensibility,[16] and that this is the hallmark of enbrained animals. Sensations, at their simplest, arrive directly from the sense organs via the nerves to the "sensory nucleus" in the brain, get distributed to the whole body, and come back again to be resent to the sensors from which they had originally arrived. Sensation is never a local process; it is always holistic. The mechanism of sensation

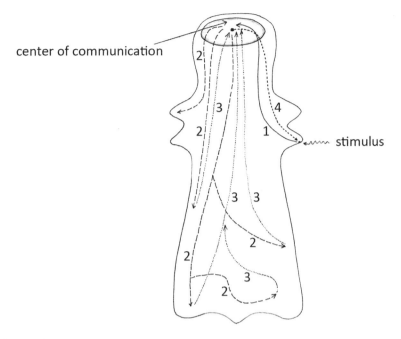

Figure 2.2

Lamarck's model of sensation and feeling. A stimulus triggers electrical conduction through the nervous fluid—along one of the nerves—through route 1. The signal arrives at a "center of communication," the sensory nucleus. It affects its activity, and as a result, signals are sent from the center and travel throughout the entire nervous system to the whole body, routes 2, and from all parts of the body back to the center of communication, routes 3. Only after reintegration at this center is a signal sent back to the original site of stimulation, through the same sensory nerve 1, to produce sensation, which the organism feels as if it were located in the actual point originally stimulated. *Source*: Based on Lamarck 1809/1914, pp. 325–326.

is illustrated in figure 2.2, which, we hope, does justice to Lamarck's rather complicated proposal.

In this account of sensation and feeling, Lamarck did not include motor responses. In a later chapter, he linked the sensory response with a motor reaction, claiming that sensations usually lead to actions that then affect sensations. Sensations from external sources and from internal sensors deep within the body become integrated in the center of communication and form a basic feeling, which Lamarck called the "inner feeling," an overall sensory state that has the faculty of exciting and controlling movement and that is itself activated by physical sensations or, in animals with cortical hemispheres, by acts of intelligence. The inner feeling is one of the most interesting concepts Lamarck developed, one he was particularly proud of, and one that was almost totally neglected for the next two hundred years. At its most elementary, it is simply a feeling of existence,[17] the result of the integration of internal sensations:

> The feeling in question ... results from the confused assembly of inner sensations which are constantly arising throughout the animal's life, owing to the continual impressions which the movements of life cause in its sensitive internal parts.... the sum-total of these impressions, and the confused sensations resulting from them, constitute in all animals subject to them a very obscure inner feeling that has been called the feeling of existence. (Lamarck 1809/1914, pp. 333–334)

Lamarck's inner feeling is linked to some intrinsic values that can stimulate adaptive, reinforcing, or neutralizing motor behaviors. These values are expressed as needs (*besoins*), one of the most misunderstood Lamarckian notions—and one that has been responsible for the ridicule that Lamarck's theory has received over the years from the many biologists who never took the trouble to read the original text or who, like Georges Cuvier, waged a holy war against his ideas. Since the notion of needs is used in the book, and since it is crucial for understanding Lamarck, Spencer, and Darwin, here is how Lamarck characterized "needs":

> Now the needs of animals with a nervous system vary in proportion to their organization, and are as follows:
> 1. The need of taking some sort of food;
> 2. The need for sexual fertilization, which is prompted in them by certain sensations;
> 3. The need for avoiding pain;
> 4. The need for seeking pleasure or well-being.
>
> For the satisfaction of these needs they acquire various kinds of habits, which become transformed in them into so many propensities; these propensities they cannot resist nor change of their own accord. Hence the origin of their habitual actions and special inclinations, which have received the name of instinct. (Lamarck 1809/1914, p. 352)

These needs are elicited by the conditions of life (e.g., lack of food or the presence of predators) and organize the activities of animals by negatively or positively reinforcing their activities so that the needs are satisfied. This is the ancient pleasure-pain principle, to which Lamarck added hunger and the sexual drive, believing that pain and pleasure are complex values— needs that emerged (relatively) late in evolution. The basic emotions are those integrated manifestations of the inner feeling evoked by the fundamental motivations that lead to action: anger, fear, joyous excitement, and so on. To these almost universal motivations/needs, he added "moral needs," such as moralistic anger, which are the products of thought and are characteristic of man.

Lamarck's "list" of characteristics of a state of consciousness or subjective experiencing are therefore: (1) a centralized nervous system with a sensory nucleus (where all sensations converge) and a rich array of afferent and efferent fibers through which electricity flows; (2) physiological integration processes that occur in the center of communication, relating the internal state of the whole organism to sensations arriving from the external world; this integration constitutes the inner feeling, which leads to interactions with the motor center(s) and the induction of actions; (3) value systems, or needs, the satisfying of which determines how the animal behaves. A point or zone of transition to consciousness was not spelled out by Lamarck, and the detailed scenario of evolution is unclear. It *is* clear, however, that Lamarck assigned some rudimentary ability for subjective experiencing to invertebrates possessing head ganglia and obvious sensory organs, although he thought that the "radiata" (e.g., sea urchins) do not experience, since they lack a head ganglion. According to Lamarck, once the cortical hemispheres started to evolve, higher intellectual functions, including reasoning and the ideas, emerged. This obviously happened in vertebrates and was a slow process, with man being its present culmination.

The mechanisms that drove the evolution of animal mentality, from sea urchins that merely coordinate their movements to humans who can reflect about evolutionary processes, were, Lamarck suggested, the already familiar processes underlying the transition from habits to instincts: in novel conditions, needs generate new positively or negatively reinforcing behaviors; new behaviors lead to new habits; and new habits become, in time, new instincts, supported by new anatomical structures. Instincts, Lamarck maintained, are neither automatic, passive reactions like those of plants or sea urchins nor "innate ideas," a notion he thought ridiculous; rather, they are compelling tendencies to behave in ways that satisfy the needs that in past conditions led to the persistent habits of the ancestors.

From feelings, instincts, and skills (which in invertebrates are just trains of instincts), Lamarck went on to derive higher mental faculties like thinking, imagining, and judging. Lamarck suggested that in order for thoughts to exist, there must be a supporting organ: the cortex of the brain or, as he called it, the "hypocephalon." This is necessary because it is in this highly complex and differentiated organ that the sensations and perceptions can be engraved; it is on this special and soft neural material that they leave memory traces. The simplest thought was, for Lamarck, the recalling of sensations, when as a result of an "effort of the inner feeling," an effort that is an act of attention, the traces of past sensations are activated. Lamarck's emphasis on attention as a prerequisite for having ideas may appear to have modern overtones, and it is indeed one of his many intuitive leaps. However, it is important to note that attention was for him a result of neural fluid dynamics. It was the effect of the inner feeling transporting and directing the available part of the nervous fluid toward the cortex, making it ready to retrace and thus reawaken old ideas or create new traces, new ideas.[18] Complex ideas, he suggested, arise when different traces become coactivated and thus associated, and the animal combines and manipulates them.

Lamarck's assumption about the dynamics of the subtle fluids in the nervous system led him to the conclusion that thought, both simple and complex, is an action and process[19] and that some feelings (e.g., moral feelings) are the natural results of thought, although in animals without a cortex, feelings exist without thought. However, thought always depends on sensation because by definition it is only when sensations become engraved in the cortex, and the engraved pattern is reactivated, that thought arises. Moreover, for Lamarck this continuity between sensation and thought in encephalized animals meant that the mental evolution of these animals followed the same associationist principles as those of simpler cortexless animals. He suggested that the ideas produced in the cortex can become associated when the fluids pass through the cortical areas in which they are engraved at the same time or in succession. Moreover, complex mental habits can develop too. For example, in man, education creates the habit of thinking: fixing attention on many things, comparing, combining, and so on, become habits of thought. The evolution of man's mentality is a result of the accumulation of many such complex, open-ended mental habits, which were greatly reinforced and sophisticated by language. Will is an intellectual act, dependent on the ability to judge, and hence there is no free will, claimed Lamarck: the illusion that we have free will is the consequence of the great number of factors that impinge on our judgments, many of which we are hardly aware.

Lamarck's ideas about the evolution of mentality received very little attention, even from those nineteenth-century evolutionists who were interested in the same questions, such as Herbert Spencer, Charles Darwin, and William James. When they did refer to Lamarck, it was his general evolutionary theory, especially use and disuse, that they discussed, not his ideas about the evolution of the mind. Perhaps this is understandable because his eighteenth-century neurophysiology was behind the times, and he left open some big and obvious questions, such as the origins of the first nervous system and the evolutionary origins of the cortical hemispheres. But Spencer, Darwin, and James were people who delighted in open questions. Did they read part III of Lamarck's *Philosophie zoologique*, or did they lose patience as they struggled through his outmoded neurophysiology? As we show in the next sections, there are some remarkable parallels between some of Lamarck's ideas about the nature and evolution of feelings and those developed later, especially the ideas put forward by Spencer and James, although both went, in many ways, far beyond Lamarck.

Herbert Spencer and the Evolutionary Principles of Psychology

Herbert Spencer (figure 2.3, *left*) was undoubtedly the most influential evolutionary philosopher in the nineteenth century—as important as Charles Darwin in the dissemination of evolutionary ideas and with far wider popular appeal. He was read widely by both professional academics in various disciplines and by lay people and had a huge influence on people's domestic life, on legislators, on education, on the establishment of the disciplines of psychology and sociology, and on the curricula of universities. Serious young lovers, like the young Ivan Pavlov and his fiancée in czarist Russia,[20] cemented their relationship by reading Spencer together (figure 2.3, *right*), factory workers discussed his work, and civil servants organized their lives in line with his philosophy.

The dramatic decline in Spencer's influence in the twentieth century—a decline to the point of near extinction among laypeople and contempt among academics—is an interesting and somewhat distressing story in the sociology of knowledge, a field that Spencer himself helped to establish. To a large measure, Spencer's decline can be traced to changes in the social and political milieu (including political-sociological thought), but it is also partly due to his adherence to unfashionable Lamarckian notions and his idiosyncratic "outsider" personality, which recoiled from any institutional associations.

We know quite a lot about Spencer's life and about his personality. He not only wrote an autobiography but was also associated with superb writers and observers of human nature, and his great fame ensured that people recorded their meetings with him and reported on his many eccentric behaviors. Spencer, completely focused on his work and on his own affairs, was a hypochondriac walking around with earmuffs, checking his pulse and retiring to rest his delicate nerves even when he invited guests to his

Figure 2.3
Spencer and his influence. *Left*, Herbert Spencer (1820–1903). *Right*, reading Spencer together as described by Chekhov's (1891/1916) *The Duel*, p. 7: "To begin with we had kisses, and calm evenings, and vows, and Spencer, and ideals and interests in common."

rented house (he never owned one). He refused most of the many honors that were offered to him, was obsessively concerned with issues of priority and with real and imaginary misunderstandings of his work, and always voiced his opinions, however unpopular.

The Law of Progressive Evolution

Evolution, according to Spencer, is a process of progressive complexification—a process inherent in any physical system, from atoms to societies. It is not extrinsically teleological: it is the result of obedience to physical principles, as inevitable as the fall of an apple to the ground because of the force of gravity. Spencer started from the reasonable assumption that a state of nature is, in the beginning, simple and relatively homogenous. The condition of homogeneity is, however, unstable because environmental conditions, which are never perfectly symmetrical, impinge on living matter, and therefore the original homogenous state cannot last. Because physical entities are dynamic and inevitably interact, the state of the system changes, and since any single cause leads to more than one interaction and hence to more than one effect, it follows that an increase in heterogeneity is inevitable.[21] Since Spencer assumed that the heterogenous state is more stable than the homogenous one, he concluded there was a universal tendency toward increased heterogeneity and complexity—for progressive evolution in the broad sense. The arrow of time and the arrow of increasing complexity were coincident.

Spencer provided a general definition for this universal law of progressive change: "Evolution is definable as a change from an incoherent homogeneity to a coherent heterogeneity, accompanying the dissipation of motion and the integrations of matter."[22] How motion is dissipated and how integration occurs differ for different systems. In the bodies of living organisms, just as in animal and human societies, the dissipation and integration take the form of increased division of labor, exemplified by Karl Ernst von Baer's law of embryological development—the progression from the general and simple to the more complex and specialized. Individual organisms and groups of organisms can respond to challenges, such as diminished resources, with a more efficient division of labor.[23] Spencer's response to Thomas Robert Malthus's gloomy predictions about overpopulation was that when populations grow and resources run out, the group reorganizes, new specializations are formed, and the better use and production of resources that follow can support the increased group size. Moreover, he argued, complex animals tend to reproduce less than simpler ones because there is a trade-off between the fertility and complexity of organization, so the problem of overexploitation of resources further diminishes. Interestingly, toward the end of his early paper discussing this, Spencer (1852) suggested that differential survival of the better adapted was another result of the challenge of diminished resources, but he did not develop this suggestion. When Darwin's On the Origin of Species was published in 1859, Spencer recognized the power of this idea and regretted that he had overlooked its importance. From then on he regarded Darwinian natural selection as one of several factors driving evolution, usually assigning to it the negative role of weeding out the unfit. Lamarckian adaptation through use and disuse still remained central to his view of evolutionary change, especially with respect to complex ("higher") neural animals, which used their behavioral-neural plasticity to cope with the environment during their own lifetime, something that eventually led to the formation of new heritable habits.

The Evolutionary Emergence of Feeling

Feeling, a Spencerian term (with a capital *F*) that is roughly equivalent to what we call "subjective experiencing," was for Spencer the first manifestation of a conscious state. However, in order to develop a theory of mentality based on Feeling, Spencer knew that he needed more than general evolutionary principles. The way he incorporated the trade-off between fertility and organizational complexity into his analysis of the Malthusian problem shows that he was aware that in addition to the general principles that apply to all evolutionary systems, something more is required for explaining a particular evolutionary problem. When discussing the evolution of mentality, he added information based on the neurobiology and the associationist psychology of his time, as well as some philosophical arguments about the nature of mental activities. In fact, it seems that one of the driving forces for *The Principles of Psychology* was to provide a philosophical critique of the associationism of John Stuart Mill (1806–1873), the most influential nineteenth-century British philosopher. Mill added two components to previous associationist ideas: First, he argued that the association of two ideas can lead to a result that is different from the sum of their components, as water is different from the oxygen and hydrogen from which it is made. Mill's "mental chemistry" could therefore account for the complexities of sophisticated types of reasoning and emotions that were difficult to explain on the assumption of a simple addition of sensations. Second, Mill added motivation and action to the passive accounts of English associationists. Although he claimed that pleasure is the prime motivator of action, for humans pleasure is not a simple gratification: ideals of justice and freedom have become associated with pleasure, and human welfare requires more than the simple fulfillment of physical needs. Although Spencer accepted the basic tenets of Mill's mental chemistry as well as the central role of the pleasure principle, he argued that Mill failed to recognize the deeply embedded neuromental foundations of all cognition. He believed that the exploration of these neural foundations could lead to an understanding of the evolutionary complexification of cognition and explain how the basic "axioms" underlying all thought had formed.

At the core of Spencer's critique was what he called the Universal Postulate: the idea that knowledge the negation of which is inconceivable is the knowledge that has the highest validity (is most likely to be correct).[24] Such knowledge includes not only logical abstractions but, crucially and fundamentally, some types of sensual-perceptual knowledge, such as our intuitions about space, time, and causality. For Spencer, certain knowledge is not logical a priori knowledge; it is the naturally constructed knowledge on which all our conceptions (including logical postulates) are founded because it is the product of experiences and relations that have persisted over eons of time. Mill's empiricism and associationism cannot account for the process of cognition without recognizing evolutionary history, which provides the foundations on which processes of association can occur. Spencer's critique of Mill remained central to his view of psychological and cognitive evolution, and he believed that his approach provided a reconciliation between the views of Locke and Immanuel Kant and their followers Mill and William Whewell (1794–1866), respectively. The Kantian a priori (Kant's "first principles"), claimed Spencer, is an evolutionary, Lockean a posteriori. Spencer thus put forward a logico-psycho-neural identity theory, uniting the logical, the mental, and the neurophysiological and showing their continuity. Associated feelings and ideas, he suggested, are embodied as neural connections, and these connections have evolved and have gradually become more complex, leading to the deep properties of the mind we see in neural animals; in man they lead additionally to logically compelling axioms.

Given the Universal Postulate, how did consciousness and intelligence actually emerge during evolution? To answer this question, Spencer defined the "abstract law of intelligence," pointed to the basic unit of neuromental action, characterized consciousness, and then proceeded to show how, from humble beginnings, a Spencerian intellect could evolve. His abstract law of intelligence states: "The strengths of the inner cohesions between psychical states must be proportionate to the persistencies of the outer relations to which they answer; and the development of intelligence into conformity with the law, being, in all cases of which we have positive knowledge, secured by the one simple principle that the outer relations produce the inner relations, and make the inner relations strong in proportion as they are themselves persistent."[25]

Spencer called the inner relations (between circuits in the brain) that correspond to or map perceived relations in the external world ("outer relations") "representations,"[26] and for the time being, we use the term "representation" in this sense. The representations that are the most coherent and most adjusted to outer relations, said Spencer, persist, while the less coherent disintegrate. Adaptation to the external conditions is the basis of this process.

Spencer suggested two sources for the production of associations. In the first edition of *The Principles of Psychology*, he suggested that "the growth of intelligence depends on the law that when any two psychical states occur in immediate succession, an effect is produced such that if the first subsequently recurs there is a certain tendency for the second to follow."[27] He illustrated this principle with the example of a shadow that becomes associated with touch. A touch normally elicits a reflex contraction response from an animal, and if a shadow is systematically associated with the touch, in time a link is formed in the nervous system, and the shadow alone will elicit a contraction response. In the second edition of *The Principles of Psychology*, Spencer introduced an additional type of associative learning, suggesting that successful (pleasurable) effects lead to the reinforcement of the actions that brought them about: "On the recurrence of the circumstances, these muscular movements that were followed by success are likely to be repeated; what was at first an accidental combination of motions will now be a combination having considerable probability."[28] This idea is the same as that suggested by Alexander Bain (1818–1903), one of the founders of modern psychology and a close associate of Mill, whom Spencer knew personally and whose books he read. Bain followed in Hartley's footsteps and explored the relations between neurological and psychological processes, stressing the role of early development and learning in the formation of complex behavior, although he did not introduce evolutionary principles. In his books *The Senses and the Intellect* (1855) and *The Emotions and the Will* (1858), Bain stressed the continuity between reflex actions, spontaneous activity, and learned behaviors. He suggested that beginning very early in development, shortly after birth, motor acts, which are at first spontaneous and chaotic, become organized: some actions become reinforced because they are pleasurable, while others are suppressed because they are unpleasant.[29]

Spencer did not attribute reinforcement learning to Bain, possibly because he regarded his own (critical) endorsement of Mill's associationism and his own discussion of learning by association in the first edition of *The Principles of Psychology* as covering this case. In the second edition, however, reinforcement learning, which in time became known as the Spencer-Bain principle (and in the twentieth century, as instrumental or operant conditioning), was given center stage. Spencer not only employed this principle to explain how complex behaviors and habits are formed during ontogeny but he also speculated about the neural basis of associative learning, something that Bain refused to do because he knew little about neurology. Spencer explained that as a result of pleasure accompanying particular motor acts (e.g., suckling), the neural routes of communication become active, facilitating the subsequent passage of neural discharges leading to the same action. If the association is very stable and persists for several generations, a new instinct will emerge. Moreover, he explained that the categories of pleasure and pain are themselves products of evolution: beneficial and deleterious experiences have become associated with feelings that were categorized as pleasant or painful, respectively. This categorization, Spencer claimed, resulted from natural selection. He considered natural selection to be especially important during the early stages of mental evolution, although it always accompanies and reinforces evolutionary changes in the nervous system that are brought about through the inherited effects of use and disuse.

Spencer suggested that the unit of neural evolution was the "reflex action," a term describing nerve-muscle interaction, which had become well known following the neurophysiological studies of Marshall Hall in 1831. However, although he considered it to be a core building block of neural activity, Spencer did *not* consider the reflex as a unit of consciousness; it was a basic *neural* process from which other more complex processes evolved. Reflex action, which he described as a single contraction following a single sensory irritation, already presupposes a differentiation between afferent sensory and efferent motor nerves, a center of communication, and a contractible muscle. The sensory stimulus is carried to a ganglion (the communication center) through the afferent nerve and is reflected from it through an efferent nerve that elicits a muscle contraction. An instinct, which may be quite elaborate, is a compound reflex action: it is the result of several reflexes occurring in parallel and in succession following relatively simple stimulation. Complex instincts are, in turn, the platform on which the first vague and preliminary manifestations of consciousness appeared. The processes of composition and coordination that are involved in eliciting and executing highly complex instincts become less determined; conflict between different neural connections may occur, the system may "hesitate" as the most coherent representations become established, and as a result the elicitation and the response cease to be entirely automatic. These neural processes take time, and the ongoing activity of integrated sensory stimulations leads to Feelings, the basic units that constitute all forms of consciousness and include both felt sensations and perceptions: "It [the mind] consists largely, and in one sense entirely, of Feelings. Not only do Feelings constitute the inferior tracts of consciousness, but Feelings are in all cases the materials out of which, in the superior tracts of consciousness, Intellect is evolved by structural combination."[30]

Basic feelings were, however, just one facet of conscious activity. Spencer suggested that feelings, memory, reason, and will are different facets of the same psychical activity that constitutes the conscious (Feeling) state, which is characterized by being highly complex and hence inevitably nonautomatic and temporally "extended." For Spencer, as for Lamarck, "memory" means not merely the reactivation of neural traces (neural internal relations) of past responses but active recollection, a mental search process involving conflicts and competition between different neural representations (the "inner relations" that correspond to the "outer relations"). Because these activities take time, the activity can be rendered conscious; Spencer firmly believed that for an inner neural state to become conscious, the neural activity has to have some minimal temporal persistence. The strongest links between neural representations are those that in the past occurred most frequently, and hence the "winning" relations are likely to be those best adapted to these circumstances. Spencer noted that "an [internal] action thus produced, is nothing else than a rational action,"[31] thus equating rationality with adaptation. Will, too, is a facet of the same process of neural reactivation, association, and searching among competing neural paths. Will involves the reactivation of an internal neural representation of a motor change: when the represented act is reactivated (recalled) and is in the process of becoming an actual motor act, the experience of willing emerges. Needless to say, with such a neurological notion of the will, the idea that there is free will was for Spencer a nonsensical, meaningless notion. There is a feeling of free will due, just as Lamarck had said, to the multiplicity of factors, few of which we are aware of, that lead to actions. Moreover, as Spencer emphasized, though we are free to do what we desire, we are not free to desire or not to desire. Free will is an illusion, a reification of a Feeling, which is then believed to have an independent existence.

Spencer's evolutionary psychology is rich and complex, and in this chapter we are merely pointing to some elements that are of particular relevance to his views about the evolution of Feeling. However, like Lamarck, Spencer did not provide a particular scenario for the evolutionary emergence of Feeling. What he did do was provide a detailed functional account of how, in general terms, subjective feelings and related conscious faculties progressively evolved from reflexes, their combinations and coordination, the coordination of coordinations, and so on. The account of how nerves and ganglia, the tissues involved in reflex actions and instinctive acts, themselves originated and evolved was presented with ample speculative detail in the 1870 second edition of *The Principles of Psychology*. We can infer from this account that he believed that some invertebrates that have complex sense organs and manifest complex behavior (e.g., bees, cephalopods) probably have some elementary ability to subjectively experience; that all vertebrates manifest basic consciousness; and that some vertebrates (those with a cortex or its equivalent) also have the ability to form representations of representations—that is, abstract ideas.

Spencer, like Darwin, was interested in the expression of emotions, in the "language of emotions," as he called it, and this interest seems to have stemmed from his love of physical beauty and of music, which were for him the supreme manifestation of the expression of complex, and in this case utility-free, emotions. Like everything else he ever considered, he analyzed the expression of emotions, including the social emotions, from an evolutionary point of view. Some of these analyses were published in articles he wrote in the 1850s and 1860s, but a summary with further extensions formed a chapter in the second edition of *The Principles of Psychology*, which appeared shortly before Darwin's *The Expression of Emotions* was published. In that chapter, Spencer suggested that the expression of emotions—facial expression, vocalizations, gestures—result either from "diffuse neural discharge," such as that seen when overwhelming joy or overwhelming anger overtakes one (when the neural discharges flow along the least resistant neural paths), or from "restricted discharge." Restricted discharge, which is far more specific, can be either undirected—the result of evolutionarily established instincts and their underlying neural structures—or directed—the result of deliberate voluntary activities. As we discuss in the next section, Darwin had very similar ideas. According to Spencer, the evolution of the diffuse and unguided restricted discharges followed the use and disuse principle. He explained, for example, that expressions of fear are reactions that first appeared, for good adaptive reasons, in combat situations, and because of their functional adequacy they became innate. Nevertheless, although he thought that the expressions of many emotions are underlain by innate dispositions, he believed that humans possess a considerable ability to control some of them. Human emotional expressions are a complex mix of voluntary and involuntary discharges and motors actions.

Spencer's general evolutionary theory, including his evolutionary psychology, is largely forgotten today; when mentioned, with the exception of a few historians, it is treated with derision. Ernst Mayr, the most dominant evolutionary biologist in the second half of the twentieth century, devoted only three paragraphs to Spencer in his massive and hugely influential 1982 book about the history of biological thought because, he claimed, Spencer's "positive contributions [to evolutionary theory] were nil."[32] We beg to differ.

Charles Darwin and the Mental Continuity between Man and Animals

In sharp contrast to Spencer, Charles Darwin (figure 2.4) is idolized by biologists, philosophers, and historians, and everything that he wrote is treated with an almost religious reverence, an attitude that we suspect would have

Figure 2.4
Charles Darwin (1809–1882), with an unmistakable expression of tenderness and
love, and his son William Erasmus Darwin in 1842. Reproduced by permission of
Cambridge University Library.

deeply embarrassed and disturbed him. There is, however, little doubt that
Darwin's ideas are crucial to any evolutionary account we may offer, so
although Darwin himself dwelled neither on the evolutionary origins of the
ability to experience nor on the evolution of the nervous system, he did
contribute to our understanding of the evolution of mentality. First, he
developed and firmly established the idea of descent with modification.
This idea was not original to him: as already indicated, both Lamarck and

Spencer had endorsed it. But Darwin argued so forcefully in its favor and marshaled so much evidence from so many fields that he convinced the biological world of its validity. His second major contribution is his powerful idea that natural selection can explain the design of complex functional traits, including mental ones.

Charles Darwin's Selection Theory

Darwin believed that evolutionary change can result not only from the natural selection of blind variations but also from the hereditary effects of use and disuse and the direct effects of the environment. But in whatever way variation is generated, natural selection was crucial in shaping the variations into complex adaptations.

Many summaries and generalizations of Darwin's idea of natural selection are available, but the one we use here is based on John Maynard Smith's summary.[33] The properties that any population of entities must possess in order to evolve by natural selection are

1. multiplication (one entity begets others);
2. heredity (entity A usually begets A descendants, entity B usually has B descendants);
3. hereditary variation (entity A sometimes begets A', which then begets A'); and
4. some hereditary variations affect the chances of multiplication.

These properties make evolution in a population inevitable. In fact, it can be said that natural selection is an emergent property of the interactions between them. The logically simple algorithm of selection hides endless possibilities and multiple realizations, and its explanatory power is immense. The generality of the idea allows it to be applied to disciplines as different as cosmology, economics, culture, and ethics, as well as to processes occurring in the brain.

Mental Evolution and the Expression of Emotions

> I must premise that I have nothing to do with the origin of the primary mental powers, any more than I have with that of life itself.
> —Darwin 1859, p. 207

As the above quotation makes amply clear, Darwin consciously avoided the questions of the origin of life and the origin of consciousness and quite firmly (with some private digressions) stuck to this decision all his life. He also never provided an account of the evolution of mental abilities from the simplest animals to the most complex. Nevertheless, like Lamarck and Spencer, Darwin believed that animals, including some invertebrates, not only feel hunger and pain but also have quite complex emotions, and he considered human rationality as a cumulative product of the evolution of animal mentality. In this way, he narrowed the gap between man and the animals as much as possible.

Darwin (1871, 1872) developed his ideas about mental evolution in two books, published within the span of just over a year, which became immediate best sellers: *The Descent of Man, and Selection in Relation to Sex* and *The Expression of the Emotions in Man and Animals*. In the first part of *The Descent of Man*, Darwin argued for evolutionary and mental continuity between man and other animals, with an ape ancestor in the human lineage. He made his motivation for writing the book crystal clear: the evolutionary process leading from ape to human mentality can be understood without introducing external powers or mysterious principles; the evolutionary principles expounded in *The Origin of Species* are sufficient to explain the evolution of man. Even the "lower animals, like man, manifestly feel pleasure, and pain, happiness and misery."[34] He granted joy and happiness to insects, which, he claimed, play together, with ants having been observed chasing each other playfully "like so many puppies." The higher animals are not only similar to man in their senses and basic instincts: Darwin claimed that they also have "similar passions, affections and emotions, even the more complex ones; they feel wonder and curiosity; they possess the same faculties of imitation, attention, memory and imagination and reason though in different degrees."[35] Even human-defining traits, such as a moral sense, language, and religiosity, have their precursors in mammals, most notably apes. These are all complex social dispositions that are half voluntary and half instinct and are products of a long evolutionary history.

In order to explain the evolution of human faculties, Darwin mobilized his suite of evolutionary mechanisms—natural selection, use and disuse, and the direct action of the environment. However, in *The Descent of Man*, he highlighted the importance of sexual selection (competition for mates) and group selection, which he believed are of special importance for explaining some human mental faculties. For example, human communities whose members have dispositions toward altruism will be ready to make sacrifices for the welfare of the group and can therefore outcompete human communities with members who have less altruistic dispositions. Through selection among communities, altruistic, cooperative dispositions spread in the hominid lineage—and with them anything that enhanced cooperation, such as emotions of shame and guilt, which in turn further reinforced human cooperative social life. Sexual selection, too, helps to explain the evolution of enhanced mental faculties. For example, Darwin explained that human males, who compete for females and are chosen by them, evolved to be ever more strong and intelligent; because of male choice, females evolved to become more beautiful.

Darwin left no doubt that both sexual and group selection are found in other animals. Selection among family groups, for example, is used to explain the evolution of social insects' caste organization and altruistic behavior, and the whole second part of *The Descent of Man* is devoted to sexual selection in animals, starting with insects and ending with humans. Sexual selection, especially mate choice, explains the evolution of certain secondary sexual characteristics and differences among members of different groups ("races") that cannot be explained by natural selection for survival. Sexual selection also suggests that the "lower" animals possess mental powers: intelligence is useful when competing for mates, and mate choice implies a mental preference. Darwin wrote in the first chapter of *The Descent of Man* that "with respect to animals very low in the scale, I shall have to give some additional facts under Sexual Selection, shewing that their mental powers are higher than might have been expected."[36] When he found evidence for sexual selection in a taxon, it meant, to Darwin, that its members are endowed with will, desire, and choice. Darwin found evidence for sexual selection (and the associated mental powers) not only in all vertebrates but also in several groups of insects, notably the Homoptera, Orthoptera, and Coleoptera. He found no evidence for it in the protists, coelenterates, echinoderms, and annelids, an absence that confirmed his belief in the "lower mental powers" and the "imperfect senses" of these animals. However, Darwin did not regard the absence of sexual selection as definite evidence for a lack of mental powers, for he found no evidence for sexual selection in cephalopods, whose high intelligence he recognized.

In *The Expression of the Emotions in Man and Animals*, Darwin discussed a somewhat less controversial and far more empirically accessible topic—the similarity in the way emotions are expressed in animals and humans. The emotions that are expressed, unlike the emotions one feels, are in the public, observable ("objective") domain; they are not inferences from behavior, they are behaviors. They can be systematically studied and compared by (1) observing infants and blind people, who cannot imitate from others the expression of the emotions they feel, thus indicating which expressions of emotions are innate; (2) observing the expression of emotions in animals, especially the higher apes, which can demonstrate the similarity and evolutionary continuity between apes and man in this respect; (3) observing the insane, whose voluntary control of their emotions is reduced and who therefore express emotions (relatively) unmodified by social norms; (4) scrutinizing works of art, which, because of the talent of artists, can highlight wide ranges and aspects of expression; and (5) comparatively studying the expression of emotions in widely different cultural groups using photographs and questionnaires. The use of photographs is a novel methodology that Darwin introduced for comparing the way emotions are expressed and, with modern modifications (videos), is still in use today.

The principles that, according to Darwin,[37] underlie the expression of emotions all stem from the associationist ideas that a response to a stimulus that is spatially or temporally contiguous with, or contrasts to a stimulus that elicited the original expression, becomes in time, after much repetition, innate. Darwin suggested three principles: "The principle of Serviceable Associated Habits.—Certain complex actions are of direct or indirect service under certain states of the mind, in order to relieve or gratify certain sensations, desires, &c.; and whenever the same state of mind is induced, however feebly, there is a tendency through the force of habit and association for the same movements to be performed, though they may not then be of the least use."[38] Expressions of aggression (anger, hate, spite), which are not only grounded in the physiology of the animal but, like the baring of the canines, were (and may still be) useful in situations of conflict, are good examples of this principle at work. The similarity to Spencer's "undirected restricted discharge" is obvious. The second principle is "The principle of Antithesis.—Certain states of the mind lead to certain habitual actions, which are of service, as under our first principle. Now when a directly opposite state of mind is induced, there is a strong and involuntary tendency to the performance of movements of a directly opposite nature, though these are of no use; and such movements are in some cases highly expressive."[39] Examples are the expressions of submission in dogs, or affection in cats. Here Darwin used the associationist law of contrast, extrapolating from the association of contrasting ideas to the association of contrasting expressions of emotions. He suggested that expressions that are very different from those in the first category and are perceived by observers (mates, members of the same social group) as manifestations of *opposite* states of mind will elicit contrasting actions. Although the communicative function of the expression of emotion is not stressed by Darwin, he suggested that the principle of antithesis is best explained by assuming that, although at first it was the by-product of directly serviceable expressions, it later evolved for communication.

Darwin's third principle, "The principle of actions due to the constitution of the Nervous System, independently from the first of the Will, and independently to a certain extent of Habit" was influenced by Spencer's principle of "diffuse neural discharge." Examples include overwhelming emotions, such as horror, that lead to trembling, secretions from glands, defecating, and more, which never had any functional, selectable significance. Of course, Darwin, like Spencer, was fully aware that observed expressions are often the result of a combination of these different principles.

Darwin suggested that the processes that led to the evolution of the various human expressions of emotion (figure 2.5 illustrates some of them) were first and foremost the inherited effects of use and disuse—and to some extent also natural selection. *The Expression of the Emotions in Man and Animals* is the most "Lamarckian" book Darwin ever wrote. Paul Ekman, in his afterword to a 1998 reprint of the third edition, suggested that this is one of the reasons for the book's decline in popularity in the first two-thirds of the twentieth century. The other reasons are, according to Ekman, Darwin's reliance on anecdotal evidence; his relatively scant attention to the communicative function of the expressions; the rise of behaviorism, which focused on learned behavior and ignored mental states; and the dominance of cultural relativism in the first half of the twentieth century, which rendered universalistic claims about the expression of emotions or any other human attributes suspect.[40]

In spite of Darwin's refusal to speculate about the origins of animal mentality, we want to stress four points that are relevant to our topic. First, Darwin believed that many invertebrates are endowed with feelings and mental states, some exhibiting will, aesthetic preferences, and desires. Thus, for him the evolution of consciousness has deep phylogenetic roots. Second, he believed that the expression of emotions and the feelings that constitute emotions are related. "He who gives way to violent gestures will increase his rage; he who does not control the signs of fear will experience fear in a greater degree," he wrote on the last page of *The Expression of the Emotions in Man and Animals*, thus paving the way for James's theory of emotions. Third, natural selection, as one of the processes driving mental evolution, played a role. Fourth, although himself unwilling to tackle the subject, Darwin relegated the study of the evolution of mentality to his young successor and protégée George Romanes, who wrote three books about it.[41] Following

Figure 2.5
Expressions of emotions.

Darwin, all psychologists, philosophers, and biologists who considered mental evolution and the evolutionary origins of mentality explained it in terms of natural selection. William James's view of consciousness, which is the basis of twenty-first-century ideas on the topic, incorporates Darwin's theory of evolution by natural selection and challenges Spencer's account of the evolution of consciousness.

The Psychological Investigations of William James

Modern psychology sprang from the introspective psychophysics that developed in Germany in the second half of the nineteenth century, from the associationist philosophical psychology of John Stuart Mill and Alexander Bain, and from the evolutionary psychology of Spencer and Darwin. It reached a magnificent and idiosyncratic peak with the work of William James (figure 2.6), who integrated the three strands, but the study of mentality fell into sharp decline for sixty years with the reign of behaviorism in North America. It slowly came back to life and entered through the back door with the advent of the cognitive revolution and became more assured following the convergence of several paths of inquiry in the early 1990s. Here we focus on James's contribution to the understanding of consciousness—on *The Principles of Psychology*, which is a challenge and alternative to Spencer's identically titled book. James's book is not only full of remarkable introspective insights—it also provides, despite its many omissions, the foundations for a twenty-first-century science of consciousness.

A charmer—imaginative, artistic, rebellious, generous, self-centered, and often exasperating—William James was loved and admired by many people, and several excellent biographies have been devoted to him. His first biographer and former student, Ralph Barton Perry, pointed to his defiant spirit: "A natural poacher, with the poacher's characteristic dislike of the gamekeeper."[42] He was described by his sister Alice as "a creature who speaks in another language as H [Henry James, their brother] says from the rest of mankind and who could lend life and charm to a treadmill."[43] His brother Henry, mourning his death, wrote to H. G. Wells of his loss: "He did surely shed light on man, and gave, of his own great spirit and beautiful genius, with splendid generosity."[44]

The Principles of Psychology: Consciousness as a Selecting Agency
James's motivation in writing *The Principles of Psychology* was very different from that which guided Lamarck's and Darwin's writings on psychology. *The Principles of Psychology* is *not* a book about mental evolution: although

Figure 2.6
William James in the 1890s.

evolutionary principles are very important, evolutionary history is not. The book is focused, almost entirely, on human consciousness and makes ample use of James's personal introspection; the results of observations and studies of animals serve mainly to explain human consciousness. For James, human consciousness was an incontestable fact. The "thinking" that each person experiences encompasses "every form of consciousness indiscriminately,"[45] by which James meant all forms of feeling, passion, perceiving, and reasoning.

James wanted to understand human consciousness through what it accomplishes, not through its origins and historical roots, and he was certainly not interested in a preconscious polyp or worm as a starting point because he did not believe the polyp would enable him to understand human consciousness with its genuinely new teloi. So although he had a good grounding in biology, especially in human neurophysiology and neuroanatomy, and he made ample use of this knowledge in *The Principles of Psychology*, he did not have much to say about the natural history of consciousness, or even about the evolution of human consciousness from ape ancestors. James's *The Principles of Psychology*, like Spencer's, is far more philosophically oriented than the treatises of Lamarck and Darwin. James and Spencer covered some similar ground, but the two books are very different because James stressed the active role of the subject in the construction of experience, emphasizing the role of attention, the active effort that is one of the determinants of what a subject experiences and wills. Agency, the sense of having a self that acts in the world, which James managed to convey so well, permeates his book.

In spite of his stress on the consciousness of humans rather than on the evolutionary history of consciousness, two general evolutionary principles were of central importance to James. First, if consciousness has been evolving and growing in complexity, as the growth in animal behavioral complexity during phylogenetic history suggests, then it must have a function, and if it has a function, it must have causal power. This was crucial to James's view of man, so the evolutionary argument for the *causal effects* of consciousness was of utmost importance. Second, Darwin's theory of natural selection and his ideas about variations had an element of spontaneity and creativity that resonated with James's view about the nature of reality and the active nature of life and of human consciousness. The central role of selection in James's thinking led to his conception of consciousness as a selective agent, "loading the dice" of precarious complex brain states. These evolution-inspired aspects of *The Principles of Psychology* are our main focus here.

The Function of Consciousness: Loading the Dice

The argument against consciousness as an epiphenomenon of material organization, foam on the waves of neural life, was the main topic of one of the first chapters of *The Principles of Psychology*. The epiphenomenal position, which James attacked, posits that mental states are the by-products of automatic neural mechanisms, and it is only the latter that have causal power: just as the shadow of A and the shadow of B can affect neither A's and B's bodies nor each other, so mental states can affect neither material states nor one another. The rationale of this position was beautifully expressed by Thomas Huxley and quoted in *The Principles of Psychology*:

> The consciousness of brutes would appear to be related to the mechanism of their body simply as a collateral product of its working, and to be as completely without any power of modifying that working as the steam-whistle which accompanies the work of a locomotive engine is without influence on its machinery.... It seems to me that in men, as in brutes, there is no proof that any state of consciousness is the cause of change in the motion of the matter of the organism. If these positions are well based, it follows that our mental conditions are simply the symbols in consciousness of the changes which take place automatically in the organism; and that, to take an extreme illustration, the feeling we call volition is not the cause of a voluntary act, but the symbol of that state of the brain which is the immediate cause of that act. We are conscious automata. (James 1890, 1:131)

Though it sounds convincing, Huxley's reasoning, James argued, is philosophically problematic. First, the argument from continuity can work both ways: since we humans are aware of feeling, of focusing attention, of willing and of thinking, and these mental states, James believed, guide our actions, we can extrapolate and attribute this causal efficacy to the brutes' mental states. Second, the assumption that feelings and thoughts do not have causal power, but muscles and nerves do, is philosophically weak: ever since the days of Hume and Kant, James argued, we have been aware that we do not have a very good idea of what causality—physical or psychic— actually is, so making dogmatic and confident statements about material causality while denying causal power to feelings and thoughts is rather hasty: "As in the night all cats are gray, so in the darkness of metaphysical criticism all causes are obscure."[46]

More importantly, according to James, there are positive reasons for believing that consciousness has causal power. Higher animals have more complex brains than lower animals and seem to be more intelligent and more conscious. However, brain complexity breeds problems. James argued that an increase in brain complexity makes it unstable and prone to mistakes, and this vulnerability points to the function of consciousness: it is consciousness that "loads the dice" and guides and increases the efficiency of the unstable brain:

> Loading its dice would mean bringing a more or less constant pressure to bear in favor of those of its performances which make for the most permanent interests of the brain's owner; it would mean a constant inhibition of the tendencies to stray aside. (James 1890, 1:140)

James advanced a series of arguments to support this view. First, the distribution of consciousness and the fact that higher animals are more conscious than lower ones reinforces the idea that consciousness evolved gradually through natural selection because of the guidance it provides to animals with increasingly large and unstable brains. No doubt Huxley would have answered that the neurophysiological machinery had progressively evolved to be more complex and, inevitably, so had its epiphenomenal, mental "shadow." A second and more convincing observation supporting the causal effects of consciousness is that it disappears with habit. Habit, for James, was a result of the brain's plasticity, which was for him, as it was for Lamarck before him, a result of the flexibility of the material of the brain, a physical property of the matter of brain tissues that allows currents to carve and shape them. The fact that habit simplifies the movements necessary to accomplish an action and diminishes the attention needed for it supports the view that when neural reactions become, through habit, simple and reflex-like, consciousness disappears—it is no longer needed to guide action. A positive argument for the causal power of consciousness is that we are aware of being conscious when we are learning something new or are faced with a difficult decision. Consciousness is most conspicuous when we dither and deliberate (a very Jamesian state of mind), which is a major characteristic of the conscious effort to decide and act upon the decision. The goal-directed nature of consciousness is particularly evident when there is an obstacle or impediment in an animal's (including human's) way—for example, when a human becomes handicapped and finds alternative ways to achieve her goals. Our felt needs, especially our conscious goals, direct our actions. Finally, the fact that the mental state of pleasure usually comes with activities that improve the chances of surviving and leaving offspring (fitness-increasing activities), while the mental state of pain accompanies states and activities that decrease fitness, suggests that these mental states are not mere epiphenomena but have a clear adaptive function. All these suggest that we are not passive mirrors, merely reflecting the relations that exist in the environment. We select among them and make them relevant. The agency of organisms, especially of humans, is central to their everyday existence as well as to their evolution.

How does this "loading the dice" actually happen? James suggested a general principle: selection, in the case of consciousness, is the result of differential stabilization among the endless variable activities in the nervous system; the telos of consciousness is to satisfy, through elicited and stabilized feelings and thoughts, the goals of the animal. Just as reproductive success is both the outcome and the telos of Darwinian organisms, so the manifestation of human consciousness is both the outcome and the telos of feeling and thinking. The ultimate telos is still survival, but there is now a mediating telos, a felt desire to survive:

> Every actually existing consciousness seems to itself at any rate to be a *fighter for ends*, of which many, but for its presence, would not be ends at all. Its powers of cognition are mainly subservient to these ends, discerning which facts further them and which do not. (James 1890, 1:141)

James did not tell us when and how consciousness emerged during evolutionary history—at what point the brain becomes too complex to work without the "guidance" of such selection and what the neural activity that enables it is. Nor did he tell us whether the notorious polyp has, by extrapolation from humans, a measure of consciousness. But the idea of selection in the brain as a necessary aspect of what we may call "mental life" is dominant in *The Principles of Psychology* and is, as we discuss in later chapters, a recurring theme among late twentieth- and early twenty-first-century philosophers like Dennett, neurobiologists like Changeux and Edelman, and theoretical biologists like Szathmáry and Fernando. There is something compelling about the idea, for, like natural selection, neural selection occurs without an external designer and can lead to complex adaptations. However, James, unlike modern scholars, never grounded the analogy in *intrinsic* neural selection, in a choice without a chooser; the selector for him was selective attention, which he regarded as a mental selecting activity that expressed itself most clearly when attention and action-selection required an effort. Consciousness is therefore not a stuff, material or transcendental, nor is it a neural mechanism or a property of neurophysiological activity, although it seems to be constituted by and composed of neural activity. It is a very special kind of activity in a whole animal (rather than just the brain of the animal), something the enbrained body does. James would have been firmly opposed to brains in vats and to conscious computer software.

Like his psychology, James's philosophy stresses the creative and unanticipated element in evolution, which is only judged a posteriori. In his course on the philosophy of evolution, James contrasted his position to that of Spencer.[47] While Spencer focused on physical laws and the "adjustments of inner relations to outer relations" through use and disuse, James highlighted the natural selection of chance variations, of agency, and of choice. In the last chapter of *The Principles of Psychology*, there is a strong endorsement of the neo-Darwinian view of evolution by the natural selection of chance variations, a view developed by August Weismann, who denied, on theoretical and empirical grounds, the inheritance of acquired characters and the role of use and disuse in evolution.[48] Although for James the undirected origin of human-specific logical and aesthetic faculties was of greater concern, undirected variation also held an important place in his general views about the nature of external reality and evolution and was closely related to his notion of consciousness as a creative "voting" element.

Chance and Creativity

In order to distinguish and evaluate the role of directed (environmentally induced) and undirected (stochastic) variation in the construction of mentality, James distinguished between internal and external constructive factors. He admitted that relations in space and time can be learned and that learned behavior can become a habit, so in this sense inner relations may be said to adjust to outer relations. However, it was equally clear to him that certain aptitudes can have either external *or* purely internal (and unlearned) causes: a boy can become a musician because he practices a lot (external influence) or because of a lucky accident in the ovum that made him musically talented with little need for exercise (internal influence). Therefore, both internal (innate) and external (learned) factors can affect brain activity.

But even if we grant external relations some stamping power, how does this happen? The blue of the sky that we see is not a copy of a blue sky up there—there is no blue up there, as we well know. Similarly, pain and pleasure are internal mental categories and have no existence outside us. What we experience as external factors and the relations among them depends on the inner structure of our brain, which at some time during evolutionary history, as a result of a lucky accident, enabled us to make better discriminations than our ancestors and was therefore established through natural selection. The world is not given, nor is it orderly. As John Stuart Mill, whom James cited, put it: "The order of nature, as perceived at a first glance, presents at every instant a chaos followed by another chaos. We must decompose each chaos into single facts. We must learn to see in the chaotic antecedent a multitude of distinct antecedents, in the chaotic consequent a multitude of distinct consequents."[49] James was undoubtedly also influenced by his friend Chauncey Wright's conception of the world. Wright described the order of the world with a weather metaphor, a "cosmic weather" view, as his friends called it. The weather is a physical system we cannot predict, although it is subject to deterministic physical laws. The complexity and unpredictability of the ever-changing initial atmospheric conditions—and (as we would say today) nonlinear interactions—preclude any universal and predictable long-term regularities. Except for the incessant change brought about by heat and gravitation, everything is transient, directionless.[50]

Given the ceaseless activity and instability of the brain, given the chaos of the world, it is not clear how a baby or a puppy learns to see objects as stable features of the world. James was aware, of course, that Spencer and Mill might agree that once an ability to impose some order on the plenitude of existence (to discriminate between colors, or feel pleasure and pain) becomes established by the natural selection of blind variations, then habit and learning could build on this and lead to the evolution of new and complex faculties. But this, he argued, could work only for simple sense impressions: "The only cohesions which experience in the literal sense of the word produces in our mind are ... the proximate laws of nature, and habitudes of concrete things, that heat melts ice, that salt preserves meat, that fish die out of water, and the like."[51] Once we go further—to the human intellectual and emotional faculties; to the ability to compare and abstract; to logical laws, metaphysical generalization, and aesthetic judgment—there is no way that we can deduce these features of our understanding from our sensory impressions. On the contrary, what science and logic do is extract order out of sensory chaos. As every scientist knows, it requires great ingenuity and hard work to create the artificial conditions in which a regularity or a natural law becomes apparent.

Of course, James is right: abstract human categories and discovered laws of nature are not the result of a simple, cumulative combination of sensory associations, a position he attributes, unfairly, to Spencer and to John Stuart Mill.[52] We need something else, and for James this something else was a mode of creative consciousness that can be reduced neither to the external world nor to the brain as a responsive organ. At the phylogenetic level, he suggested that creativity can be best explained biologically as the consequence of random variation and natural selection. But it is not at all clear how random variation in the ability for abstraction can become established in a population. What is the initial function of the selected variation? Our human mental faculties do not simply and directly serve survival and reproduction; James (1878) rebelled against this vulgar and simplistic idea in his anti-Spencer paper of 1878. What led to human-specific faculties must be a by-product of a variation that was useful for other, more mundane reasons. James did not, however, tell us the by-product of what these faculties might be.[53]

It is not just in the case of the evolution of human reflective self-consciousness that James preferred the selection of chance variations to the inheritance of learned associations. James did not subscribe to the

Lamarck-Spencer-Darwin assumption that ontogenetically acquired traits can become inherited. Habits do not become instincts, he argued; habits are *enabled by instincts*. Instincts arise through the accidental generation of variations that enhanced survival because they happened to make organisms better able to respond to the causal relations in their external environment. He referred here to Weismann's *Essays Upon Heredity*, where Weismann contended that the inheritance of changes brought about by use and disuse is empirically unfounded and conceptually problematic. James therefore concluded that there are big problems with "experience-psychology" based on the use-disuse principle and that the evolution of consciousness is shrouded in mystery:

> The more sincerely one seeks to trace the actual course of psychogenesis, the steps by which as a race we may have come by the peculiar mental attributes which we possess, the more clearly one perceives "the slowly gathering twilight close in utter night." (James 1890, 2:688)

In spite of this somewhat discouraging conclusion—this is the very last sentence of *The Principles of Psychology*—James opened new routes for studying consciousness. His theory of the embodied nature of the stream of feeling and willing, of instincts as springboards of mentality, and of the role of past experience and learning in the causal construction of future goals can be (and some facets have been) redescribed in modern neurophysiological terms and tell us how the dice could become loaded.

Embodied Subjective Experiencing

"The stream of consciousness," one of the phrases coined by James, is so much a part of common parlance today that it seems to be a common-sense description of what everyone goes through as they experience. Yet it was not the way many psychologists regarded consciousness in the past. For the associationists, ideas were made up of sensations, and complex ideas were made of combined simple ideas, with the parts preceding the whole. This view of consciousness, starting with simple parts and ending with complex wholes, seems to be based on the way we build artifacts—a house is made of simple units, bricks, which must preexist prior to the house. However, this is certainly not the only way that a process of biological complexification can occur. Spencer, following von Baer, thought that wholes always precede parts—that differentiation into tissues and organs is the complexification process par excellence, with a simpler homogenous entity ending up as a more differentiated, complex whole made of parts. Nevertheless, the result of development was for Spencer a more or less modularly organized mental system with a clear division of labor—with distinct parts that could inter-act, recombine, and become more refined during subsequent processes of differentiation. James modified this view and injected energy into it. He added time-consuming neural integration processes to his conception of the manner in which consciousness emerges from previously unconscious processes and made the conscious process continuous and dynamic—part of a flux, of a stream. A feeling or a thought, according to this view, starts like a wave, with an initially nebulous state of meaning becoming more and more differentiated (distinct and focused) as the stream proceeds: "The stream is made higher at its end than at its beginning, because the final way of feel-ing the content is fuller and richer than the initial way. As Joubert says, 'we only know just what we meant to say, after we have said it'."[54] Thoughts, he suggested, were decomposable only a posteriori.

Although, according to James, there are parts of the stream that can be differentiated, some flowing rapidly, others more slowly, the stream is, by definition, an activity. With another beautiful metaphor, emphasizing the different rates of flow in different sections of the stream, James com-pared the stream of consciousness to a bird's life "made of an alternation of flights and perching,"[55] but there is flux and fluidity in this bird's life, even in its perching.

The stream metaphor is central to James's view of consciousness and has profound implications for the most basic attributes of subjective experienc-ing. First, it makes it clear that consciousness is not a state but an activity. Second, it means that we never undergo the same experience twice and that except for the very first sensations just after birth (and as we now know, for some sensations even in utero), there are no "pure sensations." All later experiences involve memories and past associations, which change from moment to moment. Third, every stream of consciousness contains echoes of the immediate past and reaches into the immediate future. Consequently, subjective experiencing must have a minimum duration—sensations and perceptions must endure for us to actually experience them:

> The practically cognized present is no knife-edge, but a saddle-back, with a certain breadth of its own on which we sit perched, and from which we look in two direc-tions into time. The unit of composition of our perception of time is a *duration*, with a bow and a stern, as it were—a rearward- and a forward-looking end. It is only as parts of this *duration-block* that the relation of *succession* of one end to the other is perceived. (James 1890, 1:609)

This extended present with its remnants of the immediate past and its anticipation of the immediate future, which James, following E. R. Clay, called the "specious present,"[56] is the basis of our intuition of time and is based on a view of consciousness-as-activity. Although this activity depends on the brain, the brain is not sufficient. It is the animal, not the brain, that feels. The sense of self, most fundamentally of the bodily self, is basic to the animal's feeling as an agent, and this feeling of self becomes broader and more intricate as mental development proceeds from birth to maturity.

Instincts, Emotions, and the Will

Instincts are the foundations on which James, like Spencer and Lamarck before him, built his theory of consciousness. However, the nature of instinct and its relation to learning, reasoning, and complex emotions was, for James, very different from that of his predecessors. In *The Principles of Psychology*, he started his chapter on instinct with a definition that makes it clear that instinct is *any bias* for neurally mediated action, an inevitable accompaniment of the bodily construction of neural animals:

> Instinct is usually defined as the faculty of acting in such a way as to produce certain ends, without foresight of the ends, and without previous education in the performance. That instincts, as thus defined, exist on an enormous scale in the animal kingdom needs no proof. *They are the functional correlatives of structure. With the presence of a certain organ goes, one may say, almost always a native aptitude for its use.* (James 1890, 2:383; emphasis added)

Looked at in this way, instincts are everywhere and, as James stressed in this chapter, humans are particularly well endowed with instincts.[57] Following Spencer, he suggested that conflict and competition among different instincts (e.g., the baby's instincts to both approach and recoil from an unknown person) may lead to choice and voluntary action. However, instincts for James are not rigid and fixed. They can almost always be modified by learning, which leads to the formation of habits that sometimes result in the inhibition of the initiating instinct. Thus, the rat's instinct to approach and eat food when hungry can be inhibited if the food is associated with a trap. The learned association, if very traumatic or repeated, will make the rat avoid food in certain conditions. Instincts can be modified by learning in another way: they can become associated with a very narrow range of contexts. For example, a rabbit always deposits its feces in the same corner as that in which it first deposited them. Furthermore, many instincts (for example, the baby's suckling instinct) are transient. James built on the concept of "imperfect instincts," which had been developed by Douglas Spalding (1841–1877), the brilliant Scottish biologist who was the first to examine instincts experimentally, and George Romanes (1848–1894), who laid the foundations of comparative psychology. Both are often cited in James's chapter on instinct. Spalding emphasized the condition-dependent nature of many instincts: sixty years before Lorenz studied and made famous the behavior of young chicks that persistently follow the first moving object they see after hatching (a behavior called "filial imprinting"), Spalding had experimented on this following behavior. He showed that following depended on the chicks' stage of development when they saw their first moving object, thus emphasizing the importance of maturation for triggering certain instinctive behaviors. Romanes, who performed cross-species fostering experiments in birds, also regarded instincts as "imperfect" and educable. James saw this modifiability of instincts as a basic property: far from being a sign of "imperfection," the modifiability of instincts is their very raison d'être because it enables the development of habits through learning, thus fulfilling the instincts' role and making it redundant:

> Most instincts are implanted for the sake of giving rise to habits, and that, this purpose once accomplished, the instincts themselves, as such, have no raison d'être in the psychical economy, and consequently fade away. (James 1890, 2:403)

The role of instincts as the scaffolds on which habits are built is related to James's view of the development of motivations and desires. On its first appearance, an instinct is blind and is elicited by a very simple stimulus, but when experienced for the second time, it comes with past-based expectations that lead to desires and create motivation.

James's view of instincts, especially the way in which initial instincts construct mental life through learning, is intimately related to his theory of emotions (a term he often used interchangeably with feelings). This famous Jamesian theory, which was developed independently by the physiologist Carl Lange,[58] suggests that emotions (e.g., fear) are not the causes of bodily changes; rather, emotions are made up of (or constituted by) bodily (physiological) changes and follow or accompany them. This theory is based on two assumptions: first, that the whole organism is the sounding board for whatever excites the nervous system[59] and second, that these bodily changes are felt as they occur. From these assumptions James came to his theory, which he contrasted with conventional wisdom, which regarded emotions as the causes of actions and bodily changes:

> My theory, on the contrary, is that the bodily changes follow directly the perception of the exciting fact, and that our feeling of the same changes as they occur IS the emotion. Common-sense says, we lose our fortune, are sorry and weep; we meet a bear, are frightened and run; we are insulted by a rival, are angry and strike. The hypothesis here to be defended says that this order of sequence is

incorrect, that the one mental state is not immediately induced by the other, that the bodily manifestations must first be interposed between, and that the more rational statement is that we feel sorry because we cry, angry because we strike, afraid because we tremble, and not that we cry, strike, or tremble, because we are sorry, angry, or fearful, as the case may be. Without the bodily states following on the perception, the latter would be purely cognitive in form, pale, colorless, destitute of emotional warmth. We might then see the bear, and judge it best to run, receive the insult and deem it right to strike, but we should not actually feel afraid or angry. (James 1890, 2:449–450)

Although James defended his emotion theory against possible objections, giving many supporting examples from cases of neural pathologies, there are, as many scholars have pointed out over the years, some problems with it.[60] James argued against dedicated emotion centers in the brain, neglected the feedback between different brain regions, and overlooked the role of meaning. As Ludwig Wittgenstein wryly noted, "A man would say he grieves in his soul, not in his stomach, because he wouldn't expect to be cured of grief by relief for the unpleasant feeling in his stomach."[61] However, James did see memories and other contextual facets of emotional situations as building blocks of emotional experiences, so Wittgenstein's criticism is not entirely fair. James focused on the primacy of visceral and somatic reactions because he saw them as the early developmental scaffolds on which adult, highly person-specific and situation-specific emotions are constructed.

In spite of its imperfections, many aspects of James's theory of embodied emotionality and cognition are accepted today, albeit with qualifications, and one of its great advantages is that it lends itself readily to an evolutionary interpretation.[62] The evolution of the expression of emotions, and therefore of the emotions that are constituted by their corresponding expressions, follows, according to James, the evolution of simpler sensorimotor behaviors, just as Spencer, and more clearly Darwin, had suggested. Another useful feature of James's embodied emotion theory is that it highlights and prevents the common confusion between two types of knowing: knowing through acquaintance (perceiving a red rose, having a toothache) and second-order reflective knowing (knowing about seeing the rose, knowing about feeling the pain). James strongly objected to the idea that in order to experience one must have a second-order representation of the state of acquaintance, a notion that was seen by him as an unfortunate fallacy; a result of a linguistic and logical muddle.[63]

Will and emotion are intimately related, according to James. Emotions are constituted by feedback between the sensory and the motor systems and often, but not necessarily, lead to motor acts. When a motor act is elicited, the animal feels it as a will to act. "Wanting" is nothing but the recalled motor representation of an instinctive act or its experiential modifications. On this view, the basis of voluntary behavior is automatic ideomotor action, as first suggested by Carpenter (1852) and elaborated on by James: recalling a motor action leads to automatic (nondeliberate) actualization as long as nothing inhibits it. More complex, truly voluntary actions occur when there are conflicts among different representations of actions, and the action representation that gets the most attention is "selected," leading to an act being experienced by the actor as voluntary. The role of attention in the process is central. In fact, for James, attention is a necessary condition for conscious acts. All conscious life depends on some focusing of attention, from the simplest triggering of attention by a sensory input such as a loud voice, through the triggering of instinct, ideomotor acts, voluntary acts, and free choices. James was acutely aware of the continuity and gradual merging of reflexes, instincts, effortless acts of will, and effortful acts of will, which give one a sense of choice and free will, but it seemed to him that in order to save the causal power of free will, he must accept nonphysical mentalism. His position, as presented in *The Principles of Psychology*, was not coherent, and the impression one is left with is one of intense and tormented intellectual wavering.

Although gaps and inconsistencies abound in James's theories, several aspects of his view of consciousness are widely shared by those who study it today and who subscribe to a "weak" emergentist position. These include the view of consciousness as an organismal, global, ever-changing, and unified activity, something that animals and humans *do*; the related and necessarily embodied nature of private subjective experience, which depends on the feedback interactions between the external world and the animal's embodied brain; the requirement for temporal depth, the specious present, necessary for subjective experiencing; the primacy of a representation of a bodily, unified, and owned self; the central role of selection in the process of subjective experiencing and the related central role of attention; the role of learning and the formation of habits; and the practical importance, for educators, of learning by consequences. In the terms of chapter 1 of this book, these are some of the major items in James's "list" of features of consciousness, and some of them point to aspects of the dynamics of the system, such as feedback relations and selection processes, that James regarded as crucial. Putting together these aspects of consciousness and using the language and the insights of a post-Jamesian neurobiology and philosophy of mind provides, as the next two chapters show, a way of moving toward a better understanding of the dynamics that instantiates subjective experiencing. Such an approach has had to wait for nearly a hundred years from the publication of James's *Principles of Psychology*. During much of this time, especially in the United States, psychology ceased to be seen as the science of psychic life; consciousness, as a legitimate subject for neurobiological-psychological study, effectively disappeared.

The Waning and Slow Reemergence of Consciousness Studies

> I have heard a most intelligent biologist say: "It is high time for scientific men to protest against the recognition of any such thing as consciousness in a scientific investigation."
> —James 1890, 1:134

Shortly after James's death, a young, intelligent American psychologist, John B. Watson (1878–1958), redefined psychology as the study of behavior. His redefinition was based on Ivan Pavlov's studies of the conditioned responses of dogs; on the assumption that what is true of dogs and rats in the lab is true of humans; and on the conjecture that conditionable reflexes are the basis of all learned behavior. For a short period, Watson led the behaviorist movement, which became widely accepted (with inevitable minor variations) among most North American psychologists:

> Behaviorism … holds that the subject matter of human psychology is the *behavior or the activities of the human being*. Behaviorism claims that "consciousness" is neither a definable nor a usable concept; that it is merely another word for the "soul" of more ancient times. (Watson 1925, p. 3)

How could this intellectual upheaval have happened? It has been argued that behaviorism fit the antihereditarian stance and the wish to improve the human condition through behavioral manipulations, as well as the ideal of methodological rigor and the positivistic philosophy to which North American scientists adhered. Although James's mighty struggle with the mind-body problem opened up (in retrospect) interesting and important possibilities for research, *The Principles of Psychology* did not lead to a clear research program and did not fit the pragmatic, down-to-earth, and to some extent anti-intellectual trend in North America at the turn of the twentieth century. James's magnum opus, although admirably written, did not yield tangible "fruits."

The behaviorist era in the history of psychology has been described and analyzed many times and from many points of view.[64] It was very much a North American phenomenon: psychology in Europe and in the Soviet

Union developed along different lines, pursuing research on animal mentality and animal and child development. Although in the Soviet Union, as in the United States, there was a strong focus on behavioral analysis, the Russians were interested in understanding the mental world of animals and humans through the physiological measurement of learned responses. Because the research on the associative learning in which the behaviorists engaged is important for themes we develop in later chapters, and the cognitivist agenda redefined the study of psychology and learning in ways that enabled the resurgence of consciousness studies in the 1990s, we need to briefly discuss these approaches.

Conditional Reflexes and Behaviorism

In the first half of the twentieth century, associative learning was generally seen as synonymous with behaviorism and was described in terms of conditioning: the formation of a conditional (if-then) association between external sensory stimuli and reflex physiological and motor responses or between actions and their reinforcing outcomes. The nature of the stimuli that enter into conditional associations can include "neutral" stimuli (stimuli that under ordinary conditions do not trigger a response), biologically important stimuli such as those that are typically linked to the maintenance of basic homeostatic and reproductive functions, the animal's own responses, and the contexts in which particular stimuli and responses occur.[65] Although most of the behaviorist studies were performed on mammals (mainly dogs and rats) and birds (pigeons), conditioning, and also some of its compound modulations, have been found in all vertebrates, from fish through amphibians to humans, and even in many invertebrates.[66] As we argue in later chapters, associative learning covers a broad set of learning processes and conditions, which need to be differentiated if the evolution of learning and its relevance to minimal consciousness is to be understood.

Classical and Instrumental Conditioning

Two major types of conditional associative learning—classical (Pavlovian) conditioning and instrumental or operant conditioning—are usually distinguished. The best-known example of classical conditioning, which was studied by Ivan Pavlov (figure 2.7A), is that of a dog that normally salivates (unconditional reflex response, UR) at the smell of food (the unconditional stimulus, US, the reinforcer); if the dog hears a buzzer (neutral conditional stimulus, CS) just before smelling the food, it will learn to associate the

A

B

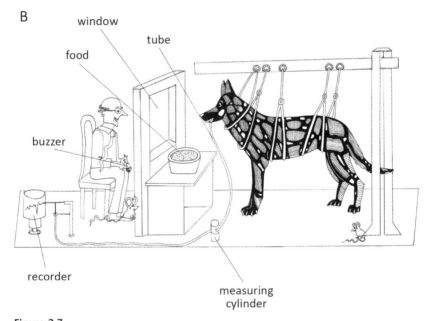

Figure 2.7

A, Ivan Pavlov (1849–1936); *B*, a Pavlovian experimental setup. Pavlov used an apparatus that included a tube inserted into the ducts of the salivary glands of a dog so that when saliva was produced, it could be collected in a measuring cylinder and measured. When presented with food (US), the dog salivates (UR), which is a natural physiological response that is not learned. In this experiment, Pavlov activated a buzzer (CS) just before he presented the dog with food; following repeated pairings of the sound of the buzzer and the presentation of food, the dog would start to salivate whenever it heard the buzzer (CR), even if no food came.

sound with being fed and salivate when it hears the bell (conditional response, CR; figure 2.7*B*).[67]

It is important to note that a UR can be a motor action the animal performs, not just a "passive" reaction such as salivation. For example, an animal can learn that a sound predicts the presence of a predator and retreat in a more or less stereotypical manner.

Many types of classical conditioning and many combinations of different types of learned associations have been described in the specialized literature, which was masterfully reviewed by the Russian experimental psychologist Gregory Razran, who rendered the monolinguistic English-speaking psychology community a great service by introducing it to the rich Soviet tradition.[68] As the studies he reviewed show, although an animal (typically, a dog) can learn many associations, not all associations are equal. The effect of different CS-US pairings depends on the type of stimuli applied and on how "dominant" the reaction is, the information content of the CS (how predictive a CS is of the US), the number of CSs applied and their salience, and the relation between the timings of the paired stimuli and the intervals between consecutive trainings. Importantly, for Pavlov and his colleagues, the reflex salivation response that became linked to various excitatory and inhibiting stimuli was a measure of neural changes in the cerebral hemisphere, a quantitative indicator (the number of saliva drops was always carefully counted) of the neural-mental processes that occurred in the dog's cortex.

The other major type of conditioning is known as instrumental or operant conditioning. It can be said to demonstrate Thorndike's law of effect,[69] where the probability that the animal executes a certain action changes as a function of its consequence: when it is followed by a positive result, it becomes more likely to occur in the same conditions, whereas when followed by a negative effect, it is less likely to occur again in the same conditions. The American psychologist B. F. Skinner (figure 2.8*A*) studied this form of learning extensively by using an apparatus known today as a Skinner box (figure 2.8B). He showed that animals, whose behavior was spontaneous and not dictated by the experimenter, learned the relation between their actions and the consequences of these actions. American psychologists explored every aspect of this type of learning by consequence.

Different types of instrumental conditioning are classified according to the type of reinforcement (desirable or aversive) and the effect of the response (preventing or enhancing a desirable or an aversive response). The sensory biases and the motor constraints of the particular species of animal,

A

B

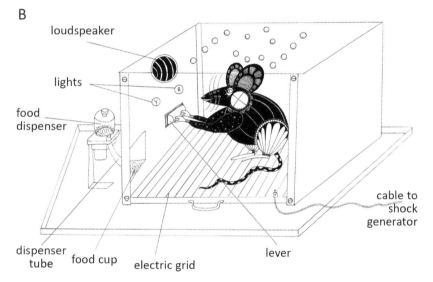

Figure 2.8

A, Burrhus Frederic Skinner (1904–1990); *B*, a Skinnerian experimental setup. The animal (a rat, in this case) is placed in a box, which is often soundproof and light-proof to avoid distracting stimuli but contains its own controllable sources of auditory and visual stimuli. The box contains an "operandum" (a device on which the rat exerts its responses, in this case a lever) that automatically registers use and then delivers reinforcers (e.g., food pellets). A hungry rat that accidentally presses the lever gets a food pellet and rapidly learns to associate its lucky action with receiving the reinforcing food.

as well as the specific organization of its nervous system, are all important factors in determining the ease, the rate, and the complexity of learning. Another distinction relates to the type of motor behavior that is involved in such learning. The motor act can be fixed and stereotypic, like the spontaneous pecking behavior of a chick, yet the bird can learn what and how to peck if some of the pecked grains are tasty and adjust its motor behavior to the task; this type of learning by consequence is known as instrumental conditioning. Behavior, such as pressing a lever or jumping through a burning hoop, can also be novel and nonstereotypical (that is, it is *not* part of the preexisting behavioral repertoire of the animal), with the animal showing some control over its actions following learning. In such cases, when the animal learns to select among its exploratory motor acts and reorganize them, conditioning is described as operant rather than instrumental.[70]

The American twentieth-century behaviorist research program purged psychology talk of mental representations and centered on the control and manipulation of behavior. When Skinner developed functionalist behaviorism (known as radical behaviorism), he excluded from his analysis everything except observable behaviors. He analyzed behavior in terms of reinforcement and defined a reinforcer as any stimulus (e.g., food) that will reinforce the operant response that preceded it (e.g., pressing a lever, pecking a key). Knowledge of the learning history of the organism can, he argued, explain any observable behavior. Although the stimuli and behavioral responses that are observed are mediated by neural processes and by the brain, these were not part of the functional behaviorist analysis. Humans were no exception to this scheme. The relationship between stimulus and response, when understood in terms of operant conditioning, was believed to enable the full prediction and control of human behavior. Skinner applied his behaviorist analysis even to language and regarded human verbal reports as "verbal behavior." The basic rules, he believed, were the same.

The Rise of Cognitivism
Behaviorists, especially radical behaviorists, were successful in discovering many things about learning and in reconditioning some pathological human behaviors, a mode of psychotherapy that is still used (with modifications) today. The limitations of behaviorism were, however, growing more obvious from the mid-1950s onward. The rise of ethology, with its focus on inborn behavioral predispositions, such as the filial imprinting studied by Konrad Lorenz (1903–1989) and the innate pecking behavior of seagull chicks explored by Niko Tinbergen (1907–1988), led to the appreciation of the species-specific constraints and affordances imposed on animal and human learning by evolutionary history.

An important influence on the study of behavior—one that transformed all aspects of twentieth-century science—was the development of cybernetics and information theory, which were associated with the construction of computational and communication systems during the Second World War. The rise of computer science, pioneered by people like John von Neumann and Alan Turing, helped to construct a new framework for the study of psychology. It became computational and "cognitive"—described by algorithms that transformed, stored, and recovered the sensory inputs that the system was able to process. A new and powerful metaphor appeared, replacing the old metaphor of a telephone exchange system: the brain was now compared to a digital computer, the most sophisticated machine then in existence. (Today we have robots, and the metaphor is changing yet again.) The computer metaphor immediately added specific internal semantic states and multiple rules of processing to the association of external inputs and behavioral outputs that the behaviorists had suggested.[71] The internal "machine" states were now thought of in terms of neural symbols equivalent to words or numbers, and the computational algorithms that operate on those representational states were seen as thinking. Artificial intelligence (AI) was born.

The cognitive approach addressed many of the questions that the behaviorists left open. For example, George Miller showed the limitation of human short-term memory (we are able to recall 7 ± 2 items) and demonstrated that in order to overcome this limitation, we organize items into groups, a process requiring the encoding and decoding of information. Other studies showed that the meaning of sentences affects memory and that unconscious states (such as seemingly forgotten memories or emotionally suppressed ones) affect how sentences are interpreted. In linguistics, Noam Chomsky demonstrated that stimulus-response chains cannot explain linguistic performance and competence and that rules for making rules must be assumed when explaining the way people construct and comprehend sentences. The studies of major European psychologists, who were nonbehaviorists, eventually found their way into the American scene: the gestalt psychology that had developed mainly in Germany, the developmental child psychology of Jean Piaget, and Bartlett's work in Great Britain on constructive memory and attention became part of a greater and more diverse scientific landscape.[72] It became clear that organizational principles must be the basis on which associations are formed. A major aspect of the cognitivist research was therefore the construction and testing of computational models that were intended to simulate mental operations such as deductive reasoning, concept formation, and memory consolidation and to conduct experiments in which humans (usually psychology students) were engaged in processes of deduction, searching, and matching. The brain, with all its complexity and intricacy, rather than the peripheral reflexes that non-Skinnerian behaviorists focused on, became the relevant research arena for studying the neurobiological foundations of cognition.

In the mid-1970s, cognitive science became recognized as an integrative approach embracing psychology, artificial intelligence, philosophy, neuroscience, linguistics, and anthropology. It seemed to be the "only game in town," as the cognitive psychologist Jerry Fodor put it, and "behaviorism" became a dirty word.[73] But, as happens with any thriving and successful science, challenges and alternatives were soon presented. In the early 1980s, connectionist models challenged the assumption that symbol-like structures manipulated by algorithms existed in the brain. The connectionists constructed simple artificial networks operating among neuron-like units that were linked by connections whose strength (weight) was determined by positive and negative reinforcement and that obeyed simple learning rules. They showed that these simple models can simulate visual pattern recognition or the production of speech sounds from written texts. They saw animal and human cognition in terms of the workings of such neural-learning networks in the brain.

The digital computer–based and connectionist models influenced philosophers of mind, who in turn influenced cognitive scientists. Daniel Dennett regarded the metaphor of the brain-as-computer as extremely useful, believing that computations are the functional equivalents of thoughts.[74] Since functionalist models are substrate independent,[75] cognitivists-functionalists regarded the ways in which the nervous system instantiates cognition and consciousness as a special and rather messy case of implementing the mental, which, they believed, could be much better understood by studying information processing by artificial intelligence algorithms in computers.

The computer metaphor inspired other philosophers by raising their hackles and making them formulate challenges to it. John Searle, in his famous Chinese room thought experiment, argued strongly against the view of the mind as a digital computer, emphasizing the difference between syntactic information processing (something a digital computer does very well) and subjective experiencing and attribution of meaning, something a digital computer cannot do.[76] Others, like Paul and Patricia Churchland, also regarded the digital computer metaphor as misleading, but for different reasons: they argued that the neurobiology of the brain and the molecular and physiological details of neural networks should be the basis of thinking about consciousness. They claimed that, in time, neurobiological-connectionist language (not AI computer language) would replace everyday mental language, a view known by the forbidding label "eliminative mate-rialism."[77] Another approach was taken by the theoretical psychologist Nicholas Humphrey, who discussed the origins and stages in the evolu-tion of the mind in terms of ever more complex and integrative feedback loops.[78]

In spite of the new view of the mind that AI and connectionism offered, consciousness remained out of bounds. Although the cognitivist research program no longer prohibited the use of mental terminology, perceptual qualia like seeing red or mental states such as the feeling of fear were simply seen as irrelevant and were not discussed. The models could not account for subjective experiencing or consciousness. Animal consciousness remained a taboo, in spite of the strong advocacy of Donald Griffin, a prominent zoologist who was one of the discoverers of bat echolocation and coined the term "cognitive ethology."[79]

The taboo around the scientific study of consciousness started to break in the late 1980s and the early 1990s, when many new approaches began to challenge accepted dogmas, and new interactions among disciplines and schools of thought started to emerge. The reawakened interest in the neurobiology of emotions, the ethological studies of animal cognition, the contributions of philosophers of mind and of cognitive psycholo-gists, and the development of robotics, which showed the importance of motor-sensory-motor feedback for the accomplishment of complex cogni-tive tasks, all contributed to the shift. Ideas about the importance of agency, of active, niche-constructing interactions of animals with the environment, began to be developed, emphasizing the embodied and dynamic nature of consciousness.[80]

A very important technological driving force for the study of con-sciousness was the development of brain-imaging techniques that enabled scientists to look at the brain while humans were engaged in specific perceptual, emotional, and cognitive activities, thus forging connections between reported mental activities and changes in the brain. Finally, a great push that legitimized the scientific study of consciousness began with vig-orous assaults on the neurophysiology of consciousness by two charismatic Nobel Prize laureates, Francis Crick, codiscoverer of the structure of DNA and one of the biologists who helped crack the genetic code, and Gerald Edelman, a codiscoverer of the structure of antibody molecules. Conscious-ness studies, a now respectable but still unsettling domain of inquiry, had at last materialized.

3   The Emergentist Consensus: Neurobiological Perspectives

Consciousness used to be the crazy aunt in psychology's attic. Behaviorists and cognitive scientists alike practiced denial, but the squeaking floorboards troubled our dreams of a truly scientific discipline. Now, the old lady has been given pride of place in the parlor, with all the respectable scientific furnishing of societies and journals. But let's face it—she's still weird.

—Gopnik 2010

What, from a neurobiological perspective, is consciousness? Is neurobiological research leading to a consensus about its nature, and if so, what kind of consensus is it? Although differences of opinion exist among the neurobiologists who study consciousness, they are all committed to the view that it has a naturalistic explanation, and all produce a similar "list" of its general attributes. Where they differ is in their approach to discovering the organization and dynamics of the neural structures and processes that constitute it. Here we discuss four major approaches that are based mainly on the investigation of consciousness in mammals. The first starts from the neural correlates of specific perceptual experiences, such as seeing red or feeling a pain in the chest. The second is a quest for brain-wide neural correlates such as those that make the difference between dreamless sleep and the wakeful state. The third attempts to locate and characterize the neural arena in which neural selection-evolution processes construct consciousness, and the fourth focuses on the emotional aspects of consciousness. We ask whether these approaches enable us to identify and measure consciousness and understand how it is constructed. Do they converge to form a coherent, single model? Have the neurobiologists solved, in principle, the old mind-body problem?

Toward the end of the twentieth century, the irresistible combination of revolutionary methodologies in neural and computer sciences and a new aura of respectability drove more and more philosophers of mind, psychologists, and neurobiologists back to the formidable mind-body challenge. Their attitudes varied, but the general mood was assertive and exuberant. It was somewhat reminiscent of the optimism that swept through the origin-of-life community after the synthesis of amino acids from inorganic material in the 1950s, when the de novo production of living entities seemed, for a moment, within easy reach. Daniel Dennett's (1991) book *Consciousness Explained*, in which the author, a philosopher, claimed that consciousness is a mirage, fully reducible to the functional equivalents of brain activities, is a good example of this confidence. Equally optimistic sentiments were expressed by two authoritative leaders of the neurobiology of consciousness. In an interview about a robot constructed on the basis of his theory of consciousness, Gerald Edelman spoke of robots as free agents, and Francis Crick famously declared:

> You, your joys and your sorrows, your memories and your ambitions, your sense of personal identity and free will, are in fact no more than the behavior of a vast assembly of nerve cells and their associated molecules. (Crick 1994, p. 3)

The claims about consciousness that are being made today, at the beginning of the third millennium, are in some ways less provocative, but they are by no means less confident. In fact, not just Edelman and Crick, but each of the prominent neurobiologists whose work we describe in this chapter believes that he (they are all male) and his group have conceptually solved the great enigma. They all readily admit that many tough problems remain, but all are sure that insofar as consciousness can be explained, it has been explained through their work, or will be explained through similar research. Moreover (although this is a point they do not always stress), their different explanations are convergent: however different their backgrounds and favorite methodologies, they share the same broad, naturalistic, philosophical framework. They agree that consciousness is a weakly emergent, intrinsic process in which the whole is different from the sum of its parts, but no new force of nature needs to be added. They regard consciousness as the outcome of the self-organizing, dynamic interactions between the low-level parts of a hierarchically structured neural system, which are constrained by the higher levels of organization and give rise to global, novel, and coherent patterns of percepts or actions.

This philosophical commitment is shared not only by neurobiologists but also by most psychologists and cognitive scientists. The difficulty they all face is how to embark on the experimental study of the emergent dynamic organization. Some (very few) researchers try to identify a unique factor that, once present, can serve as both a necessary and a sufficient condition for consciousness. Quantum coherence in the microtubules of neurons, certain electromagnetic wave patterns, or a particular molecular signature are examples of physical and chemical indicators that have been hopefully suggested as unique correlates of consciousness. Unfortunately, although not surprisingly, no such reliable, unique, necessary, and sufficient correlate has been found. At the other end of the spectrum of approaches are the psychologists and cognitive scientists who try to identify behavioral correlates: to find elaborate, flexible, and complex-enough behaviors that are sufficient conditions for consciousness. However, they do not make clear what the minimal behavioral complexity should be: Are the behaviors of honeybees and red ants complex enough? And if not, why not? Do existing robots exhibit complex-enough behavior? When and why shall we decide that a robot is a conscious being?

Another approach is the one followed by most neurobiologists: the characterization of the neural differences in information processing between unconscious and conscious subjects in different states. For example, they compare patterns of activity in the brains of normal people who are in aroused, attentive, wakeful, sleeping, dreaming, meditative, or hypnotic states and examine how experimental conditions can trigger changes in these states. In addition, they investigate the effects of brain lesions, anatomical irregularities associated with specific psychiatric disorders, and peculiarities such as synesthesia and compare reports of awareness/unawareness to identical stimuli while the attention of the subject is being manipulated. Neural activities in various brain areas or in single neurons are monitored using single-cell electrophysiological recording, magnetoencephalography (MEG), electroencephalography (EEG), transcranial magnetic stimulation (TMS), imaging techniques such as positron emission tomography (PET), functional magnetic resonance imaging (fMRI), and optogenetic techniques that allow the manipulation of single neural circuits.[1] The range of techniques is growing—and with it the ability of neurobiologists to manipulate and study conscious states in humans and other mammals.

Through these comparative studies of patterns and processes, neurobiologists have attempted to capture the neural correlates of consciousness (NCCs).[2] Although the term "neural correlates" of mental phenomena did not become popular among neuroscientists until the 1990s, it was already in use in the early twentieth century. The American philosopher Grace de Laguna was one of the first to use the term in the modern sense, as a system property of the nervous system. She wrote:

> What the empirical evidence points to as the *neural correlate of the sensation "red,"* is not the occurrence of specific processes in the visual center, but the functioning of that center as a member of a complicated system. (de Laguna 1918, pp. 539–540; emphasis added)

The modern notion of a neural or physical correlate follows but also qualifies de Laguna's conception. NCCs are supposed to both identify consciousness and to be a constituent of its neural dynamics. However, it is far from clear how to delimit an NCC. First, NCCs can be specific or global: an NCC can relate to a specific content of experience, like seeing red, or it can be global and characterize states of arousal, such as focused attention while one is wide awake or the consciousness of the dreaming state. Second, NCCs are supposed to be more than merely reliable system signatures of consciousness (such as consistent by-products or consequences, necessary background conditions, or parallel but unrelated aspects of brain activity). Rather, they are assumed to specifically cause consciousness or, even more ambitiously, to fully capture consciousness—to be a complete causal-constitutive description of it.[3] But can NCCs be precisely delimited? The ability to distill the NCC and distinguish its special essence from that of its prerequisites and consequences has become an explicit goal of some consciousness researchers. However, it is not clear that the elusive essence of the NCC is not in the relation between certain prerequisites and consequences, in the process of transition from prerequisites processing to consequence processing.[4]

Using the strategy outlined in chapter 1, in this chapter we first list the attributes of consciousness as seen by neurobiologists and cognitive scientists and then describe the intrinsic system dynamics that different leading neurobiologists have proposed for it. Their common emergentist philosophical commitment, combined with their common ambition to identify specific and global NCCs, leads to a shared list, many elements of which go all the way back to James.[5] The order of the seven entries in the list does not reflect their importance: although different neurobiologists highlight certain aspects more than others and the different attributes partially overlap, all are important and all are considered crucial for the emergence of conscious states. They are individually necessary and jointly sufficient for consciousness.

1. *Global activity and accessibility*: It is widely agreed that consciousness is not localized in specific brain parts; although its initiation or "ignition" can be detected and localized, consciousness is characterized by widely distributed recurrent brain activity that makes information globally available to different cognitive processes that are otherwise computationally isolated. Information is assumed to be transmitted between and across multiple neural processing circuits, allowing them to communicate and to influence one another. The brain states characterizing conscious states are integrated,

distinct, and metastable. Like an ocean landscape, the neural activity land-scape of the conscious brain keeps changing.

2. *Binding and unification*: The unified, integrated nature of subjective experiencing—for example, the awareness of a ripe banana, involves the "binding" of various features. We experience the banana as a totality that is yellow, elongated, and has a typical fragrance. Many theories suggest that binding is associated with the synchronized firing of ensembles of neurons dedicated to processing sensory inputs, although sequential processes that lead to loops of back-and-forth activity are crucial to binding. The unity of experience that is brought about by binding seems to involve the construc-tion of isomorphic sensory maps, in which the features of the percept (the banana in our case) are successively combined and mapped at ascending layers of the brain hierarchy. The unified nature of a particular experience always involves, as James noted, the exclusion of alternatives. At the phe-nomenological level, the unity of perception and the absence of aware-ness of the neural activity underlying it lead to the naïve belief that the world perceived is identical to the world independently of perception. This belief in the *transparency* of the world and body is called "naïve realism" by philosophers.

3. *Selection, plasticity, learning, attention*: During the construction of conscious experience, selection among neurons, neural connections, and neural networks is fundamental. Neural circuits in the brain are plastic, constantly changing, with synaptic connections among their neurons forming or dissolving, strengthening or weakening. Some biologists use the language of selection and competition when they describe the role of processes of neural exclusion and suppression in the generation of con-scious states; others talk about selection among replicating neural patterns of activity, but the idea that selection processes are involved is generally accepted. Neural-behavioral plasticity is not only a necessary background condition for the development of consciousness, with plasticity itself being plastic and controllable; such plasticity also requires differential exclusion and selection mechanisms. A prominent feature of conscious experience is its serial nature: the perception of one item at a time. For example, when each of the two eyes is presented with a different picture, only one of the two pictures is seen by the subject at any one time. Seeing only one percept at a time involves processes of selection: differential inhibition and differ-ential amplification.

Learning is a prominent way in which plasticity is manifest. Memory, a facet of learning, involves stabilization processes, whereby a memory trace is not only established but reliably maintained. Some complex forms of learning are assumed to require conscious attention, which entails a great deal of lateral inhibition. The process involves updating already existing neural patterns through the use of information that is "newsworthy" and different from what the system already has in store (this means that information that is not new is excluded). This selection includes behavioral target selection (what the organism attends to) and action selection (the choice of one of several action programs and the exclusion of others). Consciousness is considered necessary for some forms of action selection—for example, in a situation of conflict that involves the suppression of compelling action patterns such as breathing or voiding. Amplification, exclusion, and selection are closely related facets of consciousness because a notion of selection is implied whenever focused attention is considered: such attention is, by definition, always selective. A close connection between focused attention and a certain type of consciousness—reportable, access consciousness—has long been suggested, and consciousness is considered by some scholars to be equivalent to, or entailed by, focused attention. However, phenomenal consciousness (subjective experiencing) may not always be readily reportable and need not be linked to focused attention because any specific perception or feeling is integrated within a general fuzzy feeling, which is part of *every* type of subjective experiencing. There are also experiments that decouple phenomenal consciousness and attention.[6] Nevertheless, when an animal displays focused and reportable attention, it is universally considered to be conscious.

4. *Intentionality (aboutness)*: Conscious states are special because unlike other objective phenomena they refer to the world or the body, so one can always ask what a certain conscious state is about. Conscious mental states are what philosophers call "intentional." The notion of mental representation (and representation more generally) is tightly related to the notion of intentionality, as is the related notion of mapping external relation to (neural) internal relation. Hence, intentionality can refer to both conscious and nonconscious states. Intentionality is also discussed in the context of intending to reach a goal, with "goals" being represented by dynamic and complex neural states that drive behavior.

5. *Temporal "thickness"*:[7] Consciousness cannot occur if the neural effects of a stimulus do not persist; they must last some time to be "captured" and become conscious. Although stimuli that are not consciously perceived can reach deep into the brain, these neural effects are localized, very variable, and short-lived. It is thought that back-and-forth reverberations among

groups of neurons and/or the consecutive activation of dedicated circuits are necessary to render the temporal persistence needed for phenomenal consciousness. The dynamics that different scientists describe—"reentrant loops," "recurrent processing," "positive feedback loops within large networks," "strange attractors"—all reflect this basic intuition; because these reentrant interactions (back-and-forth interactions among groups of neurons) involve many neural networks at several scales, activity is maintained, thereby allowing the percept to be retained in what is known as "working memory."

6. *Values, emotions, goals*: Experiences have subjectively felt valence; they are felt, directly or indirectly, as being positive or negative and may have specific emotional dimensions, depending on the sensorimotor interactions involved. Felt values (pleasure and displeasure, although not all values need to be felt) guide the organism's behavior and its ever-changing internal states and actions so that a homeostatic, fitness-promoting state is achieved. Striving to satisfy felt needs gives rise to what one calls "motivations" and is manifest as goal-directed behavior.[8] Consciousness is therefore seen as an interface where motivations become available for regulating and driving goal-directed behavior.

7. *Embodiment, agency, and a notion of "self"*: Feedback between the brain and the rest of the body (and the external environment that the organism actively constructs and responds to) is necessary for the developmental construction of consciousness and the sense of agency that we call "self." "Self" entails a stable distinction between the body and the exterior environment, made possible by their neural mapping as well as the mapping of their relations. Consequently, although the architecture of the brain and its internal dynamics are crucial, the brain alone is not sufficient: no consciousness is possible for a brain in a vat, or for that matter, for a body in a vat (unless the body and the world are simulated). Consciousness is therefore conceptualized as a special kind of ongoing, bodily sensorimotor activity, in which the brain has an organizing, integrating part. It leads to the localization of some feelings, such as pain, and to the feeling of ownership and agency that we call "self," which implies that experiences are projected from a particular point of view.

There is nothing in this list that relates to the evolution or phylogenetic distribution of consciousness. Yet it goes without saying that all the leading neurobiologists of consciousness whose work we describe embrace evolution, and some make creative use of the selection principle. They also all agree that consciousness has a functional significance related to the integration of information within and between modalities, which enables flexible, adaptive, goal-directed behavior. However, since they are not evolutionary biologists, they address only a small range of evolutionary questions. They do not suggest the type of detailed scenario for the origins and refinement of consciousness that an evolutionary biologist would deem satisfactory (even if speculative). They do not ask: What is the distribution of consciousness in the living word? When, and in what geochemical and ecological conditions, did it arise? What were the biological, functional-structural preconditions (the building blocks, the enabling conditions) for its emergence? What were its original function/s, if any? How, when, and why were these structures and functions modified, and through what stages did the process pass? Was consciousness lost in some lineages? Did it emerge just once or more than once? What were its evolutionary effects? The questions these scholars address pertain to the functions and evolutionary stages of consciousness—usually, a progression of two or three stages from simple consciousness to a more complex one; the other questions are hardly ever mentioned. This is probably due not only to different professional commitments but also to the exclusive reliance of most consciousness researchers on investigations using highly evolved animals, the mammals. Since they all agree that mammals are phenomenally conscious (have subjective experiences) the distribution question—how is consciousness taxonomically distributed—is not their main concern. They sometimes acknowledge in passing that birds may be conscious, and those that stress the role of the brain stem also suggest that lower vertebrates (reptiles, amphibians, and fish) may be minimally conscious. However, these possibilities are not explored in their work because their studies are focused on the mammalian nervous system. The result is that the comparative method, which is central to evolutionary biology, is hardly ever applied beyond the mammalian sphere by the major neuroscientists of consciousness, although a comparative evolutionary perspective does exist and is actively pursued by a minority of scientists in the field.

Even though the studies described in this chapter rely on a very narrow range of animals, these are the most detailed and extensive neurobiological investigations available to date; the models and frameworks based on them are inevitably an important point of departure for theories of consciousness. Because some knowledge of the anatomy and functions of different parts of the human brain and of "brain waves" is necessary for grasping the neurobiological frameworks we discuss, in box 3.1 we summarize basic information about the major parts and functions of the human brain and briefly describe the types of electrical oscillations it produces.

We have divided the neurobiologists' approaches to consciousness into four main overlapping frameworks. All share the assumptions of the emergentist consensus, but they emphasize different aspects of it, so we have tried to highlight those facets of their work that the scientists believe capture the dynamics that constitute consciousness. These frameworks do not cover all the current approaches to consciousness—there are other views based primarily on *cognitive* considerations, which, we believe, are no less fundamental to the understanding of minimal consciousness and are central to our own evolution-of-learning approach. Two frameworks are particularly important for our own perspective and are discussed in subsequent chapters. The first is the viewpoint-dependent perspective of the cognitive scientist Bjoern Merker, which focuses on the construction of a "self" and goes beyond the neurobiologists' mammalian-centered approach; the second is the Bayesian, broad, hierarchical predictive processing (HPP) framework developed by the cognitive scientist Karl Friston and his colleagues.[9] However, although we recognize that distinctions between neurobiological and cognitive frameworks cannot be sharp—most cognitive scientists (and many philosophers of mind) are engaged in experimental research and vice versa—in this chapter we consider only the views of influential biologists who have been engaged in attempts to uncover the specific and general neurobiological correlates of consciousness. Four main frameworks can be

Box 3.1
Brain parts and brain waves.

Figure 3.1*A* illustrates major areas of the human brain and some of the functions attributed to them. The correlation between some structures and the control of conscious arousal is shown in figure 3.1*B*.

Some of the ideas about consciousness that we discuss are based on the functional interpretations of observations involving the rhythmic electrical patterns that are detected in the brain using electroencephalography (EEG), magnetoencephalography (MEG), and similar noninvasive techniques. These oscillations are the result of the synchronous activity of large numbers of neurons and are generated in two ways: (1) spontaneously, as a result of the electrical properties of neurons and the ionic channels in their membranes; (2) in response to excitatory and inhibitory inputs from the external environment and from other neurons. The five types of brain oscillations that are commonly recognized and the different states of wakefulness and sleep associated with them are shown in figure 3.2.

(continued)

Box 3.1 (continued)

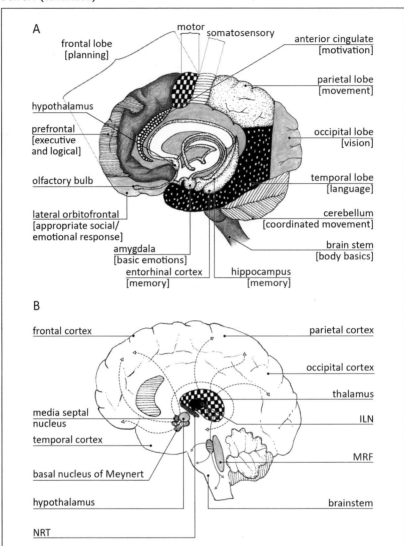

Figure 3.1

The human brain. *A,* sensory, motor, cognitive, and affective functions of different brain areas. Note that the language and movement areas labeled here are not the only brain areas involved in these functions. The basic body functions regulated by the brain stem include, for example, heart rate and breathing. *B,* midline structures in the brain stem and thalamus that regulate the level of brain arousal; bilateral lesions in many of these structures cause a global loss of consciousness. ILN, intralaminar nuclei; MRF, medulla reticular formation; NRT, thalamic reticular nucleus. The dotted and full lines with the arrows mark the spread of excitation from the thalamus and brain stem to other brain areas.

Box 3.1 (continued)

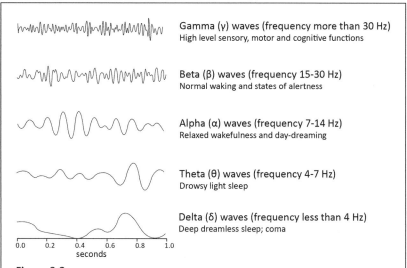

Gamma (γ) waves (frequency more than 30 Hz)
High level sensory, motor and cognitive functions

Beta (β) waves (frequency 15-30 Hz)
Normal waking and states of alertness

Alpha (α) waves (frequency 7-14 Hz)
Relaxed wakefulness and day-dreaming

Theta (θ) waves (frequency 4-7 Hz)
Drowsy light sleep

Delta (δ) waves (frequency less than 4 Hz)
Deep dreamless sleep; coma

Figure 3.2
EEG frequency bands and their functional correlates.

Gamma waves seem to accompany reportable conscious states and have been hailed as essential components of them. It should be noted that some researchers doubt whether gamma waves (and other events in an EEG) are reliable markers of consciousness. (See Tononi and Koch 2015; Koch et al. 2016.)

distinguished, which, though overlapping, differ in their emphasis of particular facets of consciousness.

The first framework, exemplified by the work of Francis Crick and Christof Koch, focuses on the correlates of specific perceptual experiences, such as seeing the blue of the sky. The philosopher John Searle called this kind of approach the "building-block approach."[10] It stems from Crick and Koch's assumption that every conscious experience has content—that it is a specific experience about something (philosophers call it "content consciousness" or "transitive consciousness") and that identifying how a subject becomes aware of a particular experience is the key to understanding consciousness in general. For example, when we say about our friend that she was so absorbed in her thoughts that she was not conscious (in the transitive sense) of the wall and bumped into it, we mean that she did not gain *access* to the visual input afforded by the wall, not that she was generally unconscious. The second framework, reflected in the different methodologies of Rodolfo Llinás and of Walter Jackson Freeman III, can, following Searle's classification, be labeled the "unified field approach" and involves studies of the state of being in a conscious state as opposed to being unconscious (philosophers call it "intransitive consciousness"). For example, when one says "as the anesthesia was wearing off, she slowly regained consciousness," one is using the intransitive sense of consciousness. Neurobiologists who study intransitive consciousness seek to characterize the general features shared by *all* conscious states. For these scientists, specific states of consciousness can be seen as modifications in the global consciousness field, as "bumps and valleys," as John Searle has put it, that generate specific consciousness patterns within the general field. Those using the third framework describe a workspace or dynamic core in which consciousness becomes manifest, and they highlight neural selection. This is why we call it the "loading-the-dice" approach. Two groups of scientists are prominent promoters of this framework: one includes Jean-Pierre Changeux, Stanislas Dehaene, Lionel Naccache, and Bernard Baars and their colleagues; the other includes Gerald Edelman and Giulio Tononi and their collaborators. Antonio Damasio is representative of those adopting the fourth framework, which emphasizes emotions rather than perception, highlighting the embodied aspects of the conscious self.

Of course, our characterization is somewhat misleading because there are overlaps and interrelations among the different approaches. In fact, *all* frameworks adopt a mix of "building-block" and "unified-field" approaches, albeit to different degrees and using different types of argumentation and methodologies. There are also, as we explore in more detail in chapters 5–9, other ways of carving up neurobiological knowledge about consciousness, some of which incorporate studies of nonmammalian taxa. Nevertheless, the four frameworks we describe, which are based on a substantial body of research in the neurobiology of consciousness, have been fundamental to current ways of thinking about consciousness.

The Neural Correlates of Specific Experiences

Crick and Koch defined NCCs as "the minimal set of neuronal events that gives rise to a specific aspect of a conscious percept."[11] Rather than trying to figure out the nature of the general neural enabling factors of all forms of consciousness, they looked at specific experiences and a specific system. Because the visual system had been investigated in detail, they concentrated on visual perception, asking which events produce a particular experience such as perceiving the shape, or the movement, or the color of an object.

They deliberately stayed away from what they considered more complex issues, like emotions and self-awareness.

The investigation of bistable visual percepts (gestalt switches) serves as a good example of Crick and Koch's approach to uncovering the NCC. Bistable percepts occur when the constant stimulus of an ambiguous composition, such as a Necker cube, results in two different perceptions that alternate over time—two flipping interpretations (see figure 3.3*A*). Binocular rivalry is another instance of bistable perception: in this case, one image (a face, for example) is presented to the left eye, and a different one (such as a sunburst pattern) is presented to the other (figure 3.3*B*). The two stimuli are presented simultaneously and continuously, but the images that subjects report seeing alternate every few seconds from one (face) to the other (sunburst pattern); most of the time, subjects do not report seeing an amalgam of the two images.

One can follow the brain changes (NCCs) that occur during shifts between different perceptual experiences such as gestalt switches or alternative percepts in binocular rivalry. The cognitive-neurobiologist Nikos Logothetis trained macaque monkeys to report when they were seeing either a face or a sunburst pattern by pulling a left lever when perceiving a face and a right lever when perceiving the sunburst pattern. As the monkeys reported, he recorded the electrical activity of identified neurons in their brains using single-unit extracellular recording. The same cells were tested when the monkeys were then presented with these pairs of images simultaneously, each image to a different eye, following the binocular rivalry paradigm.[12] The results showed that very few neurons in the primary visual cortex (the sensory area where the visual information first arrives) modulated their activities when the monkey reported a particular image, whereas correlations with the reported image were observed in higher cortical visual areas, notably, in the inferior temporal lobe (the IT cortex). The response of nearly all the neurons in this region depended on the perceptually dominant stimulus (the one the monkeys reported seeing). For instance, a neuron that was normally active only when presented with a face in the binocular rivalry experiment fired impulses only when the monkey reported perceiving a face and not when it perceived the sunburst pattern. From these experiments, Crick and Koch concluded that neuronal activity in the IT cortex is a neural correlate of visual experiencing.

Crick and Koch collected a mass of data from this and other investigations of visual consciousness and concluded that conscious perception requires global feedback from the frontal regions of the cortex back to the sensory areas. This sustained, reverberating neural activity builds up until

A

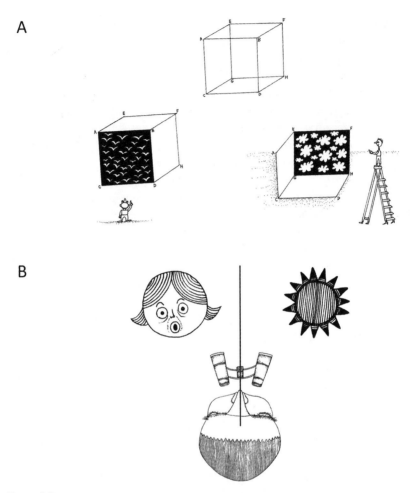

B

Figure 3.3

Bistable percepts. *A*, an example of a gestalt switch—Necker cubes and their flipping interpretations. *B*, binocular rivalry. Each of the eyes of an experimental subject sees a different picture (in this case a face or a patterned sun). Although the two eyes receive the visual inputs simultaneously, the subject perceives them as alternating, not amalgamated.

it reaches a critical threshold, which, once exceeded, propagates to additional cortical regions and to the thalamus, claustrum, and other structures involved in the integration of sensory modalities and in memory and planning.[13] Moreover, back propagation—from the frontal cortical regions back to the primary visual areas—is also necessary for the percept to be conscious/ accessible/ reportable.

On the basis of their understanding of the neural activities underlying visual perception, Crick and Koch summarized some of the ideas and assumptions they thought should guide the search for a general theory of consciousness.[14] First, they suggested that the much-criticized idea of an internal homunculus—a little man inside the head who receives information from the senses and decides what to do with it—may be appropriate if we think of the front of the brain as "observing" the sensory systems, which are mostly located at the back. Their second suggestion was that the brain has both "zombie" (nonconscious) and conscious modes. Many zombie behaviors are rapid, transient, stereotypical, and unconscious, even though some may be rather complex and require substantial integration, such as the very rapid motor decisions a skilled tennis player makes. These zombie behaviors differ from conscious modes of action, which emerge more slowly and purposefully and lead to less stereotyped behavior. In zombie mode, Crick and Koch suggested, information propagates only from lower to higher cortical areas, whereas in conscious mode information flows in both directions.

Crick and Koch's third assumption was that awareness emerges from the activity of coalitions of neurons: neurons that are spatially distributed form a transient coalition lasting for the duration of the specific perception (at least 20 ms to 200 ms); a particular neuron may be part of different coalitions at different times. They assumed that for conscious perception, activity in the coalition must reach a threshold. Normally, a coalition involves continuous interactions among widely dispersed pyramidal cells and emerges as a result of synchronous firing, which contributes to reaching the threshold.

The fourth of Crick and Koch's suggestions recognized that consciousness is related to attention and to neural selection processes.[15] In their view, several events or objects may be perceived simultaneously provided there is no overlap in the cortical or thalamic networks involved. If they do overlap, different coalitions of neurons may be said to "compete." The coalition that "wins" is the one with activities that reach the threshold, producing the perception of particular events or objects at the expense of other percepts. Attention biases competition among rival coalitions, particularly during their formation, by modulating the degree of firing synchrony among the coalition neurons. In a true Jamesian spirit, they suggest that attention may be driven by the salience of a feature relative to other features in the vicinity ("bottom-up attention"), in which case it is automatic and rapid; or, it may be controlled by volition ("top-down attention").

Finally, Crick and Koch dealt with binding. Different features or attributes of an object—the sound it emits, its motion, its shape, and so on—are explicitly represented by different cortical neuronal circuits, yet they contribute to the formation of an integrated, unitary experience. Neurons that bring this about can be wired up through experience to encode an object such as the face, voice, and mannerisms of a famous figure, such as Charlie Chaplin. In addition, groups of nerve cells may be developmentally modified to respond to particular combinations of inputs. Crick and Koch built on previous suggestions and studies, such as those of Wolf Singer and his colleagues, who showed that binding may be the result of synchronizing the activities of neurons that are related to different features of an event or object around the same frequency (30–70 Hz gamma waves; see figure 3.2).[16] For example, a unitary perception of a simple form—a green triangle—is assumed to require that two groups of distributed neurons, one specialized in signaling the stimulus shape (a triangle) and the other the color (green), would (1) increase their firing rates upon presentation of the stimulus and (2) synchronize their output responses, so their activities oscillate hand in hand, "locked" in the same frequency. However, this hypothesis meets with difficulties: first, there can be binding without temporal synchronization, and second, it is not clear that the temporal synchronization, when it does occur, actually constitutes binding rather than merely signaling that binding has occurred (by other means).[17] Crick and Koch initially regarded synchronized oscillations as a diagnostic mark of consciousness, but they changed their mind shortly after 2000 and suggested that the role of these oscillations is to resolve the competition among coalitions of neurons—a competition that they and some other researchers saw as the essence of conscious activity.[18]

Although Crick and Koch did not offer a list of conditions that must be met for consciousness to arise, their theory does share most of the points in the emergentist consensus we listed earlier—namely, integration and binding and a focus on selective attention and global activity. After Crick's death, Koch continued their joint venture to pinpoint the NCC. His memoir of their work together and his latest thoughts on consciousness and projections about the future are documented in his autobiography.[19] The wealth of publications employing neuroimaging techniques to study NCCs attest to the ongoing popularity of their building-block approach. Current research continues to focus on visual awareness, for which the experimental manipulation of perception is relatively easy, but imaging methods are also used to investigate the NCCs underlying emotions and human behaviors involving learning, memory, and the use of language.[20] Parallel work is also conducted using animals, although not all of the researchers involved in this accept that nonhuman animals have consciousness (i.e., subjective experiencing).

The Field View of Consciousness: Oscillators and Attractors

Several neurobiologists, including Rodolfo Llinás and Walter Jackson Freeman III, investigated the nature of consciousness by first identifying its global correlates. Their methodologies and concepts differ, but they share the same global "unified-field" focus, highlighting the importance of interacting brain-wide circuits rather than specialized brain areas and emphasizing the high degree of connectivity in the brain.

Coupled Oscillators

The neuroscientist Rodolfo Llinás has been studying the brain for half a century.[21] Like other neurobiologists, he has recognized both the global aspects of brain activity and the more specific processes that render experiences unique, coherent, and unitary. Unlike Crick and Koch, however, he started with the global NCC. He therefore has focused his studies on the neural differences between wakefulness, dreaming (a subjectively experiencing state), and dreamless sleep. These differences are, in his opinion, the key to uncovering the global NCC. He also has referred to the NCC as the neural correlate of *cognition*, perhaps demonstrating by this his tendency to equate consciousness with perceptual rather than affective states.

Llinás and his coworkers used lesion data and MEG to compare the interactions between specific cortical and subcortical regions during wakefulness and sleep. They found vigorous gamma waves spreading from the front to the back of the brain during the wakeful and dreaming states, but these waves were highly attenuated in sleep, disappearing altogether during deep sleep. They agreed with the view of Singer and others that the rapid integration and binding of sensory information relating to different features of an object or event are associated with the synchronization of neuronal activities around an oscillating frequency of about 40 Hz. However, they recognize that rapid binding does not explain how the represented unified features of a given object or event become conscious: other objects and events may be represented in the brain and synchronized without reaching awareness, so consciousness requires something more. Llinás addressed both the binding problem and the question of how a particular representation becomes conscious and accessible by proposing that a conscious experience depends on the action of coupled thalamocortical oscillators of two types. Each type of oscillator consists of a population of neurons that connect the cortex and the thalamus via distinct, specific connections.

The first type, the "nonspecific" oscillator, is made up of thalamic neurons within the intralaminar nuclei (ILN, figure 3.1B) that connect to wide areas of the cortex. The cell bodies of these neurons have long axons that send projections to the entire cortex, and the cortical neurons that receive them send their projections back to the intralaminar nuclei; this system is responsible for producing the 40 Hz waves that sweep through the cortex, providing a general "context" for conscious perception. The second type are "specific" oscillators, which contribute to the binding of the different features of an object or event into a unified whole. They too are made up of thalamic neurons, but in this case the nerve cells of each oscillator are part of thalamic nuclei that process modality-specific information (auditory, visual, somatosensory, and more). The neurons of each specific oscillator connect to a relatively small area of the cortex. Thus, for example, one of the specific oscillators consists of nerve cells of the thalamus that receive input from the retina and connect to neurons in the corresponding cortical area, the primary visual cortex, passing on information to them. The modality-specific oscillators supply the content of consciousness (e.g., visual, auditory, and somatosensory), and the nonspecific oscillator provides the integrating glue. Put a different way, neural populations that represent a given content of consciousness are cell assemblies that oscillate in phase not just with one another (supplying binding) but also with the nonspecific oscillations that sweep through the cortex (see figure 3.4). Consciousness is marked by the synchronization between specific and nonspecific oscillators. Llinás and his colleagues summed up their idea succinctly: "The specific loops give the content of cognition, and a nonspecific loop gives the temporal binding required for the unity of cognitive experience."[22]

The 40 Hz activities of the two types of oscillators are coupled via excitatory and inhibitory connections and feedback loops. These connections, and the cellular mechanisms contributing to the oscillations, have been described in great detail by Llinás and his colleagues. During wakefulness, while the subject is active in mental tasks, strong global 40 Hz oscillations can be recorded by MEG, and intracellular recording shows that specific oscillators are also active, their networks responding to external signals. It is the activities of specific oscillators, coupled to the global oscillator (i.e., the synchronized combination) that give rise, they suggest, to consciousness. While dreaming, oscillators are also coupled, but the neurons in the specific thalamic nuclei are not electrically responsive to external signals; hence, in

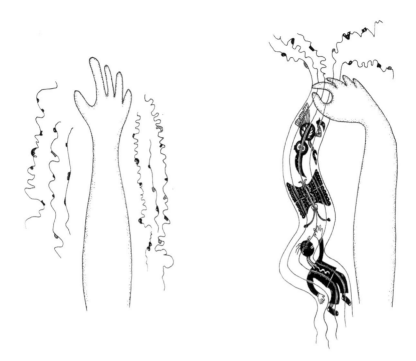

Figure 3.4
Synchronizing global and specific oscillations. *Left*, each thread denotes a separate, *specific* oscillator that has disjointed perceptual effects (*black dots*); in this state the threads are not coupled to the *nonspecific* oscillator, represented by the hand. *Right*, the nonspecific oscillator is coupled (*hand in a grasping state*) with the specific oscillators tying them together in time, so they start oscillating in phase, thus producing coherent images (*percepts, black shapes*) representing conscious experiences.

dreaming states, the content of consciousness is drawn from internal signals only—from scrambled memories. In contrast, during deep sleep, the two types of oscillators are decoupled, and there is no conscious content at all.

Many predictions based on these ideas have been corroborated. Lesions affecting modality-specific thalamic nuclei result in the loss of specific awareness: for example, visual awareness is lost by lateral geniculate nucleus lesions without affecting the subjective experiencing of sounds and touch; lesions to the intralaminar nuclei result in the total loss of subjective experiencing—there is no response to sensory stimuli and extremely little initiation of behavior; if the lesions are massive, they can lead to coma. Newer findings, such as the observations that dyslexia and aging alter gamma oscillations, and additional theoretical work have lent support to this aspect of Llinás's hypothesis.[23]

Global Attractors

Another "field" approach is that of the neuroscientist and philosopher Walter Jackson Freeman III,[24] who used EEG to study the olfactory experiences of the rabbit. Freeman explored what he called "meaning-making"— the formation of organismal states that are unified and guided by values (such as pleasant and unpleasant). Meaning-making involves exploratory behaviors and the selection of stimuli that create dynamically changing patterns of neural activity.

Following the pragmatist tradition, the point of departure for Freeman is action and its consequences. Therefore, the way he thought about the neural dynamics associated with consciousness does not track the usual causal chain, which starts with sensory stimuli impinging on the animal and is followed by the animal's response. Rather, he began with the endogenous chaotic neural activity that leads to exploratory motor behavior. Through this exploratory activity, the animal selects those environmental stimuli whose effects match or transiently reequilibrate its internal states and lead to perception or directed movement. These neural processes alter the dynamic "neural landscape" of the brain, which in turn leads to new sensorimotor states and new changes in the neural landscape. Such behavior is always goal-directed or, as Freeman preferred to call it (using the philosophers' term) "intentional"; it is guided by arousal and expectancy, which are aspects of emotions. This is a very Jamesian approach to consciousness.

Freeman's model of consciousness is based on findings from EEG recordings of rabbits trained to react to different odorants. He found that the olfactory bulb, which responds to odorants coming from the sensory nerves in the nose, has a continuous, self-stabilizing background activity. When the rabbit inhales, there is a burst of neural activity with oscillations in the gamma range, which ends when it exhales. The oscillations have the same frequency at different locations within the bulb, but the amplitude of the waves differs. According to Freeman, a particular, recognized odor is not marked by specialized neurons firing at a certain frequency; rather, the entire olfactory bulb fires synchronously at a common rate, with the specific "signature" of an odor represented by regional differences in amplitude—the "amplitude modulation" (AM) pattern. AM patterns depend on the individual rabbit, the present context, the learning history, and the rabbit's state of arousal. They change over time, drifting slowly when there is no obvious change in conditions and changing rapidly and profoundly with learning. When a new odorant is introduced and there is no negative or positive reinforcement (it is neutral), no new AM pattern is formed (the animal habituates), but when new odorants are reinforced, a new AM pattern is added.

Freeman's account of the olfactory experiencing of the rabbit is very different from the account of odor discrimination in the mammalian (mouse) and fly (*Drosophila*) brain, for which Richard Axel received the Nobel Prize. Axel and his colleagues showed that the discrimination among different odorants is based on a great diversity of odorant receptors, with each sensory neuron expressing only one receptor. The sensory cells project to particular glomeruli (spherical structures located in the olfactory bulb) and form a sensory odor map so that an olfactory stimulus composed of several odorants creates specific topographic patterns of activity within the olfactory bulb.[25] This beautiful research still leaves open the question of how the *experiencing* of odors occurs. As Axel (2004) put it:

> Who in the brain is looking at the olfactory image? Who reads the map? How are spatially defined bits of electrical information in the brain decoded to allow the perception of an olfactory image? (p. 19)

Freeman tried to answer these big questions by using the formal mathematical tools and the language of dynamic systems theory. Although his suggestion is highly abstract,[26] we need to understand what his dynamic system approach entails.

A dynamic system is described in terms of an abstract phase or state space whose coordinates define the state at any instant and a rule that specifies the immediate future of all state variables given their present values. For example, the motion of a pendulum, a simple dynamic system, is determined fully by its initial velocity and position. The dynamics of this simple system can be described by plotting the pendulum's circular velocity against its position, producing an orbit (a path) in a plane (figure 3.5A, *left*). At any moment in time, the velocity and position are represented by a single point on the circle. In this example, a state space is any point in a plane whose coordinates are velocity and position, and Newtonian mechanics serve to describe how the state changes over time. The one-dimensional circular orbit is a limit cycle; it is a periodic attractor because if the pendulum is gently perturbed, it may shift a little but will settle back into its usual motion, as though the circle attracts it back to its orbit. The right-hand side of figure 3.5A shows the dynamics of a pendulum that is losing its power: the orbit spirals toward a point, the point attractor, and comes to rest. In general, attractors are a point or group of points toward which dynamical

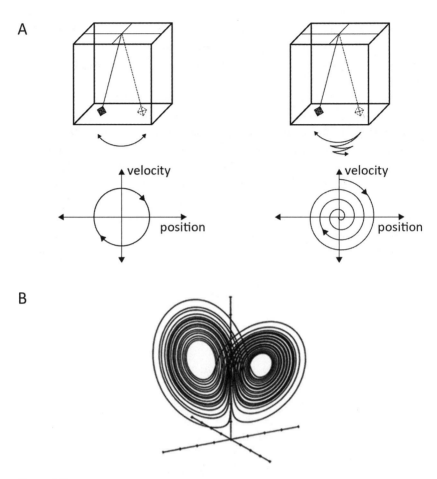

Figure 3.5
Attractors and their trajectories of change. *A*, *left*, an ideal pendulum, whose orbit is a limit cycle (a periodic attractor); *right*, a pendulum with friction, whose orbit is a spiral moving toward a point (a point attractor). *B*, the Lorenz attractor developed by Edward Lorenz as a simplified mathematical model for atmospheric convection (although it can be applied to other complex systems). The Lorenz system, which is nonlinear, three-dimensional, and deterministic, is described by three ordinary differential equations. The butterfly-like shape presents a graphic plotting of a solution to the equations.

systems evolve over time, provided that the system starts out in what is called a "basin of attraction" (the set of initial conditions that lead to long-term behavior that approaches that attractor).

The dynamics of nervous systems are of course more complex and less predictable than pendulum motion, and Freeman suggested they can be described as what in the jargon of systems theory are called "strange attractors." A famous example of a strange attractor, the Lorenz attractor, is shown in figure 3.5B. This three-dimensional attractor has the unpredictability of a chaotic system: the trajectories are highly sensitive to initial conditions and wander inside the two diverging wings in an apparently erratic manner, never returning to the same place. The geometric form shown in the figure portrays the behavior of a system in state space over long periods.[27]

It is the kind of spatial and temporal dynamics shown by strange attractors that Freeman suggested underlie conscious olfactory experiences in the rabbit, and he argued that olfactory experiences can be generalized to other modalities and other animals.[28] The EEGs recorded from the brain in different states of consciousness during different perceptions always consist of patterns of oscillations that follow the rules of chaotic systems. Thus, although the EEG patterns may seem arbitrary, they reflect *regularities* in brain activities. Freeman suggested that the observed oscillations (which, he claimed, are chaotic) are the result of the electrical perturbations of the primary cortices when the brain responds to changes in the perceived world. In these conditions, ensembles of neurons in the primary cortex jointly and synchronously alter their activity pattern in response to a sensory stimulus that may be very weak. Then, following the dynamics of chaotic systems, these perturbations spread to other remote brain areas, where they trigger further synchronization of the activities of other neuronal ensembles, resulting in a single, transient, overall activity pattern spreading through the whole brain—a process that takes time.

Freeman concluded that every individual brain has a unique history that depends on its learned experiences, and meanings develop within it based on the specific activity patterns it has generated, which are manifested as particular attractor dynamics. Consciousness is the process of change and interaction among high-level attractors. It takes time for activity to become organized and thus increases the chances that an action will be adaptive. It is always related to emotions, creating an experience involving the body as a whole.

Freeman was not alone in believing that strange attractors in the brain are significant for understanding consciousness.[29] Peter Walling (an anesthesiologist) and Kenneth Hicks (a biomedical engineer) have taken up

Freeman's ideas and suggested that the strange attractors of consciousness require increasingly multidimensional description as mental activity gets richer.[30] They followed the changing dynamics of EEG patterns as humans in deep anesthesia emerged gradually to the fully conscious and alert state and interpreted the patterns observed with progressive awakening as manifesting increasing "strangeness" and increased dimensionality. They also suggested that the dimensionality of attractors can be a criterion for mental complexity in the animal world. A point has a dimension of zero; a line, one; a plane, two; a wavy line, between one and two; a solid, three; and additional abstract dimensions may be ascribed to systems to uncover their symmetry and otherwise hidden regularities. Using the EEG dynamics of eleven animal species—sea anemone, starfish, earthworm, moth larva, crayfish, minnow, perch, catfish, frog, dog, and human—Walling and Hicks found that the dimensionality of the attractors was zero for the first four in this series and then increased.

A description of experiencing in terms of attractor dynamics is interesting and captures an important intuition—the systemic nature of experiencing and its active exploratory facets—but it needs to be integrated with molecular, developmental, and neurobiological research like that of Axel. We now turn to the proponents of our third approach, who suggested that the dynamics of neural selection, with their molecular and developmental underpinnings, can be used as the Archimedean point for the study of experiencing.

Loading the Dice

In the previous chapter, we described William James's idea that selection both contributes to the constitution of consciousness and serves as its main function. He thought that consciousness emerges as a result of selection in a noisy, labile, complex brain, a process that he called "loading the dice." Selectionist views of learning and culture, particularly of the workings of the brain in memory and thinking, became more common in the second half of the twentieth century.[31] In the 1970s, J. Z. Young, a neurobiologist, offered some of the earliest ideas about selective processes in the formation of memory:

> We may thus define the memories of an individual as the particular ordered states selected by a living system during its lifetime from among the range of possible states opened to it by its inherited DNA. Learning is the process of making that selection. (J. Z. Young 1979, p. 801)

In the late 1980s, another neurobiologist, William Calvin, suggested that decisions about what to do (action selection) depend on what he called "activated memory codes" (neuronal pathways of activity) that replicate and compete with others in the brain.[32] The idea that multiple representations ("drafts") compete within the brain was also put forward by Dennett, who offered it as an alternative to the view that consciousness occurs in some central "theater" within the brain.[33] However, the most neurobiologically explicit selectionist way of thinking about how the brain works—one that specifies the units of variation and targets of selection and stresses the cumulative, developmental effects of selection—was developed independently by groups led by Jean-Pierre Changeux and by Gerald Edelman.

Adaptive Evolution in the Brain

Changeux and Edelman both regarded neural selection as essential for explaining the central features of brain organization, cognition, and consciousness. Both suggested that Darwinian-like evolution occurs within and among populations of neurons and among numerous labile populations of neural connections that are formed during early development. At later stages and following learning, increasingly more sophisticated patterns of activity are formed on the basis of the earlier ones, and these later patterns, which occur at higher hierarchical levels than the early ones, constitute our experiences. However, the neural processes the two biologists envisaged differ in an important way from classical Darwinism: their notion of evolution in the brain does not involve replication and multiplication. It is based on selection during embryogenesis from the spontaneously formed (though genetically constrained) vast repertoire of nerve cells and their labile connections. It is the richness and enormity of this early repertoire that enables interesting, cumulative evolution: it makes it possible to have many rounds of selection and fine-tuning, culminating in functional cognitive states and behaviors. The brain is thus a developmental-evolutionary organ. It is extremely plastic and able to cope with many contingencies, thanks to progressive cycles of positive and negative reinforcements, which depend on the internal selection criteria the organism employs.

Changeux and his colleagues were the first to publish a theory about neural selection.[34] Their ideas were influenced by Changeux's reading of Jacques Monod's book *Chance and Necessity*,[35] in which Monod showed the explanatory power of the neo-Darwinian principle of accidentally generated variations followed by selection. Changeux applied this principle to the nervous system, suggesting that during embryogenesis a huge repertoire of nerve cells and synaptic connections is generated spontaneously, and the stabilization or elimination of the synaptic connections is the result of local positive or negative reinforcement, something that he sees as a developmental-selection process (figure 3.6A). Such stabilization processes continue throughout life; they are shaped by both the neural architecture and the constraints typical of the species and by the specific experiences of the individual. To account for increasingly complex forms of learning and cognizing, Changeux developed the idea further, adding more levels of stabilization within rich neural circuits. For example, his model shows how a crude and fuzzy version of a swamp-sparrow song crystalizes into its mature melodious form through interactions with the external world and the negative and positive reinforcement the sparrow receives.[36]

Edelman proposed a similar theory of neural selection in the late 1980s.[37] His theory was inspired by his research in immunology, for which he received the Nobel Prize, and especially by the insight that it is selection among diverse immune system cells that generates the incredible specificity of the immune response. Edelman applied this idea to the nervous system, arguing that an animal's nervous system also changes dynamically during development and throughout life: the neural circuits that process information are subject to many rounds of selection, and this is what leads to both the flexibility and specificity of animal activities. Edelman called this idea the "theory of neuronal group selection" (TNGS); it is also known as "neural Darwinism" (ND).[38]

The main features of Edelman's selection model are shown in figure 3.6B, which summarizes the three selection processes he posited. Each process involves large collectives of neurons. The first, "developmental selection," occurs during early embryogenesis, when neuronal cells divide and migrate and extend large numbers of branches in many directions toward other neurons. Some neurons die, and some branches are eliminated; in addition, synaptic connections are strengthened or weakened, depending on individual neurons' histories of electrical and chemical activity. Thus, developmental somatic selection forms an enormous number of variant neural circuits, which constitute the "primary repertoire" in each individual brain (figure 3.6B, left).

Edelman's second process, "experiential selection," begins as the early anatomical connections are being established and continues throughout life (figure 3.6B, right): the activity of neuronal groups is continuously selectively modified by synaptic changes that result from learning and experience. This selection occurs among collections of synapses, without large anatomical changes, on the basis of strengthening certain synapses and weakening others. The outcome is the formation of secondary repertoires

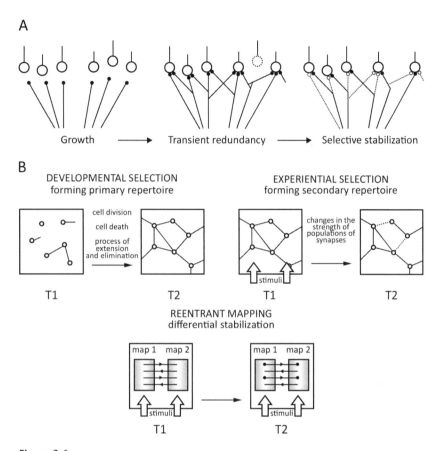

Figure 3.6

Neural selectionist schemes. *A*, the development of neural connectivity through the selective stabilization of connections, according to Changeux. During early development, multiple connections begin to form (*left, solid lines*), leading to the transient redundancy of connections (*middle*); a cell that is not connected does not survive (in this illustration, only five of the six cells survive). Some connections are later strengthened, but many are eliminated (*right, dashed lines*), so surviving cells end up having one connection. *B*, major features of Edelman's theory of neuronal group selection. *Left*, developmental selection involving differential cell division, cell death, and branch elimination or extension leads to the primary repertoire of neural circuits; *right*, experiential selection involving the strengthening and weakening of synaptic connections leads to the secondary repertoire of neuronal groups; some survive (*circuits in black*), while others dissolve (*circuits with dashed lines*); *bottom*, the spatiotemporal coordination of maps by ongoing signaling across reciprocal connections (reentrant signaling); black dots signify synapse strengthening. *Sources: A* is loosely based on Changeux 1983. *B* is reproduced by permission of Yale University Press.

of neuronal groups in response to particular stimuli. Because of the changes in synaptic efficacies, when signals of the same type occur, previously active pathways will be favored over others.

The third process in the TNGS is "reentrant signaling" (figure 3.6B, *bottom*). Reentrant signaling is the back-and-forth activity that links and coordinates interconnected series of neuronal groups that selectively respond to a particular category of input such as movement or color, thus in a sense "mapping" them. Through reentrant signaling, connections among simultaneously activated groups of neurons are strengthened. Synchronization and strengthening result in binding and are the basis for perceptual categorization. It is therefore at this level that the integration necessary for consciousness occurs.

Reentrant connections also link cortical regions with the thalamus and other areas of the brain. According to Edelman (and many neurobiologists agree with him), reentry is essential for consciousness. Without the backward flow from the higher brain level, where integration occurs, to the lower level, where the stimuli were received from the world, there can be no consciousness.

Edelman emphasized that reentry is distinct from feedback. Feedback diminishes or amplifies signals across a single fixed loop using "previous *instructionally* derived information for control and correction."[39] In contrast, reentry occurs across many parallel reciprocal connections, involves several neural regions and multiple back-and-forth iterations, and is constructive, leading to new neuronal responses.[40] Crucially, stabilized reentrant relations are *not* the outcome of prespecified instructions; they are the product of selection within and among neuronal networks. The huge amount of neural variation and selection ("differential stabilization" might be a better term than selection) in all three of the processes in Edelman's TNGS means that both the primary and secondary repertoires are what he called highly "degenerate": a given percept or a motor output can be generated by several alternative, distinct networks.[41]

Another necessary part of Edelman's TNGS is value, which is analogous to fitness in evolution, and for the organism marks events or objects as "good" or "bad." Edelman defined values as phenotypic aspects of an organism selected during evolution to constrain somatic selective events that occur during development and learning, and they do not necessarily involve mental states.[42] Value, however, is determined not only by evolutionary history but also by experience. It is shown in pain, hunger, fear, and more and is underlain by brain networks involving the neuromodulatory systems (e.g., dopaminergic, cholinergic, and noradrenergic).

Edelman and his coworkers' view of the brain as an evolutionary organ with selection responsible for the generation of conscious content has many features in common with that of Changeux and his collaborators. But for both groups, this is only the first step in the explanation of the relation between the brain and consciousness. They both ask: How is conscious experience generated by neural (selective and other) processes? What is the neural "arena" in which it occurs—which functions and brain areas participate in it? The two teams have provided the most detailed answers to these questions that are presently available, so we will look at the work of each in turn.

The Arena of Consciousness: The Dynamic Core

For Edelman and his colleague Guilio Tononi, primary consciousness comes from an adaptive connection between momentary perceptual categorization, past memories, and future needs. This connection persists in time, leading to a "remembered present" (similar to James's "specious present"). Because the construction of primary consciousness requires integration and temporal persistence, the animal possesses a greater ability to discriminate, more flexibility of action, and a stronger capacity for planning. This gives it an evolutionary advantage over nonconscious animals.

As Edelman and Tononi see it, upon receiving a new stimulus, several brain maps are activated, each coding for a particular aspect of an object or event; primary consciousness stems from the rapid reentrant interactions among them. They describe these neural processes in terms of a "dynamic core":

> We propose that a large cluster of neuronal groups that together constitute, on a time scale of hundreds of milliseconds, a unified neural process of high complexity be termed the "dynamic core," in order to emphasize both its integration and its constantly changing activity patterns. The dynamic core is a functional cluster: its participating neuronal groups are much more strongly interactive among themselves than with the rest of the brain. The dynamic core must also have high complexity: its global activity patterns must be selected within less than a second out of a very large repertoire. (Tononi and Edelman 1998, p. 1849)

Higher-order consciousness, in its full linguistic glory, emerged only during human evolution and depends on reentrant loops among large-scale connections in the cortex, especially in cortical regions associated with language and abstract concepts.

Like all the other frameworks for consciousness we are considering here, that suggested by Edelman and Tononi gives importance to the multilevel integration of information within the brain. They suggest that integrating a large amount of information very rapidly, so that discrimination and goal-directed activities are possible, is the main role of consciousness. Conscious phenomena are, on the one hand, global, unified, and indivisible into independent components, which means they must involve underlying integration processes. On the other hand, each conscious experience is distinct and does not merge with others: individual conscious states are experienced serially, one at a time. Integration and differentiation are hallmarks of consciousness that must be accounted for. Edelman and Tononi argue that the interconnectivity of the thalamocortical system (i.e., the thalamus and many regions in the neocortex that are reciprocally connected) offers a good basis for integration and differentiation and is therefore well suited to form the "dynamic core" for primary consciousness. Other brain structures, such as the cerebellum, which contains as many neurons as the cortex, are devoid of this type of interconnectivity. The types of connectivity crucial for primary consciousness are shown in figure 3.7.

Tononi believes that this view of consciousness offers a way of comparing conscious states, and he suggests a measure based on the degree of complexity underlying neural systems. According to his "integrated information theory" (IIT), consciousness corresponds to the capacity to integrate information. Integrated information, denoted by the Greek letter phi (Φ), is the amount of information generated by a complex, dynamic system of elements over and above the sum of information contained in the individual elements. In other words, it is a measure of the statistical dependence between a subset of a system and the rest of the system. It depends on the extent to which the components of a neural system are both integrated and differentiated. A system's complexity is zero if the elements are randomly dispersed and statistically independent (like gas molecules in a chamber in equilibrium, figure 3.8A); it has low complexity if the elements are uniformly coupled, as in a crystal, so the whole is an additive sum of the repeating parts (figure 3.8B). Only when there is coupling among the elements that systematically deviates from linearity (figure 3.8C) does integrated information have high complexity. In this case, the amount of information in the system as a whole is greater than the sum of its parts.

A brain state with high phi implies that a specific experience is occurring, and a huge number of alternative states are excluded. In a biological system, a high phi—a high level of both integration and differentiation—suggests that the integrated states have functional significance and that the system exerts causal power on itself. Although we do not know how to

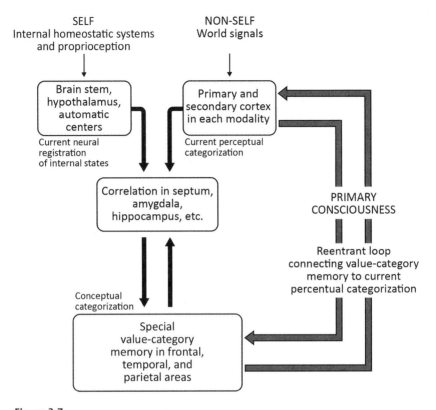

Figure 3.7

Reentrant pathways leading to primary consciousness. Two main kinds of signals are critical: those from self, constituting value systems and regulatory elements of the brain and body, and those from nonself, which are signals from the world that are transformed through global mappings. Signals related to value and categorized signals from the outside world are correlated and lead to memory and conceptual categorization. This value-category memory is linked by reentrant paths (*heavy lines*) to the current perceptual categorization of world signals. This reentrant linkage is the critical evolutionary development that results in primary consciousness. When it occurs across many modalities (sight, touch, and so forth), primary consciousness can connect objects and events through the memory of previous value-laden experiences. The activity of the underlying reentrant neural systems results in the ability to carry out high-level discriminations. This ability enhances survival. Reproduced by permission of The National Academy of Sciences, U.S.A.

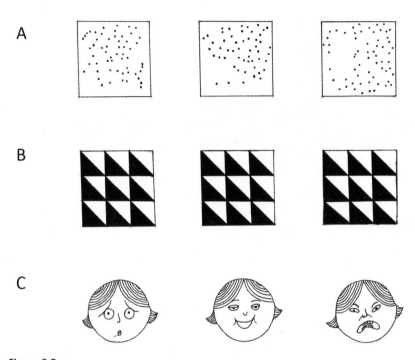

Figure 3.8
Degrees of complexity. Three cases are shown: *A*, random organization (of gas molecules); *B*, uniform organization of units (crystal); *C*, a mixture of local and global connections that systematically deviate from additive linear interactions and form integrated wholes. Complexity is high only for the last case.

measure phi directly from neurological data, one can use simulated neural networks with different types of connectivity (and different corresponding values of phi) and compare them to neural patterns in brains at different stages of development (immature cortex, degenerate cortex, mature normal cortex) and at different levels of arousal. From such comparisons coupled with inferences from simulations, it seems that there is a correlation between levels of conscious arousal and phi (figure 3.9). Moreover, what we know about mammalian brain regions that are important for consciousness also seems to be in line with the idea that nonlinear interactions between neuronal groups, which lead to extensive integration of information, are associated with consciousness: the cerebellum, though rich in neurons, does not engage to any large extent in integrating information and can be removed without destroying the ability to experience; conversely, the thalamocortical system exhibits a high degree of integration (and is therefore

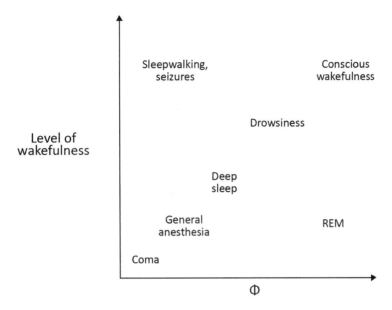

Figure 3.9
Correlation between the level of arousal (wakefulness) and phi (Φ) according to the IIT theory.

assumed to have the highest phi), and lesions in this system lead to loss of consciousness. Tononi's proposal, however, goes beyond the measure of arousal. He suggests that the neural mechanisms that lead to arousal enable the making of many nested discriminations in the attentive brain and that specific conscious contents can be described in terms of the geometrical shape of the informational relationships.[43]

Another way of quantifying consciousness has been proposed by Anil Seth, whose studies of consciousness encompass many disciplines, including computational modeling. He suggests a measure that he calls "causal density," which identifies the density of the interconnections that causally contribute to the organization of a system. Simulations using artificial neural networks show that causal density reaches a peak as a simulated neural system reaches a level of sustained yet differentiated activity, suggesting that causal density captures an important facet of the dynamics that are assumed to characterize conscious activity.[44]

As both Tononi and Seth recognize, neither the current model of phi nor the causal density model captures the dynamic instability of neural systems. Moreover, these measures cannot give us any idea as to the value above which an animal can be deemed conscious. Does a spider have a high-enough value? A crab? An octopus? A tuna? As we discuss in chapters 5 and 8, we share Tononi's and Seth's focus on intrinsic integration processes as central to consciousness, but we also emphasize learning processes that update integrated neural states when there are mismatches between incoming information and "predictions" based on prior states (neural "models" or neural representations).

Edelman's theory and the attempts at quantifying consciousness that stemmed from it led to speculations about the taxonomic distribution of consciousness. Edelman and Tononi believe that consciousness appeared in the lineage that gave rise to birds and mammals (because of the homology between the brains in the two taxa), but beyond this short and somewhat cryptic suggestion, not much is said.

In spite of its shortcoming, Edelman's TNGS adheres to the emergentist consensus in virtually all of the attributes listed at the beginning of this chapter: this theory accounts for globalization and differentiation and emphasizes plasticity, binding, attention, selection, embodiment, temporal thickness, value, and goal-directedness.

The Global Neural Workspace

The framework suggested by Changeux, Dehaene, and their colleagues at the Pasteur Institute in Paris is somewhat different from that suggested by Edelman. Their model is based on the global workspace (GW) concept developed in the late 1980s by Bernard Baars.[45] Baars, a cognitive scientist, viewed information processing as the basis of behavior. He focused on integration and the processes that lead to selective attention, suggesting that their outcomes become conscious once they gain entry into a GW. He used a theater metaphor to explain this idea (figure 3.10):

> GW theory may be thought of as a theater of mental functioning. Consciousness in this metaphor resembles a bright spot on the stage of immediate memory, directed there by a spotlight of attention under executive guidance. Only the bright spot is conscious, while the rest of the theater is dark and unconscious.... Once a conscious sensory content is established, it is distributed widely to a decentralized "audience" of expert networks sitting in the darkened theater, presumably using corticocortical and corticothalamic fibers. (Baars 2005a, pp. 45–46)

Baars assumed that instinctive responses (and those based on acquired, automatic, skilled actions) require little or no attention, whereas flexible patterns of behavior (for example, responses to novel challenges and unpredictable events) mobilize focused attention. He suggested that the contents of consciousness are generated within the system of working memory, which

Figure 3.10
A depiction of GW theory. The bright spot corresponds to the transitive conscious state on the "stage" of working memory. Conscious states respond to and activate regions that process memories, language capacities, and procedural activities (nonconscious processes, depicted by the audience sitting in the dark) and also respond to internal and external signals processes by sensory and motor areas (nonconscious, behind the curtains). The part of the stage that is not illuminated contains items in working memory.

holds on to transient information and enables it to be used and manipulated (for example, during mental arithmetic). When the specific results of nonconscious information processing carried out in several subsystems are pooled together into the GW, they become conscious, whereas information processed by the subsystems in isolation is unconscious. It is assumed that the output of this GW can reach wide regions of the brain and can recruit many subsystems to the task in hand at a particular time. The automaticity and high capacity of rapid parallel processing within the "unconscious" subsystems may be transformed into or traded for the limited-capacity, flexible, slower, serial control of conscious processing that occurs when using the GW.

Baars conjectured that the anatomical structures and/or the neural ensembles that form the GW may consist of the reticular formation and the thalamus: the former because this brainstem structure is crucial to arousal at any moment, the latter because of its widespread cortical projections, which may serve as the mechanism for broadcasting whatever appeared in the GW.

Jesse Prinz has moved the idea of GW and working memory in another direction, viewing consciousness as "attended intermediate representation" (AIR). He suggests that it is attending to a particular family of mental representations of the world—namely, intermediate-level representations, such as the view of a particular chair from a particular angle, that amounts to consciousness. Prinz proposes that phenomenal experience always stems from such integrated representations of several features of a stimulus, perceived from a particular point of view. Representations of elementary properties (e.g., the dis-unified features of the visual stimulus generated by the chair as these hit the retina) or representations of higher-level abstractions (e.g., the chair from all possible points of view) do not materialize as conscious experiences. Consciousness arises when an intermediate-level representation becomes available to working memory. These conscious experiences are expressed as synchronized gamma wave oscillations.[46]

Stanislas Dehaene and Jean-Pierre Changeux have suggested a detailed neural implementation of the global workspace—the global neural workspace (GNW):

> Our proposal is that a subset of cortical pyramidal cells with long-range excitatory axons, particularly dense in prefrontal, cingulate, and parietal regions, together with the relevant thalamocortical loops, form a horizontal "neuronal workspace" interconnecting the multiple specialized, automatic, and nonconscious processors.... A conscious content is assumed to be encoded by the sustained activity of a fraction of GNW neurons, the rest being inhibited. Through their numerous reciprocal connections, GNW neurons amplify and maintain a specific neural representation. The long-distance axons of GNW neurons then broadcast it to many other processors brain-wide.... Conscious stimuli would be distinguished by their lack of "encapsulation" in specialized processes and their flexible circulation to various processes of verbal report, evaluation, memory, planning, and intentional action, many seconds after their disappearance. (Dehaene and Changeux 2011, p. 210)

In their GNW model (figure 3.11A), Dehaene, Changeux, and their colleagues suggest that five main types of parallel processors are connected to the GW.[47] The perceptual systems give the workspace access to the present state of the external environment; motor programming circuits enable the representations of the workspace to be used in guiding future behavior; long-term memory circuits give the workspace access to past events

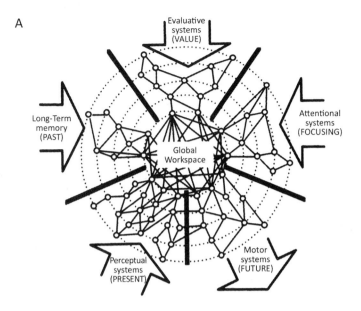

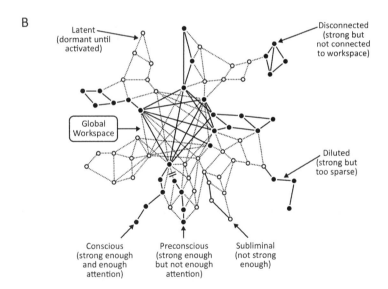

Figure 3.11

The global neural workspace (GNW) model. *A,* major categories of parallel processors connected to the GW; local processors have specialized operations, but when they access the GW, they share information, hold it, and disseminate it; the information reaching the GW is conscious. *B,* schematic representation of the GW model, showing the conscious state that reaches the GW and five nonconscious states (see text for details). Adapted by permission of the National Academy of Sciences, U.S.A.

and percepts; the evaluation circuits allow workspace representations to be allocated negative or positive value; and attention circuits enable the workspace to mobilize neural circuits independently from the external world (that is, lead to top-down attention). The five major categories of processors interact via their mutual connections, which include the pyramidal neurons that are activated during conscious processing and become inhibited in nonconscious states, such as coma or under anesthesia. The integrated activation ("broadcasting") renders sensory information conscious.

On the basis of many cognitive experiments and various types of stimulation, using data collected by brain-imaging techniques, EEG, and TMS interferences, Changeux and Dehaene suggest that in addition to conscious processes, there are several nonconscious types of information processing—disconnected processing, diluted processing, processing of latent memory traces, subliminal processing, and preconscious processing; a schematic diagram of these is shown in figure 3.11B. Disconnected processing is automatic processing, such as that involved in breathing or controlling heartbeats. Neural firing in the circuits responsible for these activities is strong and extended in time, but the circuits are not connected to the GNW, and hence their outputs are not, and cannot be, conscious. Diluted processing is characterized by patterns of connectivity that may be strong but are too sparse to become amplified and make it into the GNW, while latent processing is that occurring in dormant circuits in which the memory traces of past experiences are stored but that are nonconscious until activated by appropriate stimuli. Latent memories can become conscious, although some may never become conscious because they stem from experiences that occurred very early in development and cannot be accessed.

The fourth and fifth types of unconscious process, subliminal and preconscious processing, have received a great deal of experimental attention from Dehaene and his colleagues. Subliminal (below threshold) activity is too weak to become conscious and is most apparent in local circuits, notably neural assemblies in the primary and secondary sensory areas of the brain. Such activity dampens down very quickly, without amplifying and spreading widely to other neural circuits. Subliminal activity can be observed with masking experiments. For example, a subject is presented with a target picture (a picture of a face) flashing every 33 ms. Before and after the presentation of the target picture, a picture with a repeated pattern (e.g., black and white triangles) is flashed. Because the target picture is preceded and followed by the patterned pictures, all the subject sees are flashing patterned pictures. The focal target face picture, which would otherwise be perfectly visible, is masked under these conditions. The visibility of the target picture can be manipulated by altering the length of time for which it is presented. Usually, if shown for less than 40 ms, the masked picture is invisible, and no amount of voluntary focused attention can lead to its unmasking (i.e., being perceived), whereas above 60 ms it is visible. Although invisible, a masked picture can affect a person's decisions, making it more likely that a particular action will be taken or a particular emotion will be expressed (e.g., discomfort if the masked face is aggressive).

Unlike subliminal processing, preconscious information processing occurs when the stimulus strength is high but attention is distracted away from it, and therefore the information cannot penetrate the GW. A preconscious stimulus may become conscious if attention is voluntarily shifted to it (top-down attention) or if a strong and unexpected stimulus leads to the release of suppression (bottom-up attention). When attended to, the information-processing outcomes that represent the stimuli penetrate the GNW, activating its widely distributed neural circuits in the anterior cingulate cortex and the parietofrontal cortex. This activation leads to an explosive reverberation throughout the GNW, which continues for some time after the stimulus has subsided. While it lasts, the conscious activity inhibits many local circuits, preventing "rival" stimuli from being processed consciously, and propagates to specialized subsystems that make use of the information and activate themselves.

The nature of preconscious processing is revealed by inattentional blindness, which was dramatically illustrated by a famous experiment by Dan Simons and Christopher Chabris.[48] People watching a video were asked to count the number of times a basketball was passed among the players wearing white, a task that demands considerable attention and focus. While the ball was passed, a person disguised as a large black gorilla strode leisurely into the scene, looked around, beat his chest, and then slowly departed. Remarkably, the majority of the people counting the throws did not see the gorilla, although many of their brain circuits responded to it. Once the gorilla is pointed out, the person who rewatches the video sees it, to her amazement and alarm. This is an example of a preconscious nonaccessed brain state that can undergo the transition to consciousness by relieving the suppression imposed by the highly focused attention on a specific task. A short-term variation of the same process is seen with a manipulation of attention known as the "attentional blink," in which the first target stimulus prevents the conscious perception of a second one. Detection of the first target makes the subject more likely to miss the second if the latter appears within 200–500 ms of the first.

With both the masking and attentional blink tests, by varying the context, strength, and duration of the stimulus, the person can be made to either consciously perceive it or remain unaware of it. The concurrent changes in the brain can be monitored. Four major NCCs (or "signatures of consciousness," as Dehaene prefers to call them) have been identified:

> Conscious perception results from a wave of neuronal activity that tips the cortex over its ignition threshold. A conscious stimulus triggers a self-amplifying avalanche of neural activity that ultimately ignites many regions into a tangled state. During the conscious state, which starts approximately 300 milliseconds after the stimulus onset, the frontal regions of the brain are being informed of sensory inputs in a bottom-up manner, but these regions also send massive projections in the converse direction, top-down, and to many distributed areas. The end result is a web of synchronized areas whose various facets provide us with many signatures of consciousness: distributed activation, particularly in the frontal and parietal lobes, a P3 wave, gamma-band amplification, and massive long distance synchrony. (Dehaene 2014, p. 140)

The beautiful experiments manipulating attention and ascertaining from the subjects' reports (not necessarily verbal reports—macaques report by pressing buttons and pigeons by pecking keys) whether or not the specific contents were perceived use several different techniques. A neural "signature" of reported, visual conscious recognition is seen, even at the level of single neurons, as an increase in gamma strength 150–200 ms after stimulus onset that outlasts the length of the stimulus.[49] All these experiments still leave open the question of whether or not recollection and reportability capture all facets of consciousness. Maybe the person did have a brief conscious experience, which was then totally forgotten? Maybe some events can be reported only as they occur, "online"? Maybe the ability to report is not a necessary part of the experiencing and is far more limited than the ability to experience? What are the neural signatures of experiences that are difficult or impossible to report? What does a neural signature that is identical to a conscious reportable one, although the subject does not report a conscious experience, mean? People like the cognitive neuroscientist Victor Lamme think that phenomenal experience is much richer than what can be recalled and reported. They seek a neural definition of consciousness that separates it from attention, working memory, and reportability. As Lamme sees it, gaining access to the GW may be relevant for only one type of consciousness within a broader consciousness field.[50]

Whether or not we accept this caveat, Dehaene and Changeux's GNW model clearly falls within the emergentist consensus. As with the other models we have discussed, it highlights the dynamic, reverberating, and synchronous neural activity occurring in widespread brain areas and tries to account for the binding that brings about consciousness; it also suggests that there are complex, long-ranging, reentrant loops between various brain areas that have either generalized or specific functions and that nonconscious states are necessary for conscious functioning. Attention, selection processes, and temporal thickness are central to this scheme. As with the Crick-Koch and Llinás frameworks, Changeux and Dehaene focus mainly on the subjective experiencing of perceptual information, but their model, like those of Freeman and of Edelman and Tononi, also adds an explicit evaluative dimension: it introduced the possibility that emotions come into play in the construction of subjective experiences. However, embodiment is not central to the cognitive-oriented approaches. In fact, Dehaene remains committed to "the brain in the vat" philosophical fabrication—to the idea that a person can be equated with her brain and that subjective experiencing can emerge from cleverly applied stimulation to a disembodied brain floating in a vat.

We now turn to our fourth neurological framework for consciousness. Rather than focusing on perception, the focus here is on affective states: on emotions, motivations, and feelings.

Emotions and Embodiment

It may seem obvious that neuroscientists who study emotions in animals assume that their subjects are conscious beings with subjective feelings. However, although the neurological study of emotions and the revival of consciousness studies have been linked since the 1980s and 1990s, the relation is by no means straightforward. Even today, some neuroscientists studying emotions such as fear are totally agnostic about whether or not the macaque, monkey, dog, cat, or rat that they study has the subjective experiences or feelings of fear. For example, Joseph LeDoux, who has contributed much to the understanding of the brain pathways involved in fear responses, believes that the subjective aspect of fear appeared much later than the physiological and behavioral fear response. Fear-behavior in the face of danger, he argues, is more ancient than the feeling of fear. Since bees, snails, and worms, which LeDoux supposes are nonfeeling animals, can be conditioned to show fear responses, and since there are various pathologies that can decouple fear behavior from the feeling of fear in humans, he assumes that animals that exhibit fear behavior can be nonconscious and nonfeeling—and indeed many possibly are. He believes that whether or not nonhuman mammals are feeling beings is something we can hardly speculate about because feelings as we know them in humans depend on highly evolved cortical areas, possibly even on the specific linguistic areas in humans.[51] Edmund Rolls, a cognitive neuroscientist, follows the same line, explicitly arguing that only animals endowed with the ability for symbolic representation have "emotional feelings" that are subjective, felt mental states.[52]

The view that LeDoux and Rolls represent is rooted in twentieth-century studies of emotion and may be one of the reasons for the singular way the term "emotion" is used by some neurobiologists, who equate emotions with their behavioral-physiological correlates alone, without a subjective-feeling facet. In the 1920s it was recognized that earlier studies of the physiology of emotions were problematical because they centered on human subjects in the artificial conditions of the laboratory and their introspection-based, highly personal, and highly context-dependent reports. A solution that was supposed to solve all these problems and release "pure" emotional expressions—physiological and behavioral—was introduced by the famous physiologist Walter B. Cannon in the form of the decerebration of experimental animals (dogs and cat). Removing the cortex unleashed long, context-free, and easy-to-study trains of intense behavioral and physiological expressions of emotions; it was assumed not to be cruel because animals without a cortex were believed to feel nothing—they are just "emotional machines" without affect. We know today that the cortex may not be necessary for feelings, rendering these practices nightmarishly horrifying.[53] Ever since, the term "emotion" has been used by emotion researchers to mean the physiological and behavioral expression of emotion or to refer to the action programs that elicit the behaviors. In addition to this historical tradition, it is far from clear that different emotions have much in common—it is not clear that emotion is what philosophers call a "natural kind." The problem of finding a defendable common denominator for different emotions has led LeDoux to avoid it altogether.[54] He prefers to use instead the neutral term "survival circuits"—brain circuits that are involved in defense, the maintenance of energy and nutritional supplies, fluid balance, thermoregulation, and reproduction. Since we refer to literature that employs the term "emotion" in several nonidentical ways, we use the terms "feelings" or "emotional feelings" to denote the subjective aspects and "emotion" (in quotes) or "expression of emotions" when talking about the physiological-neurological correlates, including action programs or brain-survival circuits.

Although the study of "emotions" and the study of feeling and consciousness have largely been independent, they have become intimately linked in the work of Antonio Damasio. Unlike the work of most of the other neurobiologists we have considered in this chapter, which is centered on perception, Damasio's main interest is in the affective aspects of consciousness and their close interactions with the cognitive and perceptual facets that constitute the self. His aim is to account for the processes that give rise in the brain of an embodied organism to "mental images" of objects; events; actions; and, most importantly, feelings. Damasio has gone back to James's theory of emotions, where the feedback between the soma and the brain constitutes emotions and drives, but unlike James, Damasio includes in his scheme the brain stem, not just the cortex, and also within-cortex feedback. Damasio explains his view of consciousness (subjective feelings and perceptions) as a relation between the neural maps of the animal's body and its environment and argues that these neural mappings lead to subjective feeling:

> Feelings are mental experiences of body states. They signify physiological need (for example, hunger), tissue injury (for example, pain), optimal function (for example, well-being), threats to the organism (for example, fear or anger) or specific social interactions (for example, compassion, gratitude or love). Feelings constitute a crucial component of the mechanisms of life regulation, from simple to complex. Their neural substrates can be found at all levels of the nervous system, from individual neurons to subcortical nuclei and cortical regions. (Damasio and Carvalho 2013, p. 143)

Damasio sees "emotions" as the back-and-forth interactions between the sense organs that receive inputs from the outside world, the brain, and the body, which lead to motor and other physiological changes in the body. They are what he calls "action programs," those triggered by changes to the internal or external environment that have evolved to restore or maintain homeostasis. For Damasio, "emotions" are initiated by external perceptual stimuli (e.g., the sight or smell of a predator or a mate) while drives such as hunger and thirst are action programs elicited by internal stimuli. Feelings are the subjective states that may accompany action programs—both the action programs leading to "emotions" and the action programs that lead to drives. Damasio suggests that the sensing of the integrated construction in the brain of body changes are the "feelings of emotions."[55] The distinction between feelings and "emotions" is very important for Damasio, although, as we see it, the recurrent and ongoing nature of the sensory-brain/motor-brain/brain-brain feedback renders this distinction somewhat arbitrary. Damasio (like Freeman) does provide an important extension to James's theory by adding to the soma-brain feedback the feedback within the brain, stressing the crucial role it plays in the construction of emotional experiences and in the ability of subjects to simulate emotions in the absence of external stimuli.[56]

For Damasio, the notion of consciousness and self are intimately related, almost interchangeable. Since consciousness is constructed through the neural processing that generates the ever-changing represented relations between the animal and the world, consciousness entails a sense of self. This self has a hierarchical structure that reflects its evolutionary history: proto-, core-, and extended-consciousness or self.

All consciousness, Damasio emphasizes, is rooted in the perceptions that the organism has of its own body—the continuous monitoring of the internal state(s) of the body by the brain. "Protoconsciousness" (or the protoself) is the perception of the internal states, constructed by neural patterns that represent, from moment to moment, the changing physical state of the body that enables the maintenance of homeostasis. Although protoconsciousness is the evolutionary basis of consciousness, it does not, according to Damasio, require subjective experiencing: the representation in the brain of internal physical states of the body, which are involved in homeostasis and the regulation of these states, can go on nonconsciously. The monitoring of these states generates "primordial feelings"—overall inner feelings of pleasure or displeasure, satisfaction or need, which, he believes, are not conscious. Damasio's views seem somewhat reminiscent of Lamarck's ideas discussed in chapter 2, but the nineteenth-century biologist would certainly not have endorsed the idea that primordial emotions are nonconscious. For Lamarck, integrated "inner feelings" were already a sufficient basis for the minimal subjective experiencing capacity of animals with sensory organs and head ganglia; what Damasio describes as protoconsciousness would therefore have meant conscious feeling—albeit very simple—for Lamarck. The sensory categorizations that we call states of pleasure or displeasure cannot, according to Lamarck, be totally unconscious.

The second level in Damasio's hierarchy of consciousness is "core consciousness," or "core self," which appears whenever an object in the environment modifies the present protoself, and this change is, in turn, represented. The core self provides the organism with a sense of the experience of the "here and now" and with the sense that the experience is "owned" by the animal; that it is "its" experience. The representation of the way in which

Figure 3.12
Losing the sense of owning one's experience.

the interaction with an object in the world affects the body state entails a model of the body (self-model) and the world (world model) and the ever-changing relationship between them. The representation of the change is regenerated all the time on the basis of the current, changing states of the body interacting with environmental stimuli. This ongoing regeneration and updating gives rise to a continuous "stream of consciousness." It is a feeling of what happens here and now. Recognizing the relationship of the organism and the object is "knowing oneself." Brain lesions may destroy aspects of core consciousness, and in these cases, patients lose the sense of owning their present experiences (figure 3.12).

"Extended consciousness" is, according to Damasio, the highest level of subjective experiencing, the capacity to connect the remembered past and the anticipated future to the present. It is based on the recall of episodes and on anticipation of the future, and in humans it is related to the auto-biographical self; it is the ability of an animal to see itself as the protagonist of the narrative of its own life history. People with certain brain lesions or temporary maladies lose their sense of autobiographical self and cannot see themselves as extended over a period of time, although their core consciousness remains intact. Transient global amnesia, which often accompanies migraine headaches, is an example of a transient state in which extended consciousness is lost (without affecting core consciousness). In this state, which may last for several hours, short-term memories are lost, including memories of events that have occurred to the person minutes or days before. The person cannot recall anything that happened outside the last few minutes, although she has a sense of self, recognizes her relatives, and appears mentally alert. In other words, due to the lack of updated autobiography, although the "here and now" experiences are intact, the temporary loss of the past makes these experiences discontinuous, broken, and partially incomprehensible.

In his 2010 book *Self Comes to Mind*, Damasio argues that "being awake, having a mind and having a self are different brain processes, concocted by the operation of different brain components."[57] We shall not go into the detailed anatomical and other evidence that Damasio offers to support this argument, but, in a nutshell, what Damasio claims is that human consciousness requires the cortex, the thalamus, and the brain stem. He thinks that each structure influences or forms certain aspects of consciousness, the self, and wakefulness.

According to Damasio, both protoconsciousness and core consciousness are innate, or as he puts it, "installed by the genome." The protoself and protoconsciousness are primarily served by brain stem nuclei and the hypothalamus, while the prominent structures that participate in core consciousness are the cingulate cortex, the superior colliculi, and the thalamus. Extended consciousness, the system of memories that is consciously accessible to the organism and that leads to the construction of future scenarios, may be uniquely human, and its processing takes place in working memory with many cortical structures participating. As Damasio sees it, the internal integrative models of the body increased in sophistication during evolution, going from mere homeostatic representations to extended consciousness. Damasio's views fit into the emergentist consensus since they are focused on integration, globalization, embodiment, and selection. His evolutionary perspective is still a rough draft, but he does offer an evolutionary explanation of different levels of consciousness and the structures that subserve them.

The ideas developed by Damasio show interesting convergence with the conclusions reached by Allan Hobson, an eminent investigator of dreaming and affect in warm-blooded mammals and birds.[58] These animals show rapid eye movement (REM) and non-REM phases during sleep. Hobson suggests that REM sleep, which is associated with dreaming and predominates in fetuses and neonates, is, at these early ontogenetic stages, a preconscious phase that prepares the brain for the primary consciousness seen in wakeful animals and prelinguistic babies and which, in mature humans, culminates in extended, symbolic consciousness. Hobson, unlike Damasio, stresses the early *ontogenetic* stage of protoconsciousness, and the neurobiology of sleep rather than wakefulness is his point of departure. However, both researchers stress the affective basis of consciousness and the major role of the brain stem and suggest a very similar series of stages and brain structures (Hobson during ontogeny, Damasio during phylogeny) underlying proto, primary/ core, and secondary/extended conscious processing.

The centrality Damasio gives to the body challenges the cognitivists' view that the brain and its symbolic or network algorithms are all-important for consciousness. This approach is related to a more radical framework, one that regards cognition and consciousness not only as embodied but also as enactive, agential, and extended into the environment. In this framework, consciousness typically involves sensorimotor activities in which the brain, the peripheral nervous system, and the nonneural parts of the body are active participants, and the interaction with the internal and external environment is crucial. An individual actively constructs his or her environment and experiences, and the environment (most obviously the social environment) plays an active role in the construction of subjective experiencing.[59] Although, if asked, all neuroscientists would concur with the centrality of the body-world interactions, the enormous focus on the brain often obscures this common-sense view. For example, because humans can experience when totally paralyzed, when large sensorimotor areas of their brains are destroyed, when their sensory capacities are inhibited, or when they dream or imagine without current sensorimotor stimulation, Tononi and Koch conclude that "consciousness—in the sense of having a subjective experience—does not require sensorimotor loops involving the body and the world."[60]

This claim, which resonates with Dehaene's subscription to the brain-in-the-vat fantasy, is deeply misleading. All humans, whether their present state is normal or pathological, develop a functional nervous system during their ontogeny, and what we know about normal neural development suggests that this requires motor and sensory stimulation and feedback, even in utero. It is certainly remarkable and significant that, once formed, representations that humans and other animals acquire through interactions among the sensory, perceptual, and motor systems can persist in the absence of exogenous sensorimotor stimulation. However, this does not mean that consciousness is possible without sensorimotor loops involving the body and the world because such loops are always part of an individual's developmental history.[61] Taking an explicitly developmental perspective (evolutionary as well as ontogenetic) makes the necessity of a body-and-agent-based view of consciousness self-evident. Yet, as our summary of the different frameworks used by neurobiologists has shown, although all are committed Darwinians and employ the neural-selection principle, evolutionary history is not central to their discussion.

An Overview

The different frameworks of research into consciousness that we have considered in this chapter have many things in common. In the terminology we are using, the neurobiologists who constructed them share similar "lists" of assumptions about consciousness and its attributes. In table 3.1 we have tried to summarize the role that these are given in the various frameworks and point to the differences in emphasis. Like all schematic outlines, the table does not do justice to the nuanced approaches of the scientists, but we hope that it provides an adequate overall view. The entries in the table broadly follow the list of attributes of consciousness with which we started this chapter and add information about the scientists' views on the evolutionary stages, the brain areas involved, and the overall constitutive dynamics.

The consensus to which the table points is acknowledged by some of the neurobiologists, who regard their views as overlapping and complementary. As we have seen, Tononi and Koch joined forces to synthesize and extend the views they had developed with Edelman and Crick, respectively. Similarly, Gerald Edelman (with his colleague Joseph Gally) and Bernard Baars recognized the complementary nature of their dynamic core and the GW models. They argued that the dynamic core provides a mechanism—reentrant signaling throughout the cortex—for GW events, explaining how and why the limited working memory capacity of momentary conscious content is compatible with the vast repertoire of long-term memory. They had no doubt that such an embodied system, in constant interaction with a complex world, provides a satisfactory answer to the mind-body problem:

> Consciousness consists of a stream of unified mental constructs that arise spontaneously from a material structure, the Dynamic Core in the brain. Consciousness is a concomitant of dynamic patterns of reentrant signaling within complex, widely dispersed, interconnected neural networks constituting a Global Workspace.
>
> The contents of consciousness, or qualia, are correlates of discriminations made within this neural system. These discriminations are made possible by perceptions, motor activity, and memories—all of which shape, and are shaped by, the activity-dependent modulations of neural connectivity and synaptic efficacies that occur as an animal interacts with its world. (Edelman, Gally, and Baars 2011, p. 5)

Table 3.1

Major approaches and assumptions of the neurobiological theories of consciousness discussed in this chapter.

	Crick, Koch	Llinás	Freeman	Changeux, Dehaene, Baars	Edelman, Tononi	Damasio
Main experimental approach	Studying specific states of consciousness (e.g., visual consciousness during binocular rivalry) in humans and other mammals (e.g., macaques)	Comparison of deep sleep, the awake state, and pathological states; EEG patterns; mainly studies of humans	Analysis of global features of brain activity, using dynamic models based on EEG; main model the olfactory system of rabbits	Study of multiple states and levels, with special attention to subliminal and preconscious states; mainly in humans and other mammals	Study of multiple states and levels of consciousness, with a focus on feedback relations within the dynamic core in mammals and humans	Study of emotions, mainly in normal and sick human subjects, through imaging and cognitive testing
Main brain areas involved	For visual consciousness: region V of visual cortex, IT cortex, prefrontal cortices, thalamus, and the claustrum	The thalamocortical system (involved in consciousness and in the generating of "self")	Brain stem, midbrain, cortex, and their connections	Thalamocortical interactions with cortical areas subserving long-term memory, perception, motor control, attention, and evaluation, particularly pyramidal neurons from the prefrontal cingulate and parietal regions	Basal ganglia, thalamus, and the cerebral cortices, forming reentrant interactions	Brain stem nuclei, thalamus, hypothalamus, amygdala, cingulate cortex, superior colliculi

(continued)

Table 3.1 (continued)

	Crick, Koch	Llinás	Freeman	Changeux, Dehaene, Baars	Edelman, Tononi	Damasio
Integration and differentiation through feedforward and backward loops (reentry)	Essential	Essential	Essential	Essential	Essential	Essential
Binding of multiple features of percepts	Essential	Essential	Essential	Essential	Essential; occurs at several levels, coupled with reentry	Assumed
Attention	Important, although top-down attention may not be necessary for consciousness	Important, but not central to the discussion	Since learning involves selective exclusion, attention is important	Essential; necessary for consciousness; leads to accessibility of experiences, which can be experimentally manipulated	Important; modulates states of consciousness from diffuse to focal	Essential for the formation of the core self
Neural selection and competition	Central; competition among coalitions of neurons	Involved; however, processes of inhibition and exclusion are not discussed in selectionist terms	Not discussed in these terms, but phase-transition from one state to another are shifts between different selectively stabilized states	Central; selection and selective stabilization during embryogenesis and several stages of learning	Central to the theory of neuronal group selection; selection among neurons within synaptic populations and among neural groups	Assumed but not specifically discussed

Plasticity and learning	Plasticity is essential; little discussion of learning	Plasticity is essential; little discussion of learning	Both plasticity and learning are essential; discussion centered on olfactory learning	Plasticity is essential; learning is important but not central	Plasticity is essential; learning is important, with emphasis on discrimination and action selection	Plasticity is essential; learning is involved since emotions are evaluation systems but not discussed much
Temporal thickness	Central; related to binding	Important	Essential; establishment of attractors' states necessitate temporal thickness	Essential for amplification of neural signals	Essential; a precondition for consciousness	Not stressed, but mapping interactions are reciprocal and must have temporal persistence
Value systems and goals (intentionality); emotions	Value systems and goals assumed but not central	Value systems select actions	Central; value systems lead to action selection, driven by goals	Values are essential inputs into the GNW; required for neural selection; goals assumed	Value systems are the basis of the neural selection processes; goals assumed	Central; emotions and drives are value systems that constitute consciousness
Embodiment, self	Implied but not central	Brain-centered; however, motor control is central, so embodiment is assumed	Central; stress on the primacy of action that involves bodily affordances and a notion of self	Important but not central; Dehaene believes in brains in vats	Important but not central	Central; the interactions between representations of body and external environment constitute the self

(continued)

Table 3.1 (continued)

	Crick, Koch	Llinás	Freeman	Changeux, Dehaene, Baars	Edelman, Tononi	Damasio
Evolutionary origins and stages	Primary consciousness evolved in vertebrates such as birds and mammals and into the higher order consciousness of humans	Seems to believe that consciousness evolved early in the history of life; no details provided	Primary consciousness evolved in "low" vertebrates (salamander); there was gradual sophistication of consciousness following the evolution of the frontal lobes, culminating in humans	Consciousness evolved in mammals, birds, and possibly fish and into high-order human consciousness	Consciousness evolved from primary (birds and mammals) to high-order (human) consciousness	Consciousness evolved from protoconsciousness (in lower vertebrates? In invertebrates?) to core (in birds and mammals) into extended (human) consciousness
System dynamics	Sustained synchronization of oscillations involving global feedback among cortical and thalamic areas	Synchronization between general (arousal-related) and content-specific oscillators generates consciousness	Attractor dynamics; consciousness is the process of change and interaction among high-level attractors in the brain	GNW dynamics; amplification of activity in higher cortical areas, P300 wave, amplified gamma oscillations, and multiple long-range reentrant connections	Core dynamics involving the brain stem, the thalamus, and the cortex; TNG; consciousness emerges above an unspecified threshold measured by phi	Recurrent mapping relations among different areas in the brain and feedback with the body

This conclusion is typical of the neurobiologists—not in its neurological details, but in its spirit. Although they all readily confess that they are still far from providing a full neurobiological characterization of the processes involved, they are all confident that the dynamics they describe constitute the subjective and private aspects of consciousness. To put it plainly and simply, they are all sure that they have dissolved or solved (in principle) the mind-body problem. They totally reject Colin McGinn's pessimistic view that we just do not have the cognitive tools that would equip us to comprehend how the brain—"mere meat," as he puts it—can lead to consciousness.[62] For the scientifically optimistic neurobiologists, the dynamic organization of "neural meat" interacting with the rest of the meat in the body does provide an adequate, albeit not yet full, answer.

Why then are laypeople and some philosophers less convinced? Is it the result of ignorance, a failure in communication, or the lack of a convincing-enough model? Are they the victims of stubborn habits of thought, held back by an epistemological gap that has already been bridged? In the next chapter, we try to understand whether the neurobiological frameworks that we have described in this chapter are compatible with the philosophers' views and how they influence central philosophy of mind problems like Nagel's query and Huxley's wonder.

4 A Biological Bridge across the Qualia Gap?

We discuss some long-standing questions in the philosophy of mind as seen through the lens of biology and neuroscience. Can Mary, the all-knowing vision neuroscientist, enclosed in a black-and-white world, be said to acquire new knowledge when she first sees red? What can we learn from Daniel Kish, a blind man who perceives distant objects by making clicking noises, like a bat? In what sense do mental states cause changes in neurophysiology, and can the evolution of the disappearing shadow of a squid assist us in answering this type of question? Are zombies and Turing robots, which are supposed to be functionally identical to humans yet lacking in consciousness, at all possible, or are they the impossible products of the peculiarities of language and logic? How is the process of consciousness dynamically constructed, and can the concept of an "enabling system" help in understanding it? What, from a biological perspective, are qualia and what is the self? Is the idea that consciousness has a function coherent, or is consciousness a teleological system that provides the framework for the development and evolution of functions that support its teloi?

Some scholars would regard the suggestion with which we ended the last chapter—that neurological data and theories may force a reformulation of philosophers' ideas about the mind—as deeply misguided. Bennett and Hacker, for example, regard the philosophy of mind and the biology of mind as neatly separate though complementary enterprises, rightly noting: "PET and fMRI can scan brains, but not concepts and their articulations. Neuroscience can investigate synaptic connections but not conceptual ones."[1] However, they draw from this the far-reaching conclusion:

> A new neuroscientific theory of anything cannot give philosophy a new lease or turn, but only a new array of conceptual puzzles to resolve and knots to disentangle. Neuroscientific discoveries (e.g., blind-sight) may pose *new* conceptual problems—they may provide grist for philosophical mills, but not solutions for philosophical problems. (Bennett and Hacker 2003, pp. 405–406)

The argument is that work on the neurobiology of consciousness cannot shed any light on the philosophical *concept* of consciousness, and the ideas of philosophers can contribute nothing to *scientific theories* about the neural basis of consciousness.

As theoretical biologists, our view is different. Conceptual analysis is not the exclusive province of philosophers, and the borderline between theoretical science and philosophy—for example, that between theoretical biology and the philosophy of biology—always was, and still is, blurred. Scientific concepts are in constant flux, and new empirical findings can fuel conceptual changes, reframe them, and help solve long-standing conceptual problems, such as the nature of life. In her seminal 1986 book, *Neurophilosophy*, Patricia Churchland clearly put forward the rationale for an interaction between philosophy and neuroscience:

> For neuroscientists, a sense of how to get a grip on the big questions and of the appropriate overarching framework with which to pursue hands-on research is essential—essential, that is, if neuroscientists are not to lose themselves, sinking blissfully into the sweet, teeming minutiae, or inching with manful dedication down a dead-end warren. For philosophers, an understanding of what progress has been made in neuroscience is essential to sustain and constrain theories about such things as how representations relate to the world, whether representations are propositional in nature, how organisms learn, whether mental states are emergent with respect to brain states, whether conscious states are a single type of state, and so on. It is essential, that is, if philosophers are not to remain boxed within the narrow canyons of the commonsense conception of the world or to content themselves with heroically plumping up the pillows of decrepit dogma. (P. S. Churchland 1986, pp. 3–4)

Churchland's endeavor has been a striking success. Today, many philosophers of mind are convinced that the mind cannot be comprehended if neurobiology is ignored, and neurobiologists take seriously concepts such as "intentionality," "self," and "free will," which traditionally resided exclusively in the philosophical realm: they are now inventing specific techniques and designing experiments to fill these concepts with scientific content. Moreover, philosophers are involved in empirical (mainly cognitive) research, and neurobiologists and cognitive scientists write philosophical papers. The boundaries between cognitive science and neuroscience are fuzzy, as we pointed out in the last chapter, and so are those between philosophy and the cognitive and neurobiological sciences. This is reflected in this chapter, where the traditional philosophical questions we address are discussed from the points of view of cognitive scientists and neurobiologists, as well as philosophers.

We call philosophers who use scientific discoveries about neurobiology to sustain and constrain their philosophical framework "natural philosophers of mind." These philosophers of mind do not regard neural processes as exterior causes of mental states but as the building blocks of consciousness (just as liver cells are the constituents of the liver and not causes external to it). This componential view of consciousness—and we are using componential here and throughout the book in the sense of forming a part or component of a dynamic system—leads natural philosophers to hold, just as neurobiologists do, a weak emergentist position. We believe that they all subscribe to a version of John Searle's suggestion that "consciousness is to neurons as the solidity of a piston is to metal molecules"[2] (with the caveat, held by many, that the neurons have to be embedded in a body that is embedded in a world) or to paraphrase Cabanis, "Consciousness is to neurons as digestion is to cells in the digestive system" (see chapter 2). They all regard the processes that lead to mental states as highly distributed yet integrated and in this sense are committed to some version of the "field" view embraced by Searle, and they integrate the intransitive and transitive views of consciousness. The philosopher Thomas Metzinger, for example, suggests that there are global neural correlates of consciousness (NCCs) underlying overall subjective experience, as well as local NCCs specific to the redness of poppies or the smell of a perfume.

In spite of the dominance today of the naturalistic view of the mind, there are many variations in the current philosophical approaches to the study of consciousness. Philosophers adhering to the phenomenological tradition treat consciousness as "a field of lived experience" and describe its deep structures (such as the gestalt structures of perceptual experiences, the intentionality or aboutness of experiential states, and the "temporal thickness" of the experienced present) mainly at the level of the subjective experiencing itself rather than the underlying neural level. However, within this tradition too, Varela's neurophenomenological approach—a methodology combining first-person reports by individuals trained in introspection with neurobiological (e.g., EEG) measures—has become more influential.[3] In addition to the naturalists and the phenomenologists, there are some philosophers (usually to be found in theology departments) for whom mind is immaterial. And even among the natural philosophers there are those, such as David Chalmers, who think that the understanding of mind requires new, as yet undiscovered (natural) laws of physics, or like Colin McGinn, who believe that it requires cognitive powers far surpassing those of humans.[4] Most natural philosophers, however, hold a more optimistic view.

Differences in perspective are important and have interesting research implications, but what we want to emphasize in this chapter is the general consensus among most philosophers of mind about the major attributes of basic consciousness and the remarkable correspondence between the views of natural philosophers and neurobiologists. To illustrate this we chose the different lists of consciousness criteria, or properties, compiled by John Searle and Thomas Metzinger, which they believe characterize the structure of human and animal mental states.[5] In Searle's scheme these include: (1) qualitative nature, (2) ontological subjectivity (first-person ontology), (3) unity, (4) intentionality (aboutness), (5) mood (an overall flavor), (6) distinction between the center and the periphery, (7) pleasure and unpleasure, (8) situatedness (9), active and passive consciousness structure (as in, for example, voluntary and involuntary actions), (10) gestalt, and (11) the sense of self. Metzinger's preliminary list (which he said will be enriched and modified by further research) describes the representational structure of phenomenal experience and includes: (1) global availability, (2) activation within a window of presence (temporal thickness), (3) integration into a coherent global state, (4) convolved holism (nestedness, a property of hierarchical systems that have integrated entities of smaller scale enclosed within a larger scale), (5) dynamicity, (6) perspectivalness (having a point of view), (7) transparency, (8) off-line activation, (9) representation of intensities, (10) "ultrasmoothness": the homogeneity of simple content, and (11) adaptivity. Metzinger describes these properties at the phenomenological, representational, computational, functional, and neurobiological levels and argues that they provide a preliminary tool kit for the construction of a theory of phenomenal consciousness, a phenomenal self-model.

The attributes suggested by Searle and Metzinger, two philosophers who approach consciousness from very different perspectives, correspond to the seven characteristics in the list we extracted from neurobiologists' work, and the correspondence is summarized in table 4.1 (the terms used by Searle and Metzinger are italicized). Other philosophers of mind also see these seven attributes as crucial for consciousness.

The table shows the overlapping concerns of philosophers and neurobiologists, but there is more to this communality than interest sharing. It suggests that major problems in the philosophy of mind can be reformulated and discussed in biological terms, and vice versa, although there is no general consensus on what form such reformulation should take.

Although, as we have made clear from the outset, this is *not* a book about the philosophy of mind, and we are *not* trying to give an overview of this vast and forbidding subject, an evolutionary approach to minimal consciousness has to address the concerns of philosophers of mind.[6] So rather

Table 4.1

Correspondence between the lists of features of subjective experiencing given by neurobiologists and by philosophers of mind.

Characteristics of mental states according to the neurobiologists' list	John Searle's features of consciousness[a]	Thomas Metzinger's characteristics of consciousness[b]	Agreement of other natural philosophers[c] with the neurobiologists' list
Global accessibility and activity	*Unity; qualitative nature of mental states.*	*Global availability and integration into a coherent state, accessible to action and/or cognitive reference.* Experiences of the ever-changing states of self and world are differentiated but unified. Global neural correlates of consciousness are probably based on neural synchrony.	All; highlighted by P. S. Churchland, P. M. Churchland, Dennett.
Binding and unification	*Qualitative nature; unity; gestalt perception.*	*Integration into a unified state.* Unity leading to transparency (naive realism) and to homogeneity. Feature binding and the ultrasmoothness of simple content is emphasized, as well as sensory and motor gestalts.	All; especially Zahavi, Gallagher, Thompson.
Selection, plasticity, learning, attention	Discussed in the context of active and passive consciousness structures and free will. Plasticity is discussed in the context of memory and the construction of the sense of self. Attention is considered as an aspect of the *center and periphery* dimensions of consciousness.	*Dynamicity, adaptivity.* Discussed in the context of agency, which allows the selection of things to attend to and to do; adaptivity may involve off-line activation. Attention is discussed in the context of *attentional agency* and the ability for controlled perspective changing.	All; highlighted by James, P. S. Churchland, P. M. Churchland, Dennett. Plasticity and learning are highlighted by Dretske, P. S. Churchland, P. M. Churchland, Dennett. Attention is discussed by all with different emphases (e.g., contrast Block and Dennett).

(continued)

Table 4.1 (continued)

Characteristics of mental states according to the neurobiologists' list	John Searle's features of consciousness[a]	Thomas Metzinger's characteristics of consciousness[b]	Agreement of other natural philosophers[c] with the neurobiologists' list
Intentionality	*Intentionality* is a central attribute of conscious states, manifest in goal-directed behavior and in the notion of representation.	*Intentionality* is a central (representational) attribute of any self-model, which must be applicable (as a level of description) to any property of consciousness. Off-line activation is a distinct manifestation of intentional systems.	All; emphasized by Clark, Zahavi, Gallagher, Thompson.
Temporal thickness	Discussed in the context of the construction of identity and "self."	Activation within a window of presence, temporal sense of present.	Emphasized by James, Dennett, P. S. Churchland, P. M. Churchland, Gallagher and Zahavi, Thompson.
Values, emotions, goals	*Pleasure and displeasure; mood.*	*Values* discussed in the context of intentionality, *emotions*, and the bodily and social self.	All.
Embodiment, agency, and a notion of "self"	Related to intentionality. *Situatedness. The sense of self.*	Integration into a coherent global state. *Perspectivalness. Agency. A sense of embodied self.*	All.

[a] *Source:* Based on Searle 2004, chap. 5, pp. 134–145.

[b] *Source:* Based on Metzinger 2003, 2009.

[c] Characteristics based on the views of James 1890; Dretske 1988; P. M. Churchland 1989; Dennett 1991; P. S. Churchland 2002; Block 2007; Thompson 2007; Clark 2008; Gallagher and Zahavi 2012.

than trying to analyze and do justice to the different views of consciousness held by these philosophers (an impossible task within the framework we have adopted), we have chosen four major philosophical issues to illustrate the interpenetration of the philosophical and biological problems:

1. The disparity between first-person subjective knowledge and third-person knowledge.

2. The question of "mental causation": How can first-person subjective mental states (like anger) possibly affect physical processes like actions and their neurobiological underpinnings?

3. The componential question: How are phenomenal experiences (qualia) and the sense of self instantiated by neuronal dynamics? And how is it that mental states are intentional (*about* something—ourselves, the world)?

4. The question of function: What is the biological function of consciousness, and does it provide advantages that are beyond and above behaving/doing?

What we hope to show is that biologically oriented discussions of these central philosophical problems can help to illuminate both the philosophical and the biological aspects of the shared list. More specifically, we want to show that one biological approach, the evolutionary approach, is particularly helpful in framing philosophical questions.

Ways of Knowing: Mary, Fred, and Daniel

> I am acquainted with many people and things, which I know very little about, except their presence in the places where I have met them. I know the color blue when I see it, and the flavor of a pear when I taste it; I know an inch when I move my finger through it; a second of time, when I feel it pass; an effort of attention when I make it; a difference between two things when I notice it; but about the inner nature of these facts or what makes them what they are, I can say nothing at all. I cannot impart acquaintance with them to any one who has not already made it himself. I cannot describe them, make a blind man guess what blue is like, define to a child a syllogism, or tell a philosopher in just what respect distance is just what it is, and differs from other forms of relation. At most, I can say to my friends, Go to certain places and act in certain ways, and these objects will probably come.
>
> —James 1890, 1:221

The distinction between *acquaintance* and *knowledge about* is the distinction between the things we have presentations of, and the things we only reach by

means of denoting phrases.... All thinking has to start from acquaintance; but it succeeds in thinking *about* many things with which we have no acquaintance.
—B. Russell 1905, pp. 479–480

The difference between acquaintance knowledge and knowledge about was a topic of philosophical discussion for a long time, as the two quotes above clearly show. This distinction became identified with that between subjective (first-person) and objective/public (third-person) knowledge[7] and gained a renewed and urgent interest, which was spelled out following the introduction of some imaginative formulations and beautiful thought experiments. Thomas Nagel gave one thought-provoking formulation of this problem: "What is it like to be a bat?" Another influential discussion was promoted in the early 1980s by the thought experiments of Frank Jackson, which starred two hypothetical characters, Mary and Fred. Mary, the protagonist of most discussion on the topic, is a brilliant neurobiologist confined to a black-and-white world. However, she has the neural machinery (intact retinal receptors and brain areas) necessary for color vision and, crucially for the story, perfect knowledge of the physics and neurobiology of color vision, including the use of the technologies associated with this domain of knowledge. What, asks Jackson, will happen "when Mary is released from her black and white room or is given a colour television monitor? Will she learn anything or not?"[8]

The answer to the question seems intuitively to be "yes." When Mary is taken out of her black-and-white world and is exposed to a freshly picked bright red poppy then, according to Jackson, something is added to her perfect knowledge of the neural correlates of seeing red. She now *experiences* red, and thus, it is assumed, she acquires new knowledge. To drive home the point, Jackson introduces us to the second character, Fred, who has first-person visual experiences that are inaccessible to the rest of us (although the neural and behavioral correlates of Fred's experiences are, according to the story, perfectly describable and accessible). Fred can reliably and effortlessly discriminate between shades of red that seem to be identical to the rest of us (possibly because he has, like a goldfish, tetrachromatic vision).[9] Jackson assumes that even if we had full theoretical understanding of Fred's vision, we would not experience the world with the enhanced richness Fred enjoys. Having full knowledge about the neural or behavioral correlates of visual experience (or any other experience) is not the same as actually, subjectively experiencing seeing. This is so even if we can detect, by using futuristic functional magnetic resonance imaging (fMRI)-like technologies (much more sensitive and reliable than those we have now) that Fred is discriminating among shades of red and even if we are able to reliably relate different fMRI patterns to different hues of red. Similarly, Mary, who uses the most advanced futuristic technology, may distinguish red from green (and from all other colors) by analyzing the subtle patterns in the brains of people looking at colors, correlating them with wavelengths and integrating them with eye movement patterns, but she herself does not experience color. Or so Jackson claimed in the original paper, and so common sense suggests.

The Mary thought experiment spawned many debates, and its most basic premise has been challenged by Paul Churchland and by Daniel Dennett. They argue that an ability to make accurate and detailed visual discriminations that are *functionally* exactly equivalent to those of a color-seeing person renders the two types of knowledge, public knowledge and experiential knowledge, identical. A related claim—the assertion that Mary *knows everything* about the physical aspects of vision—means for Dennett that she can translate her perfect theoretical (third-person) knowledge into phenomenological experience and that the claim that this is impossible is simply the result of a failure of the imagination. If, when coming out of her black-and-white room, Mary was presented with a blue banana, Dennett suggests that she would tell her companions to stop pulling her leg, for she knows perfectly well that bananas are yellow.[10] Paul Churchland, too, doubts that it is only after leaving the room that Mary will recognize colors. He thinks that, contrary to intuition, Mary's complete theoretical knowledge may actually enable her to imagine and know what it is like to see red within her black-and-white room.[11]

It is not easy to share the intuitions of Dennett and Churchland. Human experience in general owes very little to our recent knowledge of neurobiology, and humans are subject to many experiential surprises. On the basis of such experiences, most people would argue that something new—whether we call it new knowledge or a new ability (is not the exercising of a new ability a form of knowing, they will say)—has happened to Mary and would happen to us if we were color-blind and suddenly became endowed with color vision. It seems to them that discriminating (e.g., by using a spectrophotometer), which appears to be the biological function involved in this case and is what the all-knowing Mary is able to do, does not exhaust the knowing of red, which is based on some more basic, technologically unaided perceptual processing. This is the gist of the objection Michael Beaton raised against the views of Paul Churchland. His main point is that knowing in the third person, which he regards as a high-level process, cannot lead to the causally relevant instantiation of knowing in the first person, which is based on what he sees as a lower-level process.[12] We disagree with Beaton and believe that it is, in principle, possible to move from a third-person to a first-person experience and agree with Paul and Patricia Churchland that this kind of "translation" may require putting into practice the theoretical (third-person) knowledge in the perceptual and affective parts of one's brain.

Thinking about other types of experiences may help in clarifying this issue. There are aspects of other people's experiences that are available to us in a more experiential and intimate manner than their visual experiences. If instead of wondering about Fred's red-color experience we were interested in Fred's emotional experiences (his fear experiences, for example), we could be in a better position to share in his first-person emotions by observing his behavior, especially if Fred was a close friend or a lover. Second-person knowledge of other people's emotions and motivations, expressed through their facial expressions, vocalizations, and body gestures, seems to be available to us as immediate emotional experiences. There are layers of neural mechanisms that are not fully understood, including, so it seems, ancient mechanisms of emotional contagion, which enable a naïve individual in a group to "catch" the emotions of others (examples are contagious fear in fish, hunting excitement in wolf packs, and hysteria in human mobs).[13] It is only recently that we have begun to understand what is involved at the neural level in such interactions. The discovery that some neurons fire both when one performs a motor act, such as grasping an apple, or facially expressing gustatory disgust, and also when one sees others make the same grasping act or expressing the same disgust, is an exciting neurobiological clue to second-person neurobiology.[14] The question is whether the kind of indirect public knowledge that we gain from science can enable an access to experience so that Mary, by virtue of knowing (conceptually) everything about pain or the color red, can feel pain or see red. Second-person knowing and we-experiences can start to bridge the knowledge gap by showing us how circuitous the route to direct acquaintance (first-person knowledge) may be.

Second-person knowing may be closest to first-person knowing when there is the physical-sexual contact of reciprocated love. In biblical Hebrew, the word "know" was used to describe not only the ability to discriminate good from bad but also the act of sexual love, as in "and Adam knew his wife Eve" (Genesis 4:1). In sexually shared love, the lovers' feelings seem to merge, and a lover gets as close as one can to first-person "knowing" the experiences of the loved one. The general point about the embodiment and situatedness of all forms of knowing, including interpersonal knowing, has been forcefully made by the neurobiologist Francisco Varela and by philosophers Evan Thompson, Alva Noë, and Dan Zahavi.[15] Moreover, the second-person way of knowing is related to first-person-plural (collective) knowing, the "we mode," in which affective and cognitive states are partly constituted by the feelings of others, and emotional contagion seems to play an important role.[16]

"Knowing" about Fred's color vision, knowing how Fred feels about a sudden unanticipated danger communicated to us through his behavior, and knowing how Fred feels love by loving and making love to Fred are somewhat different types of knowing. And all these ways of knowing differ—to different extents—from our own private knowledge of seeing colors, feeling fear, and enjoying love. The word "know" is used in all cases (first person, first person plural, second person, third person) in rather different albeit overlapping senses. Like James, Russell called the first-person knowledge "acquaintance," and he called third-person knowledge "knowledge of something," a type of knowing that is quintessentially public and reportable (although, like all forms of knowing, it does have an acquaintance aspect too). In between these we can interpose the second-person and first-person-plural knowing. When we examine the different types of knowing carefully, even on the basis of our limited knowledge of their neural basis, it seems that not only can a process of acquaintance (e.g., feeling hot) inform *knowing about* the experience (knowing about the science of heat and about the neural processes involved in response to heat) but the symbolic-based levels can inform the sensory-experiential ones.

There can be an even more direct link between first-person and third-person types of knowing, and to understand this we need to return to Thomas Nagel's "what it is like for a bat to be a bat" question.[17] Imagining batness on the basis of our third-person knowledge of bat behavior and neurobiology is clearly very different from the intimate acquaintance knowledge of batness, and intimate second-person knowledge of bats does not seem like a viable option.

Learning to Perceive: The Nonimaginary Case of Daniel Kish

We believe that it is possible to get closer to acquaintance knowing even in such difficult cases. What we are going to suggest is similar to Churchland's and Dennett's proposal. In order to explain how well (although of course not perfect) third-person knowledge can generate first-person knowledge, we introduce here a third (and real) character, Daniel Kish, who has a special insight into bathood. Daniel Kish, who was born in 1966, lost his eyesight to retinoblastoma when he was thirteen months old. Yet he navigates his way through city streets, rides mountain bikes (figure 4.1), plays basketball, can find a small ball on a large golf course, and teaches other blind people—mainly youngsters—to do the same. He does all this because he

Figure 4.1
Daniel Kish riding a bike.

echolocates, just as some bats and dolphins do, by making clicking noises that return as echoes and inform him about the shape, position, size, texture, and motions of the distant objects around him. Daniel started echolocating when he was a young child, and he is very good at it, but blind people can learn to echolocate at a later age, and even sighted people who practice echolocation for one to two hours a day can begin to echolocate after a relatively short period of training.[18] fMRI work on Kish and another echolocator has shown that part of the visual cortex, the calcarine cortex, is specifically activated when the echoes are picked up and as the echolocators report what they perceive on the basis of the echoes.[19]

Of course, Daniel Kish is not a bat, and many aspects of bat experiences are as inaccessible to him as they are to us; it is likely that his experiences are somewhat more "visual-like" than those of bats. But there is something about his subjective, first-person, private echolocation experience that must help him know a bit better about what being a bat is like. It is an embodied and situated type of experience, using his senses in a new way and training his body and his brain to perceive distant objects. In this case, expert third-person knowledge about bats and their way of perceiving could help blind people improve the use of their senses and enable the development of artifacts to help them acquire better first-person knowledge. At the most elementary and obvious level, what kind of clicks to utter, and when and how to use them, could be informed by knowledge of plasticity in the relevant areas of the brain. If, in the future, we have greater knowledge about brains—both bat and human—and better technologies for developing echolocation, people like Daniel might be able to identify the auditory stimuli that are most efficient for human brains and find effective ways of generating them. If so, it is highly likely that future blind echolocators will be able to experience the world in an even more bat-like manner. Conceptual knowledge can be used to generate new ways of first-person knowing.

Returning to Mary, with her vast expertise in neurophysiology (and inevitably also in endocrinology, physics, and biochemistry), it is clear that her perfect, theoretical third-person knowledge may be put to an empirical test by manipulating the world and her own body and brain in ways that can bypass the physical limitations imposed on her. If, upon stimulating her own brain and body to produce fMRI patterns characteristic of seeing red, she cries out, "WOW, I am having a RED, new experience"—while still sitting in her black-and-white room—we would be justified in saying that her third-person knowledge was put to use to produce first-person knowledge.[20]

Once the distinctions and relations between ways of knowing are made clearer, one can attend to what Mary-type questions teach us about first-person knowledge. As Patricia Churchland has noted, it is only when learning techniques based on a good theory are implemented in one's own brain and body that a specific subjective experience, such as seeing a red poppy, will become manifest.[21] Hence, Mary will indeed gain new knowledge when she stimulates her brain in the right way, even if she does not see red by being exposed to red objects. Her access to her experience of red has to be embodied, just like Daniel's echolocation-based experience of the distant objects around him. Moreover, even when some limitations exist in one's brain or in the external milieu, as in Mary's case, a good theory coupled with inspired learning practices (through subtle biofeedback and through highly informed experimenting with her brain) may enable Mary to do new things with her brain and see colors in her black-and-white room. This is the gist of Daniel Dennett's robo-Mary thought experiment, in which he describes a sophisticated robot (robo-Mary) with the same history as that of flesh-and-blood Mary but with the added ability to scan the brains of individuals who experience red and to implement these states (with or without a lengthy process of learning) in her own silicon-based brain. Maybe the biological or robotic Mary could even imagine (without complex technological assistance) what red is like on the basis of her fantastic knowledge.[22] After all, with our much inferior abilities, we can imagine what flying elephants with five tails may look like and imagine, while reading a novel, how an eleventh-century Japanese woman lived and felt. We must add a cautionary note, however: the fact that Mary can implement her vast knowledge and engage in a controlled process of imagining that leads to a RED experience does not mean that she fully understands why the correlations she found and the practices she used led to the subjectivity of her experience. Answering this question requires a causal-componential description based on convincing theoretical assumptions.

Why Square Pegs Do Not Fit in Round Holes and How a Shadow Can Evolve

Thought experiments can help us grapple with the overlapping problems of mental causation and mental constitution: How can mental states like anger or love change physiology? How can mental states arise from biological factors and processes (how are mental states constituted)? The in-principle neurobiological answer presented in the previous chapter and accepted by all biologist-philosophers is fairly simple: mental states are what brains in animals do—mental states are embodied and situated neurophysiological activities.[23] High-level neural processes that have emergent mental properties (e.g., anger) are composed of the activities of the neurons and neural networks that are their building blocks and at the same time constrain them. This constraining "top-down" effect imposed by high-level representations is inevitable since the top-down effects, just like bottom-up effects (from neurons to emergent mental states), are mediated by molecules such as neurotransmitters and hormones. Therefore, top-down effects are not biochemically mysterious.

Although based on the same biochemistry, it is still far from clear how exactly top-down neural effects work. As Helmholtz pointed out long ago, the brain has to infer, on the basis of altered bodily states, the sensory inputs it receives through the animal's senses, and it must do so by putting each input in the context of other inputs so that the causal relations among the inputs from the signal source can be tracked. Many cognitive scientists have argued that the brain does this by forming a flexible high-level model or map of reality, which is updated as inputs from the world and the body arrive. This is compatible with the observation that for visual inputs to become fully conscious and accessible, it is necessary for information to be exchanged back and forth among the visual cortex and other cortical areas, among higher and lower areas within the visual cortex, and among circuits within these circuits.[24]

The interpretation of mental causation in terms of neural mechanisms can be criticized, however. For example, Gerald Edelman argued that a mental state such as anger, denoted by C, is "the phenomenal transform," a relational property of brain activity C', which in his scheme is the result of dynamic-core activity. According to him, C is a property of C' and is entailed by it, but C itself lacks causal power, which can be attributed only to C'.[25] Although we believe that mental states are composed of neural (and nonneural) physiological processes, and in this sense they are not only entailed by these processes but also entail them, an explanation in terms of the composing mechanisms cannot alone fully account for the emergent processes that lead to mental causation. It is necessary to explain why high-level properties like anger impose constraints on low-level ones like neural firing. It seems to us that the causal, top-down effects of high-level states on the neural and bodily networks that constitute them should be broken down to two closely related but nonidentical problems. The first concerns the type of theoretical model that can describe how high-level properties can have nonreducible causal effects—the need to recognize not only Aristotle's efficient causes (explanations in terms of mechanisms) but also his formal causes (explanations based on a model); the second is the causal efficacy of what Aristotle called the "final cause."[26]

We start with formal causes. A famous argument given by the philosopher Hilary Putnam showing the advantage of formal explanations for some types of problems is based on the fact that very similar macroscopic causal states (for instance, a mental state like exasperation) may have multiple realizations at the micro level. Putnam gave a very simple example: when trying to understand the cause of a square peg not fitting into a round hole, the properties of squareness and roundness are better explanations for the fitting failure than the details of the quantum mechanical molecular arrangement and properties of the specific peg and the specific hole. Geometrical explanations in terms of squareness and roundness provide a more economical and general explanation of the peg's failure to fit the hole, since many microstates other than the specific molecular arrangement (the peg could be made of plastic, of wood, of bone, and so on) that realized these properties in the specific hole and peg can explain the same failure to fit. Similarly, your decision to participate in a demonstration against the passing of a racist law depends on the macroscopic properties of the neural dynamics in your brain, which can change over time at both very fine- and somewhat coarse-grain levels yet preserve the decision to go to the demonstration. These properties can therefore be regarded as the more essential cause of your action than any specific, fleeting microstate.[27] One can rephrase this formal cause-type explanation in terms of functional *information*: the geometrical pattern, rather than the specific underlying microstates, explains the "difference that makes a difference,"[28] which leads to the mismatch between the hole and the square peg, and the patterns in the brain rather than the neuronal microstates make the difference that leads to your decision to attend the demonstration. The organizational pattern in the case of square pegs is the information embedded in the geometry. The pattern must be described in terms of information and not just in terms of the transfer of energy and matter. Unlike energy and matter, information is not conserved, and this nonconservation explains why biological information can be amplified or reduced, something of central importance for the explanation of causal processes in biology, psychology, and sociology, including, of course, those processes that underlie consciousness.[29] Moreover, the effects of patterns of organization provide feedback on the low-level properties of the system. Consider Gánti's chemoton, which we described in chapter 1: the lipid molecules that result from the chemical processes within the chemoton lead to the construction of a membrane that encloses the metabolic cycle and the polymer cycle; the membrane has the ability to selectively allow or prevent chemicals from entering and leaving the system, thus changing both the internal environment of the chemoton and the external microenvironment with which the chemoton interacts (through the elimination of waste products and selective permeability). Hence, the closed membrane has properties resulting from its three-dimensional organization that provide feedback on its own synthesis through its effects on the system and the external world with which it interacts. The membrane cannot be fully explained in terms of the building-block components in the chemoton without considering its role as an enclosing and selectively permeable structure affecting the synthesis of the specific components of which it is made. The explanation of the membrane requires an understanding of the system as a whole, which includes the feedback from high-level products to the synthesis of low-level components, setting the boundary conditions of the system.[30] So although we agree with Putnam that there may be several neurobiological realizations for a single mental state, just as there may be variant chemotons, the idea of multiple nonbiological realizations existing does not follow. We are committed only to multiple neural-physiological realizations of mental states in *biological organisms*, which, we stress again, are the only entities that we know can be (but need not be) conscious.

The second, final (or in modern parlance, teleofunctional) account of causation posits that high-level mental states have functions that cause low-level activities through the mediation of neural and/or genetic selection. This is a developmental-evolutionary answer to the question of how and why the "mental" can cause the "physical." To see how this functional-selection argument works, we will take a closer look at the causal efficacy of shadows, since a shadow is a paradigmatic example of an emergent, high-level property. Although the mental has been compared to a shadow, our goal here is not to reify this metaphor but to point to the way in which inevitable, high-level, emergent products of structures or activities, which cannot easily be decomposed into microconstituents, can have selective functional effects that exert causal influences on the processes that generate them.

The small bobtail squid *Euprymna scolopes* swims in the ocean waters around Hawaii, searching for and feeding on shrimp at night. On the ventral side of its body is a light organ inhabited by luminescent symbiotic bacteria (*Vibrio fischeri*). The organ contains reflective plates that intensify and direct the light produced, preventing the squid from casting a shadow and having a visible silhouette on moonlit nights.[31] This camouflage is beneficial and necessary for the squid's survival because it helps to hide the squid from fish that might eat it (figure 4.2). This process of counterillumination has been observed in other cephalopods, crustaceans, and fish living in the twilight depths of the ocean, where predators evolved to have upward-looking eyes adapted for spotting the silhouettes of shadow-casting prey.[32] It seems that the squid's silhouette was the target of selection since it "caused" the evolution of counterillumination mechanisms, albeit its causal power was functional rather than directly mechanical. And the silhouette and shadow can be said to have evolved, just like any other organismal trait, from a full to a barely observable silhouette.

Evolved, phylogenetic adaptations are not the only way in which emergent properties can have new functions that are causally relevant for an organism. Developmental adaptation through open-ended learning, for example, can have a similar outcome. An animal casting a shadow that makes it vulnerable to predators can learn, through trial and error, to behave in ways that minimize or eliminate the shadow it casts—for example, by altering its time of activity, by moving to locations where its shadow merges with those of the local vegetation, or by moving in ways that change the size of the shadow it casts. Through ontogenetic learning, the shadow can alter the behavior, physiology, and gene expression of the animal, and through evolution it can lead to changes in the DNA sequences of relevant genes. In both ontogeny and phylogeny, such high-level causation is mediated by selective processes, but this does not render its causal, top-down efficacy any less impressive. It does require, however, that we specify the level of description of both the cause and the effect so that the cause will

Figure 4.2
The evolution of the disappearance of a shadow. The silhouette of the shadow-casting squid makes it visible and therefore vulnerable (*left*); eliminating the shadow and camouflaging the silhouette by light-producing bacteria confers an obvious advantage (*right*).

give a good prediction of the effect, and the effect can lead to an inference about the type of macrodynamics that were its cause.

In summary, and as we have argued in previous chapters, saying that high-level neural processes that constitute mental states are in some sense emergent does not mean that they have no causal efficacy. In most cases, the "formal" way of describing causal efficacy—in terms of the behavior of the animal or of the neurobiological principles involved, rather than in terms of the microstate of billions of neuronal connections—is the most general and economical causal explanation, at least for the purpose of explaining why one type of action rather than another was taken, which is usually what one is interested in. Moreover, the macroscopic products of the high-level processes have properties (the selective permeability of the enclosing membrane for the chemoton case) that affect the low-level processes (the components engaged in the metabolic cycle), so the chemoton is, as Kant would have put it, both a cause and an effect of itself. Importantly, the causal feedback interactions are brought about by selection processes at the ontogenetic and/or the phylogenetic level, so a selective-functional explanation may often be adequate for the "why" question. Usually, natural selection acts on phenotypic high-level traits through their effects: *sharp* teeth, *good* sight, *devoted* maternal care, *executive control* of emotions, or a *large* or a *small* shadow. Just like the squid's shadow, mental states have functional causal power, often determining life and death—we need only think of human history to come up with endless examples of the causal power of anger, hate, and jealousy. Importantly, mental states can not only evolve but also coevolve with other cognitive processes, through their functional effects. Like the formal explanation, the functional evolutionary perspective presents a coherent way of answering the mental causation question and can help clarify how consciousness is instantiated.

The Componential Problem: Consciousness's Magic Show

How mental states arise from biological factors and processes (how mental states are constituted) has been a topic of much philosophical debate. A recurring theme in the debate is the philosophical zombie. Philosophers usually depict zombies as entities that are atom-for-atom identical to humans or animals but nevertheless lack phenomenal consciousness. Like most natural philosophers, we regard the idea of such zombies as incoherent. The properties attributed to them are, as with many logical paradoxes, an artifact of language that gives us the illusion that linguistically possible worlds (possible because of language and the logic it endows us with) are also physically possible.[33]

Turing Robots

Another related and much-discussed philosophical issue that arises is not about zombies but about their close philosophical kin, robots that pass the consciousness Turing test (we call them "T-robots"). The Turing test, which attempts to determine whether a machine can think, is based on a social imitation game in which a judge tries to guess, on the basis of printed answers to his clever questions, whether the subjects he is examining (but does not see or hear) are male or female while they attempt to imitate the opposite sex. Using the design of this game, Alan Turing, the mathematician and computer-science pioneer, argued that if a sophisticated and highly skeptical judge cannot, based on the answers to his questions, distinguish between a digital computer's and a human's answers, the computer can be considered to be thinking. Extrapolating from this test, one can ask whether a robot that behaves in a manner that can fool a very clever judge—and make the judge believe that it is a sentient, experiencing being—should be considered a conscious automaton.

Such T-robots are still a figment of the imagination, but they highlight interesting problems about the relation between performance and mental makeup. Stevan Harnad has invoked an imaginary T-robot that can do everything a normal human being can do except feel (experience, have phenomenal consciousness) to drive home the distinction between feeling and doing/behaving. But because it has been argued that a silicon robot, even if it passes a demanding Turing-like test, may possess such a different internal organization that comparison to a human is impossible, Harnad went on to imagine a feeling T-robot, made up of biological materials, and argued that it can tell us little about feeling.[34]

Enabling Systems

What is the basis for Harnad's first claim that a T-robot can be an unfeeling automaton and that complex "doing" does not entail "feeling"? And is his second claim, that if we meet a feeling biorobot, we are still no wiser with regard to what mental states are, valid? We believe that both Harnad's assumptions are dubious. In order to see why, we introduce a new term, that of an "enabling system," which we explain by using a version (with a twist) of the famous eighteenth-century thought experiment introduced by the British theologian William Paley; a thought experiment that creationists past and present have used and to whose challenge Charles Darwin rose magnificently. Paley wrote:

> In crossing a heath, suppose I pitched my foot against a stone, and were asked how the stone came to be there; I might possibly answer, that, for anything I knew to the contrary, it had lain there forever: nor would it perhaps be very easy to show the absurdity of this answer. But suppose I had found a watch upon the ground, and it should be inquired how the watch happened to be in that place; I should hardly think of the answer I had before given, that for anything I knew, the watch might have always been there.... There must have existed, at some time, and at some place or other, an artificer or artificers, who formed [the watch] for the purpose which we find it actually to answer; who comprehended its construction, and designed its use.... Every indication of contrivance, every manifestation of design, which existed in the watch, exists in the works of nature; with the difference, on the side of nature, of being greater or more, and that in a degree which exceeds all computation. (Paley 1802, pp. 1–2)

The reasoning Paley used is far too simple—and misleading in more than one way. To have a watch, we need much more than a watchmaker. We

Figure 4.3
Stumbling, amazed, upon a grandfather clock in the desert.

need to have a world in which watchmakers live and in which watches are desirable and necessary. We need a whole cultural-social infrastructure in which watches are embedded, bought, used, repaired, and constructed. This larger system, which makes the finding of a watch on a heath, or a grandfather clock in the desert (figure 4.3) possible, is the enabling system of the watch. We reconstruct, or "reverse-engineer," it very much in the way an archaeologist reconstructs a long-gone culture on the basis of few artifacts or a paleontologist reconstructs a whole animal on the basis of a bit of skeleton. Of course, we also need to think about the cultural evolution of watches and clocks, but our understanding of the current enabling system for watchmaking and watch-using will not only explain why watches are present in a particular cultural system but may also go a long way toward unraveling the history/evolution of the human-designed watch.

To come closer to biology, let us make a variation on the watch thought experiment. Imagine that intelligent aliens, equipped with scientific technologies and instruments, land on a past life-harboring, frozen planet Earth and find several long polymers made up of several different units (long DNA molecules). Their knowledge of physics and chemistry tells them that such polymers cannot be formed spontaneously—the chance of the spontaneous generation of several such large molecules verges on the statistically impossible. They therefore assume that there must have been (and maybe still is) a system that enables the synthesis and maintenance of such complex molecules. Before they investigate how such molecules evolved (a phylogenetic question), they need to discover the properties of the system of which these molecules are, or were, part. In other words, they need to answer the developmental-physiological question—they need to unravel the machinery that is generating them and making their maintenance possible. The ordered complexity of the polymers' structure suggests they had to be synthesized and actively maintained by chemical mechanisms involving specialized factors organized within a spatially distinct self-reproducing system, similar to a sophisticated autopoietic chemoton system, with an intricate maintenance and repair machinery, which is indeed a bit like a modern cell. Moreover, experimenting with these molecules suggests that their properties enable them to be copied (replicated) with variations, and the aliens may infer that the function of the polymers was information storage and that they operated as parts of an unlimited heredity subsystem that allowed the system as a whole to persist for long (evolutionary) periods. After they solved the problem of the dynamic organization of the enabling system of these molecules, the aliens would be in a better position to answer the phylogenetic question—how a system with the particular interesting parts they found had originated and evolved. The existence of the long polymers, or more specifically, the unlimited hereditary system that they instantiate, are, in fact, diagnostic of life. They are good markers of a sustainable living system.

We may now ask the parallel question: Is passing the consciousness Turing test a good diagnostic marker of consciousness? Thinking in terms of the enabling system underlying the intelligent behavior of a T-robot can help us understand both its development and its evolution. If we just explore the current mechanisms and factors that instantiate the T-robot, we are likely to find it extremely difficult to discover how its conscious behavior is generated, but if we know something about the technological evolution that led to its construction and about the developmental processes that such a robot may need to undergo, we may be better equipped to try and figure out if they are conscious. A conscious T-robot has not yet been built, and we do not know whether the fact that it is intelligently rather than naturally designed involves far simpler mechanisms, leading to intelligent sentient-like behavior, than the sentience of a biologically evolved animal. So Harnad's first claim cannot at present be evaluated, although it will be possible if a T-robot is constructed. Harnad's claim that even finding a feeling T-robot, or a biorobot, gives us little insight into consciousness because it is not clear how it helps explain the "how and why of (calling a spade a spade) consciousness"[35] is puzzling. Surely, if we knew enough about the enabling system generating a feeling in the T-robot, and about the technological evolution that led to it, we would gain some insights regarding the "how" and "why" questions of the T-robot's consciousness. Just as understanding the enabling systems for life provides insight into what life is and leads to reliable diagnostic tests for determining which systems are living, insights into consciousness could be made by studying the enabling systems for sentient beings, whether artificial or natural.

A good strategy would be to examine the enabling systems that instantiate consciousness in animals, the only conscious entities we are aware of at present. In chapters 1 and 3, we pointed to some general properties of an internal system of communication that seem to be necessary (though not sufficient) for it to show something like mental states. For example, it must transfer signals rapidly, so they do not fade away en route; it must enable the flexible integration of information coming from different parts of the body and/or from different aspects of the external world and construct coherent wholes; it must preserve the specificity of the signal's origins and targets; and it must ensure the temporal persistence of the integrated input for it to be perceived or felt. These special properties (and some others that will be discussed in subsequent chapters) were all identified in the nervous system. It therefore seems that an evolved nervous system is necessary for consciousness.

But one can go a bit further. The neurologist Todd Feinberg suggests that considering the special organization and the idiosyncratic dynamics of the nervous system may uncover major characteristics of mental systems and help reformulate some philosophical dilemmas.[36] Feinberg stresses the hierarchical organization of the nervous system, which displays both nested hierarchy (brains are made up of anatomical modules that are made up of neurons) and nonnested control hierarchy (the nervous system is the top level of control for other bodily systems, and within the brain some parts control others). Like the neurobiologists whose ideas we discussed in chapter 3, Feinberg focuses on the multiple-level mapping relations between parts of the brain. Rough guesses about the world that are formed at high levels are mapped back to lower ones, prediction errors are corrected, lower-level information is back-mapped into higher levels, and so on, leading to the formation of coherent, integrated, dynamic "maps." In addition,

Feinberg points out that there are two basic *limitations* to this system, which can explain some properties of mental states that have a distinguished philosophical history: embodiment and transparency. The embodied nature of consciousness leads to the privacy of consciousness and to the inevitable first-person/third-person gap, while transparency, the apparent "obviousness" of reality, which is a result of the fact that neural activity can be said to refer to states of the world body rather than to itself, leads to naïve realism.[37] Since both features are central to the constitution question, we will look at them more closely.

Embodied Consciousness: Brain Centered or Animal Centered?

As we have seen, embodiment was central to James's view of consciousness and is central to neurobiological approaches, notably that of Freeman, with its sympathetic leaning on the phenomenological tradition. Phenomenologists have emphasized the centrality of the body in experience—both the consciousness of the body as an object (the body image) and, more importantly, what they call "prereflective bodily consciousness," as seen in a skilled activity such as expert driving or dancing. They oppose the idea (now rarely entertained) of the early cognitivists who regarded the mind as "software" that could be understood without knowing anything about the "hardware"—the biology of the brain, the peripheral nervous system, and the rest of the situated body.[38]

The philosophers who pursue embodiment-oriented, animal-centered research stress, like brain-centered scientists, the bidirectional feedback between the body and the brain. However, they argue that all cognitive processes begin with exploratory motor activity and that recurrent sensorimotor activity on several temporal scales governs consciousness in animals.[39] They stress the affordances (action opportunities) of anatomy and movement and their effects on animal perception: for example, how a hand is shaped and is able to move determines what one does with it and therefore how one perceives hand-relevant objects. This means that the perceived structure of the environment is guided by activity; animals extract invariant features through exploratory activities and perceive these aspects of their environments as relevant to their specific needs and goals, using the multiple resources of their environment flexibly and opportunistically. That is why cognition, the term preferred by phenomenologists, is always situated and environmentally embedded: cognition is an interaction between an embodied animal and its lived world. An additional important assumption is that perceptions, thoughts, and feelings are not just caused by sensorimotor processes. They are constituted by them. The interacting body does not just affect the brain and is affected by it; cognition needs to be understood and modeled as bodily constituted.[40]

But how can one prove that the environmentally embedded and interacting body constitutes cognition (or subjective experiencing) rather than affecting and causing it without being part of it? There are many experiments showing the profound effects of sensorimotor, hormonal, and immunological activities on mental activities like perception and learning, and there are also some (few as yet) dynamic models of "embodied" cognition (i.e., cognition shaped by the activities of the whole body, not just the brain).[41] Although, as we argued in chapter 1, a brain of some sort is needed for an animal to experience, it is patently clear that a brain cannot think, dream, feel, or perceive without a web of dynamically coupled inputs and output from the peripheral nervous system, the nonneural body, and the environment. A brain floating in a saline-filled vat is fit only for anatomy lessons. A brain in a vat (figure 4.4) has to have a virtual body and world for the system to be said to experience. These common-sense reflections do not, however, clinch the pro–bodily constitution argument. The problem is the delineation of the boundaries of a biological system of experiencing, for not everything that causally affects the system's functioning can be considered to be part of it, to constitute it. In order to count as part of some whole (a system or an individual), the system must be shown to have some coherent, stable properties, to which the parts contribute.

We believe that an experimental approach focused on morphogenesis and regeneration sheds interesting light on the componential role of the body in subjective experiencing. The study of morphogenesis and regeneration has been vigorously and imaginatively pursued by Michael Levin, who has been studying these processes in animals such as flatworms, tadpoles, and mice.[42] Levin's basic idea is that biological information on various temporal and spatial scales unifies the organism. First, cells carry information about the maintenance of cell shape, so, for example, unicellular organisms such as ciliates can regenerate amputated bits of their surface structures. Second, in all multicellular organisms, the internal milieu of cells contains chemical and electrical cues that are communicated among cells through specialized intercellular connections (gap junctions) and generate a morphogenetic field. Third (and related), bioelectric gradients are found on scales ranging from the intracellular organelle level through the cell level, the tissue level, and the appendage level to the whole-animal level. Fourth, when present, the nervous system influences the body's bioelectric field and affects regeneration; for example, if the central nervous system (CNS) or the spinal cord is damaged in specific places, regeneration in tadpoles is

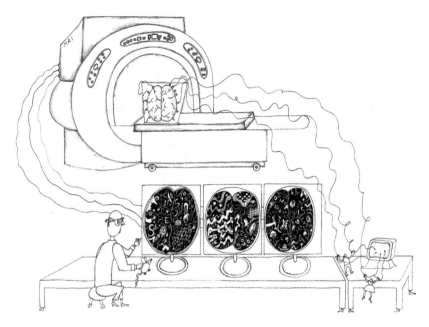

Figure 4.4
The brain-in-the-vat fantasy. A brain in a vat at the top of the picture, stimulated by a computer (*right*), is monitored by an MRI machine. The neurobiologist below observes the MRI results on computer screens, which indicate a rich internal (virtual) life.

deregulated. On the basis of the data that shows that the appendage level can affect the tissue and cell level, Levin argues that in addition to emergent and dynamic bottom-up processes, there are also top-down processes driven by bioelectric fields at the tissue and organism level.

We think that Levin's ideas are relevant to both ideas about the evolutionary origins of the nervous system, a topic we discuss in chapter 6, and the question of the bodily constitution of consciousness. Although a centralized nervous system is obligatory for consciousness and is a major determinant in the construction of a whole-body bioelectric network, it is still a *part* of such a network. The nervous system alters the chemical and electric patterns (or "codes," as Levin calls them) of the body as a whole, including the nonneural parts of the animal's body, and in turn is altered by them. From this perspective, the brain is not the sole part that constitutes mentality, and when we think about the evolution of subjective experiencing and of brains, we have to think about the way that information in the integrating brain has become part of the body as a whole. We must resist the tendency to speak of the brain as an agent that performs computations and makes predictions and decisions, a tendency that reflects the "brain in the vat" approach and ignores the ongoing feedback between sensory stimuli and motor acts.

Transparency and Selves: World Models and Self Models

While embodiment leads to irreducible subjectivity, transparency—which is the consequence of the operations of neural processes that integrate information about the world and the body and constructs models of the two—gives us the strong conviction that the things and properties that we perceive and feel are "out there," in the body or the world.[43] However, the realization that what we call the "external world" actually involves mental construction has been with philosophy for a long time and is well captured in a fifteenth-century Japanese poem:[44]

> Oh green green willow
> wonderfully red flower
> but I know
> the colors are not there

The colors, indeed, are not there. Without certain receptors and without certain brain structures and modes of neural processing, there are no red and no green experiences. Of course, red depends on a property of the physical world, but the physical objects and phenomena of the world— here the approximately 560–580 nm wavelength band of light reflected from the apple—have to interact with evolved receptors and brains that respond to it. We can see an apple as red under very different illumination, so this color constancy is independent of the wavelengths of the light that strike the apple. To insist that red is out there in the external world in the absence of an evolved or designed observer is incoherent. This does not mean that perceptions of red or green are illusions. Colors are the result of interactions, like almost anything in the world, and are as real in the realm of their interaction as are protein molecules (the outcome of interacting atoms) in the realm of organic chemistry.[45] Similarly, the feeling of pain in a woman's chest is a complex construction of an embodied nervous system, which makes the woman feel that she is in pain and that the pain is located in her chest. These perceptions, of wonderfully red poppies or of pain in the chest, are the basis of the claim that mental states are intentional—that they are directed toward the world and the body and are, in this sense, *about* the world or *about* the body.

Subjectively perceived or felt constructions like red poppies or pain in the chest are qualia. But qualia are not disconnected: for a neurally intact woman to feel that her chest is painful is to have a notion of herself as the "owner" of this pain. There is, we say, someone who suffers or who enjoys—a self. What, or who, is it?

There are more views about the "self" than there are philosophers. They include the notion that having a self requires the ability to reflect about one's feelings and perceptions and the notion that an awareness of others is a prerequisite for a conception of self.[46] There is also a distinction between different levels of self: phenomenologists distinguish between prereflective and reflective self, with the latter divided into an explicit but nonnarrative self and an explicit narrative (human) self.[47] As we described in the last chapter, Antonio Damasio makes similar distinctions, differentiating between a protoself; a core self; and an extended, autobiographic-narrative self. However, since understanding even the simplest self in neurobiological terms is a tall order, we begin with what we see as the self's neurophysiological precursor.

It has long been known that an animal's motor activities have sensory outcomes that are perceived as different from similar sensory inputs received from the external world. For example, when touched by an external rough object, an earthworm has a strong withdrawal reflex, but when crawling on an equally rough surface, it does not reflexively withdraw, thus showing that it can distinguish between self-produced and world-produced sensory inputs.[48] Similarly, we respond very differently when we tickle ourselves and when we are tickled by others, and we do not perceive the darkness caused by our eye blinks, whereas we would experience brief episodes of darkness if such "blinks" were externally imposed. It seems that the nervous system, whether it is a worm's or a human's, compensates for self-imposed sensory changes. Von Holst and Mittelstaedt suggested in the 1950s that to deal with the problem of discrimination, which any moving animal faces, animals must track their actions and inform the system that processes sensory information about their imminent movement. They argued that when the motor circuit sends a command to the muscles, it also sends a copy of this motor command (they called it "efference copy signal") to a circuit that influences processing in the sensory pathway, a process they called "reafference." Both the input received from the sensory receptors and the motor copy received from the motor system influence—via an intervening circuit—how sensory inputs are interpreted. This requires complex integration, as is clear when we remember that motor and sensory organs are made up of tissues (muscle sheets and multicell, organized sensors).[49] It seems that the animal's motor commands must entail a dynamic "model" of its own body actions. Such a "self-model" may not be sufficient for consciousness and for an experiencing self, but is it necessary?

The cognitive scientist Björn Merker maintains that it is, and he has developed a framework for studying consciousness that is focused on the self. He argues that moving animals with a CNS and several sensory modalities must make many corrections to remove the effects of self-motion from world-generated sensory information. If they are to respond adaptively to their complex world and their no less complex motions, animals must construct in their brains what he calls "a coherent and stable world-space," a reality model. The reality model that animals construct consists of a representation of the world in which a representation of the body is embedded; the latter includes decision-making mechanisms driven by motivations (figure 4.5). According to Merker, the interface among these components can be called a "self" or "ego center," and consciousness is the online, real-time working of this entire functional organization.[50]

The physical brain is part of the physical body, which is part of the world. The brain contains the conscious "reality space" (*unshaded*) that is separated from other, nonconscious (*shaded*) functional domains; the conscious part contains a dynamic model of the body embedded within a dynamic model of the world. The "ego center" is the point of view from which the world is perceived, acted upon, and evaluated (by motivational systems).

Biologically inclined philosophers approach the idea of "self" in a far more abstract manner, yet some of them employ surprisingly convergent notions. A dynamic notion of "self" as a model (a self-model) is central to Thomas Metzinger's view of consciousness.[51] He suggests that the effects of back projections from the body world to the brain and back again construct a unified model of the embodied self. We make a model, an ongoing representation, of the tiny bit of the world-body action to which we have access. He defines a minimal self-model as a representation of the entire body that is functionally available for global control and provides information about new causal properties. This representation is a process, or as he puts it, "not so much an image of reality but a tunnel *through* reality."[52] Importantly for an evolution-oriented view, Metzinger distinguishes between different levels of self-models, which involve either weak or strong first-person perspectives (1PP). While a weak 1PP "is a purely geometrical feature of a perceptual or imagined model of reality possessing a point of projection functioning as its origin in sensory and mental processing … [a] strong 1PP appears when the system as a whole is internally represented as directed at an object component, for example a perceptual object, an action goal as internally simulated or perhaps the body as a whole. A strong 1PP is exactly what makes consciousness *subjective*: the fact that a system represents itself not only as a self but also as 'a self *in the act of knowing*'."[53] The 1PP that is central to

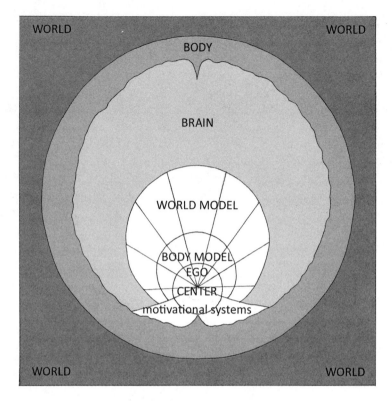

Figure 4.5
The "reality space" model. The physical brain is part of the physical body, which is
part of the world. The brain contains the conscious "reality space" (unshaded) that
is separated from other, nonconscious (shaded) functional domains; the conscious
part contains a dynamic model of the body embedded within a dynamic model of
the world. The "ego center" is the point of view from which the world is perceived,
acted upon, and evaluated (by motivational systems). Adapted by permission of
Cambridge University Press.

Metzinger's view of the minimal self is very similar to Merker's ego center
and is normally linked to a feeling of ownership and personal continuity,
the feeling of being essentially the same self over time.[54] Anil Seth developed
this idea using the framework of theoretical cognitive science, specifically of
the hierarchical predictive processing theory of Friston and his colleagues,
who propose that high-level models of the world, body, and prospective
actions are updated and upgraded in a hierarchical manner on the basis of
"prediction errors" (the mismatches between the actual effects of the input
and the predictions of the preexisting model) sent from lower levels of the

neural hierarchy. Seth suggests that subjective feelings, especially the experience of body ownership, arise from actively inferred generative models of the causes of stimulation coming from the body, which become updated as they interact with signals coming from interoceptors. The interaction of signals coming from the body with the self-model, and the interaction of the self-model with an updated world model, give rise to a feeling of body ownership.[55]

The related idea that self-referentiality is necessary for consciousness was developed in a very different (and very brain-centered) context by Douglas Hofstadter. In his 2007 book, *I am a Strange Loop*, Hofstadter explored the notion of the self through the use of a powerful metaphor, that of Gödel's self-referential mathematical incompleteness theorems. Gödel showed that for any consistent system of arithmetic axioms (such as the *Principia Mathematica*), there will always be statements about the natural numbers that are true yet unprovable within the system. The arithmetic symbols are so rich that they create subtle modes of self-referentiality that lead to paradoxes if the system is to be regarded as noncontradictory—just as language with its rich words can generate logically paradoxical sentences such as "this sentence is false." Hofstadter compares the self-referential loops among categories that create the powerful paradoxical "I" in the brain to Gödel's logical "loop" and suggests that the strangeness of the loop of selfhood is made up of two ingredients. The first is a positive property, "an extensible repertoire of symbols [that] give our brains the power to represent phenomena of unlimited complexity and thus to twist back and engulf themselves via a strange loop"; the second is a negative property, the "inability to see feel or sense in any way the constant, frenetic churning of and roiling of microstuff all the unfelt bubbling and boiling that underlies our thinking."[56] The result of these two ingredients is "an emergent entity that exerts an upside down causality on the world leading to the intense reinforcement of, and the final, invincible, immutable locking-in of this belief [in an 'I']."[57]

Different metaphors of the self may illuminate different facets of this complex notion. Dennett suggests that the self is a center of narrative gravity, an ongoing self-told plot, and a kind of narrative strange loop,[58] and the theoretical psychologist Nicholas Humphrey proposes the metaphor of a magic show that includes the magic show creator-observer (the self):

> Consciousness is a magical mystery show that you lay on for yourself. You respond to sensory input by creating a personal response, a seemingly otherworldly object, the ipsundrum [illusion-generating inner creation], which you present to yourself in your inner theatre. Watching from the royal box, as it were, you find yourself transported to that other world. (Humphrey 2011, p. 39)

Figure 4.6

Altering the body image. *A*, relieving the phantom pain of an amputated hand. A mirror is placed in the middle of a box that has two holes in the front. The amputee, who suffers great pain due to a feeling that the missing hand is clenched and frozen, is asked to insert both his arms (the actual arm and the phantom arm) into the holes in front of him and observe the reflection of his real arm in the mirror. The mirror image of his real arm creates the visual illusion that the amputated arm

Humphrey's consciousness, which includes the self, is an emergent virtual creation that has some special, evolved properties that are part of the magic (the feeling of ineffability, specialness, and irreducibility) leading to the joy of living that characterizes animals and especially humans.[59]

What all these metaphors stress is that the conscious self does not exist as an entity, as a distinct and separate "thing." Like every other conscious state, it is a special neurophysiological *process* with imperative presence, whose complex emergent and dynamic architecture is uncovered by pathologies, special brain states, and clever experiments. There are many disorders (notably strokes) that result in a person feeling that parts of her body (her left leg and left hand, for example) are not owned by her or even believing that she is dead and disintegrating (Cotard syndrome). The mirror image is the persistent integration of a lost body part within the body model—the phantom limb syndrome. With this condition, the brain is intact, but following an limb amputation, the absent limb is felt as fully present and can cause pain and suffering. Vilayanur Ramachandran showed, using simple ingenious experiments, how one can alter the body image and relieve the phantom pain (figure 4.6A), and he interprets many unusual states of consciousness as alterations in the neural body image.[60] The dream state, a normal and necessary part of the daily cycle of many animals, shows that once evolved and developed, the self-model does not require waking activity—one has a strong experience of acting and feeling while the body is lying without movement. Metzinger describes lucid dreams and out-of-body-experiences as important (albeit relatively rare) variations of the dreaming state, which can shed light on the self-model, and he links these experiences to experiments that create the illusion that a visible rubber hand (figure 4.6B) or a whole virtual body is being touched and is therefore the "real" (owned) body.[61]

---

exists. When asked to make symmetrical movements with the two hands, the amputee gets visual feedback from the mirror, seeing the frozen hand moving, and this enables the brain to reorganize and match the visual input with the representation of the phantom hand. The visual input, showing an image of the phantom hand as no longer frozen, relieves the pain resulting from the stored image of the clenched phantom hand. *B*, in the rubber hand illusion, a subject whose right hand is hidden from her view and stroked without her seeing it (under the table) observes a rubber hand on the table in front of her being simultaneously stroked. After a short time, she feels that it is the rubber hand (not her actual, invisible hand) that is being stroked, a result of the integration of visual, touch, and proprioception, in which vision dominates.

The extended applications of virtual reality in the exploration of human life open up many questions.[62] For example: Can we learn, in addition to echolocating like bats, the body-feeling of a flying bat using virtual reality construction techniques? What kind of activity will our brain show once we do that? Will some patterns of brain activity converge on those found in bats? Can we alter the self and body models of other animals? For example, how will a dog react to a "rubber paw" experiment? Virtual body experiments may allow us to investigate the richness and the limitations of the self-models of other animals. A reflex-based reafference system, such as that of a worm, will generate a very simple and inflexible self-model, while that of a mammal, an octopus, or a bumblebee, is likely to be very rich. In all cases, it is the embodied and situated enbrained animal that generates a self-model. Changes in the feeling of body ownership, even their "transfer" to an avatar, show that not only the brain is important for generating a self-model: it is the body-brain-world relationship that constructs a model of reality, and the richness of this model is proportional to this relation.

These ideas about the self and about qualia, which portray them as dynamic, recursive, and self-referential processes created by, and then operating on, sensorimotor categories, lead to some intriguing evolutionary questions. Qualia have their more-or-less distinct discriminating and motivating properties because of the way evolution built up sensory and perceptual categories and the way it led to their integration as part of the dynamics of the "self." We can therefore ask: How did the sensory categorization processes that created something like Merker's ego center, Metzinger's ego tunnel, Humphrey's ipsundrum, and the strange Goedelian loop Hofstadter identifies with the self, arise during evolution? Did selfless but qualia-experiencing creatures come first, or were creatures with a primordial sense of self and very few and fuzzy qualia primary? The fact that in highly evolved animals like us, various brain injuries can lead to dissociation between specific experiences and a sense of self does not necessarily mean that qualia preceded selves in evolution, or that qualia and self can be *evolutionarily* separated. For example, it may be the case, as Merker suggests, that a nonconscious animal, with a neural model that enabled it to distinguish between world-generated and self-generated sensory perceptions, was a precondition for the evolution of a conscious self.

In addition to questioning our all-too-ready intuitions, an evolutionary approach can point to critical biological building blocks and interactions that enabled the formation of the extraordinary sensory categorizations we call "qualia." It can help answer questions such as why a particular neural process is accompanied by qualia and another neural process is not and why different qualia, like red and green, are discriminated. Moreover, as we indicated in the previous chapter, there is more to evolution than the phylogenetic aspect. Selection within the animal, especially in the brain, is as important. For Dennett, selection among competing "multiple neural drafts," one of which gets the upper hand and achieves what he calls "fame in the brain," is the basis of distinct experiences,[63] and we saw that Changeux and Edelman developed theories of neural selection that are also based on selection among multiple parallel processes. If, as they suggest, evolution happens within brains and it is through ontogenetic neural evolution that mental states are constructed, the nature and dynamics of these evolutionary ontogenetic processes have to be elucidated. Moreover, we also need to account for the phylogeny of neural evolution within the brain.

Let us go back to Harnad's ideas about the nature of conscious T-robots. Although we do not yet have a theoretical model that will account for all the properties attributed to a conscious entity, there is no doubt that a dynamic model of minimal consciousness—something akin to the chemoton model for minimal life—would advance our understanding. In fact, a description/simulation of the dynamic system enabling a robot's conscious behavior would give us answers to a whole suite of interrelated causal questions, just as it did in the case of life. In chapter 1 we described how the reactions and factors that make up an autopoietic system like the chemoton are dynamically organized and how this organization yields a system that is alive according to all the criteria scientists have proposed. In the chemoton, the molecular building blocks are the Aristotelian "material cause," the chemical reactions between them are the "efficient (mechanismic) cause" (the source of action), and the overall relations between the reactions and subsystems are the "formal cause" (which is also an intrinsic "final cause"). Life came into being when the chemical subsystems and their activities interacted in a way that resulted in autopoietic dynamics that have to be explained in terms of *all* of Aristotle's four causes. An evolutionary approach was fundamental for constructing the chemoton model: it helped in its conceptualization and was essential to the recognition that an autopoietic system did not emerge as a result of an improbable stroke of luck, but through chemical evolution. The employment of the evolutionary approach enabled scientists to construct plausible, preliving reaction systems and scenarios that might eventually culminate in a living system. The same strategy can be applied to the construction of a model of minimal consciousness. This strategy has not been central in philosophical discussions, although the evolution question does concern most philosophers. It usually takes the form of a question about the teleofunction/s of consciousness.

The Teleofunctions Question: Consciousness Has Telos but No Function

The usual argument about the teleofunctions of consciousness is that if being conscious ("feeling" is the term Harnad prefers) has originated at some point in the history of life and has been maintained through evolutionary history in at least some lineages, it stands to reason that it has some teleofunctions. Teeth have functions, eyes have functions, maternal care has functions, and even an absent shadow may have a function. What then is the function of subjective experiencing? Does it really have one? Although it seems to us that the ability to see, hear, feel pain, and so on is wonderfully useful, many experimental results show that some things can be done without awareness of surrounding stimuli. People can navigate familiar complex routes on "autopilot," avoid obstacles while denying that they have visual experience (in blindsight), correctly answer questions when primed about perceptual inputs while denying that they ever experienced them, and so on. Nevertheless, the fact that feelings are indicators (and constituents) of departures from and returns to homeostasis and that coping with new elaborate challenges always involves consciousness suggests that consciousness is not an epiphenomenon. What, then, are its benefits? What does it contribute to?

In the late nineteenth century, the benefits of consciousness were regarded as identical to those of intelligence. George John Romanes, who studied the evolution of animal intelligence, and James Mark Baldwin, who investigated child development, suggested that the evolutionary function of consciousness was to enable animals to learn flexibly; to go beyond what reflexes can provide.[64] The focus of most philosophers and cognitive scientists today, in the early twenty-first century, is similar, although the neural and cognitive mechanisms that enable flexible, highly sophisticated, goal-directed behavior currently occupy center stage. Anil Seth has reviewed various ideas about the functions of consciousness and has shown that there is a common thread in the thoughts of both philosophers of mind and cognitivists today: the function of consciousness is to integrate information; resolve ambiguities; and enable novel, flexible, and goal-directed behavior. Seth calls this broad, shared agreement "the integration consensus."[65] However, the list of possible functions of consciousness includes additional suggestions (mainly elaborations), such as error correction, multiple object tracking, the formation of a transparent and stable world and body image, conflict resolution and coordination among subsystems, decision-making and voluntary action, skill acquisition, and learning (table 4.2). We come back to these functions in the next chapter, where we look for criteria that may mark a conscious state and help researchers to decide whether a particular animal—a snail, a bee, a tuna—is or is not conscious.

All the capacities listed in table 4.2 are undoubtedly important, and all are related to the adaptive advantages of flexible learning, but we think the question about the function of consciousness is misleading. Before we explain why, we want to present the integration consensus about the functions of consciousness from an evolutionary point of view. As Baars stresses, at least in mammals, all goal-directed activities occur during the waking, alert, and conscious state:

> Mammalian locomotion, hunting, evasive action, exploring, sensing, actively attending, learning, eating, grazing, nursing, mating, social interaction, and all other goal-directed survival and reproductive actions take place only during waking, as defined by EEG and other indices. Perceptual consciousness, as defined objectively by recent brain research, only takes place during waking periods. *It therefore appears that brain activity that supports consciousness is a precondition for all goal-directed survival and reproductive behavior in humans and other mammals. The biologically fundamental nature of the conscious waking state is beyond serious question. ... Waking consciousness involves a basic biological adaptation with many survival functions.* (Baars 2005b, p. 18; emphasis added)

According to Baars, this waking, attentive consciousness is an obvious evolutionary adaptation, allowing animals to flexibly adjust to their world through their ability to access and "report" (not necessarily through words) on their mental states. The emphasis on focused, reportable attention brings us back to a problem we encountered in the previous chapter. Are *all* consciousness states accessible to the animal in the same way? Can nonreportable and nonsemantically recalled events that occur in the wakeful state be said to have actually been experienced, as Koch, Lamme, and Block argue,[66] or are global accessibility and behavioral or verbal reportability necessary defining attributes of consciousness, as Dennett, Baars, and Dehaene claim? In philosophers' jargon, is it the case that *phenomenal consciousness* is richer than *access consciousness*? Are there some aspects of conscious states that are ineffable? As we see it, there are in fact two questions here. One is whether animals (and humans) have an overall feeling of existence that is extremely difficult to describe because it is both unified and rapidly changing, something like Lamarck's "inner feeling." We think

Table 4.2

Functions of nonreflective subjective experiencing (phenomenal consciousness) suggested by some philosophers and cognitive scientists.

Philosopher/cognitive scientists	Suggested functions for basic consciousness
James 1890	Controlling and allocating resources in an unstable complex brain by "loading the dice" and acting as a selector of physiological and behavioral reactions; enables decision-making (see chapter 2)
P. M. Churchland 1985, 1988, 2002; P. S. Churchland 2002	Coordinating all known cognitive and affective capacities
Dennett 1991, 1997, 2005	Enabling access to information; enabling cognitive functions such as discrimination, reporting, and more
Dretske 1988	Flexible, goal-directed learning and behavior
Searle 2004, 2013	Enabling all cognitive/affective capacities observed in animals
Baars 2005a	Allowing access to information that enables flexible, goal-directed behavior through facilitation of information exchange among multiple, special-purpose unconscious brain processes
Merker 2005, 2007	Enabling goal-directed target and action selection that allows flexible behavior
Morsella 2005; Shadlen and Kiani 2013; Tse 2013	Enabling decision-making (the decision to engage); attentional multiple-object tracking; conflict resolution among subsystems, allowing motor coordination and the production of skeletomotor action
Lamme 2006	Enabling skill acquisition and learning
Hofstadter 2007	Self-monitoring
Dickinson 2008	Making unconscious learning available to goal-directed behavior
Metzinger 2003, 2009	Transparently representing the world and body, thus enabling access to them and allowing multiple cognitive and affective functions
Cabanac et al. 2009	Creating world and self that lead to the motivation for living; enabling the formation of affective rules of thumb
Humphrey 2011	Creating world and self that lead to the motivation for living
Panksepp 2011a	Enabling the formation of emotion-specific, systemic reward and punishment systems
Dehaene 2014	Enabling inferences via scale tipping in a complex, changing world, thus resolving ambiguities; learning over time; planning; and access to working memory for the execution of a train of operations

Note: The functions enumerated in the table do not include cognitive and affective functions related to social life and to rational-symbolic thought because we wanted to highlight the primary, nonreflective functions of minimal consciousness.

the answer to this first question is positive and that (some) animals and humans have a general sense of being that is normally experienced as a necessary background condition, like the background of a picture, and that this sense of being can be described in terms of arousal and valence. Such a fuzzy minimally conscious state has the function of enabling an animal to distinguish between the complex effects of its own actions and the effects of the external world on its sensors. The second question concerns levels of accessibility and reportability of specific mental contents. What we know of neural dynamics suggests that the borderline between conscious and unconscious states is very blurred, and correspondingly, there are degrees of accessibility and reportability: some experiences can be clearly reported, others only partially; some experiences can be recalled though semantic memory, others only through acting out in real time, online;[67] some representations are suppressed by others at different stages of production while others are widely accessible (in the sense of connecting with many brain regions) but do not involve awareness, and so on. Understanding the evolutionary history of these mental states may give us a clue to both their causal roles within the system, their teleofunctions, and their mental "status." However, we agree with Baars that the hallmark functions that accompany consciousness, many of which are listed in table 4.2, are manifest by alert, active, and attentive animals that need to access their experiences in order to act in a flexible and goal-directed manner.

What, then, can we make of Chalmers's and Harnad's argument that since observed functional actions, like playing chess ("doings," in Harnad's terminology), can be performed by nonconscious robots, complex behavior cannot inform us about consciousness? Are we missing some special function unique to consciousness? Or is there something misleading about the function question and about the Chalmers-Harnad argument?

As is clear from our earlier discussion of T-robots, we believe (though we are by no means certain) that it may be possible to build a feeling robot, although, as we argue in chapter 10, there may be rather few ways of doing so. Even if this is not possible, the incredible cognitive and behavioral abilities of a T-robot can teach us about the affordances of naturally evolved, biological, sentient animals. Moreover, as John Searle argues, posing the question of the function of consciousness by invoking zombies (or T-robots) is mere polemics, even if we focus on just one of the many functions associated with consciousness. Searle points out that we can imagine plants producing their nourishment by ways other than photosynthesis, but this does not prove that photosynthesis has no function in evolution. In the real world, says Searle, humans, and presumably other animals, do need consciousness in order to survive.[68] In other words, although a particular function may be realized by a robot that has no consciousness, in an evolved animal, it is through consciousness that this function is realized. For example, the fact that a simple nonconscious robot can discriminate between shapes and colors does not mean that discriminating through phenomenal experiencing in an evolved animal or in a future T-robot is thereby excluded. It seems that consciousness in biological organisms entails processes that enable the development of capacities such as discrimination, motivation, attending, and flexible learning—capacities that jointly constitute consciousness. In fact, as table 4.2 indicates, consciousness seems to be conducive to multiple cognitive functions in evolved animals. Why?

In our view, the idea that consciousness is a functional trait like an internal skeleton or an immune system is wrong. Although brains have changed during evolution in ways that have led to the emergence of consciousness (just as altered morphological organization has led to the development of an internal skeleton), consciousness is not a functional trait. It is a framework for the development of cognitive and affective functional traits. We argued in chapter 1 that a change in the organization of complex chemical reactions and structures generated the first living entities—the first intrinsically teleological systems. Consciousness, we suggest, is more like life than like a skeleton or an immune system. It is neither a property nor a process—it is a new teleological, intrinsic mode of being, its teloi being the ascription of values to encountered objects or states through ontogenetically constructed desires that the animal strives to fulfill. The processes that implement desires and underlie flexible learning have functions. Just as it would be a category mistake to ask what the functions of being alive are, so it is a category mistake to ask what the functions of being conscious include.[69] However, it does make a lot of sense to ask about the intrinsic goal of living or the intrinsic goal of consciousness, for goals can be attributed to whole systems, while biological functions are defined as the parts and processes that contribute to the operation of a goal-directed, encompassing system. The emergence of conscious beings led to the generation of a new functional realm, altering the very notion of living. Consciousness, therefore, introduces an additional explanatory framework, adding one more "cause" to the four causes that Nikolaas Tinbergen listed as necessary for biological explanations.[70] In addition to his phylogenetic (evolutionary), functional, developmental, and immediate (mechanistic) causes, which apply to all living organisms, including plants and bacteria, in (some) metazoans with a CNS there is a fifth cause—intrinsic motivating subjective experiencing—that is special to them and sets them apart. The addition of a new explanatory framework—function in the case of living things, motivation in the case of conscious animals—is the hallmark of these teleological transitions. With the transition to the rational soul, yet another teleological system and another "cause" emerged, with values and truths becoming function-generating teloi, drivers, and motivators of human action.

We interpret Nicholas Humphrey's proposal that consciousness has created the joy of living and the motivation to live as a way of expressing the same basic idea. Living became a goal in a new way, a highly individualized way, a conscious way, and this can explain the evolution of multiple cognitive and affective functions. As Humphrey puts it:

> If natural selection can arrange that you enjoy the feeling of existing, then existence can and does become a goal: something—indeed as we'll see some *thing*—you *want*. And the difference between you *wanting* to exist and simply having some kind of instinct is that, when you *want* something, you will tend to engage in rational actions—flexible, intelligent behaviour—to achieve it. (Humphrey 2011, p. 85)

Wanting to do things that promote existence and positively enjoying existence are not identical, and it may be that before general joie de vivre emerged in animals, more humble and distinct wantings paved the way. However, Humphrey's suggestion, which implies that the general joy of being was the first positive experiential value, is certainly plausible. It fits with Lamarck's suggestion that the first experience is the feeling of existence, the inner feeling of living, which if experienced as positive, would drive behaviors that sustain it and lead to the evolution of what we call more specific "wanting." But who among the animals actually "wants"? Who can be said to have felt needs, which they strive to satisfy? And in what selective context did such "wanting" evolve?

5 The Distribution Question: Which Animals Are Conscious?

Which animals are conscious? We are certain that adult humans are, but what can we say about newborn babies, developmentally disabled children, chimpanzees, or gorillas? Can dogs, cats, and rats feel hunger and pain, fear and joy, or are they soft automata with sophisticated action programs but no subjective experiencing? Is human symbolic cognition necessary for consciousness, as some philosophers and psychologists suggest, or is the anatomical and physiological similarity between nonhuman mammals and humans a good enough reason for suggesting that other mammals are conscious? What about clever birds like gray parrots, with their differently evolved brains? What about the ingenious octopuses? Is a cerebral cortex necessary for consciousness? Do fish feel pain? We present some of the major arguments and criteria for recognizing the presence of consciousness: the humans-only argument, which grants consciousness only to beings with reflective self-consciousness and the neocortical structures that subserve it; the argument from analogy, which by comparing the behavior, anatomy, and physiology of the (conscious) psychologist to the analogous behavior, anatomy, and physiology of nonhuman animals grants consciousness to mammals, birds, and maybe even reptiles; arguments suggesting that affects (feelings) construct primary consciousness; and the claim that consciousness should be described in terms of a dynamic dependency between target selection, action selection, and motivation, which renders all vertebrates and maybe some invertebrates conscious and explicitly dispenses with the need for a cerebral cortex. Our own suggestion is that unlimited associative learning (UAL) marks the evolutionary transition to subjectively experiencing animals. After discussing the notion of learning, we characterize UAL as open-ended learning that enables an organism to ascribe motivational value to a compound stimulus or action and use it as the basis for future learning. We show that UAL entails the consensus list of consciousness criteria and is consistent with the consciousness models described in chapter 3. It therefore qualifies as a transition marker of minimal consciousness.

> Somewhere between medusa and human there is a transition to conscious function, and the nature of the capacity it bestows has exercised psychology, neuroscience, and cognitive studies virtually since their inceptions.
> —Merker 2007, p. 63

On July 7, 2012, an international group of prominent cognitive neuroscientists, neuropharmacologists, neurophysiologists, neuroanatomists, and computational neuroscientists, conferring at the University of Cambridge, came up with a *Declaration on Consciousness*, which ends with the following paragraph:

> The absence of a neocortex does not appear to preclude an organism from experiencing affective states. Convergent evidence indicates that non-human animals have the neuroanatomical, neurochemical, and neurophysiological substrates of conscious states along with the capacity to exhibit intentional behaviors. Consequently, the weight of evidence indicates that humans are not unique in possessing the neurological substrates that generate consciousness. Nonhuman animals, including all mammals and birds, and many other creatures, including octopuses, also possess these neurological substrates.[1]

What motivated this strange public declaration, unprecedented in the annals of neuroscience? Why was it deemed necessary? Which animals belong to the "other creatures" alluded to in the declaration? How broad a consensus does it represent? The declaration certainly points to a shift away from the anthropocentric perspective about the distribution of consciousness that had dominated scientific and philosophical thinking, particularly since the 1930s. Nevertheless, there is still no agreement about which animals are endowed with consciousness, and the question, known in the consciousness literature as the "distribution question" or the "who problem," remains the subject of considerable debate.[2] There are those (in the minority now) who still believe that only speaking humans are conscious; others allow chimpanzees and a few other select mammals to join the list; a large (and growing) number grant consciousness to mammals, birds, and some cephalopods; some argue that reptiles, but not amphibians or fish, are conscious; there are those who maintain that all vertebrates are conscious; and a small number of people (including ourselves) propose that consciousness can be attributed not only to all vertebrates but also to some invertebrates in addition to cephalopods.

In spite of the diversity of opinions, there is common ground among scholars: they all agree that having a nervous system was a precondition for the transition to consciousness. Even those who see seamless continuity between living and consciousness recognize that during evolution something qualitatively new occurred that underlies the ability of some neural animals to experience pains, pleasures, and percepts and to have a mode of being very different from that of tomatoes and bacteria—a mode of being that makes a distinction between conscious and unconscious-but-living states intelligible.

To answer the distribution question, we need to identify the criteria necessary in order to attribute consciousness to an animal. These criteria must be compatible with the neurobiologists' and philosophers' lists discussed in the last two chapters. The distribution question is an evolutionary question, but it addresses the pattern of evolution—how a trait is distributed among taxa—rather than the mechanisms involved and the ecological context of selection. One need not (in theory) know much about how a novel system originated or what drove its evolution in order to describe its phylogenetic distribution. However, as we shall see in subsequent chapters, the mechanisms and contexts of selection for minimal consciousness offer crucial insights into the "who problem."

The Argument from Analogy

> The activities of organisms other than our own, when analogous to those activities of our own which we know to be accompanied by certain mental states, are in them accompanied by analogous mental states.
>
> —G. J. Romanes 1883, p. 23

These are the words of George John Romanes (figure 5.1), to whom Darwin passed his mantle for the study of mental evolution. The two men met when Romanes was twenty-six and Darwin sixty-five, and they developed a warm and firm friendship "marked on one side by absolute worship, reverence, and affection, on the other by an almost fatherly kindness and a wonderful interest in the younger man's work and in his career."[3]

Darwin gave Romanes all his unpublished notes on animal behavior—those that were omitted for reasons of space from *The Origin of Species* and those he collected later—with the hope that Romanes would develop this still-neglected aspect of his theory. G. J. Romanes (1882, 1882, 1888) did just that. In a series of three books, *Animal Intelligence*, *Mental Evolution in Animals*, and *Mental Evolution in Man*, he addressed the question of mental evolution from a Darwinian-Spencerian point of view.[4]

Like Darwin and Spencer, Romanes was committed to the idea that mental evolution was a gradual process and that its sophistication involved slow changes from the simple to the complex through the process of natural selection. He was aware, of course, of the diverging and irregular patterns of phylogeny and noted, for example, how very dissimilar in their mental capacities were members of different rodent species. However, because, like Spencer, he was interested mainly in the evolutionary trend that led to higher mental capacities, he was not very concerned with the specific

Figure 5.1
George John Romanes (1848–1894).

ecological aspects that can explain idiosyncratic adaptations and nonlinearities in different lineages. Romanes's focus on the progressive trend of mental evolution, together with his adherence to Darwin's conviction that the inheritance of acquired characters played a significant role, were the sources of severe, and not always fair, criticism in later decades.[5]

In the first and best known of his three books (*Animal Intelligence*), which he described as a compendium of facts relating to animal (nonhuman) intelligence, Romanes marshaled much anecdotal evidence about the mental faculties of animals, based on what he regarded as reliable sources. (Darwin was deemed unfailingly reliable.) Most of the observations were about mammals, and among the invertebrates, bees and ants received the greatest attention. The second book, *Mental Evolution in Animals*, provided the theoretical and evolutionary analysis of the data assembled in the first one, and half of it was concerned with the evolution of instincts, a subject of central importance for Darwin; the appendix to the book was a chapter on instincts written by Darwin that had not been included in *Origin*. In both the first and the second books, Romanes was explicit about his criterion of consciousness (the term he usually used was "mind"):

> The criterion of mind, therefore, which I propose ... is as follows:—Does the organism learn to make new adjustments, or to modify old ones, in accordance with the results of its own individual experience? (G. J. Romanes 1883, pp. 20–21)

The criterion, then, was the capacity to learn new things, and this capacity was, for him, identical to the ability of an animal to make a choice: "Agents that are able to *choose* their actions are agents that are able to *feel* the stimuli which determine the choice."[6] This experience and learning-based choice, Romanes explained, is in fact two pronged: it involves both choosing the target and discriminating and choosing among different action options regarding this target.[7] For Romanes, such learning-based choices entail consciousness.

Romanes was well aware of the imperfect nature of his definition of consciousness. He knew that failures of memory may render a conscious animal unable to learn, so some subjective states may not be revealed using this criterion. He was also aware of unconscious learning based on some form of training of the reflexes that does not require a central nervous system ("ganglionic learning").[8] However, he argued that flexible learning—the kind of learning we see in alert animals—is accompanied by Feeling (in the Spencerian sense, with a capital *F*). But how can we know which other creatures subjectively Feel? How do we know that a reflex, such as a dog flexing a leg in response to a sudden blow, is not conscious, whereas its begging for food is? Romanes believed, as the quotation in the beginning of this section shows, that our own subjective Feelings, which accompany our own learning from experience, can be projected onto other animals that act like us and are anatomically similar to us; assuming they do have such feelings can provide the best explanation of the behaviors we observe, as well as predict other behaviors. For simplicity we call this position "the argument from analogy." The signatories and supporters of the *Cambridge Declaration* adhere to it, although in addition to the inferences from other species' behavior and neuroanatomy, they make inferences based on comparative neurophysiology, neuropharmacology, and computational neurobiology.

Before we turn to twenty-first-century opinions about the distribution question, we need to consider the views of Conwy Lloyd Morgan, who followed in Romanes's footsteps but diverged from his views in several important respects. For example, he did not accept the Lamarckian views of Darwin and Romanes, and consequently, his interpretation of the evolution of instincts differed from theirs. And although Morgan accepted Romanes's reasoning by analogy, which he called the "doubly inductive process," he suggested a more rigorous methodology when assessing the cognitive capacities of animals. Morgan's method was first to make a detailed subjective introspection of one's own experiences (e.g., the feeling of fear) and correlate it with detailed objective observations of the behaviors and physiological responses that accompanied this fear; this leads to the induction that the physiological and behavioral manifestations are causally related to the subjectively introspected feeling. Second, the observed human behaviors and physiological responses that were correlated with subjective feelings are compared to the behaviors and physiological responses of nonhuman animals. If these are found to be similar to those of humans, we may deduce that animals have corresponding mental states.[9]

Morgan was confident that all higher vertebrates (mammals and birds) had feelings, and he granted them all basic emotions and consciousness. He believed that consciousness was an emergent property of the activity of the brain but was very skeptical about the ability of any nonhuman animal to perceive abstract relations and to reason, issues about which both Darwin and Romanes were extremely liberal. He argued that even seemingly sophisticated behaviors, such as the ability of his terrier to open the latch of a gate, a behavior assumed to support the idea that dogs display abstract causal reasoning, are the result of a lower-level process of learning by consequences. He claimed that such learning, which can involve either the formation of a simple association between two external events, one of which is innately rewarded or punished, or the formation of associations between the animal's actions and the reinforcing external events that follow them, can explain all learned behaviors. And since a lower-level form of learning can be an adequate explanation, Morgan claimed that it should be the preferred explanation. This methodological stance was expressed in what is known as Morgan's canon:

> In no case is an animal activity to be interpreted in terms of higher psychological processes, if it can be fairly interpreted in terms of processes which stand lower in the scale of psychological evolution and development. To this, however, it should be added, lest the range of the principle be misunderstood, that the canon by no means excludes the interpretation of a particular activity in terms of the higher processes, if we already have independent evidence of the occurrence of these higher processes in the animal under observation. (Morgan 1903, p. 59)

The canon, including the qualifying second sentence that was added to the 1903 second edition of his book on comparative psychology, provided a good general methodological guide for research. Neglecting the second sentence, as often happened among zealous twentieth-century experimental psychologists, led to misconceptions about animals' cognitive abilities.

In addition to the famous canon just quoted, Morgan suggested a second, much less well-known one. He proposed that the ability to modify instincts through learning is the criterion of consciousness:

> Now, we may safely lay down this canon: *That which is outside experience can afford no data for the conscious guidance of future behaviour.* When we say that conduct is modified in the light of experience, we mean that the consciousness of what happened, say yesterday, helps us to avoid similar consequences to-day. (Morgan 1896, p. 131)

Positively worded, the ability to learn flexibly from experience was, for Morgan, a necessary condition for consciousness. It is, however, not clear when learning is flexible *enough*. Morgan was agnostic about the status of consciousness in animals other than birds and mammals.

The "Who Problem": Twenty-First-Century Views

The argument from analogy, coupled with the first of Morgan's canons, is an accepted heuristic among consciousness investigators, and it is used by most twenty-first-century researchers. For example, in a 2008 book titled *Consciousness Transitions*, an edited volume of articles about the evolution of consciousness, several papers addressing the distribution question all employ different versions of the argument from analogy.[10] Reasoning by analogy is also clearly spelled out in the *Cambridge Declaration* (somewhat abbreviated here):

- Studies of non-human animals have shown that homologous brain circuits correlated with conscious experience and perception can be selectively facilitated and disrupted to assess whether they are in fact necessary to those experiences. Moreover, in humans, new non-invasive techniques are readily available to survey the correlates of consciousness.

- The neural substrates of emotions do not appear to be confined to cortical structures. In fact, subcortical neural networks aroused during affective states in humans are also critically important for generating emotional behaviors in animals....Furthermore, neural circuits supporting behavioral/electrophysiological states of attentiveness, sleep and decision making appear to have arisen in evolution as early as the invertebrate radiation, being evident in insects and cephalopod mollusks (e.g., octopus).

- Birds appear to offer, in their behavior, neurophysiology, and neuroanatomy a striking case of parallel evolution of consciousness. Evidence of near human-like levels of consciousness has been most dramatically observed in African grey parrots. Mammalian and avian emotional networks and cognitive microcircuitries appear to be far more homologous than previously thought.

- In humans, the effect of certain hallucinogens appears to be associated with a disruption in cortical feedforward and feedback processing. Pharmacological interventions in non-human animals with compounds known to affect conscious behavior in humans can lead to similar perturbations in behavior in non-human animals....Evidence that human and nonhuman animal emotional feelings arise from homologous subcortical brain networks provides compelling evidence for evolutionarily shared primal affective qualia.[11]

Clearly, we now have a lot more comparative data than Romanes and Morgan had, so we have a better basis for granting consciousness to some nonhuman animals (by reasoning from analogy). However, the *Cambridge Declaration* is somewhat vague about which animals other than mammals and birds have consciousness. Moreover, although the argument from analogy makes sense, it is less convincing as one moves from mammals and birds to lower vertebrates and even less so with invertebrates such as mollusks or crustaceans. In addition, the argument from analogy can be challenged on theoretical grounds. One type of challenge takes the form of arguing that subjective experiencing can be attributed only to animals capable of reflective self-consciousness—mainly linguistically able humans.

Is Reflective Self-Consciousness a Necessary Condition for Subjective Experiencing?

> There is nothing which leads feeble minds more readily astray from the straight path of virtue than to imagine that the soul of animals is of the same nature as our own.
>
> —Descartes 1637/1976, p. 76

> They [animals] eat without pleasure, cry without pain, grow without knowing it; they desire nothing, fear nothing, know nothing.
>
> —Nicolas Malebranche, quoted in Jolley 2000, p. 42

Some philosophers and a few biologists firmly believe that in order to be sentient—conscious in the most elementary sense of feeling pain or discomfort—animals need more than to perceive colors, sounds, touches, tastes, and smells; that is, they need more than what some philosophers call *first-order representations*. They must also have representations of these first-order mental representations, that is, *second-order mental representations*. This idea is often attributed to James, who argued that the peripheral reactions sent to the brain must be integrated and "read-out," and he mentioned the cortex as the locus of interpretation. However, James had no doubt that animals such as birds and mammals have subjective experiences. As we described in chapter 2, the consciousness of nonhuman animals was not of great interest to him, but his references to the feelings and consciousness of his dog, his discussion of the relation between emotions and instincts (in the egg-laying hen, for example), and many other such allusions leave little doubt that he thought that animals other than humans have subjective feelings.

Those who believe that for consciousness there must be a high-level, neocortical read-out of sensations stress that this read-out is not a mere second- or third-order neural mapping but constitutes a *mental representation*. Mental representations are not the same as the high-level neural representations (or high-level mapping); second- and third-order neural representations may be necessary for complex sensory integrations and motor activities and are a necessary (but not sufficient) condition for consciousness, but they can exist in nonconscious animals. Mental representations, on the other hand, require reflective (e.g., linguistic) consciousness and explicit metacognition.

The animal needs to know that it knows, it has to be aware of its perceptions (e.g., pain), and it has to have thought about thoughts. It is therefore not enough to point to a neural difference between an alert and attentive cat and one presented with a subliminal stimulus, nor does it suffice to show that there is distinct and strong activity at several levels in the brain of the alert cat. Even when multilevel neural mapping occurs, even when the brain is massively active and the animal is alert and brimming with what seems to the naïve eye to be motivation, these scholars maintain that it may not be conscious. Only high-order thought (HOT), they argue, can render neural states mental, distinguishing between conscious and unconscious states such as those occurring in a sighted human and a human with blindsight.[12] Typically, those who uphold this view believe that humans and perhaps a few other animals with impressive neocortices (maybe some great apes) have consciousness-rendering HOTs. The philosopher Peter Carruthers, for example, believes that few animals other than humans are sentient, although the chimpanzee and bonobo are possible exceptions (figure 5.2).[13]

The second-order representationalists' conviction that most animals and very young human babies cannot have subjective feelings of pain or pleasure is based on their belief that without metacognition there is no self, and without self, there is no sentience (e.g., feeling). Edmund Rolls, an experimental psychologist who has carried out important work on olfaction and taste by probing the brains of monkeys and rodents, subscribes to this view.[14] His argument is that symbolic representation (mainly, but not necessarily, through language) provides the conscious awareness that one's emotions are *about* something and have a particular content. As he puts it:

> Consciousness may be the state which arises in a system that can think about (or reflect on) its own (or other people's) thoughts, that is in a system capable of second or higher order thoughts. (Rolls 2013, p. 246)

In other words, a mental state is conscious only if one also has at the same time the thought that one is in that mental state. Such symbolic consciousness is, Rolls suggests, syntactic and has the function of correcting errors in multistep plans or long trains of reasoning. This is supposed to explain its subjective aspect. Although Rolls does mention the affective states of nonhuman animals, he consistently claims that without thoughts *about* these affective states, they are nonconscious—these animals are not sentient, feeling beings. Without HOTs, what we call affect is a mere triggering of the reward-punishment system. There are no feelings accompanying or following this triggering.

Figure 5.2
Two conscious beings: *Left*, a nonspeaking bonobo, whose consciousness is questioned by some scholars although it is able to communicate, when encultured, through human-made symbols; *right*, the linguistic *Homo sapiens*, who lives in a full-blown symbolic world.

Euan Macphail, a psychologist who works within the associationist tradition and studies learning, arrived at a similar conclusion but provides somewhat clearer arguments. He explicitly rejects the argument from analogy and, on the basis of psychological data and HOT intuitions, maintains that only linguistic humans are conscious.[15] Macphail starts by pointing out that there are many disanalogies between the behavior of humans (who we know are conscious) and other animals. For example, nonmammals have a very different range of responses to tissue damage when compared to humans and other mammals (e.g., fish do not vocalize when physically damaged). Of course, as Macphail is well aware, this is not a good enough reason to deny them pain: lack of similarity of response is neither necessary nor sufficient for not attributing feelings to other animals. Nor, he argues, is similarity a good reason for the attribution of mentality. It is just a reminder that behavioral similarity may not be a good indicator of mental communality, a point that scholars adhering to the argument from analogy have acknowledged and tried to handle by including multiple similarities from as many domains as possible.

A second of Macphail's arguments is that withdrawal responses to noxious stimuli, which we interpret as pain behavior, occur in very simple invertebrates—for example, in worms, which Macphail assumes are not feeling (an assumption that he does not justify). A more serious claim is that such withdrawal responses can occur in spinal animals (animals whose spinal cord has been severed from their brain), including spinal humans, who do not report pain but do withdraw. Moreover, some simple learning can occur in spinal individuals. For example, spinal humans can learn to anticipate a strong shock after being presented with a weak shock, even though they feel nothing. This means that nonconscious learned responses that are behaviorally similar to those of conscious humans may involve no feeling whatsoever. However, as Macphail acknowledges, the responses in spinal animals are simple reflex responses, and the spinal learning that is based on these reflexes is also exceedingly simple.[16] It is also worth mentioning that such limited learning (Romanes's "ganglionic learning") has never been advanced as a criterion for consciousness. Ever since Romanes suggested that learning from experience is a good criterion for mind, only open-ended learning by alert animals, which can learn multiple contingent and often novel associations, has been put forward as a criterion for consciousness.

Macphail's third argument is that animals such as the sea slug *Aplysia*, which has a very simple nervous system, are able to learn a great deal, and invertebrates such as bees and worms, which, Macphail assumes, do not feel, provide many examples of complex preferences and complex learning. This argument is based on an a priori assumption that these animals are unfeeling, something that needs to be supported or refuted rather than dogmatically stated.

The fourth of Macphail's arguments holds that since the mechanisms of associative learning are the same in all vertebrates, complex learned behavior does not point to the presence of experiencing/feeling. Macphail seems to have based this conclusion on his conviction that every nonhuman behavior, however complex, can be explained in terms of associative learning, coupled with the assumption that flexible associative learning does not entail feelings because goldfish, as well as chimpanzees, are capable of it. However, the supposition that the associative learning displayed by an alert goldfish and a vigilant chimpanzee is based on unconscious rather than conscious responses requires justification rather than a strong anthropocentric intuition, even if the intuition seems to be shared by others, especially those associated with the recreational fishing industry. In a 2002 review, James Rose argued that fish do not feel pain, claiming that the behavioral evidence is inconclusive and that fish and most other nonhuman vertebrates (some primates seem to be the only exception) lack the neocortical brain areas that lead to the experience of pain in humans. However, the evidence in this influential review has been disputed by the eminent pain researcher Lynn Sneddon, and the arguments concerning the attribution of pain in fish are ongoing, with current studies suggesting that Rose's strong skepticism is unwarranted.[17] Studies of arthropods suggest that if the behavioral and physiological criteria that are deemed adequate for attributing pain to vertebrates are applied to invertebrates (e.g., crabs), these animals, too, probably feel. However, the fact that bees, for example, may not feel pain does not mean that they do not have other affective and perceptual experiences—humans with congenital analgesia are certainly conscious and subjectively experience affective states such as pleasure.[18]

In addition to these four arguments, Macphail also endorses the view that feeling adds nothing to behaving (to "doing"), and therefore feeling has no function. We discussed this view in the previous chapter, where we pointed out that it is based on a confused notion of consciousness and its relation to function.

On the basis of all these arguments, Macphail concludes that there is no good reason to assume that any animal other than humans actually feels anything, so we should concentrate on what it is that sets humans apart.[19] He believes this is language: only linguistically able humans feel. Macphail does not flinch from the implication that, in the period before they can understand language, human babies also do not feel and are therefore not sentient. In fact, he advances an additional proposition, which he believes supports this conclusion: he suggests that since children do not have an episodic memory of things that happened to them before they learned to speak (they manifest "childhood amnesia"), they do not have conscious experiences in the prelinguistic period. A baby may show fear when seeing a doctor who once gave her a painful injection, but when she is older and able to speak and is asked about it, she will not know why she feels the fear because she cannot recall the frightening episode with the doctor.[20] Lacking episodic recall of events that occurred at the prelinguistic phase means, for Macphail, that the sensed events were unconscious. In other words, a here-and-now experience cannot be conscious if it is not accompanied by later (episodic) conscious recall. But why not? Why is conscious recall required

for conscious sensation? How does Macphail justify this demanding criterion for basic consciousness?

Macphail's reasoning is that language allows a relation of "aboutness"
and that a "self" is formed through voluntary access to long-term and short-
term autobiographical memories, which requires linguistic competence.
Since, in his opinion, having a self requires the ability to recall and construct an autobiographical narrative, and this is something nonhuman animals and babies do not have, they are not sentient beings. He writes:

> The position taken here is, then, that the concept of self is peculiar to humans,
> and that it develops alongside the acquisition of language, both achievements
> being dependent upon the capacity of comprehending a predicate as being about
> a subject. (Macphail 2008, p. 117)

> Organisms that have not developed a self cannot, then, be sentient beings; and
> if animals and infants have no self, then they do not feel anything. They do, of
> course, learn, using the implicit associative system. (Macphail 2008, p. 118)

So it all hinges on the idea that self is needed for feeling, and that self
depends on second-order cognition, on HOT. But as we saw in the last
two chapters, there are far simpler notions of self, and there are, possibly, levels of selfhood. The aboutness that Macphail, Carruthers, Rolls, and
others espouse need not be symbolic-linguistic aboutness. Their particular
notion of intentionality is derived from a philosophical discourse based
on *linguistic intentionality*, and this narrow and specialized philosophical
notion, which was very popular among twentieth-century analytical philosophers, is applied across the board. True, we need to understand the difference between conscious and subconscious states, and this requires far more
understanding of neural dynamics than we currently have, but the present
data from studies of sleep, coma, minimal conscious states, anesthesia, subliminal states, and various pharmacological interventions in both humans
and other animals strongly suggest that there are great parallels between,
for example, conscious humans and conscious dogs and between unconscious or subconscious humans and un- or subconscious dogs. When an
animal becomes reflectively self-conscious and has the ability to represent
and communicate through symbolic language, the self becomes more complex, and consciousness is enriched in a dramatic manner, but it is not
thereby created. It seems that scholars committed to the HOT view of consciousness commit what James called "the psychologist's fallacy":

> Many philosophers, however, hold that the reflective consciousness of the self
> is essential to the cognitive function of thought. They hold that a thought, in
> order to know a thing at all, must expressly distinguish between the thing and its

own self. This is a perfectly wanton assumption, and not the faintest shadow of reason exists for supposing it true. As well might I contend that I cannot dream without dreaming that I dream, swear without swearing that I swear, deny without denying that I deny, as maintain that I cannot know without knowing that I know. ... It is a case of the "psychologist's fallacy" (see p. 197). *They* know the object to be one thing and the thought another; and they forthwith foist their own knowledge into that of the thought of which they pretend to give a true account. To conclude, then, *thought may, but need not, in knowing, discriminate between its object and itself.* (James 1890, pp. 274–275)

Little can be added to James's eloquent words.

Arguments from Homology: Mammals, Birds, and Reptiles

Dogs are (simple) People, and People are (complex) Dogs.
—Ivan Pavlov, quoted in Todes 2014, p. 406

Although the idea that subjective experiencing is confined to humans is not rare, the consensus among most students or researchers of consciousness, which is reflected in the *Cambridge Declaration*, is that some nonhuman animals are phenomenally conscious in the basic sense of having subjective experiences such as pain, hunger, fear, and lust, as well as experiences such as perceiving the image of a predator or the smell of a mate. This consensus is based on the neuroanatomical, neurophysiological, neuropharmacological, and behavioral homologies between humans and other mammals and between mammals and birds. The homologies between human and other mammals' brains are extensive: humans share in all mammalian brain structures. We therefore start with this most widely accepted view—namely, that all mammals are conscious beings, with perceptions and feelings.

Following the logic of analogical reasoning and using humans as the reference species, Anil Seth, Bernard Baars, and David Edelman compiled a list of seventeen neural and behavioral attributes that characterize the consciousness of humans.[21] As shown in table 5.1, these can be used to evaluate whether or not other mammals are conscious.

The list in the table reflects the strong influence of Gerald Edelman's view of consciousness. However, it provides only a very general guide. First, it is not clear that the list covers the most important aspects of consciousness: there is no explicit reference to emotions, which some people (to whose ideas we shall come shortly) regard as crucial. Second, it is unclear how many of the listed features, singly or combined, are *necessary* for mammalian consciousness, which are sufficient, which are facilitating, and which are human-specific. Feature 12, subjectivity, is sufficient—and indeed is

Table 5.1
Recognized properties of consciousness in humans (based on Seth, Baars, and Edelman 2005) and their occurrence in nonhuman mammals.

Features of human consciousness	Found in nonhuman mammals?
1. Typical consciousness-related EEG signature, which is irregular, low amplitude, and fast, ranging from 12 to 70 Hz	Yes
2. Extensive back-and-forth interactions between the cortex and thalamus	Yes
3. Widespread brain activation	Yes
4. Very large range of different content (many different images, emotions, and so on)	Yes (if the ability to switch tasks and learn multiple relations is taken as indicative)
5. Consciousness may fade if signals become repetitive	Yes (as measured by habituation and alteration in brain activity with habit)
6. Rapid and fleeting nature of conscious scenes	Probably yes, judging from behavioral data obtained during tasks such as hunting and EEG data during sensory experiencing
7. Consciousness is marked by a consistency constraint, which means that only one content among the many presented becomes conscious at any moment	Yes (as judged, for example, by binocular rivalry experiments in monkeys)
8. Limited capacity and seriality, with the capacity for consciousness at any given moment limited to one consistent scene	Yes (as judged, for example, by tasks requiring attentional modulation)
9. Sensory binding to form a unified whole	Yes (based on neurophysiological criteria)
10. Self-attribution, with conscious experiences always attributed to an experiencing self	Debatable; self-recognition in a mirror by some great apes may be a case of sophisticated self-attribution
11. Accurate reporting of conscious contents through a wide range of voluntary responses	Linguistic reporting is (almost) exclusive to humans; some types of accurate reporting have been demonstrated in nonhumans (e.g., parrots and monkeys)
12. Subjectivity (private, first-person experiencing)	This is what needs to be established
13. As well as focused attention, there is fringe consciousness—for example, feelings of familiarity	Probably yes, but experiments needed
14. Facilitation of learning: conscious episodes can be learned readily, whereas unconscious learning is limited	Yes, based on differences between priming, which is nonconscious, and learning during an alert state

Table 5.1 (continued)

Features of human consciousness	Found in nonhuman mammals?
15. Stability of contents: conscious contents are remarkably stable, given the variability of input encountered	Yes
16. Allocentricity: diverse frames of reference are used when neurally representing external objects	Probably; allocentricity is related to the ability for complex bound representations (episodic-like memory of where, when, and what happened is a dramatic example of allocentricity)
17. Conscious knowing and decision-making: consciousness is obviously useful for knowing about the external world and being aware of some internal processes, and for voluntary acts	Yes, *if and only if* we assume that nonhuman animals are conscious and take alertness as an indicator

what such a list is supposed to explain; it is, in fact, the "hard problem." Clearly, accurate reporting (*property 11*) through mature language is human-specific, and although humans and other mammals can report using other means, it is not clear why the ability to report (not just react but explicitly refer) needs to be part of a list of criteria for primary consciousness. Neither is self-attribution (*property 10*), a necessary criterion if it is taken to mean sophisticated metacognition such as possessing autobiographical memory. Seth and his colleagues note the different degrees of testability of different criteria and emphasize the provisional nature of the list, which will need to be updated as we learn more. In spite of these qualifications, however, they come to the conclusion that on the basis of the many homologies in brain anatomy and physiology (e.g., extensive thalamocortical connections) and the existence of many reliable behavioral and cognitive similarities, it is highly plausible that primary consciousness (human consciousness minus HOT) is present in all mammals.

Edelman, Baars, and Seth go beyond mammals and suggest that birds are also endowed with primary consciousness.[22] The case for bird consciousness is also made by Butler and her colleagues, who discuss the topic from a broad comparative perspective, with an emphasis on the sophisticated behaviors of some birds, which can rival those of apes in their complexity.[23] The achievements of Alex, the African gray parrot studied by Irene Pepperberg (figure 5.3), who was able to count, do basic arithmetic, categorize, and reason transitively,[24] are described and discussed at length, along with additional well-documented examples of elaborate bird behaviors.

Figure 5.3
Alex, the African gray parrot.

Extended parental and alloparental care, manipulation of tools, vocal imitation, complex social learning, and cultural transmission have all been documented in birds. In general, one can say that mammals and birds are no different as far as behavioral sophistication is concerned, in the sense that there is no nonhuman mammalian behavior that cannot be displayed by members of some bird species. Investigators of both groups also point to deep neuroanatomical homologies and to avian neurophysiological counterparts of mammalian conscious states, such as the similar waking electroencephalogram (EEG) patterns of birds and mammals, developmental homologies, and similar neurotransmitter systems; they point out that birds, like mammals but unlike reptiles and amphibians, are endotherms, maintaining a stable body temperature through metabolic control.[25] They conclude that although much needs to be done to consolidate this assertion, most criteria indicate that birds are conscious.

Interestingly, the conclusion that birds are conscious has led to different suggested evolutionary origins for this trait (figure 5.4). Seth and his colleagues follow the suggestion of Edelman and Tononi that consciousness

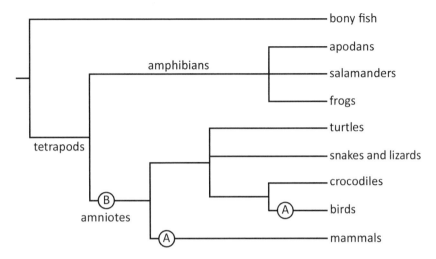

bony fish

apodans

salamanders

amphibians

frogs

turtles

tetrapods

snakes and lizards

crocodiles

B

A birds

amniotes

A mammals

Figure 5.4
Two views on the origins of consciousness. In the first, there were separate origins
in the bird lineage and in the reptilian ancestors of mammals (*A*); in the second, the
origin of consciousness took place in early amniotes (*B*) before the lineages leading
to birds, reptiles, and mammals diverged.

evolved twice, independently, after the divergence of the lineages from
which birds and mammals originated. Butler and her colleagues, on the
other hand, suggest that consciousness first appeared during the transition
between amphibians and reptiles. They argue that the reptile → bird brain
transition shows more neuroanatomical continuity than the corresponding
stem amniote → mammal transition, and it therefore makes sense to investigate the possibility that consciousness emerged in an ancient reptilian
ancestor. This idea that consciousness evolved in the early amniotes (tetrapods whose embryos are protected by several membranes and are therefore
able to live on land) has been advocated by several authors. Unlike the
perceptual-cognitive focus of those studying mammalian and avian consciousness, amniote supporters, like the supporters of even more ancient
forms of vertebrate consciousness, stress the more affective and instinctual-
emotional aspects of animal life.

Michel Cabanac has for many years been an advocate of an early-amniote
origin for consciousness. In his opinion, brain anatomy, brain physiology, and certain behaviors all suggest that consciousness arose about three
hundred million years ago, in the early amniotes that evolved from the
amphibian reptiliomorphs that are the common ancestors of present-day

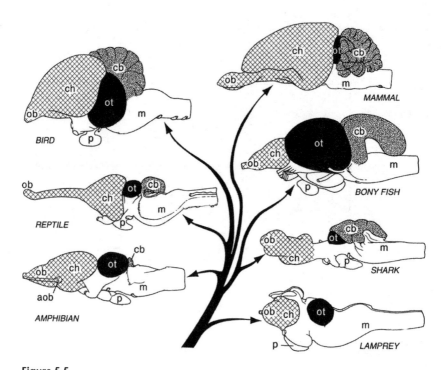

Figure 5.5

Lateral views of the brains of some extant vertebrates (not to scale) showing that most vertebrate brains have the same divisions: *aob*, accessory olfactory bulb; *cb*, cerebellum; *ch*, cerebral hemispheres; *m*, medulla oblongata; *ob*, olfactory bulb; *ot*, optic tectum; *p*, pituitary gland. Reproduced by permission of Oxford University Press on behalf of The Society for Integrative and Comparative Biology.

reptiles, birds, and mammals (letter *B* in figure 5.4). The amphibian brain is similar in morphology and function to that of a fish, although interestingly, most fish—bony fish and rays, for example—have a more complex and much larger brain (relative to body weight) than amphibians (see figure 5.5).[26] However, the amphibian brain lacks some structures and connections that characterize the amniotes' brain.

Cabanac stresses the striking similarity between reptiles and birds in brain structure and cell architecture; the existence of double pyramidal cells in amniotes, but not in amphibians; the relative abundance of dopamine-producing cells in amniotes (there are significantly fewer in amphibians); and several behavioral traits that are supposed to be good indicators of consciousness and that, he claims, are present in amniotes but not in amphibians. These include the emotion of fear, as manifested by the increase

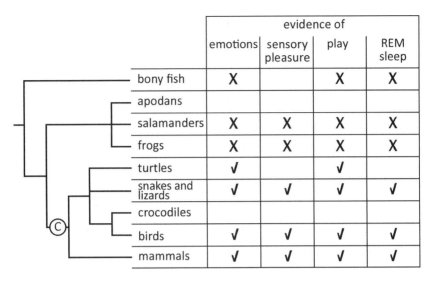

	evidence of			
	emotions	sensory pleasure	play	REM sleep
bony fish	X		X	X
apodans				
salamanders	X	X	X	X
frogs	X	X	X	X
turtles	√		√	
snakes and lizards	√	√	√	√
crocodiles				
birds	√	√	√	√
mammals	√	√	√	√

Figure 5.6
Cabanac's view of the emergence of consciousness (C) during phylogeny, its distribution in living vertebrates, and the supporting experimental evidence.

in heart rate (tachycardia) and body temperature; the ability to feel displeasure, suggested by aversion learning; the capacity for action selection based on reward and punishment (the ranking of actions); detour behavior (the ability to reach a goal that is lost, temporarily, from direct perception); and the manifestation of sleep and play.[27] According to Cabanac, the presence of these behaviors in reptiles but not amphibians is important because it suggests the emergence of consciousness in the early amniotes (figures 5.6 and 5.7).

All the behavioral criteria that Cabanac believes are diagnostic of reptilian consciousness and are absent in amphibians can be, and have been, challenged. The lack of tachycardia in amphibians and fish may have more to do with their poikilothermic, water-adapted physiology than with a lack of fear emotions; a behavior similar to play has been reported in amphibians; some amphibians (e.g., tree frogs) display sophisticated behavior (figure 5.8) and behavioral sleep; and finally, food-aversion learning has been reported in one amphibian species.[28] The learning capabilities of the more than five thousand species of amphibians are very poorly understood, and conclusions regarding a lack of detour behavior (something that is readily displayed by fish, which precede amphibians in evolution) seem premature. Moreover, all of Cabanac's suggested criteria for consciousness

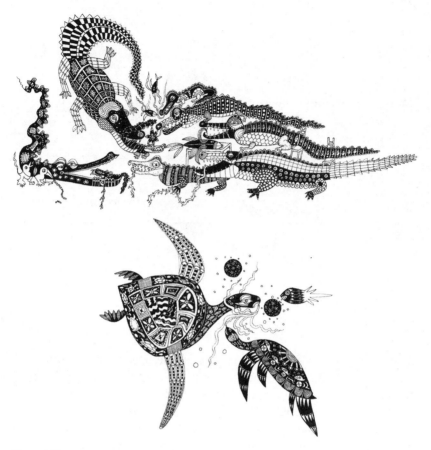

Figure 5.7
Did consciousness emerge in early amniotes? Crocodiles display elaborate maternal care and turtles play.

(with the exception of tachycardia), are seen in fish,[29] and as figure 5.5 suggests, some fish have larger brains and cerebral hemispheres than reptiles and amphibians. So did the amphibian lineage lose consciousness? Or did consciousness evolve in parallel in some fish? Whatever the answer might be, the available data suggest that there is little justification for the conclusion that amphibians lack basic consciousness and that it originated in the amniote lineage.

For us, the most important aspect of Cabanac's work is not his amniote-origins scenario but his suggestion that the criterion for consciousness is a unified and integrated representational domain for perception and action,

Figure 5.8
A nonconscious (?) tree frog finding refuge from the rain under a large leaf. *Source*: Based on http://yourshot.nationalgeographic.com/photos/1530236/.

with affect (a feeling of pleasure or displeasure) being the "common currency" for evaluating stimuli and actions. Cabanac insists that the study of sensorimotor relations and integrations should focus on reward systems that evaluate them and that are expressed as feelings. This stress on evaluating feelings is central to most suggestions that consciousness originated early in vertebrate evolution. It is seen, for example, in theories such as those of Jaak Panksepp and Derek Denton, both of whom grant consciousness to mammals; birds; reptiles; and, more cautiously, to other vertebrates. They, too, rely on explicit reasoning from homology but emphasize different aspects of neurobiology and behavior—the affective aspects.

Affective Consciousness in Vertebrates

> The raison d'être of Consciousness may have been that of supplying the condition to the feeling of Pleasure and Pain.
>
> —G. J. Romanes 1883, p. 111

Affective neuroscience has been developing rapidly since the 1990s. It is focused on basic emotions such as fear, rage, and joy; on sensations like those related to touch, taste, and smell; and on visceral drives such as hunger and thirst and the sex drive. These emotions, sensations, and drives either negatively or positively reinforce behaviors and dispositions, and the study of the mechanisms underlying their evaluative and motivating role has been at the heart of neuroscientists' research.

An important discovery made in 1954 by Olds and Milner drew attention to the systems that lead to reward.[30] They implanted electrodes in the lateral hypothalamus and nearby septum of the brains of rats and found that animals that learned the behavior that activated these electrodes would stimulate themselves again and again, to the point where they disregarded eating and eventually died of hunger. The assumption was that the rats stimulated themselves because of the pleasurable feelings they received. The brain areas that lead to such obsessive self-stimulatory behavior include the ventral tegmentum area (VTA) in the midbrain, the medial forebrain bundle and the nucleus accumbens, which are controlled directly or indirectly by areas in the prefrontal cortex. In rats, stimulation of the VTA causes the release of dopamine in the nucleus accumbens in a way similar to that seen with addictive drugs and with natural rewards such as water or food, which are needed in states of thirst and hunger. Not surprisingly, the areas in the brain that produced the response were dubbed the "pleasure centers." But are they really responsible for pleasure? If so, in what sense?

Kent Berridge and his colleagues have suggested that we often confuse liking and wanting, two affective states that may be distinct: for example, we may desire things that we do not enjoy.[31] According to Berridge, the rats in the Olds and Milner experiment actually stimulated a "wanting system" that leads to actions that usually (but not always and not necessarily) result in the fulfillment of anticipated desires. This wanting system, driven by the neurotransmitter dopamine, leads to eager excitement and arousal but is not to be identified with the sensation of pleasure. Berridge has proposed that in addition to the wanting system, there is a distinct, though tightly connected, "liking system." This liking system engages the opioid circuitry and underlies fulfilled pleasure, such as the pleasure shown by mammals (expressed as enthusiastic lip smacking) when tasting a sucrose solution. It seems that brain stem areas are sufficient to elicit reactions of pleasure and satisfaction: decerebrate rats continue to respond to chemicals that enhance liking responses. In addition to the liking and wanting systems, Berridge suggests that there is a closely linked "learning system" through which implicit and explicit predictions about future rewards based on past learned associations are represented.

Yet is pleasure conscious? What does a neuroscientist mean by saying that a rat is pleased? Berridge recognizes that pleasure is not a mere sensation. It is what he calls a "hedonic gloss" painted over sensations that makes them *liked*, just as pain makes sensations *disliked*. This hedonic gloss is something humans share with other mammals, although Berridge implies that only in humans does the gloss become transformed into subjective feeling. What a nonsubjective hedonic gloss does is not clear.

The functional neuroanatomist Bud Craig shares the view that the type of feelings that humans and their close relatives experience is very different from that of other mammals.[32] He bases his conclusion on detailed studies of the brain and has drawn attention to the anterior insular cortex (AIC), an ancient cortical structure that in humans and some primates is more developed than in other mammals and also has a unique concentration of clusters of large spindle-shaped neurons known as "von Economo neurons." In mammals, the AIC seems to play an important role in integrating sensations such as hunger, thirst, warmth, and pain with the motivational behavior that helps regulate body physiology. According to Craig, it is the human-specific aspects of the AIC and its interconnections with other parts of the brain (amygdala, cingulate and prefrontal cortices, hypothalamus, and more) that give humans their feelings of self and awareness of the outside world. Dogs and cats, which have a different anatomical organization in this region and are unable to recognize themselves in mirrors, may not have similar subjective feelings from the body.

Another believer in HOT-only consciousness is Edmund Rolls, whose ideas we have already mentioned. Rolls comes to the distribution problem from a different direction. As a learning theory–oriented researcher, he thinks about emotions and affects (e.g., gustatory affects) in terms of reward and punishment: "Emotions are states elicited by rewards and punishments, including changes in rewards and punishments. A reward is anything for which an animal will work. A punishment is anything that an animal will work to escape or avoid."[33] Because of their lack of symbolic interpretation, Rolls believes that the elicited emotions are not subjectively felt by nonhumans, but as we argued earlier, this conclusion is invalid—it is the outcome of what James called the "psychologist's fallacy." Nevertheless, we agree with Rolls that reward and punishment should be distinguished from pleasure and pain.

Jaak Panksepp, one of the authors of the *Cambridge Declaration*, who investigates emotions, promotes a conclusion that is diametrically opposed to that of Rolls.[34] He takes a decidedly evolutionary stance. For Panksepp, the fact that specific, homologous brain areas in humans and rats are involved in the generation of specific rewarding or punishing emotional responses that the animals will work to obtain or avoid means that these homologous punishing and rewarding systems lead to conscious states in both. Panksepp points out that in decorticated animals and humans, basic emotional responses remain or are even strengthened; in fact, humans whose brain stem areas have been stimulated report stronger feelings than when they are stimulated in higher brain areas. Interestingly, during extreme feelings, higher cortical regions appear to be relatively inhibited. On the basis of various comparative studies of human and animal behavior, Panksepp (2005) asserts:

> At present, the evidence is most consistent with the conclusion that our core *emotional* feelings (e.g., fear, anger, joy, and various forms of distress), motivational experiences (e.g., hunger and thirst), and sensory affects (pain, taste, temperature, etc.) reflect activities of massive subcortical networks that establish rather global states within primitive body representations that exist below the neocortex.
>
> Affective states are critically linked to the dynamics of the instinctual emotional action apparatus, which can be regulated, but not created, by higher cortico-cognitive activities. (p. 64)

Emotional feelings are for Panksepp facets of basic, innate, and instinct-bound affect programs constituted by the activity of subcortical regions of the brain. He identifies seven such core systems that generate emotional actions and the associated feelings: SEEKING (which is similar to Berridge's wanting system), FEAR, RAGE, PANIC/GRIEF, LUST, CARE, and PLAY, putting all in capital letters to point to their generic nature—they have been found in all the mammalian brains that have been studied. Clearly, Panksepp subscribes to the view that there are innate basic emotions (or affect programs), rather than general processes such as positive or negative affective evaluations and low and high arousal, on which specific, context-sensitive emotions are constructed.[35]

Panksepp argues that both anatomical and experimental evidence suggests an evolutionary trend starting with an initial stage of basic instinctual emotions and followed by a second stage in which the basic emotions become associated through learning with various facets of the world; this progression culminates in a third stage, which is specific to humans with their symbolic metacognition, and elaborates upon the second stage (figure 5.9). Panksepp's progression is similar in its basic structure to the suggestions of Edelman and Damasio, and it makes intuitive sense. However, it is not the only way to describe the evolutionary trend from fish to humans, and in later chapters we propose a different approach

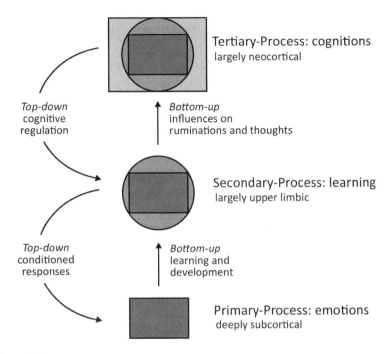

Figure 5.9
Panksepp's evolutionary view of the emotional organization of the brain. In a system he describes as two-way or "circular" causation, there is a nested hierarchy in which bottom-up processes integrate lower brain functions with those at a higher level, and higher-level functions exert top-down regulatory control over lower-level functions. Primary processes are depicted as dark-gray rectangles, secondary processes as light-gray circles, and the tertiary process as a very light-gray rectangle. Modified by permission (unrestricted open access) of J. Panksepp.

to thinking about the evolutionary growth of complexity in feelings and consciousness—one that starts with the evolution of learning processes.

Panksepp's views of affective consciousness center on well-studied emotional systems such as FEAR and ANGER, and he has relatively little to say about visceral affects like hunger and thirst. Visceral homeostatic affects are, however, central to Derek Denton's view of the origins of consciousness. Denton, who is an expert on the physiology of thirst, goes into the depths of the vertebrate body and looks at basic survival and reproductive functions. His intuition is simple and compelling: "The primordial emotions— the subjective element of the instinctive behavior subserving control of the vegetative systems of the body—were the beginning of consciousness."[36] Primordial emotions—Denton uses the term "emotions" synonymously with feelings—are those most obviously related to survival: hunger; thirst; the need to get the right balance of gases, minerals, and nutrients and to eliminate wastes; pain; and sexual urges. These primordial emotions are initiated by receptors located deep within the body (interoceptors), although with both sex and pain, externally located receptors (exteroceptors) are also involved. Most of the interoceptors signal when survival is threatened and are felt as "imperious states of arousal and compelling intentions to act."[37]

Primordial emotions, like all emotions, are subjective and have survival value because they motivate actions that relieve or reinforce the corresponding feelings and, crucially for Denton, restore physiological homeostasis. The primordial emotions differ from the secondary or "classic" emotions, such as anger and fear, on which Panksepp concentrates in several ways: (1) they are harder to suppress; (2) it is harder to imagine them in the absence of the exciting stimulus; (3) like the "emotions" Panksepp discusses, their neural basis is located deep in the phylogenetically ancient medulla, midbrain, and hypothalamus; (4) they are activated by less complex mechanisms and are generally less dependent on perception of the outside world (although sex and pain are obvious exceptions); (5) the relief of the needs underlying them is subjectively felt before the physiological state of the body returns to equilibrium; for example, relief of thirst is felt many minutes *before* the salt balance in the previously thirsty animal is restored, a fact of great survival value for a thirsty antelope. Primordial emotions (like classic emotions and perceptions) can be recovered following distraction (for example, following the intrusion of a new dangerous situation, an animal may "forget" its thirst but recover it as soon as the danger is over), suggesting that they have stable neural representations.

The view of both Panksepp and Denton that subjective feelings are not restricted to mammals is shared by Antonio Damasio. Damasio and his colleagues have described the experiences of a man who, as a result of herpes simplex encephalitis, lost his insular cortices (the structures that according to Bud Craig are the neural basis of subjective human feelings). The unfortunate man, though deeply amnesiac, retained all aspects and types of feelings—pain, pleasure, itch, tickle, happiness, sadness, apprehension, irritation, caring and compassion, hunger, thirst, and the desire to void. Analyzing this case and additional evidence from studies of human and animal cortical dysfunction that left feelings intact led Damasio and his colleagues to conclude:

> Feelings, ranging from feelings of pain and pleasure to feelings of emotions, first emerge from the integrated operation of structures in the brain stem, hypothalamus and the deep layers of the nearby superior colliculi.... Non-human species,

certainly mammals but also birds and even simpler species, clearly exhibit appetitive behaviors and emotions....While it is obvious that non-human minds are
not capable of intellectual feats of the human variety, it does not follow that nonhuman minds are deprived of body feelings, emotional feelings, and sentience.
(Damasio, Damasio, and Tranel 2013, p. 843)

Once it is accepted that some nonprimate species, and possibly even
nonmammalian species, have a form of subjective experiencing, the question becomes, Which ones? Denton explicitly includes reptiles within his
scheme, and both Denton and Panksepp mention the possibility that all
vertebrates may have basic consciousness. But in spite of their ambition to
account for the origins of consciousness, Panksepp, Denton, and Damasio
are all surprisingly vague about its phylogenetic origin and the related distribution question.

Do All and Only Vertebrates Have Consciousness, and When Did They Acquire It?

In 2007, Björn Merker, whose ideas about the self we discussed in the previous chapter, published an influential paper that, in our judgment, has
made a real difference to the consciousness discourse, rattling, so to speak,
its neocortical shackles and providing a coherent, integrated framework for
thinking about minimal consciousness. Merker (2007) came up with a bold
proposal:[38]

> The primary function of consciousness—that of matching opportunities with
> needs in a central motion-stabilized body–world interface organized around an
> ego-center—vastly antedates the invention of neocortex by mammals, and may
> in fact have an implementation in the upper brainstem without it. (p. 80)

He suggested that consciousness—in the basic sense of subjective
experiencing—appeared in moving visual animals whose world was constantly changing and who needed to adjust their body and their goals. The
selection of a target in the external world, the selection of a bodily action
directed toward that target, and the motivation for that action had to be
integrated:

> My suggestion is that the emotional, sensory, and action aspects of consciousness
> were linked from the outset by providing the functional reason for a specifically
> conscious mode of organization. (Merker 2007, p. 113)

According to Merker, there is therefore a developmental and evolutionary "triangular dependency" between target selection, action selection, and
motivation. Together they constitute primary consciousness, enabling the
animal to act in the world in a goal-directed manner, using what Merker

describes as a motivational "common currency" for selecting among both targets and actions. To achieve this feat, the brain engages in analog reality simulation of the tripartite interaction between the body, the world, and the animal's needs:

> It features an analog (spatial) mobile "body" (action domain) embedded within a movement-stabilized analog (spatial) "world" (target domain) via a shared spatial coordinate system, subject to bias from motivational variables, and supplying a premotor output for the control of the full species-specific orienting reflex. (Merker 2007, p. 71)

There is an implicit "ego center" in this coordinate space, a perspective from which the world is perceived. This ego center creates a motivated "subject," a "self" located in a body-within-world space (see figure 4.5 in chapter 4). Hence, by integrating spatially structured sensory information while differentiating world- and self-originated changes in the sensory influx, a motion-stabilized body-world interface organized around an ego center is constructed. It enables the animal to adopt a stable first-person perspective, loaded with motivational value, which is derived from its current homeostatic state and is continuously synchronizing goals with possible actions.

Merker provides neuroanatomical and neurophysiological evidence from vertebrates that supports this model. On the basis of empirical findings, he argues that the integrative structures engaged in such simulations of reality involve structures in the upper brain stem;[39] electrical stimulation of these leads to emotional responses in humans that are far more powerful and integrated than in the stimulation of higher areas. He also cites decortication experiments, especially in very young mammals (rat pups and kittens) that show that many basic functions, behaviors, and even quite complex learning are intact. Merker discusses human anencephaly, a condition in which the cerebral hemispheres fail to develop (in some cases very little cortex is left). Anencephalic children are born very seriously handicapped, but nevertheless they show unmistakable emotional responses and, according to conventional diagnostic criteria, are considered conscious. Moreover, in spite of the lack of any human-specific cortical structure, the smiles and laughter of these children are clearly human, so changes must have occurred in the upper brain stem as well as the neocortical regions during human evolution. A complementary line of evidence comes from studies of loss and recovery of consciousness following anesthesia in humans, which show that the brain stem and the midbrain are essential for their minimal consciousness.[40]

On the basis of his analysis of midbrain structures and processes in vertebrates, Merker suggests that reality simulation from a first-person perspective occurs in fish, which are therefore conscious. The cerebrum of all vertebrates is covered by the pallium, consisting of layers of gray and white matter, which in mammals comprises most of the cerebral cortex. Although fish (as well as amphibians, reptiles, and birds) lack a laminar, columnar pallium, which some people assume to be necessary for consciousness, there is increasing evidence that such a pallial structure is not necessary for sentience and that the tectum and the superior colliculus, reciprocally interacting with the immediately underlying periaqueductal gray matter, have the functional architecture necessary for sentience. The sensory maps in the tectum of fish represent the body and most of the externally arrived composite information (except olfaction), and the connectivity of the tectum includes reentrant reciprocal connections with the periaqueductal gray, which are similar to thalamocortical interactions in mammals. Although sentience can be manifest without a pallium, once the pallium is in place, it is involved in maintaining the relationships mapped in the tectum. In his book *What Fish Know*, ethologist Jonathan Balcombe shows that some fish species manifest highly sophisticated behaviors, including transitive inference, tool use, and self-recognition, which in mammals and birds have been considered as indicative of consciousness. These behaviors, as many studies have shown, are enabled through activities in both the tectum and the pallium.[41]

What about invertebrates, which show no homology to vertebrates? There have always been advocates for invertebrate consciousness because of the complexity of some of their behaviors. Darwin was in no doubt that ants are conscious beings with complex emotions (see chapter 2), and Romanes, too, granted a degree of consciousness to some invertebrates. Donald Griffin, one of the scientists who braved the consciousness taboo in the early 1970s, also argued that nonhuman animals, including some invertebrates, are conscious.[42] His conclusions, like those of Romanes, are based on the sophisticated and adaptive responses of these invertebrates when faced with novel challenges and on their sometimes amazing (as in honeybees and red ants) ability to communicate. According to Griffin, the arthropods may be another taxon in which consciousness has emerged.[43]

The idea that minimal consciousness is not unique to vertebrates is implied by Merker's proposal. Indeed, in response to the paper in which Merker put forward his ideas, Emmanuel Gilissen described the remarkable behavior of predatory jumping spiders of the genus *Portia*, which show

amazing feats of planned complex behavior rather similar to those of hunting cats and lions.[44] Gilissen describes how these spiders stalk their prey
(flies or other spiders), taking detours and indirect routes and even leaving
their prey temporarily out of sight, which points to the persistence of the
spiders' representations and to their ability to plan. The triangle of target
selection, action selection, and motivation seems to be in place, so do these
predatory spiders have some primary consciousness? Merker bites this bullet
and agrees that it is worth looking for such an interface in *Portia*, something
that would open the way for investigating the question of consciousness in
other invertebrates. The idea that insects' brains have the integrated and
egocentric representation required for minimal consciousness, solving the
basic problems of sensory reference and spatial navigation, was discussed,
with reference to Merker's original suggestion, by the biologist Andrew Barron and the philosopher Colin Klein, who recognized the corresponding
functional architecture in arthropods.[45] Slowly but surely, the idea that
arthropod species may have some form of minimal consciousness is making its way into the discourse of consciousness studies.

Other obvious candidates for consciousness among the invertebrates are
the versatile cephalopods, which include octopuses, squid, and cuttlefish
(figure 5.10). Some animals in this group display the kind of "triangular
dependency" that Merker highlights. Octopuses can classify and categorize
objects and discriminate among them using size, shape, and intensity; they
can find their way in complex mazes, rapidly learn by observation from
other octopuses, and use abandoned shells for protection. In addition, they
have a complex central nervous system with distinct regions involved in
learning and long-term memory. They also have dopamine, noradrenaline,
and serotonin receptor subtypes that are similar to those found in vertebrates, and their EEG patterns resemble those of wakeful vertebrates. They
even express FOXP2, a gene that is expressed in communicating birds and
mammals, in their chromatophore lobes, the neural structures that regulate
changes in skin color. By controlling these chromatophores, which are specialized cells with light -reflecting pigments, the cephalopods can display
waves of color and texture, which they use for social interactions and camouflage. Although not conclusive, the combination of highly sophisticated
behavior and richly complex nervous systems suggests to those who study
cephalopods that they may well be endowed with consciousness of a sort,
although it is not easy to imagine what it is like to be an octopus.[46]

The view presented by Merker has important implications for the distribution question and for the closely related question about when minimal

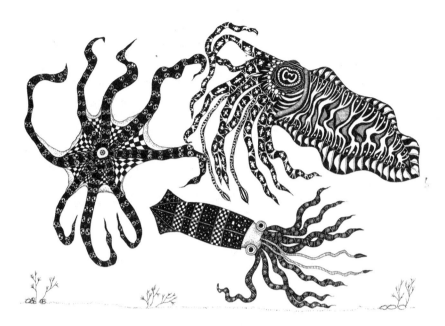

Figure 5.10
An artist's depiction of coleoid cephalopods. Octopus (*left*), squid (*bottom right*), and cuttlefish (*top right*).

consciousness evolved. The vertebrate architecture that he analyzes is present in fish, which evolved at least 520 million years ago, during the great Cambrian explosion, when most animals made their first appearance. It can therefore be inferred that the Cambrian era is when vertebrate consciousness emerged.[47] The idea that the Cambrian marks the emergence of consciousness was developed during our early analyses of the evolution of open-ended associative learning, which we recognized as the marker of minimal consciousness. Our investigations led us to suggest that consciousness appeared during the Cambrian era in the first vertebrates and, in parallel or somewhat earlier, in arthropods and possibly in other invertebrate groups.[48] This idea has since been gaining in popularity: it was applied some years later to all vertebrates by Todd Feinberg and Jon Mallatt, who argued that the ability of early vertebrates, including the jawless fish (and also some invertebrates such as arthropods and cephalopods), to represent the world and the body through neural sensory maps that are organized in three or four hierarchical layers points to their ability for conscious experiencing.[49] The evolutionary approach to consciousness is thus beginning (slowly but surely) to spread.[50]

We return to it in chapters 7–9 when we develop our model of the evolution of associative learning and propose that it was associative learning and minimal consciousness that drove the Cambrian explosion.

Suggestions about the presence of consciousness in a particular animal that are based on sophisticated behavior or on the complexity of the nervous system are always open to challenge: Why draw the line here and not there? The answer supplied by the arguments from analogy and even homology is always the same: because they are similar enough to the changes in behavior, neuroanatomy, and neurophysiology that accompany the conscious experiences reported by humans. But how similar is similar enough?

In this and previous chapters, we described several architectures and capacities that can serve as indicators of animal consciousness: thalamocortical reentrant dynamics (chapter 3), an amplification wave characterized by typical peaks in electric activity and gamma waves (chapter 3), Merker's triangle (chapter 4 and this chapter), and the arguments from behavioral-learning complexity advanced by Romanes and by Griffin (this chapter). These suggestions highlight important properties of primary consciousness, but they, like other current suggestions, all have their problems. For example, the signatures of consciousness identified by Dehaene in mammals may be different from those in other taxa,[51] and as yet no experiments have been carried out to distinguish between subliminal, preconscious, and conscious processes in nonhuman animals. Even if we adopt Merker's more general criteria, it is not clear how flexible or sophisticated action selection and target selection need to be: behavioral sophistication is a notoriously vague criterion (when is behavior sophisticated enough?).[52] We therefore adopted a different, two-pronged approach: we looked at how the seven criteria of minimal consciousness, which we enumerated in chapter 3, are distributed across the phylogenetic tree and tried to find an evolutionary transition marker to minimal consciousness that requires these criteria.

Evolutionary Transition Markers: The Evolutionary Transition Approach

Mind can be understood only by showing how mind is evolved.
—Spencer 1890, p. 291

In chapter 1 we described a diagnostic feature of an evolutionary teleological transition as a "transition marker" and noted that Gánti, and more explicitly Maynard Smith and Szathmáry, suggested that the transition to life is marked by the system's capacity to manifest unlimited heredity. In such a system, the number of possible hereditary variants is vast, the system can persist in the face of challenges, and open-ended evolutionary change is possible because hereditary variants have novel, variable, functional (selectable) effects. A biological system that shows such unlimited heredity requires that the remaining items in the list of features and mechanisms of life that have been suggested by origin-of-life researchers is in place. Unlimited heredity entails a highly complex enabling system (with metabolic pathways, closure, a high level of functional plasticity, and so on) that carries out the many processes that render it possible. A modern cell is clearly such an enabling system, and we can assume that it was preceded by precursor-enabling systems that occupied the gray area between the pre- and posttransition states. Discovering a similarly convincing evolutionary transition marker for consciousness would go a long way toward elucidating the biological underpinnings of minimal consciousness. But can we find such a marker?

Unlimited Associative Learning: A Conceptual Characterization

We suggest that the conceptual analog to unlimited heredity is unlimited associative learning (UAL), which we contrast with limited learning. In animals with limited learning, most learning is nonassociative, mainly learning by sensitization (a decrease in the threshold of a reflex response as a result of exposure to the stimulus) and habituation (increase in the threshold of a reflexive response as a result of repeated exposure). A great increase in adaptability occurs with the evolution of restricted associative learning, in which associations between simple ("elemental") stimuli and stereotypical actions can be reinforced. We describe this type of learning in detail in chapter 7 and call it "limited associative learning," or LAL. With UAL, however, the number of associations within and between modalities that can be learned and recalled is greatly increased because many different compound patterns of stimuli as well as the components of the compound stimuli can be learned. Consequently, the number of associations far exceeds those that actually form during the individual's lifetime. In terms of experimental psychology, UAL describes associative learning that is cumulative and involves the learning of novel compound stimuli and novel trains of actions. Just as unlimited heredity that entails innovations and new selection regimes enables open-ended evolution, so UAL allows open-ended behavioral adjustments and through cumulative learning can lead to complex behavior that has functional (selectable and selecting) effects.[53]

It is important to note that we suggest UAL is a positive marker of animal consciousness—its presence can tell us that an animal is endowed with minimal consciousness, but its absence cannot determine whether or not an animal is nonconscious. A human who does not manifest UAL, such as a neonate or an anencephalic person, can nonetheless be minimally conscious because she is endowed with a preexisting, evolved, and functionally active UAL-supporting organization that is in place and is exercised even when UAL itself cannot manifest for developmental or pathological reasons. A cell whose nucleus has been removed and has hence lost some or even all of its unlimited heredity system (like human erythrocytes, which live for four months without a nucleus) is still considered alive because supporting cellular systems (which had evolved in the context of unlimited heredity) are in place and are functional. Similarly, an animal devoid of a UAL capacity for developmental or pathological reasons can be considered minimally conscious as long as it possesses and exercises the preexisting, evolved structures that enable UAL. Thus, although decorticated mammals can feel, the ability to feel evolved through interactions with the pallium. The midbrain in decorticated animals can support sentience because of the properties it acquired during its coevolution with the pallium. Our argument is that the evolutionary origins of UAL and its enabling system were coupled. Moreover, just as a lineage of living autopoietic systems that has lost its system of unlimited heredity will eventually become extinct, we expect that a lineage of animals with consciousness that has lost its ability for UAL will, over evolutionary time, lose sentience.

We can sharpen the notion of a transition marker for consciousness by comparing it with the correlate of consciousness concept. In chapter 3 we showed that the term "correlate of consciousness" (usually a neural one, an NCC) is somewhat ambiguous because it does not specify whether the correlate is a necessary background condition, a specific cause, a consistent effect, or a necessary functional consequence. This ambiguity led Anil Seth to suggest that one should search for *explanatory* correlates of consciousness (ECCs)—the causal mechanisms explaining the fundamental properties of consciousness such as its unity and diversity; its subjective first-person nature; and its emotional, motivational, and intentional aspects. A transition marker is not such a causal correlate, however: it is neither necessary nor sufficient for the expression of sentience in present-day animals as long as its supporting mechanisms are in place and are functioning. In other words, an evolved system exhibiting a teleological function (life, sentience) may exist even if the transition marker that was a major driver of its evolution is not currently in place. However, an evolutionary transition marker is the evolved feature that permits one to reconstruct or to reverse engineer the enabling system of which it has been, and usually remains, a part. Just as an evolved unlimited heredity system enables us to reconstruct a protocell, and it is the protocell and not the heredity system that is alive, so, we argue, the appropriate transition marker of consciousness may allow us to reconstruct the enabling system of which it was evolutionarily a necessary part—the minimally conscious animal.

The embedding of the transition marker system within a coevolved set of supporting mechanisms (which jointly comprise its enabling system) explains why the nature of the enabling system is crucial for assigning life or sentience to the system. The capacity for UAL per se (in a robot, for example) may not be sufficient for sentience. Just as a genetic algorithm is an unlimited heredity system (realized in a computer chip) although neither the chip nor the computer is alive, so a robot exhibiting UAL (when such robots are built) may not be sentient because its supporting system is inadequate. This means that a functionalist, substrate-neutral account of life or sentience may be insufficient. Moreover, although consciousness is an intrinsic process, the marker need not be intrinsic; it can be a type of behavior. A useful marker must, however, enable the construction of the intrinsic dynamics of consciousness.

Our suggestion that UAL is a marker of consciousness follows the views articulated by Romanes and Morgan, and shared by many others since, that intelligent behavior that manifests as flexible learning is an indicator of consciousness in animals.[54] However, if not qualified, this type of suggestion is either uninformative or easy to refute because intelligence is a notoriously vague term; if defined simply as the ability to solve complex problems, then nonconscious computers can be said to be "intelligent." Moreover, as we noted earlier, some forms and facets of associative learning do not require consciousness, since very simple forms of associative learning are found in single-celled organisms and in plants. Having a nervous system is also not sufficient for consciousness: humans whose spinal cords have been severed from their brains can learn reflex-based associations (for example, that a weak electric shock predicts a stronger shock) even though they report no subjective experience of the stimuli.[55] Therefore, in order to clarify how we see the evolutionary relation between learning and consciousness, we need a general definition of learning and a characterization of UAL.

Defining Learning and Memory

> It appears to me that the conjunction of memory with sensations, together with the feelings consequent upon memory and sensation, may be said as it were to write words in our souls.
>
> —Plato 1961, p. 39a

Learning is an ontogenetic adaptation that involves a relatively long-lasting change in behavior. Neurobiologists, psychologists, and cognitive scientists have stressed different aspects of the processes involved,[56] but our own emphasis is on memory. All learning entails memory: there is no learning without the encoding, retention, and then the retrieval of information related to past experiences, and the evolution of memory is therefore a key to the evolution of learning.

"Memory" is an umbrella term that covers different relations between past and present/future occurrences. Looking at the memory of humans, which is where most of our intuitions come from, reveals that this memory is not monolithic. Over a century ago, Henri Bergson distinguished between "habit memory"—the manifestation of habits upon the presentation of cues without conscious recall—and "recollective memory"—the explicit, conscious recall of facts and of past events.[57] Habit memory, sometimes also called "implicit memory," includes the memory the body has of past learned skills (for example, riding a bike) and is typically involuntary. Recollective memory (commonly called "explicit memory" or "declarative memory") has been divided into two types by cognitive scientist Endel Tulving. The first is the recall of facts without recollection of the context—the place, time, and manner in which they were learned; for example, the learning context of one's memory that World War II started in 1939 is not remembered; this is called "semantic memory." The second is the context-rich recollection of events and chains of events, known as "episodic memory"; for example, the recollection of an encounter with a predator, which includes where, when, and how this encounter happened. In mature humans, episodic memory involves mental time-traveling that is constructed around an autobiographical narrative in which the recalled events are embedded and through which identity is constructed ("autobiographical memory").[58] In view of the multitude of memory types and the plausible supposition that the distinctions between them are based on different types of processes, it is not surprising that the concept of memory has received a great deal of scrutiny, and the supposedly explanatory metaphors used to describe it have been heavily criticized.[59]

Although the old metaphors and their presuppositions can be misleading, we find the distinctions between encoding, retention, and retrieval helpful, providing one remains mindful of the active relations between these processes. We therefore use the traditional three-operation characterization that was discussed at length and in depth by the German zoologist and theoretician Richard Semon, who coined the term "engram" and anticipated, over a hundred years ago, many of the current discoveries about the acquisition, storage, and retrieval of neural and nonneural memories. Following Semon, our characterization emphasizes and makes explicit the dependence of encoding and retention on reinforcement:[60]

- *Encoding*: A sensory stimulus that is the result of a change in a dynamic internal state or is the result of an external stimulation that typically follows exploration (e.g., the sight of a food following undirected motor activity) starts an internal process or reaction that culminates in a (usually) functional effector response (e.g., feeding behavior). The recoordination of activity and sensory stimulations occurs within the specific context of the physiology and environment of the animal and metaphorically can be said to *encode* the relation between them.

- *Retention*: The coordinated internal reaction is partially retained. By this we mean that some physical traces of the relation persist even when the original stimulus is no longer present, and the response is no longer manifest. A latent memory trace, an engram, is formed following one or more phases of consolidation. The engram can be described at several levels of organization, beginning with the epigenetic cellular level. It can, for example, be an epigenetic pattern in the chromatin of the nucleus of a single cell or induced regulatory RNA molecules and protein complexes that not only change the threshold of the reaction of the cell to the inducing stimulus but can also be transferred between cells. In addition to such cell-universal memory systems, neurons also have the specialized memory encoded in synaptic connections, and there are engrams at the anatomical level of the brain and the body. Some changes in the external environment (e.g., written words, monuments of ancestors) can be regarded as external memory traces, as exograms.[61] For retention to occur, some form of stabilization of the reaction is necessary. Although the original response is transient, and with time and in the absence of the stimulus it often decays, the encoded reaction is stabilized through value (reinforcement) systems.

- *Retrieval*: This process occurs when, upon later exposure to a similar, partial, and/or associated type of condition, the engram facilitates an internal reaction similar to the original one. Retrieval can involve more than simple triggering of the engram; it can also involve encoding, with additions, reconstructions, and reconsolidations of the engram.

This general characterization, when suitably qualified to recognize the distributed nature of encoding, the consolidation processes during storage, the reconsolidation occurring following retrieval, and the interactions among engrams during encoding and retrieval, is appropriate for all types of learning and memory systems, including nonneural learning. Indeed, Semon thought about memory in such general terms. He suggested that a common principle, which he called the "Mneme" (the Greek word for memory or remembrance and also the name of the muse of memory in Greek mythology), underlies the organisms' capacity to retain the effects of past stimulations and to interact with the environment on the basis of these engrams. The Mneme, according to Semon, is manifest in different ways at the cellular level (cell memory), at the level of interactions among neurons (neural memory), and at the phylogenetic level (heredity).[62]

An Operational Definition of UAL

UAL is a specific, demanding type of associative learning characterized by generativity, which is the result of the combinatorial explosion of learnable stimuli and actions, and by "reflectivity," which is the capacity to use the outcomes of past learning as the basis for further learning.[63] UAL requires that three conditions are satisfied:

First, for classical (Pavlovian) UAL, the animal should be able to associate a compound conditioned stimulus, a CS (e.g., a smooth, round, fragrant blue object, or a melody comprised of several notes) to an unconditioned stimulus (US) such as a tasty food or an electric shock, that elicits an innate reflex response, an unconditioned response (UR) (see figure 5.11, *A1* and *A2*). The "value" of the US is based on the release of neurotransmitters, such as serotonin or dopamine, that are negatively or positively reinforcing. Crucially, a compound CS is a stimulus that has several features that may not elicit a response when perceived separately and that can be discriminated from compound stimuli consisting of the same component stimuli but in a different combination. In other words, the production of the conditioned response (CR) must be specific to the conjunction of the features, a phenomenon known as "spontaneous configuration" or "perceptual fusion," so a single component of the whole (e.g., shape alone for a visual stimulus, or a single note for an auditory one) cannot elicit a CR, or elicits only a much reduced one.[64] This means that the percept is stably encoded and

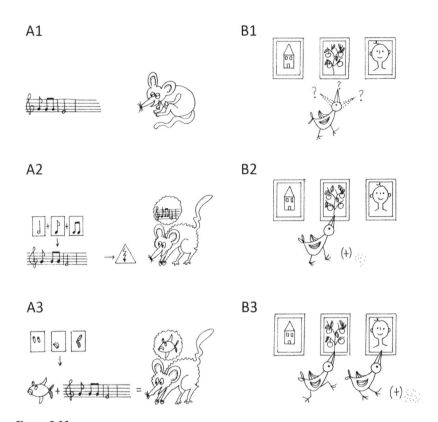

Figure 5.11

Two types of unlimited associative learning. In *A1*, three different musical notes are bound to form a compound conditioned stimulus, a melody, which has no value for the mouse. In *A2* the same melody (which we can now consider as CS1) becomes associated with negative value (punishment) through an electric shock (the US). *A3* shows how, following this learning, CS1 can serve as a US for future second-order learning and become associated with a different compound stimulus, a picture of a fish (CS2). In *B1*, exploratory, motor, nonreinforced behavior leads the bird to peck whatever it sees. In *B2*, pecking a picture of a tomato plant three times (a compound CS and a specific train of acts) becomes associated with positive value (reward), the tasty seeds. *B3* shows that the previously learned association can be the basis for learning an additional motor behavior directed toward another compound stimulus and result in the chaining of the action patterns: the seeds are obtained only after first pecking the tomato plant picture three times, then pecking the picture of a face twice.

recalled as perceptually bound, going beyond an online transient encoding of a multifeatured percept. Learning leading to memorized perceptual fusion is known as "nonelemental learning," and there are several experimental procedures for testing it. For example, nonelemental learning is assumed to have occurred when the animal can learn that two different stimuli are associated separately with a negative value (A–, B–) or are neutral (A$_{neutral}$, B$_{neutral}$), yet the compound stimulus is associated with a positive value (AB+). Another manifestation of nonelemental learning is discrimination learning, in which the animal learns to ascribe a particular value only to a specific compound pattern of stimuli—for example, it can discriminate between AB+ CD+ and compound AC– BD–.[65] With compound instrumental learning, the requirements are similar: a particular combination of action patterns is learned through reinforcement, whereas a single action or a compound action made up of the same action patterns in a different temporal order is not. Similarly, the animal must be able to learn to generate a compound sequence of motor actions rather than a stereotypical reflex action (although the novel sequence of actions is likely to be based on a combination of innate action patterns) to attain or avoid a reinforcer (e.g., food, shock), including a reinforcer that is associated with a compound perceptual stimulus (figure 5.11, *B1–B3*). The learning of such compound action sequences is incremental and involves chaining among individual action patterns (see below).

Second, the stimulus or the reinforced action must be novel—that is, it is neither reflex eliciting nor preassociated with a US or with past reinforcement. For example, prior to learning, the sight of blue flowers with yellow spots, of a musical phrase, or of a spontaneously emitted sequence of actions do not elicit any observable aversive or appetitive response.

Third, it is necessary that the learned compound CS (or the learned sequence of actions) is not only recalled as being integrated but can subsequently support second-order, cumulative conditioning, thus itself acting as a US (or reinforcement) (figure 5.11, *A3* and *B3*).[66]

When considering instrumental/operant conditioning, it should be possible to link together ("chain," in behaviorist jargon) short, single, reinforced behaviors into a longer sequence of actions, something that is dramatically manifest in circus tricks accomplished by trained animals. This would result in compound series of actions that are remembered as a whole and can become habitual. The behavioral sequence of actions is learned incrementally and can lead to compound action patterns manifest as self-control, tool use, and expectations of future events, which are usually attributed to "higher" cognitive processes.[67] Hence, although there are taxon-specific and ecology-specific constraints on the associability of different classes of compound stimuli and actions, UAL broadens the range of factors and processes that make up each of the four pillars on which learning is based (perception, memory, value, and action): the ability to form a recollectable representation of a compound (bound) stimulus entails enhanced discrimination ability; second-order learning dramatically extends the number of value-signals; and learning to link sequences of actions leads to many potentially novel action patterns. Perception of compound objects and scenes and flexible action patterns are linked, since action is guided by perception, and learning new adaptive action patterns depends on the integration of value signals about the homeostatic state of the body. This is a demanding type of learning, which we unpack further in chapter 8.

How is this great expansion in the repertoire of learned behaviors and the cognition underlying it related to minimal consciousness? Even if one finds the analogy with life and its transition marker attractive, this is not a sufficient reason to accept UAL as a transition marker for consciousness. It could be argued that other components of the list of properties of consciousness that were discussed in the previous chapters, or features of the models described in chapter 3, could have been put forward as markers of minimal consciousness: binding, attention and working memory, thalamocortical reentry loops, high Φ, and so on. Three reasons led us to suggest UAL as the transition marker: First, it is an overt, behavioral property that is neutral with respect to underlying neural implementation and hence not mammalian-avian biased. Second, it relies on several well-investigated behavioral functions (perceptual fusion, operant associative learning, second-order conditioning), and thus it is a well-defined and measurable form of learning that can be the basis of empirical hypothesis testing. Third, and most importantly for our argument, just as unlimited heredity entails an autopoietic enabling system that satisfies the characterization of a living system, UAL entails an enabling system that satisfies the list of properties characterizing consciousness. We therefore return to the consensus list that we presented in chapter 3 and examine each of the listed seven properties in the light of UAL.

1. *Global activity and accessibility of information*: According to our definition of UAL, a compound CS becomes associated with value and action and can become a US enabling future cumulative learning. This entails the integration of information (to enable perceptual fusion) followed by its widespread distribution across dedicated circuits so that value and motivation systems can be updated.[68] The associations formed between the compound CS and the US get incorporated into long-term memory, which underlies

US-UR relations, and into the organism's value system, thus enabling the "reevaluation" of the CS or of learned motor behavior.

2. *Binding and unification*: According to our definition, UAL entails the formation of a compound stimulus, which is based on activities within distributed neural networks on several different anatomical scales. This entails something like binding, which may be dependent on both synchronous and sequential-reentrant processes between neural maps.[69] UAL therefore enables the identification of and discrimination among compound stimuli that differ in their conjunction of underlying features within and across modalities. As suggested long ago, discrimination is tightly related both to consciousness and to the evolution of complex forms of learning,[70] and selection for an increased ability to discriminate probably drove the evolution of perceptual fusion, which in turn drove learning, forming a "learning ratchet." Pattern completion, whereby a partial facet of the compound stimulus induces memory of the whole compound percept, is a testimony to the original unification of the compound percept.

3. *Selection, plasticity, learning, and attention*: Learning, a striking manifestation of plasticity, is inherently selective at both the neural and behavioral levels because it always entails the exclusion of multiple possibilities. It also, by definition, includes memory, which entails stabilization. Selection mechanisms are therefore a prerequisite for UAL. However, UAL entails complex modes of selection, requiring that a compound CS can guide future behavior. This means that when the value associated with the CS is greater than that of the competing US, the CS will override the US and hence will guide behavior. Consider an example adapted from Morsella—taking a hot plate out of the oven.[71] Although the heat of the plate (US) elicits a reflexive releasing response (UR), holding on to the plate is associated with a positive cumulatively learned CS (social norms), which may override the releasing reflex. Learned bowel and bladder control by a child or a dog is another common and compelling example of reflex inhibition and action selection mediated by cumulative learning. In this case, a learned behavior is reinforced only when a sequence of actions is ordered in a particular way. Since this behavior is a consequence of cumulative and temporally ordered learning, we consider it an instance of UAL. The integration of multifeatured information, which is a characteristic of UAL, requires attention, and attention is, by definition, selective (albeit to different degrees). Strong support for this requirement is provided by a study in which participants were engaged in an attention-demanding primary task and, as a result, were unable to differentiate between red-green (red on

the right, green on the left) and green-red (green on the right, red on the left) disks that appeared in the location on which they focused their attention. The participants were, however, able to discriminate between male and female faces because facial recognition relies on the automatic, previously learned processing of stimuli.[72] However, although alertness generally increases the success of learning-demanding tasks, a low level of implicit learning may occur if the cognitive architecture includes an evolved ability for forming fused percepts.

4. *Intentionality*: UAL entails referral and aboutness because it involves behavior that is aimed at the compound CS—it is goal-directed and intentional. It also requires that sensory stimuli coming from the world and from the body are mapped and is therefore intentional in the "aboutness" sense too.[73]

5. *Temporal thickness*: Binding, or feature integration, entails temporal thickness since it seems to occur at a (relatively) late processing stage.[74] It requires more time than processing a single feature or a reflex-like stimulus (e.g., a face, which is processed by a dedicated preestablished circuitry), and it relies on temporal reverberations across distributed neural maps. The complex, time-consuming, global dynamics underlying the formation of associations thus renders UAL temporally "thick." An extension of temporal thickness occurs with trace conditioning—a form of conditioning in which the CS ends before the US begins, so there is no temporal overlap between their presentation. Although trace conditioning is not a necessary condition for UAL as we define it, it can further expand the potential associations that may be learned because the constraint of temporal overlap between the CS and the US is relaxed. In mammals, trace conditioning requires an intact hippocampus and involves working memory processes that maintain the CS-US relation during the intervening time. Since trace conditioning requires working memory and in humans is evident only when the CS-US relation can be verbally reported, it has been suggested that consciousness is necessary for trace conditioning. However, from our perspective, trace conditioning may be both too stringent and insufficient as a criterion for *minimal* consciousness: on the one hand, it is too stringent because conscious recollection of the CS-US relation is evidence of higher-order consciousness rather than minimal consciousness (one can experience the CS-US contingency while being unable to later report it); on the other hand, it is an insufficient condition because it neither requires binding nor assumes a flexible, modifiable value system (i.e., it is not necessarily cumulative). Nonetheless, we think that it is highly likely that trace

conditioning emerged alongside UAL as a facet of the temporal thickening entailed by complex binding, the exclusion of alternatives, and by the ability to remember and perform long sequences of learned motor acts.

6. *Values, emotions, goals*: Any form of learning implies a value (reinforcement) system. In simple forms of learning, such as the reflex conditioning in *Aplysia* (described in detail in chapter 7), the value system is innate and fixed: only reflex reactions can be modified or acquire value. Unlike simple forms of learning, UAL is cumulative, so the animal can learn to respond to new stimuli in new ways. The organism's value system is therefore adjustable: any stimulus may potentially serve as a US, and any sequence of actions may be reinforced. Thus, the organism's goal-directed behavior becomes much more flexible—in addition to phylogenetically inherited goals (obtained, for example, through reflex actions), new goals can be defined during ontogeny. For example, consider a mammal with an innate fear response to the sight of a snake but with no innate sensitivity to the snake's hissing, which is heard before the snake is seen. After a few traumatic encounters with snakes, the hissing sound (CS) and the sight of the snake (US) are automatically associated, and the mammal will now flee immediately upon hearing the hissing. Assuming second-order learning, if the hissing (now a US) is consistently preceded by a sudden rustling sound made by the snake's movement in the dry grass (a new CS), the mammal will learn to flee from rustling sounds, which are heard well before hearing the hissing or seeing the snake. This type of cumulative learning dramatically expands the organism's motivational landscape, rendering many novel states, stimuli, and action patterns motivationally relevant. Since the animal's activities must accommodate to its changing needs, these activities are intentional in the broad "goal-directed" sense discussed by Freeman (see chapter 4).

7. *Embodiment, agency, and a notion of "self"*: UAL is strongly related to the notion of self and is a necessary condition for it. The minimal "self" as described by Merker and by Metzinger requires a model of the integrated yet rapidly changing world in which a model of the coordinated and flexibly changing body is nested, providing a stable, updateable perspective that enables the flexible evaluation of changes in world-body relations. This notion of self is related to that of elementary agency, which stems from the ability of animals to distinguish between the effects of self-generated and world-generated stimuli. Comparing the actual sensory consequences of the animal's complex motor activity to the sensory feedback that is predicted by its self-model inhibits compensatory reflexes when the consequences of own body–derived sensory inputs and world-derived

sensory inputs match. Although this distinction can be accomplished by peripheral systems and does not require consciousness, once the animal has UAL and can learn in an open-ended manner about its own actions and the world, both the world model and the self-model can become integrated, leading to a minimal self. The formation and maintenance of complex body, action, and world models and distinguishing between self-produced and world-produced sensory inputs involves the binding and integration of world (exteroceptive) and body (interoceptive) stimuli along with models of action patterns (coming from the proprioceptive system) that are constantly updated, evaluated, and compared with past learned memories.[75] Interactions between conspecifics—for example, between mates or other social competitors and cooperators—further enrich the basic perception of the self. The self is defined not only vis-a-vis the abiotic environment but also vis-a-vis *self-similar others* (e.g., kin, mates).

The functional attributes of UAL not only fulfill all the requirements of the neurobiologists' list, they are also compatible with the main functional features of all the models of consciousness described in chapter 3. For example, both the dynamic core model and the GW model, which are the most detailed models of the organization of a consciousness-generating mammalian system, suggest that nonreflexive information must be attended to and globally synchronized to enable the integration processes that are necessary for the formation of a representation of a compound stimulus. This information must then be stably maintained, allowing it to be widely distributed to the different functional components underlying learning, such as motivational and value signals; various motor programs; and preexisting perceptual, semantic, or procedural schemes. Finally, the integrated information combined with the evaluative inputs from long-term memory must control and direct the animal's behavior.

Existing investigations of the relation between associative learning and conditioning seem to support our hypothesis that the learning of novel compound stimuli and actions is related to consciousness. Studies in humans show that masking an unconsciously perceived, non–reflex eliciting, compound and novel CS that is associated with positive or negative value, such as a picture of flowers or mushrooms that is followed by a mild shock, does not give rise to conditioning, whereas masking a stimulus such as an angry face, which elicits an innate or already learned response, does lead to conditioning.[76] Hence, during the encoding stage, the novel compound stimulus needs to be consciously processed for it to be learned, although the recall of that stimulus need not be conscious. People often do not remember why they are afraid of something or like something yet react predictably following past exposure to masked stimuli. The case of the deeply amnesiac patient who, even though she could not remember the event, avoided shaking the hand of a physician who had once pricked her hand,[77] is an often-discussed example of unaware recall, which is an extreme (and in this case, pathology-induced) example of unconscious recall. A low level of learning, when a signal is rarely recalled, could nevertheless occur for complex percepts masked during encoding if the UAL functional architecture is in place.[78] However, it would be extremely disadvantageous to use a highly unreliable signal as a second-order US. Only highly reliable signals of value should give rise to second-order learning; otherwise, many misleading, "incidental" associations could be formed. We therefore expect that if second-order conditioning to masked, compound, novel percepts occurs, it will happen only under highly unusual conditions. We found no studies showing that a subliminal, compound, non–reflex eliciting stimulus can be assigned a stable value and subsequently used as a US. More experimental studies of humans and other animals are necessary to test our hypothesis that open-ended, second-order associative learning requires conscious encoding.

Our suggestion that UAL is the transition marker for consciousness has obvious implications for the distribution question. Discovering whether or not UAL occurs in different groups could provide an answer to the question about which animals can positively be said to possess minimal consciousness. Systematic comparative studies of UAL have not yet been done, but there is information in the literature about the distribution of complex associative learning involving perceptual fusion and complex operant conditioning, as well as behaviors such as learned navigation through space and relational learning or concept formation, which can be seen as a proxy for UAL.[79] As we indicated earlier and as we show in more detail in subsequent chapters, the data suggest that UAL evolved in the arthropods and the vertebrates during the Cambrian and in the mollusks (cephalopods) approximately 250 million years later. In many groups it was lost (e.g., in parasitic arthropods). Moreover, the seven criteria of consciousness in the neurobiologists' list are clustered in these three groups. Almost all vertebrates exhibit the seven characteristics and so do many arthropods and a few mollusks (the coleoid cephalopods). These seven capacities were identified in members of these groups through studies of perceptual discrimination, attention, action selection, complex reafference, emotional expression, memory, and flexible learning abilities. The historical record thus shows that all seven criteria form a natural grouping in these taxa.

A comparative approach, including systematic experiments testing for UAL and the seven criteria in different animal groups, is crucial for an evolutionary approach to consciousness, but this is not the only way in which one can make progress. We believe that by working "backward" from a functional model of UAL to its enabling system, the existing models of consciousness can be expanded and refined. However, in order to do that we must know what the building blocks of UAL are and how they are organized. Our way of approaching this question is to follow the evolution of UAL. This should enable us to identify the type of processes that are necessary for the integration of neural information and its constant updating, investigate the kind of neural structures that implement it, and build a functional model of UAL. Although we cannot reconstruct the details of this long evolutionary process, we can try to look at the emergence of important stages in the evolution of UAL (and of sentience) and then revisit and update the list and models of minimal consciousness. This is what we attempt to do in the second part of the book.

Introduction to Part II   Major Transitions in the Evolution of the Mind

To understand the origin of some structure, one must first understand what is essential about it—what features it must have to work at all.
—Maynard Smith and Szathmáry 1995, p. 3

We ended the first part of this book with the suggestion that an evolutionary transition marker for consciousness is unlimited (open-ended) associative learning (UAL). We explained what led us to this suggestion, showing that UAL requires the list of properties of consciousness agreed upon by neurobiologists and philosophers and is compatible with the models of consciousness suggested by neuroscientists. However, a chemoton analog, a Gantianesque model of consciousness, is still not available. We believe that the construction of such a model would be facilitated by an account of how UAL, the transition marker of minimal consciousness, evolved. A good evolutionary analysis of UAL could help identify the criteria that enable a minimally conscious system to be recognized, point to gray areas, and help unravel the type of dynamics instantiating such a system. It could also provide insights into the ecological context in which UAL evolved and throw further light on the distribution of consciousness in the living world and its effects on the history of life.

We need to say a little more about the transition-based evolutionary approach because there is more than one way of thinking about evolutionary transitions. As we indicated in chapter 1, we follow Aristotle, who carved biological organization at its teleological joints, and we interpret Daniel Dennett's Generate-and-Test Tower as an evolutionary version of the Aristotelian teleological classification.[1] But as will become clear in the following chapters, we also make use of other transition-oriented approaches that order the biological world from the simple to the complex.[2] Defining

biological complexity and deciding when an increase in complexity qualifies as a major event in evolutionary history are not easy tasks, but there
is general agreement that in their influential book *The Major Transitions in
Evolution*, John Maynard Smith and Eörs Szathmáry provided a good framework for discussing major changes in biological complexity.[3]

The unifying concept used by Maynard Smith and Szathmáry when they
discussed the origins of new levels and types of individuals was *information*,
and they focused on evolutionary changes in the way that information is
stored, transmitted, and interpreted. They identified eight major transitions:
(1) from replicating molecules to populations of molecules in compartments
(protocells); (2) from independent genes to chromosomes; (3) from RNA as
both an information carrier and catalyst to DNA as the carrier of information and proteins as enzymes; (4) from prokaryotes to eukaryotes; (5) from
asexual clones to sexual populations; (6) from single-cell eukaryotes to multicellular organisms with differentiated cells; (7) from solitary individuals
to colonies with nonreproductive castes; and (8) from primate societies to
human societies with language.[4] With the exception of the last, linguistic
transition, all their transitions were associated with changes in the way
that *genetic* (DNA) information is stored, transmitted, or interpreted. In
the transition to linguistic animals (humans), they allowed a new type of
information to be incorporated—information embodied in linguistic representation and communication. Other types of nongenetic information—
cellular epigenetic information, for example, or neural information (which
is a high-level form of developmental information)—were either seen only
in terms of the genetics that gave rise to them or were not mentioned at all.
As a result, the discussion of the major evolutionary transitions remained
incomplete, and the neural transition was omitted;[5] so was the transition
to consciousness.

Maynard Smith and Szathmáry argued that a major transition occurred
when entities that had reproduced autonomously (before the transition)
became part of an assembly of similar entities that formed a new functioning whole (e.g., a multicellular organism made up of previously autonomous
cells). The lower-level entities lost their ability to reproduce independently,
and the new individual, the higher-level entity, became the unit of reproduction. Its stability was assured by the division of labor that led to the
obligatory interdependence of the component units and by the evolution
of policing mechanisms that prevented the components from behaving
autonomously. However, as they also pointed out, not all the major transitions involved an increase in such a nested hierarchy. The transition to
DNA as hereditary material and proteins as enzymes, and the transition

to symbolic language, did not result in the emergence of a new, higher-level entity made of hierarchically nested lower-level units. The result of these transitions was the reorganization of a preexisting level of individuality through the addition of new levels of control to the existing control hierarchy and the emergence of a new type of information (DNA, symbolic language). The new type of information and associated processing and control systems altered the activities of entities in ways that had profound effects on their development and evolution. Clearly, the transition to neural organisms and to consciousness belongs to this latter type of transition. The addition of a new level of hierarchical control to a system can be seen in other progressive evolutionary changes, such as transitions in learning ability that involve changes in the organization, storage, and use of neurally encoded information. Using these criteria, we identify four new transitions in the evolution of neural animals: the transition from multiple exploratory and directed reflexes to limited associative learning; the transition from limited associative learning to UAL and minimal consciousness; the transition from UAL to imagination; and the transition from imagination to symbolic representation and communication.

Our analysis of evolutionary transitions in neural organisms is based on a development-oriented (evo-devo) approach and on an expanded notion of selection. The evo-devo approach has deep roots in late nineteenth-century evolutionary biology and was revived in the 1980s by biologists like Stephen J. Gould, who were interested in the relationship between ontogeny and phylogeny. Since it is a framework for studying evolutionary biology rather than a new theory, evo-devo is difficult to define and delimit, but unlike the traditional neo-Darwinian approaches that dominated evolutionary biology for most of the twentieth century, its point of departure is not population genetics. Rather, analysis starts with the study of the plasticity and robustness of individuals' ontogenies—their adaptive developmental responses to environmental and genetic challenges, which can lead to evolutionary novelties.[6] We identify five main research themes in evo-devo. First, using the comparative method, it is developmental processes from the fertilized egg onward that are being compared, rather than the biological features of adult animals. Second, there is a strong focus on the effects of genetic variations on embryonic development and recognition that some variants can have large, saltational outcomes. Third, the role of developmental plasticity—the ability of the same genotype to generate different phenotypes in different environmental conditions—and the primacy of developmental responses in evolution are highlighted. Fourth, the generation of developmental variations—and their maintenance and inheritance within and between individuals—play a role in evolutionary explanations. Fifth, physical, chemical, and cybernetic constraints on the direction, mode, and tempo of development, and their role in evolution, are emphasized. All these (partially overlapping) approaches are placed within an ecological (and where appropriate, social) dynamic context, and together they make up the twenty-first-century evo-devo (or eco-evo-devo) perspective. In the following chapters, when analyzing the transition to consciousness, we draw on these approaches.

The evo-devo perspective encourages a developmental-evolutionary extension of the traditional view of selection, something that is particularly important for our approach to consciousness. Our extended view of selection is influenced by the ideas of George Price, who during a short foray into biology between the late 1960s and early 1970s made a great contribution to evolutionary theory by developing the most general formalization of selection, the now famous Price equation.[7] His notion of selection is clearly expressed in his 1971 paper, written shortly before his suicide and published twenty-four years later:

> Selection has been studied mainly in genetics, but of course there is much more to selection than just genetical selection. In psychology, for example, trial-and-error learning is simply learning by selection. In chemistry, selection operates in a recrystallization under equilibrium conditions, with impure and irregular crystals dissolving and pure, well-formed crystals growing. In palaeontology and archaeology, selection especially favours stones, pottery, and teeth, and greatly increases the frequency of mandibles among the bones of the hominid skeleton. In linguistics, selection unceasingly shapes and reshapes phonetics, grammar, and vocabulary. In history we see political selection in the rise of Macedonia, Rome, and Muscovy. Similarly, economic selection in private enterprise systems causes the rise and fall of firms and products. And science itself is shaped in part by selection, with experimental tests and other criteria selecting among rival hypotheses. (Price 1971/1995, p. 389)

What Price argued for is a general concept of selection that includes *both* the familiar concept of Darwinian selection (selection among multiplying, replicating entities) and what he called "sample selection" (S-selection), a process of picking up a subset from a set according to some value criterion. S-selection *does not* require reproduction or replication, and in a letter to a friend, Price gave, as an example of such a process, the selection of radio stations by turning a dial.[8]

Price's broad notion of selection applies to everyday observations of organisms coping with novel challenges. It is implicitly based on their adaptive plasticity; their ability to respond flexibly to altered, predictable, and unpredictable environmental conditions and retain or regain functional homeostatic states. When organisms, including single-celled ones, are faced with a challenge to which they do not have a specific evolved solution, they do not sit and wait passively for miracles. They mobilize internal processes of random or semirandom search, exploring and sampling the environment, and stabilize internal states that lead to the alleviation of or, if they are lucky, to the resolution of the problem. We therefore describe selection here as the sampling process that leads to a decrease in the uncertainty of reaching an attractor. From this broad Pricean perspective, classical Darwinian evolution by natural selection is a special case of the exploration-stabilization principle; a case that involves differential reproduction through multiplication and replication.

Exploration and selective stabilization mechanisms are all based on a similar principle—the generation of a large set of local variations from which only a subset is eventually stabilized and manifested. Which particular output is realized depends on the initial conditions and the states around which development can be stably organized. Such dynamically stabilized states are called by systems biologists (who borrowed the term from physicists) "attractor states," or "attractors." As we described in chapter 3, an attractor in physics is defined as a set (a defined collection of distinct objects) toward which a variable moves according to the dictates of a dynamical system. Points that get close enough to the attractor (within the basin of attraction) remain close even if slightly disturbed. Biological attractors are usually functional: the mechanisms that enable them to be reached reliably, in spite of different starting conditions, evolved by natural selection. For example, when a cell whose growth has been arrested finds, through exploration, an internal state that allows it to divide, the networks that produce this functional state are stabilized.[9] Mary Jane West-Eberhard, an evolutionary biologist who explores the role of plasticity in evolution, calls the processes of reaching a new, stable, and functional attractor state "phenotypic accommodation."[10]

Biology is full of examples of phenotypic accommodation through exploration-stabilization processes. Foraging in army ants, for example, is based on exploration stabilization: ants move in random directions, chemically marking outgoing and returning paths while doing so. A path that is productive (leads to food) is used more frequently than others and hence is more heavily marked, leading to further use (stabilization of this particular usage), and so on. The persistence of the more frequently used path through this type of positive feedback is therefore an example of selective stabilization.[11] Other examples include spindle formation within cells, the already

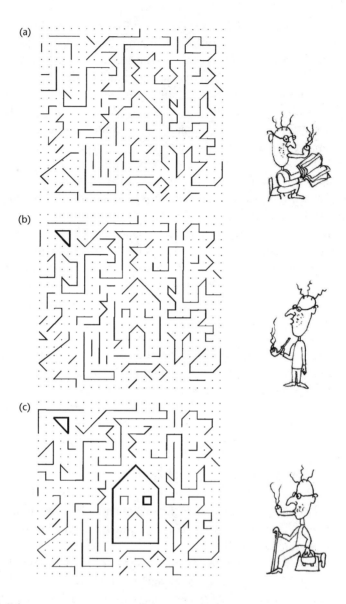

Figure II.1

An imaginary exploration-stabilization system. The rules of this system are as follows: (*a*) lines between points continuously form, disappear, or extend to join neighboring points; (*b*) the longer the line, the less it changes, but stability is achieved only when a closed shape (e.g., a triangle) is formed; (*c*) a functional closed structure (the house) has enhanced stability and increases the likelihood of closure of enclosed structures. Reproduced by permission of the MIT Press.

mentioned exploration of metabolic networks in highly stressed cells, the exploratory search for water by plant roots during drought, exploration-based perceptual and motor learning in animals, and the exploratory cultural learning and reconstruction of social structures in humans.[12]

Exploration and selective stabilization is a strategy that is employed across the board. As Gerhart and Kirschner have stressed, it helps maximize the production of adaptive phenotypic variations (which they call "facilitated variation") in evolution.[13] At both the developmental and the phylogenetic level, it is the only strategy that can adaptively cope with the unknown (unless the hand of God is mobilized). This general strategy leads, through what the developmental biologist Yoav Soen calls "adaptive improvisation," to a functional, stress-reducing state that preserves the trajectory of developmental change despite perturbations.[14] The exploration-stabilization principle is shown in figure II.1, which depicts an imaginary system and articulates the rules that lead to an improvised adaptive outcome.

The exploration and selective stabilization mechanisms in the nervous system are central to our discussion of the evolution of learning in the following chapters. They are fundamental to the nervous system's development and function and occur intracellularly and at several other levels: that of neuronal cell number, synaptic connectivity, and at higher levels of connectivity between neuronal groups and neuronal maps. First, nerve cells are overproduced and most die. They "commit suicide" unless they are stabilized by "survival factors" during development: only axons whose tips accidentally hit a region with "survival remain alive.[15] In this way, the number of nerve cells is pruned through competition for "survival factors," the location of which determines where neurons will connect and hence which neurons will remain. Second, the synapses in surviving neurons are also overproduced, and most are pruned: connections that have the highest functional efficacy persist according to Hebb's rule: "Neurons that fire together wire together." Many other factors guide the formation of stable synapses, which are determined by both the presynaptic growth cone and the postsynaptic target cell,[16] but the net result is that only certain synaptic connections among the many initially possible are stabilized and retained. As we discussed in chapter 3, Edelman and Changeux both suggested that neural selection (mostly sample selection, which does not involve processes of replication) occurs in the nervous system during ontogenetic development and learning. Such selection is based on both selection for neuronal survival and selection for synapse stability. Both Changeux and Edelman added a third level of stabilization: Edelman argued that reentry between neuronal groups (maps) constitutes the third level of stabilization, and

Changeux argued that higher levels of exploration and stabilization among neural networks occur in the brain during complex learning—for example, when a songbird learns its species-typical song.[17]

Exploration and stabilization were also central to Freeman's view of consciousness and cognition (chapter 3). Freeman suggested that endogenous exploratory and chaotic neural activity leads to exploratory motor behavior through which the animal selects those environmental stimuli whose effects match its internal states and lead to perception or goal-directed movement. He used the term "neural attractors" to describe the dynamic neural states that form and change as the animal learns though this process of exploration and stabilization. This view resonates with the description of trial-and-error learning (learning by consequences) that was described in purely behavioral terms by Bain, Spencer, and Skinner and will be discussed in chapters 7 and 8.[18]

If exploration-stabilization offers a general perspective on selection, can one develop a notion of fitness that will accommodate S-selection through exploration-stabilization processes? In other words, can one not only describe functional developmental information but also measure it? Evolutionary change through Darwinian selection is measured using the classical notion of fitness—relative or absolute reproductive success—but what is the equivalent of fitness in the case of sample selection? A very useful notion, suggested by Karl Friston, is the "free energy principle," which says that any self-organizing system that is at equilibrium with its environment must minimize its free energy.[19] Friston developed a hierarchical predictive model of cognition, with cognition understood in the broad, autopoietic sense. He suggested that, in theory, one can measure the variational free energy: the difference between the probability distribution of environmental factors that affect the animal system (for example, food items, predators) and an arbitrary distribution encoded by an internal configuration in the animal's brain and body. The enbrained animal can minimize the free energy difference by altering its state in two ways: either by changing its overt behavior, and thus the way in which it samples its environments, or by changing the way it perceives the environment—that is, the way it "interprets" what it perceives. Hence, a decrease in the variational free energy can measure how a self-organizing, goal-directed system like an animal resists the natural tendency to disorder by using regulatory, homeostasis-retaining or homeostasis-increasing processes.[20] Free energy minimization is the most general notion of how the organism can increase its adaptive fit to its environment. Darwinian selection can be seen as a special case of the free energy principle. Exploration-stabilization and adaptive improvisation are explicit in Friston's principle: estimating the change in variational free energy and using it as a measure of a change in adaptive fit depend on the use of internal models that generate neural states. Such internal models can be seen as exploratory "hypotheses" about the world, and those that increase homeostasis and homeorhesis (the stability of the trajectory of change) are stabilized. Of course, during both ontogeny and phylogeny, as new levels of biological organization emerge, there is increased and correlated variation in the relevant environment and in the internal states (such as those of a nervous system) that represent it. Using the variational free energy instead of classical fitness provides a more general concept of adaptive fit, one that is suitable for the study of the selective processes that occur as cognition and consciousness develop (during ontogeny) and evolve (during phylogeny).

In the second part of the book, we pursue this general approach and the evo-devo, transition-oriented directions of research that it entails, emphasizing the role of plasticity and of exploration-stabilization mechanisms in the evolution of learning. A theme that runs throughout our reconstruction of the evolutionary history of learning and minimal consciousness is that evolutionary innovations not only create new adaptive solutions but also open up a Pandora's box of new problems, which lead to a spiral of new selection regimes, new "solutions," new problems, and so on. We start (chapter 6) with the building blocks of learning, provide an overview of the transition to neural animals, and discuss the molecular and behavioral components found in cnidarians, from which simple forms of associative learning, and later UAL, probably evolved. Like many before us, we link the evolution of the nervous system with the evolution of mobility and muscles. We stress the problem mobility opened up: once moving macroscopic animals had evolved, they had to distinguish between the sensory effects of their own movements and those that were independent of their own actions, a difficulty that led to the evolution of new modulatory interactions between sensory and motor neural centers. In chapter 7 we describe the evolution of limited associative learning and the problem that this great adaptive innovation brought about—the problem of overlearning. This stumbling block was partially overcome by restricting learning to surprising, newsworthy discrepancies between expectations based on what has been learned and the actual, current effects of a new stimulus.

Building on the discussion in the preceding seven chapters, chapter 8 considers the transition to UAL and to minimal consciousness. We describe the functional neural architecture that constructed UAL, which, we argue, is the architecture underlying the simplest mental representations, and describe the different realizations of this architecture in vertebrates, arthropods, and mollusks. UAL led to a great increase in adaptability, but like limited associative learning, it also led to a severe problem of overlearning, which was evolutionarily solved by modulating the animals' memory and their responses to stress. This topic is discussed in chapter 9, where we position our evolutionary proposal within an ecological context. We suggest that the evolutionary emergence of limited and unlimited associative learning had dramatic effects, acting as an adaptability driver of the Cambrian explosion. Once in place, the evolution of UAL led in some lineages (notably, birds and mammals, but also in the very different cephalopods) to the emergence of "Popperian" animals, creatures endowed with imagination. Chapter 10 brings this book to an end. We discuss the continuity between life and consciousness, examine the possibility (and implications) of constructing artificial conscious beings, and outline a further stage in the evolution of consciousness: the transition to the human "rational soul," to human symbolic-based cognition, and to human abstract values. This last chapter takes the form of a dialogue, with a critical reader who questions our interpretations and who wants to understand the implications of our proposal for neural and cognitive consciousness studies, for the philosophy of mind, and for ethics.

6 The Neural Transition and the Building Blocks of Minimal Consciousness

If unlimited associative learning (UAL) is the diagnostic marker of minimal consciousness, what were the conditions necessary for its evolution? What were its building blocks? We start with the origins of a new type of individual, the neural, mobile individual. With neurons, a new unifying "language" appeared: the action potential became a currency of communication among cells. We therefore need to explore how neurons communicating across gaps generate a nervous system, why multicellular large animals that move need a nervous system, and how interactions among neurons bring about memory and learning. This opens up a whole plethora of questions: What cell type gave rise to the first neuron? In what animal groups did the neuron and the nervous system first emerge? Have present-day sponges lost their nerves? Did the nervous system first appear in medusas, in hydras, or in comb jellies, all of which have diffuse nerve nets spread throughout their bodies? Observations and experiments on cnidarians (medusas and sea anemones) show that dispersed nerve nets enable their owners to explore their world through rapid, coordinated movements and to react to stimuli in a flexible, adaptive manner, solving the problem of distinguishing between the sensory effects that depend on their own action and those that are independent of activities. Moreover, cnidarians can learn: they learn nonassociatively, through habituation and sensitization, and these simple forms of learning enable them to adapt efficiently to their complex world. Can these animals be said to have the precursors of primordial consciousness?

> The chief function of the body is to carry the brain around.
> —Thomas Edison[1]

> Behavior is initiated from within, spontaneously, actively. This is part of the organism's self-ness.
> —Heisenberg 2013, p. 9

Ever since Lamarck first pondered the question about why neural organization emerged only in the animal kingdom, the answer generally suggested has been because animals move. All animals that have neurons also have muscles, the only exception being some minute, degenerate, parasitic cnidarians, the myxozoans. The intuition that movement is a key characteristic of animals is an ancient one, and Aristotle recognized that patterned or coordinated movement requires a special type of control, an extending and contracting center (which he thought was the heart). Unlike plants and fungi, with their rigid cell walls, animals do not have to rely solely on growth by cell division to change shape and location: the elasticity and contractility of animal cells, notably of muscle cells, enable them to shorten, contract, stretch, and elongate. This type of cell-level movement seems to be a prerequisite for the mobility of all but the smallest multicellular animals.

The movement of a multicellular organism poses a big problem. A tiny multicellular animal can move through coordinated movements of cilia, brought about by localized, self-generated, rhythmic motions being transferred among connected cells, but this is impossible when organisms become bigger. When a sizeable, moving, multicellular organism alters its position in space, it moves as a unit, so changes in the shapes and functions of many different cell clusters have to be coordinated. Within a single cell, many molecules act as specific signals, conveying information from one part or organelle to another while chemical signals, such as hormones, are conveyed throughout the body by a circulation system (e.g., the xylem and phloem in plants and the blood vessels in animals). For a large mobile animal, however, this type of signaling is insufficient in terms of both coordination and speed.

The members of the two animal phyla that lack nerves and muscles, the sponges and the placozoans, do not face this problem because they are either sessile or move slowly using cilia.[2] But rapid and efficient communication among moving body parts in a large multicellular animal cannot be achieved by diffusion or the slow circulation of chemical signals. The muscles used when sizeable animals relocate do not move as single cells but as tissue sheets, and the animal's movements require integrated control that organizes the activity of muscle sheets that are spatially disconnected.[3] Such coordinated movements are necessary if macroscopic animals are to avoid predators and other obstacles and search for food and mates. While sensing threats or opportunities is possible without locomotion, moving to avoid danger or search for benefits demands well-orchestrated movements. Coordination of the activities of the muscle sheets is also crucial for the proper functioning of internal organs such as the gut, whose peristaltic contractions must be synchronized and sequentially organized.[4]

The focus on coordinated movement has led some researchers to suggest that the original driving force for the evolution of nervous systems was the need to bind together many cellular units into muscle sheets; the function of the nervous system as a general information processor and its organization around sensors and effectors were later evolutionary developments. This "movement-first" view was presented by the British zoologist Carl Pantin, who suggested that during evolution spontaneous, internally generated motor activity only later became modulated by external signals and came under the control of the nervous system. The idea that autonomous and spontaneous motor behavior is a key characteristic of animals—and the basis of their agency—was also put forward by Martin Heisenberg, who emphasized the role of stochastic behavior in the generation of flexible adaptive responses.[5] Although Heisenberg did not discuss evolution, his view, like that of Walter Freeman III, whose ideas were discussed in chapter 3, implies that the need for coordinated movement should be seen as a driving force in the evolution of neural animals. As is shown by the many examples of exploration-stabilization behaviors, from the chemotactic behavior of bacteria to the activities of humans during trial-and-error learning, living organisms employ spontaneous, internally initiated movements to explore their environment, and they organize their behavior by using sensory inputs to reinforce and stabilize some of their actions. Spontaneous, highly variable activity is beneficial because unpredictable movements enable the animal to avoid predators, find new resources, and develop new behaviors.

However, a moving animal faces an immediate problem that if not solved would render even coordinated and rapid movements maladaptive. An animal must be able to determine whether a sensory stimulus is the consequence of a change in the environment or is the outcome of its own actions. For example, is the stimulation of mechanoreceptors in the skin the result of an external obstacle or the result of one's own movement? The sensory stimulations may be identical, but the consequences for the animal could be dire if it does not distinguish between an action-dependent, self-generated (reafferent) stimulation and a world-generated (exafferent) stimulation.

Obviously, a moving macroscopic animal that can coordinate tissue and organ movements and is able to distinguish between its own and world-generated stimulation must have a nervous system adequate for the job.

There can be no "simple" nervous systems in moving animals. Such animals need neural circuits that control sensorimotor coordination, and there must be something about neurons that enables this. What is special about neurons?

The Major Neural Transition: A New Information System

> Gotch and Keith Lucas, by their analysis of the "refractory period" in nerves, gave us for the first time a clear idea of what may be called the functional value of the nervous impulse. They showed what the nerve fibre can do as a means of communication and what it cannot.... It is of the first importance in the problems of sensation, for it shows what sort of information a sense organ can transmit to the brain and in what form the message must be sent.
> —Adrian 1928, pp. 13–14

Lord Adrian, the British electrophysiologist who shared the 1932 Nobel Prize in Physiology or Medicine with Charles Sherrington, introduced the "information metaphor" into neurobiology as early as 1928.[6] At that time, the neuron (box 6.1) was already recognized as the basic unit of organization of the nervous system, and Adrian referred to the "nervous impulse" as the information currency for rapidly communicating messages within the body. Adrian showed that the electrical signal (the "action potential," the nerve impulse) was the same whether carried by sensory or motor nerves. The nerve impulse is a stereotyped electrical signal into which all modes of sensory stimuli are translated—photons, chemicals, heat, sound waves, and other mechanical types of energy. This common currency can be used to "map" the external world within the nervous system—that is, to create a "model" of the rapidly changing world surrounding the moving animal. Furthermore, nerve impulses are also used to map the internal environment—the organism's body—within the same system. Importantly, the common currency makes it possible to bind together and integrate different types of stimuli that carry information from a single modality and, even more remarkably, from several different modalities.

The nervous system can accomplish these feats because of the combination of its unique structural and functional properties: each neuron can form synapses with many other neurons, including very distant ones; electrical signaling is rapid; and there is an electric "common language."[7] Figure 6.2 shows how the many fibers emanating from a neuron can contact many other neurons, thus forming neural networks. Through them,

Box 6.1
The neuron, the synapse, and the supporting glia.

The term "neuron" was first used in 1891 by the German anatomist Heinrich Wilhelm Waldeyer-Hartz (1836–1921) in the context of the debate between two great anatomists, the Italian Camillo Golgi (1843–1926) and the Spaniard Santiago Ramón y Cajal (1852–1934), about the nature of the organization of the nervous system. Golgi invented a new staining technique, which when used by Ramón y Cajal led to a detailed description of the nervous system (for which both men received the 1906 Nobel Prize), but the two anatomists disagreed about the nature of the neural circuits they identified and described. While Golgi believed that the nervous system consisted of a continuous network (a view called the "reticular hypothesis"), Ramón y Cajal envisaged the nervous system as composed of discrete cells with no continuity among them.[8] This view was called the "neuron doctrine" by Waldeyer-Hartz, who was persuaded by the Spanish anatomist's arguments and used the term "neuron" to describe the cellular unit of the system. Once the electron microscope was developed and synaptic gaps became an unquestionable fact, Ramón y Cajal's view was vindicated, and the reticular hypothesis vanished.

As neurobiologists see it today, the neuron, a cell type specialized for communication, is an asymmetrical cell composed of several compartments that carry out information processing. These compartments are specialized to receive signals, to integrate them, and to deliver them to other cells. The three main morphological specializations of a neuron are illustrated in figure 6.1A: a cell body—the "soma"; many thin tree-like branches, the "dendrites," which are attached to the soma; and a single long branch, the "axon," emanating from one point of the cell body.

Electrical information is transmitted in one direction along the neuron: The dendrites of a neuron receive and gather chemical signals from other neurons (or from receptor cells like those found in the ear or eye), and after a transduction process, send them on as small electrical changes to the cell soma. The received signals are integrated in the soma, and if they reach a certain threshold, they are transformed into an all-or-none, large, constant-sized electrical impulse, which consists of a brief, reversible electrical polarization—a transient change in the voltage between the inside of the neuron and the outside. In a typical neuron, the impulse, an "action potential," is propagated very quickly and unidirectionally along the axon—away from the soma toward another neuron.[9]

Neurons communicate in a special way: between the end of the axon of one neuron (the nerve terminal or terminal buttons) and the beginning of a dendrite of another neuron is a small physical gap that does not allow

(continued)

Box 6.1 (continued)

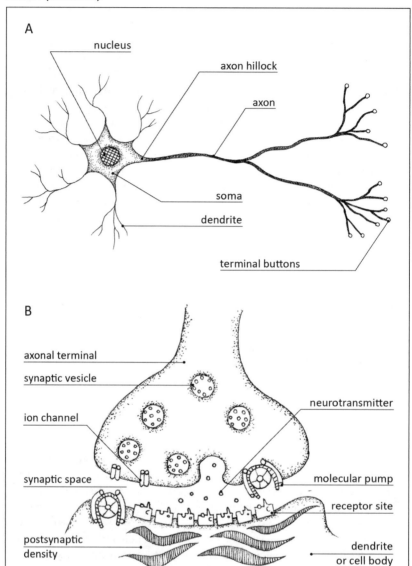

Figure 6.1

A, a schematic depiction of a neuron. *B*, the synapse between an axon terminal (*top*) and a dendrite (*bottom*). The neuron from which the electrical impulses arrive, and from which neurotransmitter molecules are released, is known as the "presynaptic neuron"; the neuron that binds the transmitter molecules is known as the "postsynaptic" neuron.

Box 6.1 (continued)

the passage of electrical signals. In this gap the electrical nerve impulse is transduced into a chemical message. In 1897, Sherrington, the British neurophysiologist who investigated reflexes in the nervous system, used the term "synapse" for the junction between the two neurons, which consists anatomically of the gap together with subcellular regions near the plasma membranes of the two separated neurons (figure 6.1B).[10] As the electrical impulse reaches the terminal of the axon of the presynaptic neuron, it causes specialized, voltage-dependent calcium channels located in the terminal's membrane to open. As a result, calcium ions, which are much more abundant extracellularly than intracellularly, rush into the terminal and trigger the fusion of synaptic vesicles with the terminal's plasma membrane; since the vesicles contain chemical neurotransmitter molecules, fusion results in the secretion of these chemicals into the synaptic gap. The transmitter molecules diffuse along the narrow, 20-nm-wide gap and reach the postsynaptic cell's dendrites, where they act as a chemical signaling communication device. The neurotransmitter molecules bind to specific receptor macromolecules embedded within the membrane of the dendrite of the postsynaptic neuron; this binding triggers changes in the postsynaptic membrane, resulting in small electrical changes that are transmitted to the soma. The synapse operates as a filter and a point of control: it allows only strong-enough chemical messages (resulting from the concerted release of transmitter molecules from many synaptic vesicles that fuse with the presynaptic membrane) to be transmitted further. When the chemical message is large enough, the depolarization of the postsynaptic cell membrane reaches the threshold necessary to elicit an action potential, thus transmitting the message further on. If the chemical message is too small, or if it is blocked, transmission from the presynaptic to the postsynaptic neuron will not take place.[11]

Glia (glial cells, also called "neuroglia") are additional nonneuron components of the nervous system. They are present in most animals but have not been found in some basal taxa, such as urochordates, hemichordates, bryozoans, rotifers, and basal platyhelminths.[12] Glia form the myelin sheath that wraps around some neural fibers, thus insulating them and increasing the conduction velocity of the nerve impulse. They have a role in the maintenance of neurons by regulating the fluid contents surrounding them and the nutrients that enter the cells. Glia create physical barriers that guide the direction taken by nerve fibers and help in pruning synapses and stimulating the growth of new neurons. At some synapses, glia actively take part in synaptic transmission by releasing neurotransmitters that can modulqate the effects of the "primary" transmitter released from the presynaptic terminal, causing excitatory or inhibitory actions at the postsynaptic cell.[13]

(continued)

Box 6.1 (continued)

Small glial cells, known as "microglia," are the primary immune cells of the brain, acting as macrophages. They protect the CNS by scavenging for damaged cells and synapses and acting against infecting molecules that manage to cross the blood-brain barrier.[14] Some researchers believe that microglial cells play a role in memory and learning by contributing to synaptic plasticity and are essential in preventing pathological states such as Alzheimer's disease.[15]

an enormous amount of information can be conveyed from virtually endless sources, which may be fluctuating in activity, to thousands and thousands of target neurons and effectors (muscle and endocrine cells). The number of messages (and bits of information) communicated within a nervous system at any given moment is simply staggering: even a simple invertebrate may have interconnected networks of several thousand nerve cells, and in humans the number of neurons is on the order of one hundred billion.

The flexibility of many of the interactions within neurons and among neurons allows variant local neural circuits to be formed. Circuits can be embedded within circuits, and circuits can communicate laterally, with both negative and positive feedback relations within and between them. Furthermore, organizing circuits into clusters or centers can contribute to even more efficient integration of information and to better control of movement and other activities. In addition, the neurons' activation thresholds may be modulated by hormones and by slow bioelectric signals from nonneural cells and organ systems, which are less precise than neurotransmitters in targeting specific cells but are better suited for controlling overall homeostatic states.

Order and functionality in such a vast, rich communication system cannot be based on precise instructions; there have to be strategies that make sense of diversity and massive stochastic variability in neural activity, as well as powerful constraints. The nervous system is a spontaneously and inherently active system: even in animals under anesthesia or in comatose states, there is ongoing spontaneous activity. All neurons, at all times, contain incessant electric discharges that can become amplified and lead to local and global activations, and in neurons, as in other cells, are ongoing changes due to the molecular turnover that accompanies and results from chemical reactions of synthesis and degradation within each cell. These stochastic activities occurring between and within nonlinearly interacting

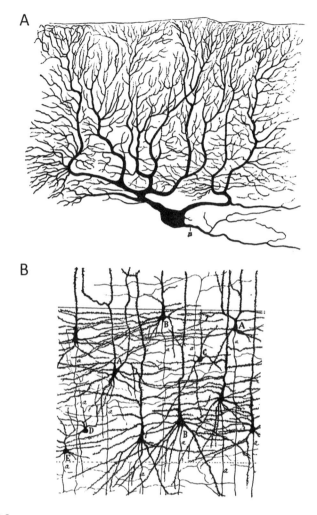

Figure 6.2
Two of Ramon y Cajal's exquisite drawings. *A*, a stained single (Purkinje) cell from a human cerebellum with a huge number of dendrites. *B*, pyramidal neurons and interneurons forming a neural net.

neurons generate great diversity. This diversity is the "raw material" that is organized, through the amplification and inhibition of neural activities, into coherent actions and coherently perceived inputs.

Once in place, the neural information system, with its vast capacity to vary, brought about a profound change in the way organisms responded to the world: neural animals could adapt very flexibly and very rapidly to changing conditions, including, importantly, conditions imposed on them by other animals. A second, no less revolutionary outcome brought about by the nervous system is related to the new type of connectivity it bestowed on animals. The structure and nature of the synapses and the types of signals that emerged in neurons enabled more than just the formation of circuits and subcircuits that could connect remote parts of the organism and take on specialized functions; because synapses and synaptic connections can change by virtue of the signals they receive, they also enable the magic of flexible learning and long-term remembering. The synapse, as noted in the previous chapter, is not the only locus of memory in neurons. Neurons also have epigenetic memory systems both in the nucleus (induced chromatin marks that function as memory traces) and in regulatory factors, such as RNAs and protein complexes, that can be transferred among cells and alter their threshold of reaction.[16] When an animal learns, connections among neurons grow and strengthen through activity and diminish with a lack of signaling. Thus, during the lifetime of the animal, whatever it senses can leave a mark on its synapses and its intracellular epigenetic memory, literally forming tracks of past environmental encounters—that is, memory traces or engrams. Furthermore, repeated encounters with the environment make certain signals and responses easier or more difficult to elicit, and linkages can form between various incoming signals from different sources, thus producing contextual learning.

As we suggested in the introduction to this part of the book, the transition to neural animals clearly qualifies as a major evolutionary transition: if one of the hallmarks of a transition is a change in the way information is stored, transmitted, and processed, then the emergence of animals that use a new type of information (neural information), which is processed, stored, and transmitted in new ways, should surely be seen as one of the most important transitions in evolution. The nervous system is a key distinguishing feature of the vast majority of animals.[17]

But what first brought about this leap in information processing and led to the neural transition? Was it the first neurons or the first neural networks? The answer is that the origins of neurons, synapses, and webs of neural connections cannot be teased apart: neurons are by definition connecting cells, and synapses are defined as connections between neurons and other cells. In a sense, it is the neural network that constitutes and defines its units (the neuron and the synapse) rather than the other way around. Therefore, although we are starting by discussing the evolution of neurons and synapses, this should not be taken to suggest that they were prior to neural *systems*.

The Advent of Communicating Neurons

The development of cheap DNA-sequencing techniques and sophisticated algorithms for comparing DNA has led to the consensus that all organisms share a great deal of their molecular machinery. The basic molecular constituents of nerve cells were clearly present prior to the advent of the first neuron, the first synapse, and the first sensory cells. Indeed, it seems that a substantial part of the molecular machinery found in neurons existed even before the first multicellular organisms evolved.

In all cells, an electrochemical gradient between the intracellular and extracellular compartments is maintained by active molecular pumps embedded within the cell membrane, which actively transport ions using cellular energy. This gradient and the flux of ions across the cell membrane is the basis for maintaining features such as cell shape, volume, osmolarity, and excitability. Ions diffuse through the ion channels—specialized protein molecules embedded in the membrane. There are several categories of ion channels: channels triggered by changes in membrane potential, channels triggered by mechanical and other physical factors, and channels triggered by chemicals. One type of channel, ligand-gated receptor channels (LGRCs), is triggered by specific molecules, such as acetylcholine, and analyses of DNA and amino acid data from many eukaryotes have revealed a high degree of amino acid sequence similarity in these channels. It appears that they originated in a common ancestor that existed before the dawn of eukaryotes and neurons 2.5 billion years ago. Some LGRCs that are prominent in the synapses of all eukaryotes have even been identified in bacteria. It is suggested that in these prokaryotes and other unicellular organisms, they functioned as chemotaxis receptors or osmotic regulators, and as multicellularity evolved their roles changed to serve in intercellular communication.

The structure and function of other types of ion channels are also similar in prokaryotes and both unicellular and multicellular eukaryotes.[18] But ion channels and electrical transmission are not the only ancient components of neurons: some elements of what we recognize as a synapse in neural

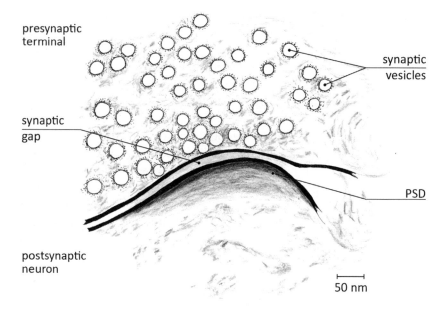

Figure 6.3
A synapse between presynaptic and postsynaptic cells. The dense area below the plasma membrane in the postsynaptic neuron is the postsynaptic density (PSD).

organisms are also found in their preneural predecessors. One example is a microanatomical structure in the postsynaptic region of the synapse that is known as the "postsynaptic density" (PSD; see figures 6.1*B* and 6.3).[19] It is a specialized part of the cytoskeleton containing many types of proteins. There are many hundreds of structural and signaling proteins in the PSD (up to fourteen hundred in mammals), but nerveless sponges also have a PSD, albeit with fewer proteins and with functions that differ from those in neural synapses. The basic structure of PSD proteins in sponges is, however, no different from that of those found in mammals.[20]

It is clearly not easy to identify the molecular innovations that mark the transition to neural organisms. Which changes are important and which are not? Are there just a few changes with large effects, or are there many changes, each with a small effect? To answer these questions, biologists need to compare the molecular constitution of neural and preneural metazoans and the organization of their gene networks. A complementary approach, which we take in the next section and which could make the identification of the key molecular features easier, is to try to decipher the cellular origin of the neuron—the type of cell from which it evolved.

A New Kind of Division of Labor among Cells: Sensors, Effectors, and Linkers Become Separate

In 1872, Nicolaus Kleinenberg, a student of the famous German naturalist Ernst Haeckel, published his discovery of neurons in the cnidarian *Hydra* and described them as polarized cells with a "sensory" side facing the environment and a second side that might have a motor function.[21] A century later, in 1970, the neurobiologist George Mackie proposed that a layer of ancestral two-sided cells (a myoepithelium) gave rise to several types of specialized cells (figure 6.4A). The end result was an epithelial sheet of cells on the surface, a sheet of muscle cells beneath it, and sensory cells and motor neurons spanning them.[22] The cells in the original layer were electrically coupled and had both sensory and motor functions; at the next stage, the two functions became distinct, and protomuscle cells (protomyocytes) separated from and sank below the "sensory" side of the original coupled cells. As the distance between the surface sheet and the protomyocytes increased, there was evolution of connecting cells, and following yet more specialization, the first synapses also appeared, with sensory cells forming at one end and motor neurons at the other, close to the muscle cells.

A related, though not identical, scenario, proposed in 2011 by Gáspár Jékely, suggests that the first eumetazoans (i.e., true neural metazoans, which means all animals except sponges, placozoans, and some obscure life forms of uncertain nature) evolved from a ciliated sponge-like larva, and the first steps of neuron evolution occurred in this stage, as a result of selection for controlling the motion of cilia.[23] Ciliated larvae are found in many animals (e.g., cnidarians, annelids, mollusks, echinoderms) and sometimes adult stages are also ciliated (e.g., in ctenophores and acoels). Jékely showed that the larvae of the annelid worm *Platynereis dumerilii* have direct neural connections between their simple eyes and the ciliated cells below and argued that these may be a heritage from a very early stage of eye and brain evolution. He suggested that epithelial-like sensory cells secreted neuropeptides or transmitters that regulated nearby ciliated cells, leading to changes in swimming speed. Greater efficacy and precision in regulation emerged when the sensory cells developed basal processes (the first axons) and secretory release sites; as the latter came into contact with the target cells, the first synapses were formed (figure 6.4B).

Yet another interesting idea is that neurons evolved in the context of defense and repair. Leonid Moroz has suggested that during the earliest stages of animal evolution, widespread triggering of gene expression occurred as a response to injury or stress, with damaged cells releasing signal molecules or transmitters and triggering repair activities.[24] Maybe, he speculated,

A

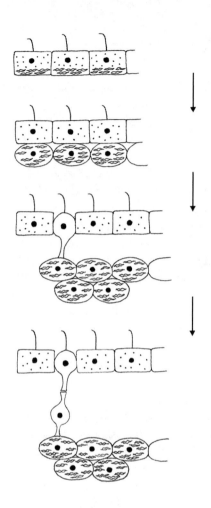

B

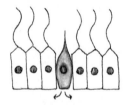 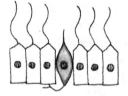 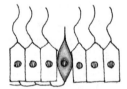

Figure 6.4

Two scenarios of evolving neurons. *A*, neurons forming from an epithelial tissue sheet. Electrically coupled epithelial cells with both sensory (*dotted*) and motor (*wavy*) functions (*uppermost level*) evolved into protomyocyte (motor) cells and sensory cells (*second level*); protoneurons (one shown—the cell with an elongated end) then evolved from sensory cells, connecting the epithelial and muscle sheets (*third level*); neurosensory and motor neurons evolved (one of each shown), with a synapse neurons evolved in ancestral metazoans as a result of selection for adaptive regenerative responses to injuries. The idea is that damage to the cell led to the release of chemical messengers that acted as growth factors and controlled transdifferentiation (differentiation into a different cell type). In response to the chemical signals released by the damaged cells and the regrowth of their processes, neuron-like cells were formed. According to this suggestion, from the outset there was a link between the nervous and repair-immune functions, and the two cell types evolved from a common, multifunctional type of ancestral cell.

Although the cellular origin of nerve cells suggested by the theories just described are different, all see the emergence of the neuron as a result of a division of labor between cells. A different approach, focusing on motor and neural *systems* and building on the ideas of Carl Pantin, has been developed by Keijzer and his colleagues. They suggested that the early evolution of nervous systems was driven by the need to organize a new multicellular effector—sheets of muscle cells—to move the animal. Control of self-induced, spontaneous movement that operated on a whole-body scale was, however, necessary, and the first nervous systems induced and coordinated the spontaneous contractile activity of extensive muscle sheets underneath the animal's sensitive skin. They therefore suggested that the first nervous system was a diffuse "skin-brain" system: the nervous system and the muscle system coevolved as connections formed across a contractile tissue (myoepithelium).[25]

The different scenarios we have just described share a major insight: the evolutionary origin of the nervous system was linked not only, or even mainly, to responses to the outside world. It evolved in the context of the necessity to coordinate the movements of a macroscopic, mobile, multicellular body and to integrate and coordinate interactions among different internal organ systems, such as the gut. There is good evidence that the nervous system is intimately involved in animal embryonic and postembryonic development and growth, and controlling the development and growth of internal organ systems may well have been one of the major functions of the first nervous systems.[26] The scenarios we have described

---

between them (*bottom level*). *Source*: Adapted from Mackie 1970. *B*, neurons evolving from sensory cells. Sensory cells (*gray*) that regulate the motion of cilia on neighboring cells by secreting transmitter molecules (*left*) evolve to form projections that reach the ciliated cell (*center*) or several cells (*right*). These projections may be the first axons, and their contacts with the target cells may be the first synapses. *Source*: Jékely 2011.

differ, however, so can we choose between them? A method known as "cell-type molecular fingerprinting" may help resolve the question of the cell type that gave rise to the neuron or to the neuron system. Every type of cell has special structural and physiological features that depend on its "differentiation signature" and "regulatory signature." The differentiation signature is the pattern of gene activity unique to a given cell type; the regulatory signature is the combination of transcription factors that switch on the differentiation genes. Together, the two signatures constitute the molecular fingerprint of each type of cell. Hence, homologous cell types, even when separated by long evolutionary periods, can be identified by comparing these molecular fingerprints. So far, peripheral sensory neurons have been discovered to be homologous in vertebrates and insects, and the motor neurons of vertebrates, insects, and nematodes have also been found to be homologous. Homologies can help in tracing the origins of neurons and, at a later evolutionary stage, the origin or origins of brains and nerve chords.[27] A complementary way to approach the question of the origins of the neurons and nervous systems is the one evolutionary biologists have used to great effect since the nineteenth century—namely, to explore in which animal taxon the system originated rather than the type of cell that served as its origin.

The First Nervous Systems and the Animals Carrying Them

> The nerve-net of the lower animals contains the germ out of which has grown the central nervous systems of the higher forms.
> —G. H. Parker 1919, p. 115

Most researchers agree that the first nervous systems appeared more than six hundred million years ago in early representatives of the kingdom Animalia. They appeared as diffuse nerve nets in which the neurons were dispersed. Later, these nets evolved into a system consisting of a central nervous system (CNS), which controls the entire body, and a peripheral nervous system (PNS), which exerts local control over body parts. In which animal group neurons first appeared is a heated topic of debate that is linked to the question of the phylogenetic origins of animals.

The animal kingdom is divided into two major groups based on body symmetry: the nonbilaterians, with either radial or indefinite symmetry, and the bilaterians, with a single plane of symmetry, dividing the animal body into left and right mirror images (see box 6.2 for additional information about these major divisions of the animal kingdom). Figure 6.5 points

Box 6.2
Bilaterian and nonbilaterian metazoans.

"Bilaterians" are divided into two superphyla, the Protostomia and Deuterostomia, which are distinguished on the basis of the body opening (mouth or anus) that their members form from the blastopore during embryogenesis. The blastopore is the opening of the archenteron, the primitive gut of the gastrula. Protostomes ("mouth first"), which are invertebrates that develop a mouth from the blastopore opening, are further divided into Ecdysozoa and Lophotrochozoa. The former possess bodies covered by an exoskeleton and hence molt as they grow; the latter are soft-bodied animals such as mollusks and annelids, whose soft tissues are frequently in contact with the environment, and many of them have cilia. Deuterostomes ("mouth second") are bilateral animals, most of which form a mouth secondarily, with the blastopore marking the region of the anus. They include vertebrates, hemichordates (acorn worms), echinoderms (e.g., sea urchins, which are secondarily radial as adults), urochordates (tunicates), cephalochordates (lancelets), and, according to some researchers, the xenacoelomorpha worms. The Xenacoelomorpha include the acoelomorph worms (so-called because they lack a conventional gut and have no anus) and the Xenoturbellida, which are made up of a few species of small, anusless and brainless worms. While xenoturbellids have very simple nerve nets, some acoel worms have brainlike structures. However, the phylogenetic position of the Xenacoelomorpha keeps changing, and some studies suggest that they are a sister group to all bilaterians.[28]

"Nonbilaterians" include the nerveless Porifera (sponges) and Placozoa and the neural Cnidaria (e.g., sea anemones, jellyfish, corals) and Ctenophora (e.g., comb jellies and sea gooseberries).

Placozoans are nerveless, flat, and irregularly shaped disk-like animals a few millimeters in diameter. They are considered to be the simplest of animals, with bodies made up of a few thousand cells of only six types. There is a debate about their taxonomic status and whether their simplicity is ancestral or secondary (i.e., whether they have lost their nerves). Although only one species has been identified, this one species is genetically very diverse.

Porifera are nerveless animals whose bodies contain canals and chambers through which they pump water to extract food. They are sessile as adults, but their larvae move by ciliary movement. There are an estimated fifteen thousand sponge species living today.

Ctenophora are transparent animals with radial symmetry, a peripheral nerve net, and an apical sensory organ used to sense gravity and light. They move by using their comb rows and capture prey by entangling it with their colloblasts (sticky cells), turning it into a sticky mass. Unlike the Xenacoelomorpha and

Box 6.2 (continued)

other nonbilaterians such as cnidarians and sponges, which lack an anus and eat and excrete through the same hole, the ctenophores have two holes (the equivalent of an anus) in their rear end, opposite the mouth, through which indigestible particles exit.[29]

Cnidaria are a large and diverse group that includes Scyphozoa (true jellyfish), Staurozoa (stalked jellyfish), Cubozoa (box jellyfish), and Hydrozoa. They are radially symmetrical around a single oral/aboral axis and have a nerve net with (usually) some degree of neural condensation. They use cnidocytes (stinging cells) to capture prey and protect themselves from predators. Their life cycle consists of two main stages: a sessile polyp and a swimming medusa, although some groups have only a single stage.

to the relationship between these two subkingdoms—without, however, ordering the various nonbilaterians.

Diffuse nerve nets are present in the Ctenophora, Cnidaria, and some Xenacoelomorpha. While the nerve nets of cnidarians and ctenophores are distributed throughout their entire bodies, in those xenacoelomorph worms that have nerve nets, the net is dispersed beneath their outer epidermal layer. The status of this group is in flux: until 2011 the acoelomorph worms were viewed as the most basal group from which all bilaterians originated, but later comparative morphological and molecular studies suggested that they are related to the xenoturbellids and form the basal phylum of the deuterostomes, from which more complex phyla (the chordates, the hemichordates, and the echinoderms) originated; newer studies maintain that they are, as originally assumed, the sister group to all bilaterians. Whatever their status, the paleontologist Douglas Erwin argued that the common ancestor of all neural animals was similar to acoels that have both a diffuse nervous system and some clustering of nerve cells at the anterior end of the animal that form a cerebral commissure.[30] Nerve nets, however, are not exclusive to these basal phyla: they are present in echinoderms such as sea urchins and starfish and (arguably) in hemichordates (e.g., acorn worms), although some researchers think that in these two taxa, they are probably secondary and simplified features of former, more centralized nervous systems. The phylogenetic relationships of "basal" Metazoa are still debated, and figure 6.6 reflects two of the proposed topologies, with the first placing Porifera (sponges) as a sister group to all other Metazoa and the second placing Ctenophora as the most basal animal group.[31] According to both

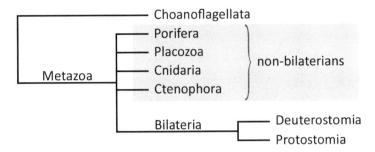

Figure 6.5
Phylogenetic relations between bilaterians and nonbilaterians. The disputed relations among nonbilaterians are not shown—the order presented here is arbitrary.
Choanoflagellates are single-celled protists from which all metazoans are assumed
to have evolved.

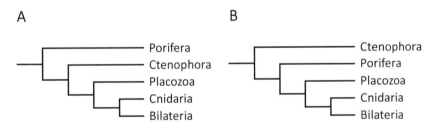

Figure 6.6
Two simplified phylogenetic trees showing possible relations of basal animals. *A*,
poriferan (sponge) origins; *B*, Ctenophora as the most basal group.

trees, the lack of nerves and muscles in placozoans is secondary (they lost
them). Tree *A*, however, suggests that the poriferans may have ancestrally
lacked nerves and muscles (although the tree is also compatible with the
possibility that they lost them), while according to tree *B*, the Porifera are
likely to have lost their nerves.

Porifera and Ctenophora: Which Is the Sister Group to
All Other Animal Phyla?

Sponges (figure 6.7) were traditionally believed to be the earliest common
ancestor of all animals, the sister group to all other animal phyla.[32] They
are sessile, although their tiny larvae swim using cilia before they settle and
develop into adults. Their bodies contain ten to twenty different types of
cells, and although they do not have a nervous system, some have cells or a

Figure 6.7
The types of adult forms found among species of sponges.

cell syncytium that exhibit the basic functions of neurons. There are three lines of evidence to support the claim that these animals were the originators of the nervous system.

First, some sponges produce and propagate electrical signals. For example, the glass sponge *Rhabdocalyptus dawsoni* filters water while feeding; when it detects sediment, an all-or-nothing impulse-like electrical voltage change lasting for five seconds travels across its body, causing the beating flagella that pump water through the sponge to shut down.[33] This "reflex" is slower than those found in animals with "proper" nervous systems, and the "impulse" propagates at the rate of about 0.2 cm per second, as compared to proper action-potential velocities of 100–10,000 cm/s. Nevertheless, the voltage change reflects an electrical excitability that is one of the hallmarks of neurons.

A second feature of Porifera that foreshadows nervous systems is the presence of "protoneurons" or "protosensory" cells in their larvae, from which, some biologists believe, fully fledged neurons and sensory cells may have evolved. Flask cells (sometimes referred to as "globular cells") in the larvae express genes that are orthologues (molecular equivalents derived from a common ancestor) of genes that play similar roles in the neurogenesis of neural animals.[34]

Finally, as already noted, most (but not all) of the structural proteins that are necessary for building the postsynaptic density in eumetazoan neurons have orthologues encoded in some sponge genomes.[35] In sponges these proteins may have a quasineural role in long-distance communication and coordinated behaviors. For example, some sponges employ peristaltic-like contractions to expel waste material from their water-circulating system, a process that requires coordination. In animals with a nervous system, the structural proteins in the PSD act as a scaffold for anchoring neurotransmitter receptors into the postsynaptic plasma membrane. The scaffolding proteins that are present in sponges, in conjunction with the transmembrane LGRCs (not present in sponges), were essential for the emergence of the PSD in neural organisms.[36]

Another feature, which may be relevant to the placing of sponges as ancestral to all neural animals, is the lack of a standardized body form. It is possible that a standardized body morphology is a precondition for the evolution of a nervous system that controls the movement of the body because nonstandardized topology cannot be effectively controlled.[37] If a standardized body form implies a nervous system in animals, then its absence in sponges suggests that sponges could be the ancestors of all neural animals.

Not everyone agrees that sponges are the progenitors of neural animals. Large-scale phylogenomic analysis of data from numerous species belonging to many phyla has led some scientists to conclude that sponges are not the earliest nor the simplest animals. They argue that their morphological simplicity is derived and that sponges lost their ancestors' nervous systems.[38] If so, in which group did neurons evolve?

On the basis of their molecular phylogenies and the fossil record, Ctenophora (figure 6.8A) and Cnidaria (figure 6.8B) are considered to be among the earliest (most basal) neurally endowed animal phyla. An additional reason for assigning them basal status is their relatively simple tissue structure. Most other animals (the Bilateria) are triploblastic, having three germ layers: the ectoderm, which gives rise to the skin and the nervous system; the endoderm, which gives rise to the intestines and digestive glands; and the mesoderm, whose derivatives include muscle, connective tissue, blood, kidneys, and skeletal elements. Ctenophora and Cnidaria have only two distinct germ layers (endoderm and ectoderm), and they have a similar neural organization, both having a diffuse nerve net (though many species have some condensed neural regions within the net).[39]

A

B

Figure 6.8
Examples of ctenophores and cnidarians. *A*, comb jellies (ctenophores); *B*, sea anemones—a group of sessile cnidarians.

Some biologists believe the Ctenophora are a sister group to all other metazoans, and the first neurons may have arisen in an extinct ancestral member of this phylum. Although their gelatinous bodies do not fossilize readily, fossils of ctenophores have been found in strata as old as 525 million years, and some suggest that they appeared in the late Ediacaran era, before the Cambrian explosion.[40] Others believe that the first neurons arose in the Cnidaria, fossils of which also date back to well before the Cambrian era, because some cnidarians, such as *Hydra*, have extremely simple nerve nets—the simplest so far found in any animal. Which group, then, "invented" the neuron and the nerve net?

Although the idea is by no means generally accepted, it is possible that neurons evolved independently in both ctenophores and cnidarians. While Cnidaria share the same neural machinery with most other animals, the Ctenophora have a unique neural makeup that sets them apart from all other neural organisms. Leonid Moroz and his colleagues studied eleven different species of comb jellies and discovered that many genes and proteins that are essential for developing a nervous system in all other animals are totally missing in ctenophores. Furthermore, most of the neurotransmitters that other animals use in their synapses (e.g., serotonin and acetylcholine) are not found in ctenophores, and one of the two transmitters that they do employ, glutamate, seems to be extremely versatile in its effects: an unusually wide variety of glutamate receptors have been discovered in comb jellies. The ctenophores use these glutamate receptors' variety (rather than neurotransmitter variety) to generate a flexible and differentiated nervous system, thereby illustrating how a similar end can be reached by different means. Moroz suggests that the nervous system of Ctenophora evolved separately, in parallel with that in other neural organisms. Since there is (again, arguably) molecular evidence that ctenophores may have appeared before the common ancestor of Cnidaria and Bilateria, and since they may have followed an independent evolutionary trajectory (and invented the all-important anus en route), the advent of neurons in comb jellies may have preceded the dawn of neurons in other animals.[41]

If the bilaterian nervous system is more similar to that of the cnidarians than that of the ctenophores, and if the phylum Cnidaria is a sister group of the Bilateria (there seems to be consensus about this, at least!), then it is the Cnidaria that may provide clues about the features that led to the evolution of animals with brains. Fortunately (from our point of view), the cnidarians are the best source of informed speculations about the functional evolution of ancient nervous systems, adaptive behaviors, and even learning.

The Cnidarians

The phylum Cnidaria is composed of two large groups, the Anthozoa and the Medusozoa. The anthozoans (sea anemones and corals) live exclusively as sessile polyps that lack a free-swimming medusa stage. Their basic structure includes a mouth (the single opening of the internal digestion sac that characterizes cnidarians), through which food is taken and waste is excreted; tentacles that radiate outward from the rim of the mouth; a pedal disc with which the animal attaches itself to the substrate; and a gonad, from which gametes are released into the water. The sea anemones (figure 6.8B) are soft bodied, whereas the corals live in compact colonies and form a hard skeleton that serves as the basis of many reefs in tropical oceans. The major subdivisions of the Medusozoa, an extremely diverse group with representatives almost everywhere in the oceans, include the scyphozoans, hydrozoans, and cubozoans, which usually have two alternating life-history phases—a planktonic medusa form and a bottom-dwelling polyp that is usually sessile, and the staurozoans, which do not have an alternation of polyp and medusa phases but live as attached medusae (figure 6.9 shows the medusoid forms of these groups).

The sensory capacities of Cnidaria are varied, but many have a stream of inputs providing them with information about their environment. For example, the moon jellyfish *Aurelia* has light receptors (though probably not photoreceptors), mechanoreceptors, chemoreceptors, gravity sensors, sound pressure wave and vibration sensors, and hydrostatic pressure receptors, which respond to light, touch, chemicals, gravity, sound pressure waves, vibration, and hydrostatic pressure, respectively.[42] Such sensory capacities enable complex behavior.

The Nervous System of Cnidaria

In 1874 and 1875, George Romanes, Darwin's devoted young disciple, performed electrical stimulation and excision experiments on the jellyfish *Aurelia* at his family summer house in Scotland. He was the first to show that cnidarians had a nervous network.[43] Today we know that despite the lack of a clear well-developed brain, many of the features of complex nervous systems are present in cnidarians, including sodium-dependent action potentials, neuron interactions through chemical and electrical synapses, and neuronal integration.[44] In addition to a dispersed nerve net in both the ectoderm and endoderm, more condensed sections of the nervous system are present in most species. For example, some Medusozoa have nerve rings that enable directed locomotion and pacemakers that enable rhythmic movements. Other specializations, such as two types of action

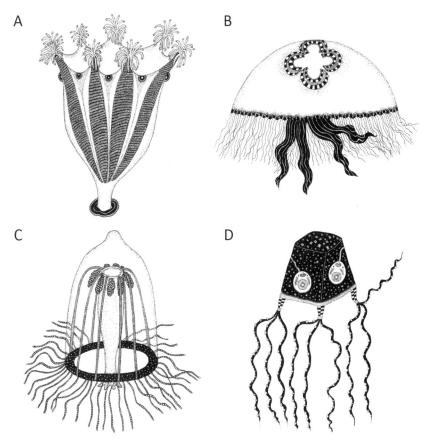

Figure 6.9
Medusoid forms of the major subdivision of the Medusozoa. *A*, a staurozoan (stalked jellyfish) modeled on *Craterolophus convolvulus*. *B*, a scyphozoan ("true" jellyfish) modeled on *Aurelia aurita*. *C*, a hydrozoan (hydra-like cnidarian) modeled on *Aglantha digitale*. *D*, a cubozoan (box jellyfish) modeled on *Chironex fleckeri*.

potentials, fast and slow, are employed for escape swimming and fishing activity, respectively, in the hydrozoan *Aglantha digitale*. These types of specialization provide a broad spectrum of strategies of neuronal integration and enable more coordination in the cnidarians than in the nerveless Placozoa or Porifera.

In cnidarians with the simplest dispersed nerve nets (e.g., *Hydra*), many of the neurons have symmetrical synapses, which may indicate an ancient origin. In such synapses, signals are transmitted bidirectionally, as opposed to the unidirectional transmission in the chemical synapses characteristic of

most animals, in which two-way signaling is achieved through reciprocal anatomical connections. In addition, some cnidarians have electricity-conducting
epithelial cells that are not nerve cells but enable the propagation of signals
in an all-or-none fashion, albeit very slowly compared to neurons. Pacemakers in some cnidarians provide ongoing rhythmical internal stimulation to
the animal. In *Hydra*, for example, the periodic alternation between a phase
of body shortening and a phase of body elongation is brought about by two
neuromuscular-pacemaker groups of cells located at the base of the polyp and
under the oral lip. Under constant environmental conditions, this contractile
behavior leads to the periodic expulsion of liquid accumulated in the gut.

It is clear that mobile cnidarians, like all mobile animals, must be able to
distinguish between sensations generated through their own movements
and world-imposed sensations. We have to assume that a neural organization
enabling sensorimotor control, through the effects of copies of motor commands modulating sensorimotor reactions, is in place. Sadly, we know
nothing about this basic control (reafference) in cnidarians.

Most of the research on the nervous system of anthozoans has been
carried out using the sea anemone *Calliactis parasitica*. Electrophysiological studies have shown that within its nerve net this cnidarian has three
separate conducting systems, with distinct stimulus thresholds and distinct
distributions. One conducting system is ectodermal, another is endodermal,
and the third is distributed everywhere.[45] We have relatively little information about the characteristics and functions of the neurons involved
because their small size makes intracellular recordings virtually impossible;
what we know comes from the less precise extracellular methods. Very few
electrophysiological studies of anthozoans have been published since the
early 1990s, which is surprising because the genome of the sea anemone
Nematostella vectensis was sequenced in 2007, and interest in the evolution
of nervous systems is steadily growing.

Almost nothing is known about the nervous system of stalked jellyfish,
but more is known for other types of medusae. The scyphozoan nervous
system consists of a condensed motor nerve net, a diffuse nerve net, and
some more specialized neural sensory structures. These include rhopalia,
which are small, pigmented, photosensitive structures; statocysts, which
are sac-like gravity receptors containing a mineralized mass (statolith) and
sensory hairs that, when touched by the statolith, signal a change in orientation that leads to a motor change restoring balance; and pacemaker
neurons that set the main swimming rhythm. The motor nerve net of scyphozoans consists of distributed bidirectional neurons that directly activate
muscle contraction upon incoming signals from the pacemakers, and their

diffuse nerve net is involved in indirectly activating tentacle contraction (by affecting pacemaker action).

Although in most cnidarians visual responses are restricted to simple phototaxis, seen as avoidance of shadows and objects, sight is well developed in box jellyfish. These cubozoans have twenty-four eyes, of four different types, which are arranged in clusters of six on four connected lobes, the rhopalia (figure 6.10). An upper-lens eye looks up while a lower-lens eye looks obliquely down; the two slit-shaped eyes aim obliquely down, and the two pit-shaped eyes are directed upward. The cubozoans that are endowed with this complex organization are the only cnidarians known to possess image-forming, camera-type eyes. Even when the box jellyfish is swimming, the rhopalia maintain a constant vertical orientation, as each rhopalium dangles on a flexible stalk and is weighted down by a heavy statolith. The intricate organization of the rhopalia led some zoologists to regard the nervous system of cubozoans as centralized. For them, the four rhopalia and the conducting nerve ring comprise a CNS, and the motor nerve net is the PNS.[46]

In the cubozoan *Tripedalia cystophora*, four of the eyes, with lenses that enable image formation, always gaze up out of the water, regardless of the way the animal body is oriented. It appears that they allow this box jellyfish to navigate its way around the mangrove swamps in which it dwells, using terrestrial landmarks seen through the water surface. The ability of another cubozoan, *Chironex fleckeri*, to avoid obstacles in a darkened tank (and in the wild) and to orient to the light of a match suggests behavior that depends on the complex coordination of movement following the integration of multiple visual stimuli. The intricate navigation that this animal displays supports the conjecture that it is able to form an image of its environment.[47]

Hydrozoans lack rhopalia, and the nervous systems of those having a medusoid stage consist of neurons arranged in inner and outer rings running along the margin of the bell. These rings contain pathways for many different sensory inputs and also pacemaker neurons. In *Aglantha digitale*, at least fourteen separate functional pathways have been observed in the nerve rings, and some researchers regard them as part of a "central circuitry," which they refer to as the cnidarian CNS.[48]

Precursors and Enabling Systems in Cnidarians: Overall Sensation and Nonassociative Learning

What types of learning do cnidarian nervous systems enable? There is ample evidence showing that cnidarians can and do learn, but their learning is limited; there is no reason to think they are capable of UAL. So what is

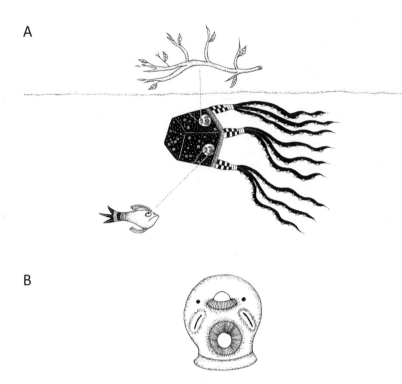

Figure 6.10
A cubozoan, its environment, its rhopalia, and its eyes. *A*, a swimming cubozoan can see the mangrove branch above and the small fish below. Two of the four rhopalia are visible. *B*, a rhopalium, showing the six eyes. The upper- and lower-lens eyes are flanked by two pairs of simpler eyes—two slit-shaped eyes lie below two smaller, pit-shaped eyes.

missing? In the terms we introduced in chapter 5, what does the enabling system that constructs and maintains the cnidarians' behavioral and cognitive functions have and, more importantly, what does it lack?[49]

Spontaneous Activity and the "White Noise" of Overall Sensation in Interconnected Nerve Nets
Constant spontaneous and induced activity, which only ceases with death or in states of "potential life" (such as a dehydrated spore), is a hallmark of life. In multicellular organisms, spontaneous activity occurs not only within the cell but also in all cell-cell interactions, and in a nervous system, it is manifest as ceaseless electrical and biochemical activity. Neurons spontaneously emit small amounts of neurotransmitter, and these generate local

patterns of electrical change that contribute to the excitation or inhibition of other neurons. The formation and dissolution of synaptic connections resulting from the summation of spontaneous and emitted discharges can be described as "neural exploration."

Animals with a ceaselessly active, interconnected nervous system and multiple sensors have what we call "overall sensation"—the kind of overall sensory buzz that was first alluded to by Lamarck when he described the internal activity underlying the "inner feeling," the feeling of existence. The overall sensation is made up of spontaneous neural activities, neural activities resulting from homeostatic maintenance like those of pacemakers, and neural activities occurring as the animal responds to contingent conditions. As a metaphor, the overall sensation can be regarded as white noise. We see it as a functionless, as yet feelingless by-product of a sensorimotor system that dynamically processes electrical and chemical signals.[50]

Overall sensation is constantly changing. Different sensations may become distinguishable from each other when persistent stimuli activate a particular local circuit. Other overall sensations may be generated when there is transient selective stabilization of altered neural states and when simple forms of learning add their temporary signatures to the overall buzz of neural activity. Although the overall sensation with its various signatures is not subjective experiencing, it is, we suggest, the evolutionary raw material from which it emerged, and the specific patterns and "signatures" it manifests are therefore of special interest to us.

Innate Behaviors, Transient Reinforcements, and Nonassociative Learning

The nervous system of a cnidarian, like any other nervous system, is not a "blank slate." Cnidarians have neural circuits that underlie species-typical motor behaviors of two types. The first is innate, default, spontaneous, exploratory (nondirected) behavior. The second type is the highly directed, stereotyped (or "fixed") kind of motor or glandular response, elicited by a specific stimulus, that is underlain by a dedicated neural circuit (such as a withdrawal response to a strong touch). It is this second type of circuit to which one refers when talking about "simple reflexes."[51]

What is the evolutionary origin of "simple" directed reflexes? There are, as we see it, two possibilities. Such reflexes may have been derived from spontaneously generated, exploratory action patterns based on nonspecific reactions to stimuli. According to this scenario, selection led to the evolution of more specific sensors that could respond to stimuli by eliciting more specific motor responses and, importantly, to lateral inhibition so that strongly excited neural clusters inhibit the activity of other clusters, thus leading to a more directed and specific response. Alternatively, some reflexes may have their origins in the division of labor between the sensory and motor functions of ancient precursors (figure 6.4A), which formerly had coordinated, one-to-one sensorimotor properties. The two possibilities are not, of course, mutually exclusive. We favor the first possibility because although some preexisting, one-to-one interactions between sensory and effector parts of the precursor cells could have been preserved, the evolution of interneurons ensured that most connections became many to many, involving motor-sensory-motor interactions. Indeed, we predict that in present-day animals, a low level of exploratory motor-neuron activity will be found even when particular motor units are preferentially linked to specialized sensory elements (as in classical reflexes). This is in line with an alternative to the classical reflex model that was suggested in the 1930s by the Russian psychologist Pyotr Anokhin, who developed the notion of a "functional system" that involves an organization with distributed neural elements that display self-organizing and nonlinear dynamics.[52] We believe that such functional organization may be applicable even to relatively simple neural systems.

Both spontaneous exploratory behavior and simple reflexes can be elicited or modulated by "positive" stimuli, such as the sensed presence of food, which signal fitness-enhancing conditions, and "negative" stimuli, such as tissue damage, which signal a persistent departure from homeostasis. We refer to states of departure from homeostasis that the animal tries to alter as "repulsor states" and see them as the mirror image of "attractor states," which are states of homeostasis that the animal strives to reach and to maintain. As long as the system is in a repulsor state, internal sensors are continuously activated, signaling an out-of-equilibrium state, until homeostasis is restored. In cnidarians, internal chemical sensors are important to maintaining homeostasis. For example, *Hydra* can be busy searching for food, semirandomly moving their tentacles for hours on end, but when they are fed to repletion, they close their mouths and stop fishing for prey. They are probably the simplest metazoans to exhibit satiety after feeding.[53] The factors (which include neurotransmitters such as dopamine and serotonin) and the processes that positively reinforce (reward) homeostasis-promoting reactions and suppress (punish) homeostasis-reducing reactions are what the neurobiologists whose work we discussed in chapter 4 call "value systems."

As we described earlier, cnidarians, like all neural animals, have many of their sensory cells organized into clusters (sensory organs). Among other things, this organization means that fluctuations and variability due to the spontaneous activity of each single sensory cell are averaged out, and there-fore the integrated sensory input that promotes or inhibits motor activity is a more reliable indicator of the external (or internal) conditions. Cnidar-ians display simple repeated rhythmic movements that are underlain by dedicated motor programs. These motor-action programs (action patterns), which are generated by relatively autonomous neural networks—the cen-tral pattern generators (CPGs)—are innate but are considered to be more complex than simple reflexes.[54] The organization of the nervous system into circuits that include sensory organs, distinct stereotypical action pat-terns, and circuits that compare the movement-generated stimulation of muscle tissues to the world-generated stimulation arriving through com-plex sense organs (by the reafference system) means that the activity of these circuits can be combined and coordinated in many different ways, making complex behavior possible.

But how rich is the actual behavioral repertoire of cnidarians? In the jel-lyfish *Aurelia*, which is one of the best-investigated cnidarian genera, fifteen species-specific innate behaviors have been documented.[55] These include swimming upward when encountering either mechanical stimulation or low oxygen; swimming downward when encountering low salinity, touch-ing the surface, meeting with turbulence, or finding themselves in the two top meters of water; swimming away (not necessarily up or down) from rocks and turbulence; remaining in areas with conspecifics; staying in areas containing the smell of prey; and altering, in various ways, their swim-ming behavior following the capture of prey. These behaviors are *Aurelia*'s response to one or more sensory inputs that are received through their many types of receptor. Some responses are complex and suggest that dis-crimination, integration, and action selection are enabled by their nervous system. For example, *Aurelia*'s response to touch depends on both mecha-noreceptors and chemoreceptors, and its reaction depends on the type of touch. While a touch by a conspecific elicits a pause in swimming fol-lowed by a resumption of the previous swimming pattern, a touch by a silicon ball initiates reorientation and swimming upward. Catching prey also elicits different behavior depending on the conditions: when prey is first caught, the speed of swimming increases for a while, but after the animal has caught several items of prey, the swimming speed decreases. The mechanisms underlying this behavior require modulations of default exploratory activity.

Consider, for example, a medusa sensing a source of food (figure 6.11). Although at first its movements are haphazard and exploratory, they have to become directional if it is to reach the food. Therefore, it cannot use

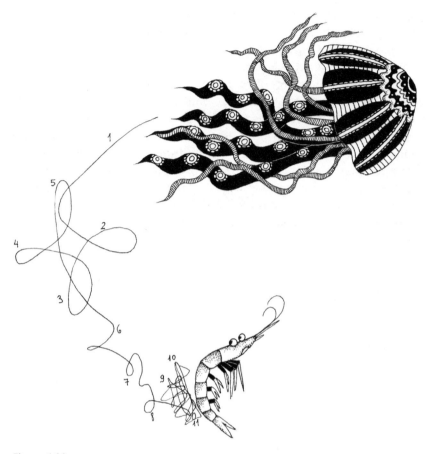

Figure 6.11
Exploration and stabilization—hypothetical exploratory behavior of a hungry medusa. The medusa starts with stochastic exploratory swimming (*1–5*) until it senses a reinforcing (positive) stimulus—a shrimp (*6*). The smell of the shrimp guides the medusa's motor activity (*6–11*). Numbers represent time points.

rhythmic motor behavior nor make use of rigid directional reflexes because the location of the food source is different at different times, and movement toward it requires the use of different muscles and neural trajectories. There has to be a mechanism that leads to consistent movement toward food, wherever it is detected. In other words, there must be a stabilization of movements along a selected trajectory, based on either local reinforcement or the inhibition of random activity. The simplest type of such behavior in neural animals is probably the phototactic behavior of the larvae of the annelid worm *Platynereis dumerilii*, a tiny and spherical 100 μm larva that is helically propelled by cilia. When one of the two eyespot photoreceptors is illuminated, it excites neighboring ciliated cells that change their beating as a consequence, resulting in locally reduced water flow and movement toward light (if not illuminated, the larvae swim in circles). This type of behavior has been described for cnidarians, but the neural mechanisms underlying it have not yet been studied.

In general, for behavior based on exploration and stabilization to be directed and specifically targeted, positive and negative reinforcement systems must be in place. The system has to evolve so that when repeated stimuli lead to reactions that enhance fitness, a reinforcing signal is released, and movement toward the target (e.g., along a smell gradient) is stabilized, while movement away from it is inhibited. Similarly, when encountering fitness-reducing conditions, movement away would be stabilized, while movement toward it will be inhibited. Over evolutionary time, the reinforcement system evolved to employ different classes of "value" signals for fitness-enhancing and fitness-reducing conditions. This provided the basis for the formation of the value-based categorization of stimuli and actions, the kind of evaluative internal states we call emotions and drives.

The neural circuits underlying such behaviors enable animals to respond to salient environmental cues without any learning. In other words, the adaptiveness of their stereotypical exploratory behavior and the responses to specific sensory stimulation have been phylogenetically fixed. The behavior is transient, and there are no stored traces of past experiences, as there would be with learning. Yet cnidarians do learn: they can modify their behavior as a result of past experience, albeit in a most elementary nonassociative manner, through sensitization and habituation.

Habituation involves a decrease in the intensity, and sometimes the disappearance, of a reaction upon repeated weak stimulation, while sensitization involves a decrease in the threshold of a response and/or an increase in the intensity of a reaction following stimulation. Short-term habituation (typically lasting up to minutes to hours) is observed when stimuli are delivered at relatively low frequencies and intensities. For example, repeatedly stimulating a sensory neuron by gentle touch leads to a decreased response of the motor neuron that innervates muscle cells and a concomitant decrease in the extent of the contraction of these muscles. The decrease in the behavioral response is caused by a short-term decrease in the amount of transmitter the sensory neuron releases at the synapse. Habituation enables the organism to ignore irrelevant stimuli, thereby minimizing energy waste. It can be reversed by a strong shock that elicits general excitation, which restores the response to its prehabituation state, a process known as "dishabituation."

Sensitization is in many ways the opposite of habituation. At its simplest, when there is a direct trajectory from a sensory neuron to a motor neuron with no other sensory units involved, high frequencies of stimulation result in an increase in the amount of transmitter released from the presynaptic neuron, which produces a more effective or intense response and lowers the threshold of future responses to similar stimuli (a process known as "facilitation"). Sensitization often takes on a more general form: a particular nonlearned ("innate") response can be affected by the general excitatory state, brought about by a strong stimulation to any of the animal's parts. Such sensitization can be seen as a form of somatosensory vigilance—the animal is prepared to face any kind of trouble.[56]

Sensitization and habituation can be either short term (minutes to hours) or long term (days to months). The anatomical and molecular mechanisms underlying short-term habituation and sensitization are accompanied by modifications of presynaptic and postsynaptic molecules and their movement to new sites, but they do not require protein synthesis; they require only covalent changes in preexisting proteins and involve alterations in preexisting synaptic connections. Long-term habituation and sensitization involve additional mechanisms. They require protein and RNA synthesis,[57] the growth of new synaptic connections (and the dissolution of existing ones), and other anatomical and structural changes in synapses. Long-term sensitization may also involve feedback between the postsynaptic and presynaptic neurons: the activation of specific postsynaptic receptors during long-term sensitization may produce a retrograde factor (such as the gas nitric oxide) that initiates an activity-dependent mechanism that enhances transmitter release from the presynaptic terminals.[58] Short-term and long-term memory characterize not only habituation and sensitization but all types of learning, and we discuss them in greater molecular detail in the next chapter.

The numerous behavioral studies of cnidarians responding to tactile stimuli, chemicals, and light all reveal that both short-term and long-term habituation and sensitization are common in this group.[59] Jellyfish and sea anemones are sensitized by a strong touch and habituate to a gentle touch, contracting less following repeated, gentle stroking. For example, an early experiment with the sea anemone *Anthopleura elegantissima* showed that repeatedly stimulating it with a jet of water reduced the strength of the contraction of the oral disc—that is, it became habituated. Unfortunately, physiological and molecular studies of cnidarian learning are scarce:

intracellular recording is difficult, and the ionic and molecular bases of their habituation and sensitization have not been worked out.[60] However, it is reasonable to assume that the basis of these forms of learning in cnidarians is the same as that found in other invertebrates.

The evolution of the ability to learn through short- and long-term habituation and sensitization presumably involved selection among animals that varied in their ability to repeat or sustain particular reflex responses. When the benefits of continuing an action were outweighed by its cost in terms of energy expenditure, animals that became less responsive to continued stimulation were favored, so selection led to habituation. Sensitization, on the other hand, would have resulted from selection in situations in which it was beneficial to respond to persistent stimulation with more of the same activity. If a strong or persistent stimulus is constantly associated with continuing danger, selection would lead to the enhancement of the animal's defensive reflexes by lowering the threshold for action and making these actions more intense.[61]

The responses that occur during general sensitization show that although simple reflex circuits can act independently, they can also interact, thus increasing behavioral flexibility. An additional simple form of interaction among different reflex pathways involves summation, which occurs when a postsynaptic cell receives inputs from several different presynaptic cells. Summation is particularly relevant for situations in which every single stimulus is too weak to produce a particular reflex response yet when occurring simultaneously sum up the effects and lead to a response (figure 6.12). There is evidence of summation in several cnidarians, including the sea anemone *Metridium senile* and the hydromedusa *Polyorchis penicillatus*, suggesting that it is one of the earliest adaptations of the nervous system.[62]

Another form of interaction between reflexes is lateral inhibition, in which the intense activation of one reflex pathway inhibits alternative pathways, thereby enabling a directional response like that of the medusa that swims toward the shrimp in figure 6.11. Lateral inhibitory relations between reflex pathways facilitate discrimination: for example, they enable sharp distinction between sensory stimulations that lead to different responses. Reflex inhibition seems to be as ubiquitous as summation (which also involves reflex inhibition through reafference), although there is disagreement about whether or not it occurs though synaptic inhibition in cnidarians.[63]

The interactions between different "reflex" paths certainly increase behavioral flexibility. However, these flexible responses are transient and are not the basis of future behaviors, so they are not learning according to

A

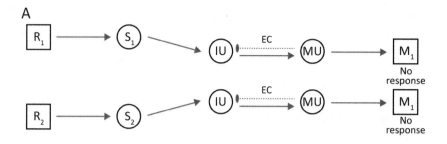

B

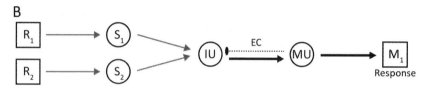

Figure 6.12
Summation in a reflex pathway. R1 and R2 are receptors. When potential reflex-eliciting sensory neural units (S_1, S_2) are weakly activated (*A*), they do not cross the threshold necessary to elicit a response mediated by the motor unit (MU), and no command goes to the muscle (M_1). However, with summation, which in this example occurs through the convergence of the two sensory units onto an integrating unit (IU) when both S_1 and S_2 are simultaneously stimulated (*B*), their contributions are pooled and the activation of MU by IU is above the threshold, leading to a response by a muscle. An efferent copy (EC) is sent from the MU to a network of integrating interneurons (IU). It modulates the activity of the IU according to the predicted action of the motor units and hence modifies the signals sent to the muscles. Subthreshold signals are shown in light gray and above-threshold signals in black. Inhibition is depicted by lines with blobs; excitation by a line with an arrowhead.

our definition (chapter 5). In contrast, true learning occurs through habituation and sensitization, but it is fixed in pattern; what changes, and what is learned and remembered, is the *extent* of the innate response—either the innate exploratory behavior or the simple reflex. That is why we regard this type of nonassociative learning as "limited": it is limited to the modulation of preexisting relations between sensors and effectors.

Adaptive modulations based on exploration-stabilization processes, summation, and lateral inhibition all seem to occur in cnidarians. In figure 6.12 we did not show the incessant stochastic activity that occurs in all neural networks, but in all cases when sensors are stimulated, many neural pathways, in addition to those specifically involved, are activated and suppressed, and some patterns of neural activity become reinforced and stabilized. The ability to transiently stabilize new sensorimotor connections is a necessary condition for the evolution of the mechanisms underlying long-term learning by association. The evolution of these mechanisms seems not to have happened in cnidarians (but see the next section for some heretical speculations). With the exception of some prespecified innate reflex pathways, the selectively stabilized, transient, and novel neural connections in these animals do not carry long-term traces of their activated states. The animal cannot build on and benefit from past encounters; it has to reinvent the wheel (the new adaptive response) time and time again.

In animals with a CNS, exploration and selective stabilization are likely to have increased behavioral plasticity because, by bringing sensors closely together and minimizing signal dissipation due to distance, new neural connections may become stronger and hence more persistent. If so, we can ask whether partially centralized cnidarians show learning that is based on the persistent formation of non-prespecified relations. In other words, can they learn by associating different stimuli? Can they learn by consequences?

Do Cnidarians Have a Limited Form of Associative Learning?

Associative learning (AL) requires remembering and evaluating new associations between sensory stimuli or between an action and reinforcing stimuli. The benefit of AL is obvious: an animal that lacks it has to depend on the continued and immediate presence of the external cue to find the adaptive (attractor) state, while an animal with AL has the great advantage that it can respond to associated cues that *anticipate* reward and punishment.

We are aware of only one study that reported AL in cnidarians. It was claimed that sea anemones learn to contract their tentacles when presented with light that predicted a shock.[64] However, the sample used was small, and further studies have not corroborated the results. Other attempts to condition cnidarians have failed.[65] Even if the experience-dependent formation of associations between highly specific and restricted stimuli is found in some cnidarians, the most generous conclusion from the negative results we have at present is this: cnidarians do not have the flexibility enabled by open-ended, multimodal AL in which many types of stimuli can become associated with many other stimuli and with the actions of the organism. Nevertheless, if limited associative learning (LAL) does occur in this phylum, where is it most likely to be found?

Good candidates are species of box jellyfish. They have a sort of centralized nervous system, an elaborate sensory system, and complex vision-based behavior that includes feeding adaptations, specialized courtship, obstacle avoidance, diurnal activity patterns, and navigation. There is no doubt that they can form images and detect shapes, although we have no idea what the extent of their visual discrimination is. However, the ability to detect images made up of multiple visual stimuli may be particularly beneficial if the animal can learn—if it can associate an image with positive or negative reinforcement. So perhaps box medusas have vision-related AL. Electrophysiological studies have shown that a significant part of the cubozoan nervous system is situated within the rhopalia, and extracellular recordings from their stalks have revealed a pacemaker signal that directly controls swimming. The complex nerve cell populations in the rhopalia, with commissures interconnecting some of the eyes, may form the structural basis for processing and integrating visual information. If visual information is processed and integrated within these structures, and if they directly control motor behavior, then associations between different sensory modalities or, more likely, between different types of cues within the visual modality, might be reinforced.

Although the details of visual processing in cubozoans are only just beginning to emerge, it seems that the different types of eye might be specialized for specific tasks.[66] But to provide more useful information (for example, about both overall shape and movement), visual stimuli from the different eyes have to be integrated. Such integration during fast swimming and predation would require enormous amounts of neural processing and may account for the need to sleep that has been observed in members of this group. The kind of sleep we have in mind is not mere inactivity or quiescence, nor torpor or hibernation (although sleep may share mechanisms with such forms of inactivity), but rather the sleep observed in neural animals that involves, in addition to reduced behavioral responsiveness, changes in brain activity and homeostatic regulation that includes sleep rebound (compensation for sleep deprivation). Evolutionary analyses link sleep in animals to the need to relieve the burden of excessive neural processing such as that imposed by vision in complex lensed eyes, although the presence of sleep in cnidarians, such as the upside-down jellyfish *Cassiopea*, suggests the origin of sleep preceded vision. Sleep in cubozoans may have acquired new functions, such as memory consolidation, following complex neural processing.[67]

Fossil evidence suggests that cubozoans evolved during the late Cambrian, when active, associatively learning predators already roamed the oceans. Did cubozoans acquire, in the highly competitive context of the Cambrian seas, not only particularly lethal poisons but also some LAL?[68] It is significant that their rhopalia, which integrate visual information, exhibit bilateral symmetry and that such symmetry in other metazoans seems to go hand in hand with AL. So do cubozoans, with their sophisticated behavior, their more centralized nervous systems, and their partial bilateral symmetry, learn through association? For example, can they learn that "neutral" shape X is associated with nutritious shape Y, or with forward, rather than backward, movement? Unfortunately, appropriate experiments searching for AL in cubozoans have not been carried out.[69] Several decades ago it was noted that "the investigation of learning in animals phylogenetically below the earthworm has been focused chiefly on the protozoans. Such animal groups as sponges, jellyfish, and ctenophores have been largely neglected,"[70] and the situation has not improved much since. If experiments investigating AL in cubozoans are eventually carried out, we think that it is probable that a limited type of learning by association will be found in this group so that signals originating from one type of eye become predictive of signals that in the past were paired with those from another type of eye. It is also possible that signals from other sensory modalities can become predictive of visual cues.

Building Blocks for Subjective Experiencing?

Many open questions linger about the learning abilities of cnidarians but there are, regrettably, very few experiments. Cnidarians clearly have nonassociative learning and cubozoans, we have speculated, may have limited, vision-based AL. According to our criterion, which requires that an animal manifests UAL for it to be *positively* considered as minimally conscious, cnidarians cannot be subjectively experiencing animals. Nevertheless, we can ask whether their neural architecture, behavioral adaptations, and learning abilities satisfy any of the features that the neurobiologists list as criteria for subjective experiencing (table 6.1).

The table shows that cnidarians do not satisfy the consciousness criteria, although there are many "not known" and "not studied" attributes. Even when they do seem to satisfy a particular criterion, the criterion has to be interpreted in an extremely general manner. For example, although we can speak about temporal persistence in cnidarians, this is not the kind of temporal thickness that accompanies perception in animals such as mammals, where it is associated with multilevel neural integration, complex discrimination, and the effects of past learning. Similarly, although a cnidarian "emotion" may be an action program elicited or stabilized by a value system, it is a very rigid action program. Clearly, the mammalian-based criteria of Seth and his colleagues (chapter 5, table 5.1) are inappropriate for determining whether or not a cnidarian may be considered conscious.

Table 6.1
Applying the criteria for subjective experiencing to cnidarians.

Criteria	Found in cnidarians
Global activity and differentiated states	*Yes*, the neural network is interconnected, and local activity reverberates through it. They have exploratory activity, specific reflexes, reflex modulations, and central pattern generators (CPGs).
Binding and unification	*Not known.* The complex reflexes of many cnidarians and the visual abilities of the cubozoans suggest that some binding in the visual modality may occur in this group.
Intentionality	*Only in the phylogenetic sense* applicable to all living organisms (including nonneural unicellular organisms, plants, and fungi); the mapping of outer relation onto inner relation seems limited.
Selection, plasticity, learning and attentional modulation	Exploratory behavior suggests that there are processes of selective stabilization in the nervous system, but no direct experiments on exploration and selective stabilization have been carried out, and the reafference system has not been studied. No action selection has been demonstrated. However, learning and selective stabilization mechanisms are manifestations of plasticity (and metaplasticity). Learning is nonassociative. *Attentional modulation has not been investigated.*
Temporal thickness	*Not known.* Persistent temporal signatures may accompany exploratory behavior and homeostatic states, but there are no experiments focusing on temporal persistence in the absence of activity-eliciting stimuli.
Attribution of values; emotions, goals	*Values are reflex bound.*
Embodiment and self	*Not studied.* Agency as manifest by the reafference system has not been studied.

The same is true, at least with our present state of knowledge, for the "consciousness signatures" (amplification of neural signals, long-range reentrant connections, specific frequency of gamma waves) suggested by Dehaene, which are also specific to mammals, and the analogs of which have not been sought in any other taxon. Using mammalian-based criteria alone clearly cannot help. If we adopt Merker's criteria (target selection, action selection, and motivation), we may grant some form of action selection and target selection to some cnidarians (although discrimination learning has not been studied), but we simply have no idea what cnidarian "motivation" might mean, other than the employment of fairly rigid value mechanisms. Nor do we know what a cnidarian self-model might be, if there is such a thing.

Maybe, then, we can answer a simpler question about cnidarians—namely, whether they may be considered as proactive, self-distinguishing agents. As we argued earlier, they must be able to distinguish between the sensory effects of their self-generated actions and world-generated stimuli. We know about such reafference systems in invertebrates such as worms and flies, as well as in vertebrates (chapters 4 and 5). In vertebrates, this ability to distinguish between world and self is assumed to depend on the existence of a self-model.[71] The varied and sophisticated swimming behaviors of medusas, in which interactions between muscle tissues and sensory organs occur, suggest that a considerable level of integration is necessary. When a cnidarian moves, its many sensors are stimulated, and these multiple incidences of stimulation must be compared to stimulation that is independent of its activity. Hence, cnidarians are definitely agents. But are their reafferent interactions generating something like a self-model? Is the first proper function of the internal, systemic "sensory buzz" a precursor of a self-model? We think that it is.

Despite the gaps in our knowledge about cnidarian agency and learning and their seeming lack of minimal consciousness, we want to emphasize three features that are crucial constitutive parts of the enabling system of their cognition. These features impart a specific signature on the overall sensation of cnidarians and are the building blocks of subjective experiencing, which appeared later, with the evolution of cephalization and UAL. They are the effects of: (1) the activation of motor neural systems (including pacemakers); (2) the short- and long-term memory mechanisms that modulate reflex paths through habituation and sensitization; and (3) distinct and persistent overall sensory states that result from the selective stabilization of new trajectories activated during exploration.

The sensory signature that different activities impose seems to construct states similar to what Denton called "primordial emotions." These are action programs based on spontaneous activities that are stabilized by sensory information continuously coming from interoceptors and are associated with departures from homeostasis.[72] Of course, to have any functional significance, action programs must lead to appropriate responses to the environment, which depend on the stimulation of exteroceptors and the reafference system. This does not, however, entail subjective experiencing. We believe that these internally and externally initiated sensory signatures should be regarded as no more than by-products of the global, consistent, and specific neural activities imposed on overall sensation. However, departures from homeostasis that have negative value are not the only by-products/precursors of feelings that one can imagine. As we speculated earlier, the ability to distinguish between the self and the world may have been the first proper function of overall sensation. When spontaneous exploratory activities leading to a distinction between self and the world had a positive value, they constituted the primordial, vague, and rough version of what would become, when subjectivity later evolved, the "joy of being" that Humphrey suggested as the most basic type of experiencing. We can put this same idea in a different way: an overall, general, positively valued sensation emanating from spontaneous activity that distinguished the animal from its milieu can evolve into the SEEKING system described by Panksepp, the basic metaemotion that explains the motivation to explore and to energetically engage with the world. Clearly, if "joy of living" is manifest in active exploratory engagement with the world, or its opposite (imperious trauma can occur), these are already minimally conscious states. However, we do not have positive reasons to attribute such subjective states to cnidarians—any more than we have reasons to attribute them to responsive garden peas, which acquire learned associations between wind (a neutral cue) and a source of light that affects the direction of their growth.[73] In the next chapters, we follow the evolution of the neural learning that eventually led, from the humble beginnings we described here, to states we can confidently recognize as conscious.

7   The Transition to Associative Learning: The First Stage

Surprisingly, the evolution of associative learning (AL), one of the most revolutionary adaptive strategies that ever evolved, has received very little attention from evolutionary biologists. The few existing accounts start with the assumption that AL, or more specifically classical Pavlovian conditioning, evolved from general sensitization and that operant conditioning was later built on these foundations. But what is the evidence for this evolutionary account? We suggest that classical and operant conditioning were from the outset evolutionarily entwined and present a functional model of limited associative learning (LAL) that encompasses both forms of learning. We highlight the observation that all animals with confirmed LAL are bilateral and have brains; they probably evolved in the Cambrian era, when most animal phyla first appeared. However, although according to our criteria LAL requires an elaborate neural architecture, animals that display only LAL are not minimally conscious. In some groups the evolution of LAL rapidly led to the evolution of more open-ended forms of learning, which entail (as we argue in the next chapter) the neural dynamics that characterize sentient, conscious animals.

Associative Learning: Distinctions and Stages

A great gulf separates animals with the nonassociative learning we described in the last chapter and those showing unlimited associative learning (UAL), which will be described in the next chapter. As with other apparent evolutionary gaps, it is a good idea to try to discover the stages necessary to bridge them, for it is clear that the transition to UAL did not happen in one miraculous quantum leap. While *what* evolved is an open question, which we address in detail in this and the next chapter, *why* AL evolved seems clear: AL is generally acknowledged to be one of the most revolutionary adaptive strategies that have emerged during the history of life. It enables animals to learn about the contingent relationships between stimuli, actions, and reinforcements during their lifetime, whereas before AL evolved, adaptive responses to such relations, if they occurred at all, were "learned" phylogenetically through the differential survival and reproduction of individuals. As we described in chapter 1, Daniel Dennett regarded the emergence of this type of intentional system as the second level of his four-tiered Generate-and-Test Tower and called organisms that can learn by association "Skinnerian organisms."

AL is not only a game changer in terms of adaptability, it is also one of the most widespread adaptive strategies used by animals. It is found in all vertebrates, from fish through amphibians, reptiles, birds, and mammals, and there is ample evidence for such learning in many invertebrate groups, including nematodes, platyhelminths, crustaceans, chordates, arthropods, annelids, and mollusks. How, then, should we study the evolution of such a fundamental adaptive strategy?

There are two main approaches to the study of the evolution of AL.[1] Mauricio Papini (who reviewed ideas about the evolution of learning in the early 2000s) called the first the "general process view." According to this view, domain-general learning processes that lead to general intelligence (e.g., the processes underlying AL) are basically the same in all learning animals, and the attempt to attribute better general intelligence and better general capacity for AL to particular taxa leads to the unidimensional and nonevolutionary notion of the *scala naturae*. This approach is most sharply and clearly articulated by Euan Macphail, who argued that with the exception of linguistically endowed humans, learning mechanisms remained essentially unmodified during animal evolution, from the fish to the chimpanzee. The only modifications that did occur were in the sensory, motor, and motivational systems, which became adapted to specific ecological conditions and imposed constraints on the learning abilities of animals that lived in those conditions. The second perspective, called by Papini the "ecological view," is shared by people who focus on domain-specific learning adaptations, such as learning to sing a species-specific song or learning how to horde food. According to these scholars, one should abandon the attempt to discover general learning mechanisms because the evolution of learning led to idiosyncratic, ecologically specific adaptations that cannot be arranged along any progressive scale, such as that of increasing intelligence. The advocates of this view think that general intelligence is nonexistent and that attempts to discover changes in general-learning processes are both hopeless and misguided. Although the two views differ in their attitude to the reality of general intelligence, according to both, a search for progressive evolutionary change in general-domain learning capacity is doomed to failure. Papini, whose analysis is inspired by an evo-devo approach, challenged both views. He suggested that changes in regulatory mechanisms and a reorganization of highly conserved learning modules could provide an explanation for evolutionary changes within the general-process framework.

Like Papini, we want to find regulatory, organizational changes involving conserved learning-related circuits. The kind of changes that we are looking for are those that are involved in the qualitative, progressive transitions in the domain-general ability to learn by association. In the introduction to part II, we described how Maynard Smith and Szathmáry used changes in the way that information was encoded, stored, and transmitted to identify major transitions in evolution that led to increases in complexity. We use this approach to identify progressive changes in learning ability, which included modifications in learning plasticity and in the organization, storage, and use of neurally encoded information. These changes required new types of learning and new supporting neural structures that augment the regulatory hierarchy in the nervous system, adding to it extra levels of representation and control. The evolutionary sophistication of AL led, we argue, to an increase in general intelligence, although, of course, motor, sensory, and motivational constraints led to ecology-specific variations in each type of AL in any particular species.

We recognize two progressive evolutionary stages, which may qualify as qualitative evolutionary transitions, that led to the increase in the complexity of AL:

1. *The transition from nonassociative learning to LAL.* With LAL, spontaneous and stochastic exploratory activities and preexisting simple reflex reactions can be combined, reinforced, and recalled. Other noncompound (elemental) stimuli such as a flash of light, or single actions such as pushing a button, which are unrelated to a particular reward or punishment, can become associated with the reinforcement and lead to a future anticipatory response.

2. *The transition from LAL to UAL.* Experimentally, it has been found that many compound or composite non–reflex eliciting features (e.g., a large, sunflower-like, yellow-blue shape) or sequences of actions (e.g., pressing a round key and then moving left toward a square key), which are made up of individually neutral components, can become associated through reinforcing rewards or punishments. They are recalled as compound stimuli or actions and can become the basis of second-order conditioning. The evolution of the increased discrimination that such integration entails has driven increasingly higher levels of control because updating past perceptual and motor neural representations in the light of surprising new information requires mechanisms that go beyond the lateral inhibition of nonreinforced

sensory processing. We suggested (chapter 5) that the emergence of fully fledged UAL involved the evolution of the recall of stimuli that were bound through "perceptual fusion," hierarchical predictive coding, and top-down attention.

Any discussion of the evolution of AL raises questions about the functional and evolutionary relationships between classical (Pavlovian) and operant/instrumental conditioning, the two types of learning identified with AL. Are both forms of AL included in each stage of the evolutionary transition, or should we tell a different evolutionary story about each? Is one form of conditioning evolutionarily derived from the other? Can we really treat classical and operant conditioning as distinct learning processes? A close look at the psychological studies supporting the distinction uncovers surprising problems, suggesting that we have to rethink this traditional classification. We therefore start by reiterating the characterization of the two types of conditioning, which we briefly described in chapters 2 and 5, and examine the problems it raises.

The Distinction between Operant and Classical Conditioning: An Evolutionary Perspective

> There is no sharp dividing line between spontaneous and stimulus-elicited behavior.
> —Hinde 1966, p. 226

As Pavlov's experiments with his dogs showed, classical conditioning entails the formation of an association between a conditioned stimulus (CS) and an unconditioned stimulus (US) that elicits an unconditioned reflex response (UR). A CS-US association is formed when the US repeatedly follows the CS in close temporal proximity, and eventually, exposure to the CS alone elicits the conditioned response (CR; often indistinguishable from the UR). For example, in the sea hare *Aplysia*, a mild, habituated touch of the siphon that does not elicit withdrawal can be considered a CS (when nonhabituated, a gentle touch *does* elicit such a reflex response). Alternatively, the CS can be a stimulus that is completely unrelated to the reflex response—for example, the sound of a buzzer that is unrelated to the salivation reflex in dogs and has never elicited this UR but does elicit an almost identical CR following the pairing of the buzzer with the smell of food. Conditioning that is based on habituated USs is called "alpha conditioning" while conditioning based on neutral CSs is called "beta conditioning" or "true (Pavlovian) conditioning." It is important to note that as animals become conditioned, they may learn not only the predictive value of the CS but also the context that cooccurs with the CS and the US. For example, rats will learn to fear not only a sound that predicts an electric shock but also the cage in which this sound and shock occurred.

With the second type of conditioning—operant or instrumental conditioning—reinforced motor behaviors initiate learning. When a certain sequence of actions (either a stereotypical action pattern such as pecking by birds or spontaneous, seemingly random acts such as pressing or pushing various objects by an inquisitive rat) is followed by positive or negative reinforcement (for example, food or electric shock), the tendency to behave in the same way under similar circumstances in the future is modified.[2]

Skinner (1981) summarized the difference between classical and operant conditioning as follows:

> Through respondent (Pavlovian) conditioning, responses prepared in advance by natural selection could come under the control of new stimuli. Through operant conditioning, new responses could be strengthened ("reinforced") by events that immediately followed them. (p. 501)

Although the difference between classical and operant conditioning seems conceptually clear, differentiating between them in actual natural or experimental conditions has proved difficult. It is clear, for example, that operant conditioning usually includes classical conditioning: the lever the rat presses is a CS no less than the reinforced behavior directed toward it. On the other hand, since a spontaneously initiated, innate, "stereotyped" motor behavior such as birds' pecking is the result of internal sensory stimulation, it can be considered a type of classical ("respondent") conditioning. These ambiguities kindled attempts to find a single formal description for both classical and operant learning and subsumed one under the other. However, others focused on the differences between the two types of learning and treated them as two distinct processes, a view that later became generally accepted.[3]

The difficulty with separating operant and classical conditioning can be partially overcome if one takes special measures to isolate the reinforcing effects of movements from those of perception. To some (rather limited) extent, such isolation was partially accomplished by both Pavlov, who tied his dogs to a stand and did not allow normal motor exploration, and Skinner, who deliberately put his animals in perceptually impoverished conditions. A more convincing isolation of "pure" operant from "pure" classical conditioning awaited studies in the fruit fly *Drosophila melanogaster*, the classical geneticists' workhorse and one of the favored model organisms in learning research. One of the fly's many advantages is that it can learn a great deal; there are also good methods to manipulate its perceptions and actions. The experimental psychologists Reinhard Wolf and Martin Heisenberg used a flight simulator (figure 7.1) in which a single fly, glued to a small hook of copper wire and attached to a torque meter, is "flying" stationary in the center of a cylindrical moving panorama (the arena). Since the experimenter can control the perceptual stimuli to which the fly is exposed in the arena and the relation between the fly's movements and the negative reinforcer (heat beam), or between the perceptual cues and the same reinforcer, it is possible to study the two types of learning separately.[4] Extending these studies, Bjoern Brembs showed that learning about the external perceptual stimuli inhibits the fly's learning about its own movements, but learning about its own movement reinforces learning about external percepts. He also showed that learning about the external percepts inhibits the retention (memorization) of learning about movements, but when movements become a habit after a lot of training, their learning can override the learning of percepts.

The conceptual and practical difficulties raised by the traditional distinction between operant and classical conditioning and the results of their experiments led Colomb and Brembs (2010) to suggest a different classification of learning:

> Today, we can propose a terminology to better distinguish what is learned (stimuli or behavior) from how it is learned (by classical or operant conditioning). We define self-learning as the process of assigning value to a specific action or movement. We define world-learning as the process assigning value to sensory stimuli. While only world-learning occurs in classical conditioning experiments, both processes may occur during operant conditioning. (p. 142)

We believe that this distinction is useful and that learning about one's own actions, about the world, and about the relations between the two is part of what constitutes all forms of AL and enables an animal to be an effective agent, able to distinguish between world-imposed stimuli that are independent of its action and stimuli that result from its own actions. However, we think that in most conditions, including most classical-conditioning experiments, animals learn both about the world and about their own behavior. Consider a case of classical conditioning, when an animal learns that a cue (e.g., a particular smell) predicts the presence of prey, and the presence of prey elicits a biting reaction, the UR. In this case, the actual response must be tailored to the specific prey that the cue predicts (its size, its texture, and so on). The animal learns both which

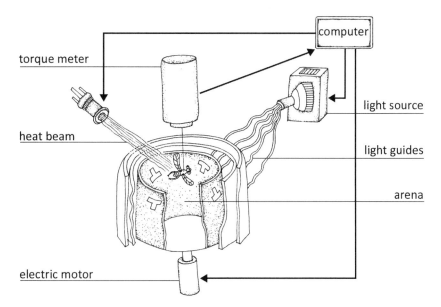

Figure 7.1
The flight simulator constructed to study conditioning. A single fly, glued to a small hook of copper wire and attached to a torque meter, is "flying" stationary in the center of a cylindrical panorama (the arena). In the flight simulator mode, the rotational speed of the arena is made proportional to the fly's recorded yaw torque (the angular momentum the fly exerts when it attempts to rotate around its vertical body axis). By altering its yaw torque and movements, the fly can stabilize the rotational movements of the panorama (i.e., "fly" straight) and adjust flight directions with respect to visual landmarks (patterns such as upright and inverted Ts). The yaw torque and flight direction of the fly are recorded continuously and stored in the computer memory. The illumination of the arena can be colored by inserting filters (e.g., green and blue) between the light source and the light guides. The patterns on the inside of the arena wall can be changed during the experiment. A laser diode provides an infrared heat beam, which is unpleasant for flies, and can be made contingent either on the fly's movements, on the various visual stimuli, or on both. Adapted (from photographs) by permission of B. Brembs.

cue predicts the prey and how to modify the UR (i.e., produce the CR) that the cue elicits. More generally, when the CR is not identical to the UR (it is a modification of it that is specific to the eliciting CS), which is often the case when the UR is a locomotor pattern, the CR is part of what the animal learns. Hence, in most conditions, classical conditioning requires both world and self-learning, although the animal's dependence on world and self-learning may vary and have different weights in different conditions.

What we know about the relation between perception and action makes separating them even more difficult because emitted exploratory actions ("operants") and sensory stimulations that trigger reflex actions ("respondents") depend on each other. Even when both the animal and the object of perception are stationary, perception involves motor actions: animals perceive the world through the active exploratory movement of the body and sensory organ movements, with both motor and sensory variables acting as players of equal importance. For example, in the visual system, spontaneous eye and head movements move the array of sensory receptors from one region to another while smaller movements scan the region of interest at finer resolution. Because most photoreceptors are activated by a change in light, if the eye rotates against a uniform background, they remain silent, and the animal fails to see. The same is true for touch—for example, with the vibrissal system of the rat, through which the rat explores objects with its whiskers.[5]

The cognitive scientist Ehud Ahissar has proposed that the relationship between exploratory motor activity and perceptual reactions can be conceptualized in terms of closed motor-sensory-motor loops (MSM loops).[6] A closed loop is one in which every signal affects it source. Since sensory organs such as eyes are associated with muscles and motions that produce changes in sensory stimulation, and since every change in a sensory input affects the sensor's motion, perception involves a dynamic sensorimotor interaction. Ahissar suggests that several MSM loops interact to construct a unified percept, forming a high-order MSM loop, with each lower-level loop typically relating to an individual feature of an object. But MSM loops, which have dynamics similar to those James suggested for the interplay between motor and sensory systems, are not the only type of closed loops linked to perception. Brains also contain internal loops that connect motor and sensory areas without involving the relevant sensory organ. Internal brain loops form models of the relations among brain, world, and movement and normally interact with high-order MSM loops.

Once the stimulation of sensors or effectors initiates MSM loop activity, a percept can form. In the context of learning, all learning (with the possible exception of sensitization that follows rapid, strong, one-time stimulation) involves closed MSM loops and motor exploration. There is no perception without motor exploration and bidirectional feedback. This notion is compatible with Anokhin's functional systems view discussed in chapter 6 and with the increasingly influential conception of the sensorimotor system as a functionally unified system.[7] Inputs to such a system, which can stem from the external world (exteroceptive), from the animal's activities (mainly proprioceptive), or from conditions within the animal (interoceptive), are selected by the animal according to their compatibility with preexisting models of the world and their contribution to fitness-promoting states.

The action-dependent conception of perception is compatible with the view presented in previous chapters, where we emphasized that spontaneous and stochastic exploration-stabilization processes and coordinated reflex reactions are involved in all neural learning. All nervous systems are inherently active and proactive, their dynamic internal states determining the effects of stimuli and constantly shaping and reshaping themselves through learning processes. In chapter 6 we argued that the sensitization and habituation of stabilized exploratory actions, as well as simple reflexes (that were probably evolutionarily derived from them) are the basis of nonassociative learning. Here we argue that the evolution of AL involved additional changes that led to genuinely novel modes of learning both about the world and about self. Our focus on exploration-stabilization and our view of the cooccurrence of different forms of learning differ from previous suggestions, which ignore exploration-stabilization processes and regard instrumental conditioning as evolutionarily derived from classical conditioning. However, since we build on some of the theories of our predecessors, we need to outline their views before we start our own analysis.

Early Ideas on the Evolutionary Origins of Associative Learning

> First, we propose that higher forms of learning may utilize the mechanisms of lower forms of learning as a general rule, and second, we speculate that this may occur because the mechanisms of higher forms of learning have evolved from those of lower forms of learning.
>
> —Hawkins and Kandel 1984a, p. 380

In the Western world, the earliest paper that proposed an explicit functional model for the evolution of AL was by Martin John Wells, a zoologist and experimental psychologist who studied the nervous system and the remarkable learning abilities of the common octopus.[8] Wells compared the learning abilities of the octopus, an animal showing sophisticated AL, with those of the annelid worm *Nereis diversicolor*, in which most learning seems to be through sensitization. According to Wells, in this worm the order of stimuli did not seem to matter: a US preceding a CS had the same effect as the opposite order. Additionally, the worm showed no cumulative effect of training, even when it continued for eight days with six trials per day. It seems that although the worm can learn through sensitization, AL (even very simple Pavlovian conditioning) is beyond its abilities.

Wells assumed that classical conditioning evolved from general sensitization. He pointed out that general sensitization is important when stimulus discrimination is limited or does not exist. For example, if a shadow can signal either a predator or food (and hence lead to two URs—approach or retreat), the only adaptive strategy is to "assume" that events in the world are repetitive: if there was food, there is likely to be more; if there is danger, it is likely to persist. Discrimination is increased when a sensitizing stimulus, such as a particular shape of the shadow, becomes more readily associated with a particular innate fixed response (e.g., approach). When the animal lives long enough and in conditions that render long-term memory advantageous, selection will lead to the evolution of a response that is remembered when a shadow with a specific shape precedes the US (e.g., food). In other words, Wells assumed that the reflex-inducing stimulus (the US) and a perceptual facet of it (e.g., shape, which can be considered a CS) had pathways that are cosensitized because the CS and the US cooccur. He believed—incorrectly, as we now know—that general sensitization is always short term and that long-term sensitization was needed to support the evolution of conditioning.[9] Although incorrect, Wells's model highlights an important point: some ability to discriminate must be in place for AL. With evolutionary improvement in an animal's discrimination, its niche expands, and more things can and need to be learned. It is therefore highly likely that discrimination between stimuli is an important driver of learning, a possibility to which we will return later. Wells's evolutionary scenario can be described as follows:

A link between an innate reinforcer (e.g., food, a US) and a sensory input associated with it (e.g., a vibration, the CS) is sensitized → long-term memory of the sensitized reaction → increased discrimination among sensed reinforcing stimuli → increased responsiveness when the CS precedes the US.

Psychologist Gregory Razran discussed the widely shared hypothesis that global sensitization led, through several stages, to true Pavlovian learning.[10] Like Wells, he argued that Pavlovian AL evolved from general sensitization, where the sequence in which the CS and US are presented is unimportant. Razran assumed that one of the stimuli (the US) that elicits an innate response (a generally arousing stimulus or a stimulus that triggers a more localized neural pathway) became linked to another reflex pathway that elicits another response. Although no new stimulus-response link is created (because the links are all prespecified and form a "reflex" network), the association between the two reflex-eliciting stimuli does have an impact on the second innate response, lowering its threshold. This kind of associative sensitization is sometimes called "pseudoconditioning."[11] The evolution of "true" classical Pavlovian conditioning from such sensitization is therefore to be sought in the mechanisms that render the temporal sequence of the stimuli significant, with the CS preceding the US and replacing it as a significantly stronger elicitor of the UR/CR. Once such mechanisms exist, a further distinction between different types of Pavlovian learning can be made. Razran, like other experimental psychologists, distinguished between alpha conditioning, where the CS (when nonhabituated) elicits the same reflex response as the sensitizing US, and beta conditioning (true Pavlovian conditioning), where the CS is a truly neutral cue, the effects of which were not evolutionarily prelinked to the US-UR circuit. He suggested that alpha conditioning preceded beta conditioning. The evolutionary progression he proposed was

general sensitization → pseudoconditioning → alpha conditioning → beta conditioning.

The functional similarity between general sensitization and simple conditioning was also the point of departure of the experimental psychologist Bruce Moore's proposal.[12] He suggests that the evolutionary progression from simple to complex modes of learning started with long-term sensitization and, following at least two intermediate steps, led to Pavlovian conditioning. The first and simplest type of classical conditioning was alpha conditioning; the second, slightly less constrained type of classical conditioning and the next step in the progression toward fully fledged Pavlovian AL was what Moore calls "Garcia conditioning." In the 1950s John Garcia showed that rats made ill by poison become aversively conditioned to various neutral or pleasurable gustatory cues, such as the taste or smell of the food they ate before they became ill. What is significant here, and what Moore underscores, is that conditioning is selective: rats that are electrically shocked are readily conditioned to audiovisual stimuli but not to gustatory ones, whereas animals made ill by giving them a poison that elicits vomiting are readily conditioned to gustatory cues but not to audiovisual ones. This makes evolutionary sense, since illness is often preceded by eating spoiled food, while painful injury is usually preceded by the sight or sound of predators. The connections between the neural circuits responding to mechanical injury and visual-auditory stimuli and between the circuits underlying digestive illness and gustatory cues were evolutionarily preorganized. Moore's suggested evolutionary sequence is

> general long-term sensitization → alpha conditioning → Garcia conditioning → beta conditioning.

The evolution of instrumental learning is a less-discussed topic than Pavlovian conditioning, but it is usually assumed that it evolved from the latter. Razran suggested that instrumental and Pavlovian conditioning are supported by the same basic mechanisms, with CS, US, CR, and UR involving interactions of different strengths, and that Pavlovian conditioning preceded instrumental conditioning in evolution. He based this proposal on the larger number of reports of Pavlovian conditioning in the learning literature, on the developmental dependence of some forms of spontaneous behavior on reflexes, and on his conjecture that instrumental conditioning seems more "advanced" or sophisticated than classical conditioning. Razran argued that what is learned through instrumental conditioning is more memorable (retained for longer periods), and the behaviors are more variable and do not easily habituate. However, the greater number of reports about Pavlovian conditioning may reflect a research bias. Furthermore, if the memorability and novelty of instrumentally conditioned behaviors are greater than those of classically conditioned ones (a generalization that seems doubtful, since memorability depends on the salience of the CS/US relation), this does not necessarily mean that instrumental conditioning is evolutionarily derived from Pavlovian conditioning.

A simple functional connection between Pavlovian and instrumental conditioning, which is in line with Razran's conception, was proposed in Moore's analysis of the evolution of learning. Moore observed that although simple instrumental conditioning and Pavlovian conditioning differ in that the reinforcer (the US) follows the CS in the Pavlovian case, whereas reinforcement follows responses (actions) in the case of instrumental conditioning, the two modes of learning have the same sorts of reinforcers and the same reinforcement-appropriate, species-typical responses. He suggested that the simplest kind of instrumental learning may be a special case of Pavlovian conditioning and provided an example from his own work with pigeons in which he observed that spontaneous innate pecking, which was rewarded when a particular key was pecked, led to the repeated and enhanced pecking of that key (figure 7.2). He called the reinforcement of such species-typical, innate reactions a "Pavlovian law of effect."[13]

Simple instrumental learning, based on the reinforcement of spontaneously elicited fixed action patterns, is found in most invertebrates. It

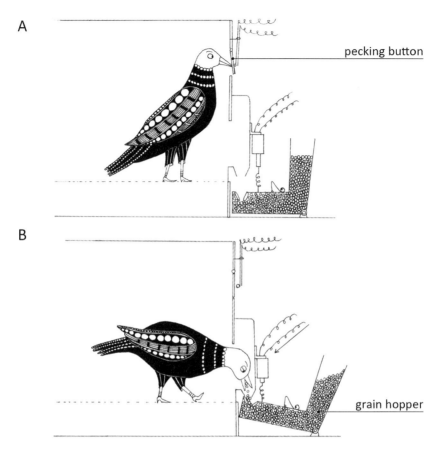

Figure 7.2
Instrumental conditioning of pigeons. *A*, through its exploratory pecking, the pigeon made contact with the pecking button. *B*, pecking the button is reinforced by the grain hopper lifting for a few seconds. Adapted by permission of Routledge.

involves, according to Moore, less constrained and more precise responses within the Pavlovian repertoire. This was the basis for his suggestion that instrumental conditioning evolved from Pavlovian conditioning. True *operant* conditioning, when the acts of the animal are novel combinations of fixed action patterns and result in behaviors that are not species-typical, is a further evolutionary stage that, he claims, is observed only in mammals and some birds. However, as we pointed out in chapter 2, even stereotypical behavior like a pigeon pecking grains is modulated by learnable variations resulting from the animal's experience with the size and texture of the grains, so the difference between instrumental and operant conditioning with regard to novelty may not be clear-cut.

The ideas we have reviewed up to this point were all based on functional-behavioral evidence and did not include cellular and molecular data. We now turn to studies of the molecular correlates of AL and the interpretation of the evolution of AL that is based on them. The biochemical basis of learning, which is described in some detail in the next sections, may be skipped by readers who are either familiar with it or find the molecular descriptions tedious. Nevertheless (and in spite of our awareness that the molecular details are likely to change as more research is done), we believe that these details provide a "feel" for the kind of biological research that is an integral part of the study of learning and consciousness.

The Evolution of Learning: Cellular and Molecular Mechanisms

Eric Kandel (figure 7.3A), who in 2000 received the Nobel Prize in Physiology or Medicine for his many contributions to understanding memory and learning, took a broad and imaginative approach that looked at learning at all levels of biological organization. Kandel's model animal was the slug *Aplysia californica*, also known as the California sea hare, a large herbivorous mollusk that can reach 75 cm in length (figure 7.3B). The animal has a relatively simple nervous system; some of its cells are very large and invariant and can be identified in all individuals. Most importantly, from our point of view, it can learn through the formation of associations.[14]

Investigations of learning in *Aplysia* have focused on the gill and siphon withdrawal reflex, in which the tender gill (the important respiratory organ) and the siphon tissues next to it are withdrawn when danger is sensed. A highly simplified neural circuit of the *Aplysia* gill withdrawal reflex is shown in figure 7.4 (where just one neuron of each type is depicted).[15] Sensory neurons from the siphon (*siphon SN*) form direct connections with the motor neurons (*MN*) that control the gill withdrawal reflex. Tail sensory neurons (*tail SN*) act indirectly on the reflex via facilitatory interneurons (*FAC INT*). These connect to the reflex pathway through synaptic contacts on the presynaptic terminals of the siphon sensory outputs (not on the motor neuron itself).[16]

Kandel and his colleagues characterized short- and long-term sensitization in *Aplysia* and then went on to characterize classical conditioning. Short-term sensitization occurs when a strong stimulus delivered to the tail enhances the gill withdrawal reflex response. This sensitization is brought about by the modulatory transmitter serotonin (5-hydroxytryptamine, 5-HT), which is released from facilitatory interneurons that synapse onto

A

B

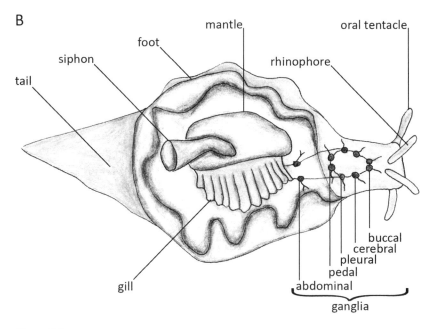

Figure 7.3
The scientist and his pet. *A*, Eric Kandel (b. 1929); *B*, the Californian sea hare, *Aplysia californica*.

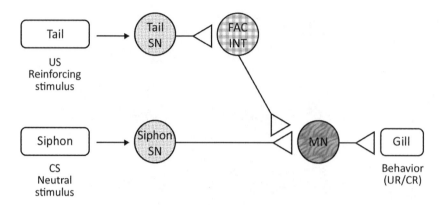

Figure 7.4
A simplified gill withdrawal reflex circuit (see text for details). SN: sensory neuron;
MN: motor neuron; FAC INT: facilitatory interneuron. Different neuron types have
different shading.

the terminals of the siphon sensory neurons, triggering a cascade of biochemical reactions. The outcome of these is presynaptic facilitation—an
increase in the synaptic strength of the connection between the sensory
and motor neurons. The cellular changes underlying short-term memory
rely entirely on various covalent modifications in preexisting proteins and
involve calcium and cyclic adenosine monophosphate (cAMP) signaling,[17]
as shown in Figure 7.5B.

Long-term memory is based on additional changes in the cellular properties of the neurons on both sides of the synapse. These changes require the
de novo transcription and synthesis of new proteins (figure 7.5C). Long-
term sensitization is based on cascades of reactions in the presynaptic cell
and its nucleus, the trafficking of receptors to the postsynaptic membrane,
and the growth of new synaptic connections. In both *Aplysia* and mammals, NMDA and AMPA receptors are involved in these processes.[18]

Having gathered the basic information about the biochemical pathways
involved in short-term and long-term nonassociative learning,[19] Kandel
and his coworkers found that the *Aplysia* withdrawal reflex lent itself to the
study of classical conditioning.[20] Normally, *Aplysia* withdraws the gill when
a noxious stimulus is applied to other parts of the body, such as the siphon
or the mantle. Weak stimulation of one pathway—say, the siphon—can be
paired with a US such as a strong shock applied to the tail. When the tail
shock (the US) precedes the siphon stimulation (the CS), the tail shock sen-
sitizes the CS reaction, but when the CS precedes the US, the response is

stronger and lasts longer and is, Kandel and his colleagues argued, a simple instance of conditioning. Conditioning was investigated by recording intracellularly from a single cell in a ganglion while one axon pathway to it was weakly stimulated electrically as a (tactile) CS, and a second pathway was stimulated as an unconditioned strong, aversive tactile stimulus.

In classical conditioning of the *Aplysia* gill withdrawal reflex (figure 7.5D), the US is "represented" in the siphon sensory neuron terminal by the action of the modulatory interneuron (*FAC INT* in figure 7.4) that releases serotonin. The CS is "represented" by the activity within the siphon sensory neuron itself. If this neuron has just fired an action potential in response to a CS, the modulatory interneuron that is activated shortly afterward by the tail US produces greater presynaptic facilitation of the sensory neuron. This phenomenon, activity-dependent facilitation, requires the same temporal sequence at the cellular level as the behavioral aspect of conditioning. Clearly, the mechanism is quite similar to the presynaptic facilitation found with sensitization of the gill withdrawal reflex, strongly implying that this type of conditioning may be an elaboration of sensitization. The important difference between sensitization and conditioning is the timing of the CS and the US: when the siphon sensory neuron has been active *before* serotonin is released, calcium concentration within the cell is even more elevated than during sensitization, and calcium binds to calmodulin, which in turn binds to the enzyme adenylyl cyclase (AC), enhancing its ability to synthesize cAMP. The increased levels of cAMP in turn enhance the activation of AC, leading the siphon sensory neuron to release more serotonin than would occur otherwise.[21] Hence, the activity of the siphon sensory neuron (before the tail shock) primes it to become more active, and AC is an important convergence site, or a coincidence detector: the CS and US are "represented" within the cell by the convergence of two different signals—calcium and serotonin—on the same enzyme.[22]

The associative-learning behaviors that Kandel described therefore rely on an elaboration of general sensitization: a synapse between two neurons is strengthened (without activity in the postsynaptic neuron) when a third, modulatory neuron (*FAC INT* in figure 7.4) acts on the presynaptic sensory neuron and enhances the release of transmitter from its the terminals.[23] This, however, is not enough: AL also requires that both the presynaptic sensory neuron and the postsynaptic motor neuron are simultaneously active and fire together (Hebb's law).[24] While the first process makes use of the associative or coincidence-detecting properties of the enzyme AC, the second employs the associative properties of the NMDA receptors. To activate the NMDA receptor and initiate long-term facilitation in the postsynaptic

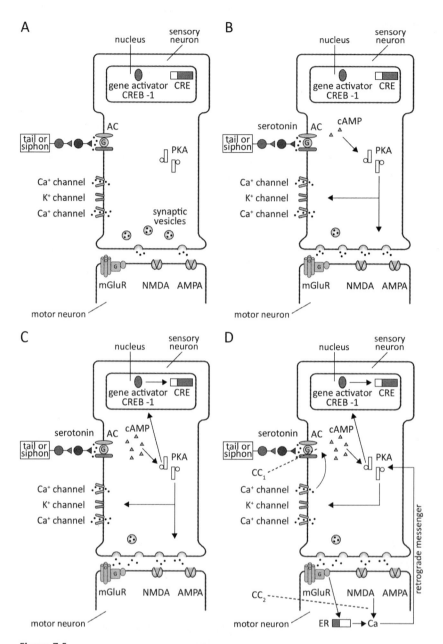

Figure 7.5

Sensitization (*A–C*) and classical conditioning (*D*) in *Aplysia*.

A, before learning. When an action potential is transmitted through the siphon sensory neuron, before learning occurs, molecules of the transmitter glutamate are released from the presynaptic vesicles of the sensory neuron terminal and bind to receptors on the motor neuron. (Circles at the bottom of the figure denote vesicles, some of which fuse with the membrane of the sensory neuron, releasing the transmitter molecules that are depicted as dots.)

B, short-term sensitization. The release of glutamate from the siphon sensory neuron is facilitated. Facilitation occurs because serotonin (released from facilitatory interneurons, as shown in figure 7.4) binds onto G protein–coupled receptors (G) in the membrane of the sensory neuron, leading to the activation of the adenylyl cyclase (AC) enzyme. This results in an enhanced level of cyclic AMP (cAMP), which triggers covalent modifications in molecules within the terminal of the sensory neuron. One modification occurs in cAMP-dependent protein kinase A (PKA), leading to its activation; activated PKA causes the phosphorylation of potassium (K+) channels in the terminal, shutting them down, thus reducing the potassium current. This in turn leads to prolongation of the action potential, to longer activation of calcium channels, to more calcium influx, and to the enhanced release of the neurotransmitter glutamate from the sensory neuron.

C, long-term sensitization. Serotonin is repeatedly released from the interneurons, leading to a persistent increase in the level of cAMP in the sensory neuron; as a result, the catalytic unit of PKA recruits another kinase, MPKA (mitogen-activated protein kinase, not shown). The two kinases reach the nucleus and trigger a cascade of gene activation and suppression: the PKA unit activates the gene CREB-1, which leads to the stabilization and prolongation of PKA signaling, which in turn enhances synaptic efficacy. The expression of another gene, CREB-2, a negative regulator of memory, is repressed. This concerted transcriptional activation and suppression involves a whole cascade of enzymes and transcription factors. Long-term sensitization may involve the growth of the synapse (not shown). Other important kinases that are involved in sensitization in *Aplysia* and other invertebrates (not shown) are PKC (Ca/phospholipid-dependent protein kinase) and CaMKII (Ca/calmodulin-dependent protein kinase II), which are present in both the pre- and the postsynaptic neurons.

D, molecular changes in classical conditioning in *Aplysia* are the consequence of the activity-dependent enhancement of presynaptic facilitation. This is the result of calcium priming of adenylyl cyclase, which leads to increased cAMP and increased activation of the protein PKA. Conditioning also involves Hebbian potentiation (a postsynaptic associative mechanism), resulting from calcium influx through NMDA receptor channels. The convergence sites leading to conditioning are designated CC (*dashed lines*). The presynaptic and postsynaptic mechanisms may interact through a retrograde messenger produced in the motor neuron as a result of elevated calcium; it can diffuse into the presynaptic sensory neuron and interact there with the cAMP cascade. Since the gas nitric oxide (NO) acts directly in both the sensory and motor neurons to affect facilitation at the synapses between them, it has been suggested that the retrograde messenger may, in fact, be NO. CC_1 depicts the involvement of a presynaptic association mechanism in the conditioning, and CC_2 shows conditioning that entails a postsynaptic mechanism.

AC, adenylyl cyclase; CRE, cAMP response element; CREB, cAMP response element-binding protein; ER, endoplasmic reticulum; G, G protein; mGluR, metabotropic glutamate receptor; PKA, protein kinase A. *A–D* modified by permission of Cold Spring Harbor Laboratory Press.

cell,[25] two events must occur simultaneously: first, glutamate released from the presynaptic neuron must bind to the receptor, and second, the post-synaptic membrane needs to be depolarized by the activation of AMPA receptors to expel magnesium from the NMDA channel. Thus, the NMDA receptor is ideally suited to act as a molecular coincidence detector. In addition, the associative mechanism may involve "retrograde signaling" from the postsynaptic motor neuron back to the siphon-sensory neuron: as a result of activating the motor neuron, the elevated level of calcium within it triggers some signal that further enhances transmitter release from the sensory neuron.[26] Kandel and his coworkers suggested that changes in these molecular components of the learning pathway led to the transition from general sensitization to Pavlovian learning. Hence, in order to distinguish conditioning from global or local sensitization, the temporal sequences of CS preceding the US must have a much stronger effect than random pairing or a US-CS pairing. Unfortunately, in many old experiments, which are currently the only source of information about learning in nonmodel animals, the experimental design does not include US-CS pairing controls, so it is impossible to know whether one is observing sensitization or conditioning. We can sum up Kandel's model of the evolution of AL as follows:

> long-term sensitization → Pavlovian conditioning, with no qualitative distinction between alpha and beta conditioning.

It has repeatedly been found that the cellular processes and molecular factors underlying learning are highly conserved in evolution. The sea slug *Aplysia*, the pond snail *Lymnaea stagnalis*, the fruit fly *Drosophila*, the honeybee, the nematode worm *C. elegans*, the mouse, and other well-researched model organisms all show that the basic cAMP pathway used in sensitization and conditioning is widely shared, as is the employment of AMPA and NMDA receptors in the postsynaptic neuron.[27] Moreover, the signaling cascade involved in CREB (cAMP response element-binding protein)-1 transcriptional activation and CREB-2 transcriptional repression is also shared (with some modifications) in diverse animal groups, including flies and mammals and nematodes,[28] and sixteen homologous learning-related genes are expressed in both the vertebrate hippocampus and the arthropod mushroom bodies.[29] Hence, the molecular "learning tool kit" is probably ancient, and it is likely that the cellular components that Kandel and others have identified were recruited in the early evolution of nonassociative neural learning.

Kandel, Wells, Razran, and Moore all assume that general sensitization is the basis on which Pavlovian learning is built, but Kandel's view differs from that of the others. While for Wells, Razran, and Moore the difference between alpha and beta conditioning is fundamental, Kandel does not see the difference as qualitative. Kandel and Hawkins, for example, argue that because *all* neural connections are prewired, there are no "real" neutral CSs, and training merely alters the strength of those preexisting connections that bring the response from below threshold to above threshold; the difference between alpha and beta conditioning is merely quantitative.[30] They do, of course, acknowledge that through natural selection some connections between specific neural trajectories (for example, connections underlying defense responses) are more likely to form than others, but they do not see it as a qualitatively important difference. However, as we argue in the next chapter, although at the level of intracellular and synaptic activity the molecular pathways of alpha and beta conditioning may indeed be similar, those forms of beta conditioning that involve perceptual fusion (e.g., seeing a round red apple) entail new *additional* functional principles and new neural structures. From both functional and evolutionary perspectives, this difference is highly significant.

What about the molecular underpinnings and the evolution of operant conditioning? Although operant and classical conditioning have much in common, some of the underlying molecular details may not be identical. In *Aplysia*, AC is the molecular convergence site for both classical and operant conditioning. However, pure operant learning depends on intact protein kinase C (PKC, which is unnecessary for classical conditioning and was not shown in figure 7.5). PKC acts in concert with dopamine to activate another type of AC (type II AC), while classical conditioning depends on intact type I AC, which is not needed for operant learning.[31] Interestingly, in *Aplysia*, the same neuron (neuron B51), a buccal ganglion cell that is part of a central pattern generator (CPG) circuit responsible for feeding, is the coincidence detector for both the CS-US association (in classical conditioning) and for the contingency between ingestive behavior and reinforcement (in operant conditioning). However, the two types of conditioning modify the membrane properties of neuron B51 in opposite directions: classical conditioning results in diminished B51 excitability, while operant conditioning results in increased B51 excitability.[32] Hence, instrumental learning is likely to have evolved through molecular changes in PKC, AC II, and the membrane excitability of particular neurons. This may suggest that world and self-learning had independent evolutionary trajectories. However, since every difference in conditions that affects behavior is reflected in *some* molecular difference, the significance of this observation is not clear. It would be necessary to compare the candidate molecules (e.g., AC I,

AC II, and PKC) and their regulators in species that *do not* learn by association (e.g., cnidarians and presumably also acorn worms, rotifers, and acoel
worms) and in those that do (e.g., most arthropods, many mollusks and
vertebrates) and between species that show substantial differences in their
ability for world and self-learning (although we are not aware of any such
animals). At present, the role of PKC in instrumental learning is known for
few species and for few learning conditions.

The existing views concerning the evolution of AL are schematically summarized in figure 7.6 (*left*), and are compared to our evolutionary scheme,
which suggests distinct, progressive changes in AL evolution (figure 7.6,
right). We are well aware that making clear-cut distinctions between evolutionary stages is impossible in the real biological world, but we believe that
our classification captures important innovations in the evolution of AL,
which are reflected in both the functional organization of animals' nervous
systems and in the taxonomic distribution of learning.

Before we describe and justify our proposal for the evolution of LAL, we
need to examine another facet of memory that is currently getting a lot of
attention from neurobiologists: cell memory and the intracellular epigenetic mechanisms that underlie it. In the last decade, epigenetic mechanisms
have been found to play a crucial role in establishing long-term memory in
neurons, so we cannot understand the evolution of learning and memory
without understanding these mechanisms. In fact, it looks as if we need to
think about different time scales and types of memory: long-term epigenetic memory at the level of the individual cell, memory at the level of synaptic interactions and neural maps, and memory at the level of the entire
animal (imparted, for example, by bioelectric fields). Memory goes all the
way down ... and up again.[33]

Going All the Way Down: The Relation between Cell Memory and Neural Memory

The idea that there is a relation between intracell memory (the active maintenance of cellular states over time) and neural memory is now a focus of
intense research. Its origins, however, can be traced back to the nineteenth
century, when cytology became an established discipline, and biologists
who sought the cellular correlates of memory conceived of cell heredity
as a form of memory. The most sophisticated and insightful view within
this tradition was developed by Richard Semon,[34] whose views of memory
were the basis of our definition of learning in chapter 5. Semon suggested
that the processes that lead to the development of new behaviors and other

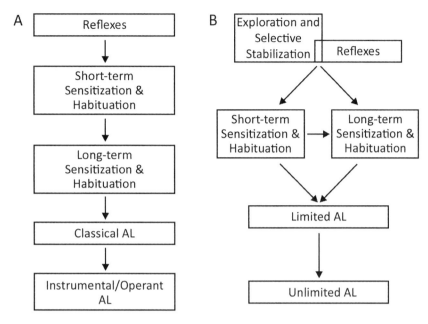

Figure 7.6
Schemes for the evolution of AL. *A*, existing models; *B*, our proposal.

features (e.g., morphological characters) that are acquired through learning or through the direct effects of the environment leave traces in the individual's biological organization. These can lead to the subsequent retrieval of characteristics. He also argued that some of these traces can be transmitted to the organism's descendants, a view that was dismissed, along with the rest of Semon's ideas, and is only recently being reinvestigated.

A molecular link between cell memory and enduring memory at the level of the whole animal was first suggested in the late 1960s, but the study of the relations between the intracellular processes of cell memory and neural memory took off around 2005. By then it had been established that cell memory is underlain by epigenetic mechanisms—cellular mechanisms that produce and maintain changes in patterns of gene expression and cellular structures in both nondividing cells, such as most mature neurons, and dividing cells, such as stem cells, which then transmit them to daughter cells (epigenetic inheritance).[35] Epigenetic cell memory seems to be crucial for long-term behavioral memory, and there are now serious challenges to the idea that long-term memory (which depends on transcriptional activation and protein synthesis) is based on the persistence of synaptic structures. Studies of *Aplysia* neurons connected into networks in a petri dish, as well as intact living *Aplysia*, suggest that long-term memory does not reside only in synaptic connections. Treatments that disrupted long-term memory led to the loss (retraction) of the synapses that were formed following sensitization, as well as to the random loss of other synapses, yet long-term memory later recovered. This implies that synapses do not store memories but rather express memories that are stored intracellularly, in epigenetic marks.[36] Other studies have shown that a single cerebellar Purkinje cell can encode the memory of a specific time delay in ferrets, again suggesting that intercellular connections cannot be the only locus of memory.[37] These studies reinforce the conclusion that *intra*cellular, epigenetic mechanisms are crucial for long-term learning and memory.

Four major types of epigenetic mechanisms underlie cell memory in all types of cells, including neurons:[38]

1. *Self-sustaining loops* involving, for example, the positive regulation of a gene's activity by its products. Such positive regulation leads to the maintenance of the pattern of gene activity. When the gene products are distributed during cell division, the same state of activity in daughter cells can be reconstructed (figure 7.7). A neuronal example of cell memory based on a self-sustaining loop is the autoactivation of CaMKII (calmodulin-modulated kinase II), whose activity depends on the levels of calcium ions and calmodulin (a calcium-binding protein). This enzyme is crucial for long-term potentiation and cell memory and is one of the most abundant proteins in the brain (it accounts for 2 percent to 6 percent of the total protein in the postsynaptic density, depicted in figure 6.3). Structurally, it is a holoenzyme, with twelve similar subunits, which in resting conditions are organized so that their catalytic segments are blocked by neighboring subunits. This autoinhibition is relieved when a strong stimulus causes an influx of calcium ions (via the NMDA receptor channels, as shown in figure 7.5), which enables the phosphorylation of the blocked sites on some of the subunits. This in turn increases the probability of the phosphorylation of the neighboring subunits so that the enzyme becomes active—and remains active— even when the levels of calcium fall again; a self-maintaining loop is thus generated.[39]

2. *Structural templating*, in which three-dimensional cellular structures act as templates for the production of similar structures (figure 7.8). Prions are infectious proteins that can lead to diseases such as bovine spongiform encephalopathy (mad cow disease) and are the best-understood examples of such self-templating proteins. More generally, proteins with identical amino acid sequences but different three-dimensional structures may become components of daughter cells or can be transmitted to other cells

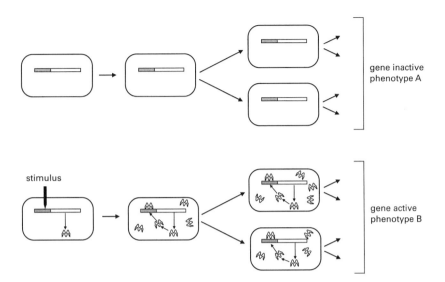

Figure 7.7
Cell memory and cell heredity through self-sustaining loops. *Top*, a gene having a control region (*shaded*) and coding sequence (open) is inactive and transmits its inactive state. *Bottom*, the same gene is transiently activated by an external stimulus and produces a product (M-like structures), which then associates with its control region and keeps it active; because the product is transmitted to daughter cells, the lineage retains the active state, even in the absence of the external stimulus. Reproduced by permission of The MIT Press.

via vesicles (exosomes). In neurons, a candidate prion is CPEB (cytoplasmic polyadenylation element-binding protein), which may be involved in synaptic long-term memory. This ubiquitous protein, which is found in the synapses of many animals (including *Aplysia*), can assume a prion form, and when it is in this conformation, it leads to the polyadenylation of local RNAs, a process needed for their translation. Kauski Si, a former student of Kandel, has proposed a model that explains how CPEB leads to local long-term synaptic memory. He suggests that CPEB is induced by serotonin. When it becomes abundant in a synapse, it assumes a prion conformation that is active and adds the necessary poly-A tails to RNA; the RNA is then translated in that particular synapse, leading to local synaptic activation.[40]

3. *Chromatin marking*, in which patterns of DNA modifications, such as the addition of methyl groups (CH_3) to some cytosines (figure 7.9) and modifications in the histone proteins associated with DNA, can be reconstructed during cell maintenance and cell division. The long-term

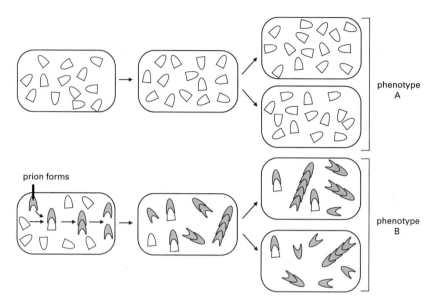

Figure 7.8
Cell memory and cell heredity through three-dimensional templating. *Top*, a cell with the normal form of a protein has phenotype A, which is maintained and inherited by daughter cells. *Bottom*, a molecule with the same amino acid sequence adopts a prion conformation. The prion interacts with normal protein molecules and converts them to its own structure. Prions are transmitted through cell divisions or via vesicles, and phenotype B is formed when all or many of the newly formed proteins have assumed the prion conformation. Reproduced by permission of The MIT Press.

memory of learned behaviors is correlated with gene-specific and genome-wide (global) changes in the neurons' chromatin structure.[41] For example, rats that learn to associate a certain experimental chamber with an electric shock (and freeze when they are introduced into it even when no shock is given) undergo epigenetic changes in their hippocampal neurons: learning-facilitating genes become demethylated while learning-suppressing genes become methylated;[42] in addition to gene-specific changes, there are also genome-wide effects. These results appear to be consistent across taxa, developmental time, brain region, and learning task. Variations in these fundamental and ancient cell memory mechanisms must have been important for the evolution of learning.

4. *RNA-mediated systems*, in which noncoding RNA molecules (ncRNA) regulate translation and transcription through interactions with mRNA or DNA to which they are complementary. When transmitted between cells,

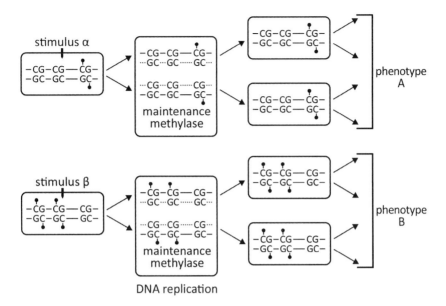

Figure 7.9
Cell memory and cell heredity through the maintenance of DNA methylation patterns. As a result of receiving different stimuli, CG sites in the same locus of two genetically identical cells become differentially methylated (•). As a consequence, they have different phenotypes, which are maintained through the activity of dedicated methylase enzymes in both nonreproducing cells and in multiplying cells. Patterns of DNA methylation are reproduced following cell division because after DNA replication the old DNA strands remain methylated, but the new strands are initially unmethylated; a maintenance methylase recognizes CG sites that are asymmetrically methylated and methylates the Cs of the new strand; the daughter cells thus inherit the mother cells' pattern of methylation (the marks). Reproduced by permission of The MIT Press.

these ncRNAs can affect translation and transcription in recipient cells (figure 7.10). In neurons, small ncRNAs interact with the chromatin-marking systems to alter transcription: in *Aplysia*, a small ncRNA (miRNA-124) is present in sensory neurons and silences CREB-1, but serotonin inhibits this reaction and leads to the relief of CREB-1 suppression, enabling the translation of the CREB-1 mRNAs needed to initiate the process of memory consolidation. Serotonin also leads to an increase in the abundance of another noncoding RNA, piRNA-F, which is a suppressor of CREB-2. This leads to the methylation and repression of CREB-2 (a repressor of memory) and the consolidation of long-term memory.[43]

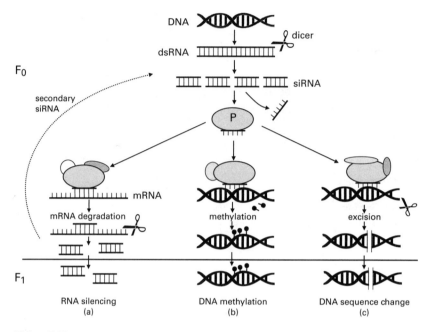

Figure 7.10
Three mechanisms of cell memory and cell heredity mediated by small regulatory RNAs. By associating with various protein complexes (P), small RNAs cut from double-stranded RNA (dsRNA) can silence genes (*a*) by causing degradation of the target mRNA with which they have sequence homology; this silencing can be inherited when the small silencing RNAs (siRNAs) are replicated by RNA polymerase and transmitted to daughter (and sometimes nondaughter) cells; (*b*) by pairing with homologous DNA sequences and methylating them; the chromatin marks are dynamically maintained and can be transmitted among cells; (*c*) by pairing with DNA and causing sequences to be excised. In some cases they can be transmitted to the next generation (F_1). Reproduced by permission of The MIT Press.

The different epigenetic mechanisms depicted in figures 7.7–7.10 often interact and lead to the formation of robust yet flexible and responsive patterns of cellular activity. They underlie developmental plasticity at both the level of the cell type and the organism's phenotype: the determined states of different types of stem cells that breed true depend on these epigenetic mechanisms, as does caste determination in honeybees, whose queen and workers have different inducible epigenetic patterns. Epigenetic processes are integral to the neural plasticity underlying learning and memory. On the basis of these new discoveries, we have suggested that changes in the epigenetic profile within neurons during learning reflect changes at the behavioral level that are described by curves depicting the way learning strengthens with repeated experience.[44] It was later shown that such epigenetic mechanisms are crucial for long-term sensitization in *Aplysia*. RNA extracted from the central nervous systems of sensitized animals induced sensitization when injected into naïve animals. Furthermore, DNA methylation was necessary for this long-term sensitization to occur.[45]

Since the persistence of intracellular epigenetic engrams (EEs) may occur on a different time scale from synaptic Hebbian memory, intraneural memory adds another layer to the formation and retention of both nonassociative and associative memories. Cell memory mechanisms, which are now recognized to be a facet of all neural processing, including neural learning, when combined with synaptic-based memory, may allow sophisticated computations and achievements that go beyond those manifest in current neural network models based on synaptic learning alone. They may have contributed to the evolution of UAL, a topic we explore in the next chapter.

The Evolution of LAL

We define LAL as conditioning that includes both self- and world learning and involves the formation of predictive relations between noncompound stimuli, actions, and reinforcers. "Noncompound" refers to features of objects or actions that cannot be fused into a single nonadditive discernable pattern, so different patterns of sensed stimuli or trains of actions made up of identical constituent parts cannot be discriminated and differentially reinforced. With world learning, CSs can be either reflex-triggering stimuli that have become habituated or are too weak to elicit the observed response or noncompound stimuli that usually elicit a different innate response. With self-learning, the stabilized exploratory actions that are reinforced are innate fixed action patterns. Unlike the general sensitization of reflexes or of exploratory behaviors, where the pairing of the CS and US or action pattern with the reinforcer is unimportant, with LAL the US that follows the CS, or the reinforcer that follows the action, is preferentially sensitized. As a result, the action pattern anticipates the reinforcement, and the CS replaces the US (it "predicts" it).

The changes in the molecular components to which Kandel and others have pointed are crucial for the evolution of LAL, but two factors that are just as essential are missing from the traditional accounts. The first is the explicit incorporation of conditioned exploratory motor behavior into the functional models describing LAL, a topic that we discussed in the previous chapter in the context of the evolution of nonassociative learning.

This includes the feedback between motor and sensory systems through sending back a "copy" of the outgoing efferent motor command (the EC), which inhibits the effects of reflex-triggering stimuli in the sensory pathway. However, in the context of LAL, the EC must also inhibit the *conditioned* pathways. The second factor we need to recognize when considering LAL is the surprising observation that all associatively learning animals seem to have brains.

A Functional Toy Model of LAL

Modulations of exploration-stabilization processes are usually only implicit in accounts of the evolution of AL. The transition from general sensitization to alpha conditioning, a type of LAL in which the CS is neutral because it is a habituated reflex response (rather than a response that is unrelated to the reinforcement), must include feedback interactions between the motor system and the reflex sensory pathway that responds to the conditional stimulus. An efferent copy (EC) from reinforced motor commands, sometimes involving several neural motor units controlling more than one muscle system, must be sent to the sensory pathways to modulate the sensory consequences of self-induced activity in motor units.

But inhibitory effects include not only reafference. AL also entails inhibiting the effects of learning-irrelevant stimuli. This is demonstrated by the "blocking" effect, which was studied by the psychologist Leon Kamin in the late 1960s and, according to some scholars, contributed to the cognitive turn in experimental psychology. With blocking, the presentation of a new CS in conjunction with an already perfectly predicted CS fails to support new learning. In this paradigm, a certain CS (X) is initially paired with the US (i.e., X is established as a reliable predictor of the reward). When combining X with a new neutral stimulus (Y) and pairing them both with the US, no association will be formed between the US and the new stimulus (Y). In this way, AL discriminates against redundant stimuli, which do not add to predicting reinforcement.[46] In other words, the extent to which the organism anticipates a reinforcement is a key factor in determining the strength of learning. What an animal already knows requires no learning, whereas a surprising (discrepant) stimulus is "newsworthy" and is worth learning if it is systematically correlated with reward or punishment. This dependence on what is called "the surprise value" of the CS or the emitted action is the prediction error (PE). One way of interpreting PE is to assume that the animal constantly generates predictions about the value of the inputs it receives. With a novel, surprisingly reinforced input, the most efficient strategy is to adjust the predictions to minimize future surprise.

Thus, a difference between the actual and the expected reinforcement positively affects the formation of associations: the bigger the PE, the greater the increase in learning success and vice versa. This is captured by a simple equation (which underwent many modifications and elaborations): $PE(t) = \lambda(t) - V(t)$, where the reward prediction error PE is the difference or the discrepancy between the received reward (λ) and the reward prediction (V) during trial t. PE reflects the surprise-induced increases in the associability of the CS and US.[47] The neuron circuits of figure 7.11 show how prediction error is generated during world and self-learning. Our LAL models are based on this scheme, which can readily explain blocking. A model of blocking posits that the effect of the new concurrent stimulus Y is inhibited because there is no difference between the reinforcing effect of X and that of XY. This makes evolutionary sense: mobile animals with LAL encounter many incidental stimuli that cooccur with an already reinforced stimulus but do not contribute to the reinforcement, and learning about these stimuli would be a waste of time and energy. It is a telling fact that blocking is found even in animals with relatively simple brains (although "simple brain" seems, increasingly, a contradiction in terms)—in *Aplysia*, in planarians, and in fruit fly larvae.[48] Both blocking and reefferent inhibition therefore enable animals to react adaptively by taking into account discrepancy signals ("prediction errors").[49]

We have developed toy models (figure 7.12) based on the scheme in figure 7.11 to describe LAL, but although the units described have a similar function (indicated by their shading) to the neurons described in figure 7.11, in figure 7.12 they depict neuronal networks that include different neuron types, rather than single neurons. Figure 7.12A is a world-learning model with S_1 representing the sensory network processing the CS (e.g., the sound of a bell) and S_2 the sensory network processing the US (food), with an EC leading to the subtraction of the effects of self-movement-generated stimulation from world-generated stimulation; figure 7.12B represents self-learning, with the activity of the animal being reinforced. In this case, exploratory activity involved sensory activation (for example, through the activation of sensory cells upon pressing a lever), and this activation has neutral effects before it is reinforced (e.g., rewarded); only after it leads to reinforcement, which is mediated by the sensory neuron delivering the effect of the reward to the integrating unit, is the motor behavior reinforced. Figure 7.12C describes both self- and world learning. (We present this byzantine figure, which is merely a combination of figures 7.12A and 7.12B, only because we want to underscore the complexity that LAL world and self-learning requires, even when presented as a highly simplified toy

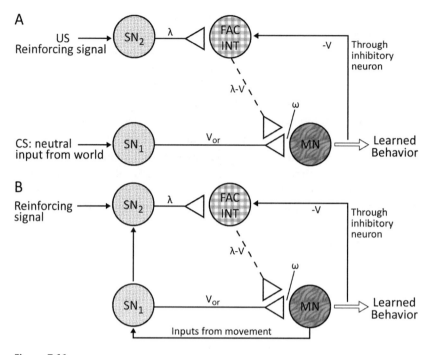

Figure 7.11

Schematic representation of a neural circuit for computing and delivering prediction error.

A, world learning. The prediction error ($\lambda - V$) is computed in the facilitatory interneuron neuron (FAC INT) as the difference between the actual reinforcement λ delivered by the sensory neuron SN_2 and the inhibitory output of the motor neuron MN (the inhibitory neuron is not shown) to the FAC INT. The prediction error changes the efficacy of the synaptic connection between the sensory neuron SN_1, which delivers the CS, and the motor neuron.

B, self-learning. The exploratory activity of the animal results in an input to sensory neuron SN_1, which sends back a signal (V_{or}), to the motor neuron MN. SN_1 also triggers the reinforcing sensory neuron SN_2, which sends the reinforcing signal λ to FAC INT. FAC INT computes the prediction error, which is sent from FAC INT to the synapse between the sensory and motor neurons, leading to reinforced behavior. V_{or}: activity of the signal unit; λ: activity of the reinforcing sensory unit; ω: weight of connection between neurons; $-V = -\omega V_{or}$. *Source*: Based on figure 7.4, and figure 1 in Schultz and Dickinson 2000.

model.) In both world and self-learning there is predictive time-sequence-sensitive priming, with the reinforcement or US following the action and/or the predictive CS. Stimulation by the motor unit, which is reinforced and becomes associated with S_1, involves the inevitable EC-mediated interactions between the motor and the sensory systems, as well as any relevant inhibitory blocking effect. Even in intact animals with simple AL, there must be neural units that integrate afferent and efferent information; that integrate discrepancy signals (an EC is such a discrepancy signal) and lead to adaptive behavior. We assume here that an EC is sent to the integrating unit (IU), where its inhibitory effects are combined with excitatory and inhibitory signals received from multiple sources. It is also possible that in such a simple system, an EC is sent directly to the sensory units S_1 and S_2, or to the sensory units *as well as* to the IU.

There are two additional attributes of learning that all functional models of LAL must accommodate. The first is second-order conditioning, whereby a learned CS can serve as a US (or as a "value-signal") in subsequent learning. Such second-order LAL has been shown in *Aplysia* using a simplified preparation that contained the relevant bits of the animal (mantle, siphon, gill, abdominal ganglion) and one thousand to two thousand neurons that included the circuit responsible for the gill withdrawal reflex. It was shown that two CSs, consisting of a gentle prod of the siphon with an iron rod on the right side (CS_1) or on the left side (CS_2), could be discriminated. Conditioning by pairing CS_1 with a shock to the mantle resulted in withdrawal, and the subsequent pairing of CS_1 with CS_2 led to gill withdrawal upon CS_2 exposure. Hence, an association was formed between CS_1 and CS_2, and the value (negative, in this case) was transferred from the US to CS_1 and from CS_1 to CS_2. The models can accommodate this type of second-order conditioning by adding more sensory and integrating units and assuming both excitatory and inhibitory relations between units, as was indeed suggested by Hawkins and Kandel.[50]

The second facet of learning that models of LAL must accommodate is the precise temporal relation between the action and the reinforcer, or the CS and the US. In natural conditions, an action and its reinforcer, or the CS and the US, may or may not overlap temporally, so a distinction must be drawn between delay and trace conditioning. In delay conditioning, the US follows the CS immediately or is presented while the US is still occurring; in trace conditioning, there is a temporal gap between the CS and the US that follows. It is noteworthy that while trace conditioning is widespread among organisms (it has been found, for example, in *Aplysia* and

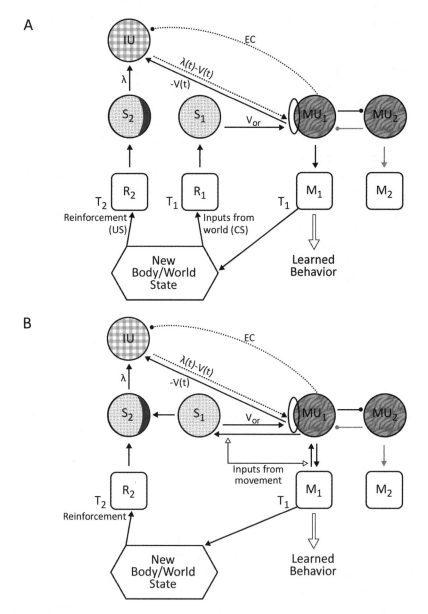

Figure 7.12

Toy models of simple learning. Neural units are depicted as circles with different types of units indicated by different textures; sensors and muscles are depicted as squares. The dark area within a sensory unit indicates that it elicits a reinforcing signal; white ellipses indicate conjunction circuits between the sensory and motor units where synapses are formed following stimulation. Arrowheads denote excitatory interactions; blobs denote inhibitory interactions; light-gray lines depict weak stimulation; dashed lines with arrows show top-down signals (e.g., the PE). R_1 and R_2 are receptors; M_1 and M_2 are muscles; S_1 and S_2 are sensory units; MU_1 and MU_2 are response units

C

(which include sensory neurons and interneurons and send motor commands to the muscles); EC is the efference copy; PE is the prediction error, $\lambda(t) - V(t)$; ($\lambda(t)$ is the expected reinforcement at time t, and $V(t)$ is the actual reinforcement at time t.

A, world learning. At time T_1, a change in sensations coming from the world elicits a change in the receptor R_1, leading to the activation of the sensory neural circuit S_1. This stimulus (e.g., change in illumination) precedes the innately reinforcing US (e.g., food) that is received via mechanoreceptor R_2 at time T_2 and is processed by the reinforcing sensory unit S_2. IU represents an integrating unit connecting the reinforcing signal from S_2 to the response unit MU_1 (a motor unit connected directly to S_1). After receiving inhibiting inputs from the sensory unit (not shown) and EC from MU_1, the IU computes the PE at time t and delivers it to the response unit MU_1. MU_2 is another response unit, connected to effector M_2; when the MU_1 unit is activated, it strongly inhibits MU_2.

B, self-learning. Before reinforcement, both MU_1 and MU_2 are spontaneously and stochastically active, with each motor unit transiently inhibiting the other when active; switches between them lead to default exploratory activity. Through activity at time T_1, which is the outcome of the activity of MU_1, the animal accidentally encounters an inherently reinforcing stimulus (food) at T_2; MU_1 is strongly activated, and this strongly inhibits MU_2. The sensory effects of the action activate S_1 and are sent as an input to S_2, the sensory unit delivering the reinforcement (the rewarding effects of food). As in world learning, an EC is sent to IU and leads to the inhibition of reflexes that would prevent motor activity.

C, combined world and self-learning. We assume that MU_1 is required for the output of world learning and as input for self-learning and that S_1 mediates both self- and world learning. This is a highly simplifying assumption; it is likely that different MUs and Ss are involved. We also assume that the signal from S_2 to IU reinforces both self- and world learning (this is likely, although each facet of learning can be separately reinforced).

Drosophila), no trace conditioning has been found in *C. elegans*, although not for lack of trying.[51] However, the temporal interval between the CS and the US that supports conditioning varies substantially across species. The models presented in figure 7.12 can accommodate trace conditioning if we assume that the molecular factors in the synapses following the CS or the emitted action pattern are not very short-lived. However, this explanation cannot account for trace conditioning in mammals and birds, where the interval may be minutes, which suggests that the transition from relatively brief trace conditioning to more prolonged forms depends on the evolution of attentional and working-memory mechanisms,[52] a topic to which we return in the next chapters.

The toy model presented in figure 7.12 highlights the basic structural unity of world and self-learning. It accommodates both types of learning in a single schema: at the modeling level, the only difference is whether the change in the world depends on learnable actions or is independent of them. Although the model is simple and does not explicitly include motor-integrating units (which coordinate stimulations from *several* independent muscle sheets), it makes clear that even with LAL there must be feedback interactions between different integrating neural units to account for inhibitory interactions such as those seen with blocking and reafference. As we noted in the last chapter, *all* animal nervous systems involve integrating units and feedback that make use of multiple sensors in sensory organs and multiple effectors (organized as muscle sheets). LAL requires an even more complex organization. The construction of motor-sensory-motor loops that control the relation between self- and world-generated stimuli—and that enable the animal to discriminate between the same *learned* self-generated and world-generated stimuli—requires additional circuitry. Even the relatively simple escape response of the crayfish, which is a reflex response, requires a complex circuitry with several types of integration circuits intervening between the sensory and motor systems (including a reafference system).[53] A brain, which we define loosely and liberally as an anterior or cerebral ganglion or several ganglia integrating multimodal sensory information and coordinating motor activities,[54] is not required: as Kandel and others have shown, some forms of LAL can be accomplished by a mere petri dish "preparation" or through ganglionic learning. However, such preparations and simplified systems avoid the complexity of multiple integrations across moving muscle sheets and integrating stimuli from one or more sense organs. Is a brain, then, necessary for building an integrating reafference system and a blocking-enabling system that can generate associatively learned behavior in an intact animal? Since LAL is seen conclusively only in animals with a brain, it seems that a brain may be evolutionarily necessary for LAL.

Why Is a Brain Required for LAL?

The apparent need for a brain is the second aspect of the evolution of AL that we want to highlight (figure 7.13 and table 7.1). Table 7.1 draws attention to information about groups in which some form of learning has been reported and the nervous system organization associated with it, focusing on the presence or absence of a brain. What constitutes a brain and distinguishes it from a mere anterior ganglion or a ring of ganglia is not always clear, and different authorities interpret various anterior ganglia in different ways, using different terms. We rely here on the most common interpretations. We indicate the arbitrariness in the usage by adding "brain-like?" when there is an anterior ganglion or nerve ring that is only occasionally referred to as a brain.

The apparent absence of AL in brainless animals that is shown in table 7.1 and figure 7.13 does not prove that a brain is necessary for AL. A lack of evidence of AL is not evidence of lack because information about learning is either extremely limited, vague, or, in most cases, completely lacking. As we indicated in the previous chapter, even our knowledge about the better-researched cnidarians is limited,[55] and for otherwise well-researched groups such as echinoderms, information about their learning is scant and inadequate. Experiments purporting to show that starfish can learn by classical conditioning did not control for the possibility that the learning observed is the result of global or local sensitization because the effects of the US-CS versus the CS-US contingency were not compared.[56]

It is also important to note that although a brain seems a *necessary* condition for LAL, it is not a *sufficient* condition. We have no data about the learning abilities of many animal groups, but table 7.1 suggests that an animal that has a brain (notably, microscopic animals) may nevertheless lack the ability for AL. Moreover, some species of sea slugs, which, like their relative *Aplysia*, do have a brain, do not show the same extent of AL and have different types of sensitization.[57] Although the biochemical features (for example, serotonin receptor sensitivity) and the modulatory interneurons that prime the presynaptic sensory neuron need to be in place for global sensitization to occur, how the nervous system is anatomically constructed may be a crucial factor in determining what kind of sensitization the animal shows—whether it is global as well as local, or only local. The ability for LAL and the scope of LAL—how many and in what manner different reflex paths are connected—may depend on brain architecture.

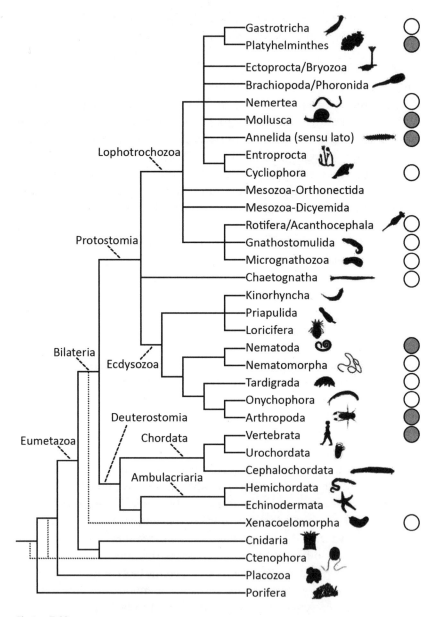

Figure 7.13
Phylogenetic relationships of major animal phyla, their "braininess," and their ability to learn associatively. *Dotted lines*: Alternative phylogenetic positions. *Open circles*: Taxa with what is referred to in the cited literature as brains or cerebral ganglia. *Shaded circles*: Animals showing AL. Modified with permission of Elsevier.

Table 7.1
Type of neural organization and types of learning in animal phyla.

Phylum	CNS or NS	Type of learning
Gastrotricha	Two ganglia, one on either side of the pharynx (*brain*), connected by a commissure; a pair of nerve cords run along the sides of the body	NK
Platyhelminthes	Bilobed ganglia under the eyespots (*brain*) and two nerve cords connected by commissures	NAL, AL[a]
Ectoprocta/ Bryozoa	A bilobed ganglion at the base of the feeding organ and a nerve net (*brain-like?*)	NK
Brachiopoda/ Phoronida	Two ganglia, one above and one below the esophagus (*brain-like?*)	NK
Nemertea	A ring of four ganglia at the anterior end (*brain*) and ventral nerve cords	NK
Mollusca#	No CNS in some (e.g., bivalve mollusks). A cerebral ganglion (*Aplysia*) or more differentiated *brain* (terrestrial mollusks, cephalopods)	NAL, AL[b]
Annelida	*Brain* consists of a pair of ganglia above and in front of the pharynx, linked to nerve cords on either side of the pharynx	NAL, AL[c]
Entoprocta	A ganglion (between mouth and anus) and nerve net (*brain-like?*)	NK
Cycliophora	A pair of ganglia (*brain*) connected by a commissure and a pair of longitudinal nerves	NK
Rotifera/ Acanthocephala	A cerebral ganglion (*brain*) and paired ventrolateral nerve cords	NAL[d]
Gnathostomulida	One anterior ganglion (*brain*) and one pair of longitudinal nerve cords	NK
Micrognathozoa	Cerebral ganglion (*brain*) and a pair of ventral nerve cords	NK
Chaetognatha	Six ganglia in the head region (*brain*), one ventral ganglion in the body, and nerves connecting these ganglia	NK
Kinorhyncha	An anterior nerve ring surrounding the pharynx and a ventral nerve cord	NK
Priapulida	Nerve ring around the pharynx and a nerve cord with ganglia	NK
Loricifera	A ring of ganglia around the pharynx and a ventral nerve cord	NK
Nematoda	An anterior nerve ring (usually referred to as a *brain*) and four nerve cords	NAL, AL[e]
Nematomorpha	Nerve ring near the anterior end of the animal (referred to as a *brain*) and a ventral nerve cord running along the body	NK

(continued)

Table 7.1 (continued)

Phylum	CNS or NS	Type of learning
Tardigrada	Brain with lobes, attached to a ganglion below the esophagus that is linked to a double ventral cord	NK[f]
Onychophora	A bipartite *brain*	NK
Arthropoda#	*Brain* consists of three (in most taxa) paired ganglia in the head; paired main nerve cords run along the body below the gut	NAL, AL[g]
Vertebrata#	*Brain* consists of forebrain, midbrain, and hindbrain; the spinal cord is part of the CNS	NAL, AL[h]
Urochordata (Tunicata)	Ganglion between siphons from which nerves issue directly	NK[i]
Cephalochordata	Anterior blister (*brain-like?*) and dorsal nerve cord	NAL, NK[j]
Hemichordata	Subepidermal nerve net thickened in the region of the proboscis and two nerve cords; also an anterior nerve ring at the proboscis base (*brain-like?*)	NK[k]
Echinodermata	Nerve net with condensed central ring around the mouth	NAL, AL?[l]
Xenacoelomorpha	Nerve nets; some species have anterior ganglia (*brain*)	NK[m]
Cnidaria	Nerve nets; condensed rings in some species; centralized zone in eye-carrying cubozoans	NAL[n]
Ctenophora	Nerve net with some condensations	NK[o]
Placozoa	No nerves	NK
Porifera	No nerves	NK

Notes: CNS: central nervous system; NS: nervous system; NK: not known; NAL: nonassociative learning; AL: associative learning, including both LAL and UAL; #: groups in which UAL has been demonstrated, in at least some species.

The tiny degenerate or simplified mesozoans, which consist of an outer layer of ciliated cells surrounding one or more reproductive cells, are not included in the table since their status is not certain. The presence of the brain as interpreted by most authorities is indicated by explicit mention of this organ (in italics). "Brain-like?" (also in italics) refers to ganglia that are found in the anterior part of the animal and that are sometimes, but not often, referred to as "brain."

Sources: Information about the nervous systems of the different phyla and their behavior is taken from several authorities (see Corning, Dyal, and Willows 1973; Macphail 1982, 1987; Torley 2007; Heyes 2012; Perry, Barron, and Cheng 2013). Information on learning is based on studies summarized in the respective notes.

[a] For examples of learning in Platyhelminthes, see Halas, James, and Knutson 1962; Nicolas, Abramson, and Levin 2008.

Table 7.1 (continued)

[b] For learning in *Aplysia*, see Hawkins, Greene, and Kandel 1998; for terrestrial snails, see Loy, Fernández, and Acebes 2006; in *Nautilus*, see Crook and Basil 2008; in octopuses, see Hochner, Shomrat, and Fiorito 2006.

[c] For learning in annelids, see Wells 1968; Bain and Strong 1972; Sahley and Ready 1988.

[d] Habituation in rotifers was shown by Applewhite (1968); according to Terry Snell (personal communication, December 14, 2015) and Robert Wallace (personal communications, December 20, 2015, and December 21, 2015), nothing more sophisticated has been shown in these animals.

[e] For learning in nematodes, see Saeki, Yamamoto, and Iino 2001; Ardiel and Rankin 2010.

[f] No studies on learning in tardigrades have been published, according to Georg Mayer (personal communication, December 11, 2015) and Karin Hochberg (personal communications, December 11, 2015, and January 26, 2016).

[g] For examples of the learning capacities of insects, see Bhagavan and Smith 1997; Brembs and Heisenberg 2001. For a review on learning in crustaceans, see Tomsic and Romano 2013. For a general review, see Perry, Barron, and Cheng 2013.

[h] For learning in vertebrates, see Macphail 1982, 1987.

[i] No studies on learning in tunicates have been published, according to Yasunori Sasakura (personal communication, December 9, 2015).

[j] According to Thurston Lacalli (personal communication, December 15, 2015), nothing is known about other learning capacities in this phylum.

[k] No studies on learning in hemichordates have been carried out, according to Christopher Cameron (personal communication, December 10, 2015), Linda Holland (personal communication, December 10, 2015), and Christopher Lowe (personal communication, December 11, 2015).

[l] See Willows and Corning 1973; Perry, Barron, and Cheng (2013) review the studies of learning in echinoderms. Although past studies reported associative learning in echinoderms, they did not fully control for sensitization (e.g., McClintock and Lawrence 1982). John Lawrence (personal communication, November 30, 2015) is not aware of more recent work on learning in any echinoderm species.

[m] Pedro Martinez Serra (personal communication, November 30, 2015) and Gabriella Wolff (personal communication, December 12, 2015) say that studies on learning in acoel worms have not been carried out.

[n] As described in chapter 6, while nonassociative learning has been found in several cnidarians, there are no trustworthy reports on AL in this phylum.

[o] According to Leonid Moroz (personal communication, November 29, 2015), there are no reliable studies on learning and memory in ctenophores.

If the connection between the presence of a brain and the capacity to learn through conditioning is not merely an artifact due to ignorance, we must try to account for this intriguing correlation, since it may be an important clue for understanding the evolution of AL. If, as we argued earlier, the evolution of both world and self-learning was entwined and involved interactions between motor and perceptual learning, something like a "center of communication," to use Lamarck's prescient phrase, may have been necessary to coordinate muscle movements with information arriving from multiple external and internal sensors. Of course, once the mechanisms allowing the memorization of such integration are in place, they can be discerned at the levels of isolated neural circuits, in neural "preparations," or through "ganglionic learning." However, LAL did not evolve in isolated ganglia. We suggest that it evolved in moving bilateral animals, with heads and brains. To justify this, we need to look at the evolution of bilaterality in animals and the benefits of a brain-containing head.

Did bilaterality evolve once from a radial or an indefinitely symmetrical ancestor? Can one talk about a single "Ur-bilaterian," or did bilateral symmetry emerge several times? Opinions on this issue, like almost anything else about the evolution of basal animals, are divided.[58] Fortunately, the answer to this question is not crucial for our purposes. What is important is, as Holló and Novák have convincingly argued, that once bilaterality evolved (probably more than once) in tiny, slow-moving ciliated animals, it was maintained in macroscopic animals and is now present in 99 percent of them.[59] According to Holló and Novák, bilateral symmetry maximizes the ability to swiftly change direction because changing direction requires the generation of instantaneous "pushing" surfaces, from which the animal can obtain the necessary force to depart in the new direction. This allows hugely improved maneuverability and therefore confers great benefits on an animal living in a world full of predators. The evolution of bilateral symmetry went hand in hand with the evolution of a single, forward direction of locomotion, with the anterior parts of animals becoming the first to meet or seek various stimuli in the environment. This led to the concentration of sensory cells at the anterior end of the body, which later became a head. It also entailed the aggregation of neurons into condensed nerve cords running along the dorsoventral axis and a motor integration center. These two changes amount to the consolidation of a central nervous system (CNS): a well-defined nervous-tissue structure containing distinct clusters of functionally specialized neuronal cell bodies (nuclei) interconnected by axon tracts (neuropil); these clusters of neurons integrate and process sensory information coming from the periphery and initiate body-wide

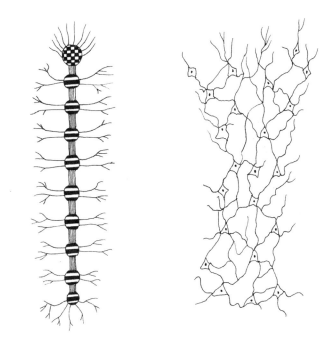

Figure 7.14

The centralized nervous system of a segmented worm (*left*) and the noncentralized nerve net of a cnidarian polyp (*right*).

responses via the neural activation of muscles and neurosecretion into the body fluids.[60] A schematic illustration of a CNS and a nerve net is shown in figure 7.14.

The neural connections in the CNS seem to be organized in a way that is consistent with a trade-off between wiring costs and information-processing efficiency: on the one hand, metabolic costs need to be kept low by minimizing wiring length, and on the other hand, integration of information should be high.[61] Figure 7.15 illustrates different possible network topologies in a theoretical brain model. The "small-world" topology provides the optimal compromise between the minimization of wiring length and an ability to integrate information. This is the topology found in the brain of the worm *C. elegans*, the only animal for which we have the necessary detailed information.[62] Bilaterality, cephalization, and a demand for the efficient and rapid processing of sensory stimuli and coordination of motor responses therefore constrain and shape brain organization. Moreover, processing and coordination occur in an intrinsically and spontaneously active nervous system, not a system that passively awaits the sensory stimuli it

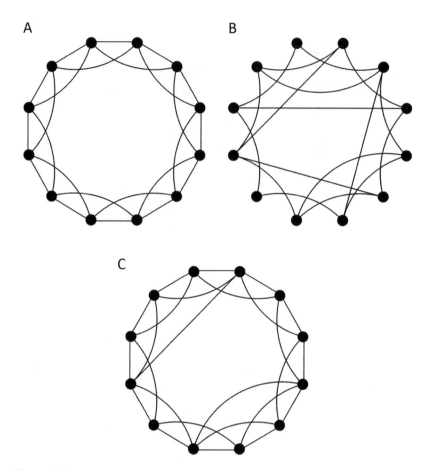

Figure 7.15

Theoretical network topologies. *A*, a regular lattice topology with each node connected to its spatially nearest neighbors minimizes wiring length and hence minimizes metabolic cost; this topology, however, does not enhance the global integration of information processing. *B*, a network with random topology (where each node connects to many nodes) would maximize processing efficiency but has a high wiring cost since it involves many long-distance connections. *C*, the complex "small-world" topology lies between the other two extremes: it contains clusters of lattice-like connections that decrease wiring costs and some high-cost, long-distance shortcuts between connector hubs (nodes connected to many other nodes) that increase processing efficiency.

receives from the world. The brain is "Lamarckian": like any living system, and like the rest of the nervous system, it is, as Lamarck noted long ago, a proactive, self-organizing organ.[63] But do all these dynamic constraints and affordances explain the deep similarities among brains across phyla? Can they tell us something about the evolution of the CNS?

The answer to these questions is relevant for deciding between two hotly debated interpretations of the observed patchy distribution of brains (table 7.1). Brains either evolved from ancestral nerve nets several times in parallel or evolved in a common Ur-bilaterian ancestor and were lost several times.[64] The case for a single origin is based on developmental, functional, and molecular analyses. These show that there are two neural centers of integration in cnidarians, ctenophores, and the larval stages of other invertebrates. The first neural cluster, which remains separate in cnidarians and ctenophores, is located at the apical region and integrates information for the global control of body physiology and motor activity; the second cluster, which integrates motor activity related to feeding, is located at the other pole of the animal, in the blastopore region. According to the "chimeric brain hypothesis" developed by the neurobiologist Detlev Arendt and his colleagues, in the common ancestor of all bilaterians (the Ur-bilaterian), the anterior end of the blastoporal nervous system fused with the apical nervous system to generate the new bilaterian brain, and the posterior part of the blastoporal system was transformed into a pair of bilateral nerve cords, which coordinated locomotion.[65]

An additional observation pointing to the single origin of the brain is that complex integrating brain structures in chordates, arthropods, and annelids (and even acoels) are surprisingly similar. For example, patterns of gene expression during development of the vertebrate pallium resemble those of the mushroom bodies in the annelid worm *Platynereis dumerilii*. The pallium is the most highly developed part of the forebrain (in vertebrates, the cortex), while mushroom bodies are invertebrates' sensory-associative brain centers that store multiple engrams. Remarkably, they develop in a similar way, and their development produces similar neuron types. The multiple developmental-molecular similarities suggest a deep homology of pallium and mushroom bodies that dates to a common bilaterian ancestor.[66] Subsequent studies seem to confirm and extend this conclusion. A comparison between the brains of insects and crustaceans, chordates, annelids, flatworms, ribbon worms, and acoelomorphs has shown that they not only have similar patterns of expression of learning-associated genes in the brain areas responsible for memory and learning (the mushroom body in annelids, the hemiellipsoid body in crustaceans, and the hippocampus in vertebrates) but that they share a similar anatomical organization of these association-memory areas. Even some of their developmental programs share striking similarities: for example, neurons divide in both the vertebrate hippocampus and the insect mushroom bodies.[67]

Yet these molecular, developmental, and morphological similarities do not clinch the matter, and the debate about the single or multiple evolutionary origins of the brain is far from over.[68] Since most of the molecular kit of "learning genes" is found in all animal phyla and since the development of a head with a brain is imposed by bilaterality and forward movement, these strong developmental, molecular, and ecological constraints may account for the similarities of brain structures. Leonid Moroz, who also suggested that the neuron evolved independently in ctenophores (see chapter 6), strongly advocates the case for parallel evolution. Moroz argues that brains originated several times in parallel not only in different phyla but also within a single phylum, the Mollusca,[69] which has an estimated two hundred thousand species and several morphologically different classes, some with brains and some without. The convergent evolution of another important part of the CNS, the nerve cord in the bilaterian trunk, which is found at different dorsoventral levels, is supported by a study showing that representatives of Xenacoelomorpha, Rotifera, Nemertea, Brachiopoda, and Annelida do not have the molecular signature that a homologous dorsoventral regionalization of their nerve cords predicts. Hence, the observed similarities between the molecular patterns along the nerve cords of vertebrates, flies, and annelids may indicate convergent evolution rather than common ancestry.[70]

Whether the brain and nerve cords evolved in parallel in several taxa, had a single origin, or were lost and modified multiple times is still an open question.[71] It is related to other questions that are also shrouded in mystery: Was there a common neural-brainy ancestor to mollusks and annelids? Can we envisage an Ur-bilaterian ancestor from which *all* mollusks, arthropods, annelids, and many other phyla radiated during the early Cambrian? Did the CNS originate independently in all these phyla, or was there a common ancestral Ur-bilaterian for several phyla (e.g., arthropods, annelids, and chordates) and another one for others, such as mollusks? How many times was the CNS lost and radically modified? Although we believe that the accumulation of more data (including more data about the evolution of learning) and a more sophisticated understanding of the physics of development will eventually resolve these issues, we are not there yet.

Regardless of the single or multiple origins of the brain and the nerve cords, what is important for us here is that brains, even those of "simple" animals, seem to have not only sensory and motor regions but also what researchers in the field call "centers of integration,"[72] a term that has some affinity with Lamarck's "center of communication" (although for Lamarck, the whole brain was such a center). Developmental, structural, and functional data show that brains have a sensory integration center and a center for directing and selecting among motor actions, although such centers may vary greatly in complexity.[73] *Aplysia*, our main source of knowledge about basic LAL, has a brain (an anterior ring of ganglia) that is the "center of communication" for the integration of incoming stimuli and the coordination of movements. We know that much of the neural integration occurs within ganglia that are part of this sea slug's brain (figure 7.3*B*). For example, the buccal ganglion that controls spontaneous and learned biting behavior is functionally complex, with several neuronal types within each functional subcircuit within it.[74] And the cerebral ganglia, which can be considered the most "brainy" part of the animal, do not appear to be specialized for any sensory, motor, or physiological function, but they contain some "higher level" interneurons that integrate information coming from several sensory and motor neurons.[75]

Another type of animal showing LAL are the planarians,[76] such as *Dugesia Japonica*, a flatworm that has a CNS with a two-lobed brain and two ventral cords connected by transverse nerves, with a pair of eyes and branches forming the sensory organs projecting from each brain lobe (shown in figure 7.16). The brain is organized into four structurally distinct and functionally diverse domains—mechanosensory, chemosensory, light sensory, and interneuron. The interneuron domain is assumed to have an integrative function, so even in this seemingly simple brain, something like an integration center seems to be present.[77]

Nematodes such as *C. elegans* have a centralized nervous system and can learn about the world, associating a wide range of reflex-eliciting stimuli, something that may be related to their highly compact brains. Nevertheless, in some respects their world learning is more limited than that of *Aplysia*: they show conditioning only when exposed simultaneously to both the CS and the US for a long time. Hence, they do not have delay conditioning or trace conditioning,[78] and they cannot learn about the spatial location of food.[79]

Clearly, LAL can take different forms and can be subject to different constraints in different taxa, and there are important limitations to what can be learned and how learning occurs. But how is all this relevant to the evolution of AL? If a brain is necessary but not sufficient for LAL, under what conditions did LAL evolve? Since the evolution of LAL required few

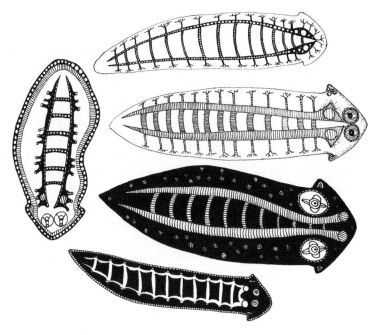

Figure 7.16
Planarians and their nervous system.

changes in an already existing molecular tool kit, how did each change become evolutionarily advantageous?

Selection for LAL
We assume that the common ancestor of the deuterostomes and protostomes was a minute, morphologically plastic animal with a simple nervous system and an integration center, possibly similar to that seen in acoels. The first roles of such an anterior integration center were presumably the coordination of motor activity and the transient stabilization of interactions between sensory information coming from simple sense organs and motor information coming from muscles. Once a brain evolved, animals could search for food and avoid predators more effectively. The avoidance of predators and active predation became more urgent and improved when animals grew in size, which probably occurred when there was a small (but biologically significant) increase in oxygen concentration during the era that preceded the Cambrian.[80] Large size led to improved sensory discrimination (due to larger sensory organs) and speed of movement (due to

larger muscle sheets). Increased body size is correlated with an increased life span—initially, for the simple reason that it takes time to grow a large body, which makes learning by association worthwhile. We assume that it was at this stage that AL made its evolutionary debut.[81] The distribution of AL (figure 7.13) strongly suggests that it emerged during the early Cambrian era, when animal phyla with brains and the potential for AL made their appearance.

If a brain with a center of association was, as we assume, a necessary condition for the evolution of LAL, we predict that LAL may be found in other uninvestigated free-living animals with brains that show basic differentiation into sensory, motor, and integration centers, such as velvet worms (phylum Onychophora).[82] Furthermore, we should also be able to determine whether for some taxa, brains have been lost or never evolved. If the brainless state is the result of secondary loss and the animals have a learning-related molecular kit similar to that of brainy animals, we can infer that LAL (similar to that seen in ganglionic learning) will be found and that the lack of a brain is secondary. If echinoderms lost their brains (the data, as we indicated earlier, are inconclusive), we expect that a molecular comparison of genes supporting AL will show that echinoderms are more similar in this respect to brainy animals than to ancestrally brainless ones. Through a comparative analysis of eight hundred genes that are involved in neural plasticity and AL, we have found that echinoderms are similar to vertebrates and other AL animals in their genetic profile, supporting the proposal that the lack of a brain in this group may be secondary.[83]

Another facet of the evolution of AL is probably the increased importance of sleep. Sleep has been reported in sessile cnidarians,[84] so it presumably had ancient origins, but it appears to have acquired additional roles in animals with AL. In these it seems to be part of the suite of characters related to neural plasticity and may be obligatory unless some compensating mechanisms can provide the same benefits, which are memory consolidation and the unlearning of insufficiently reinforced information. Almost all vertebrates sleep, as do many arthropods, the nematode *C. elegans*, and some mollusks (including, in addition to cephalopods, which sleep deeply, *Aplysia* and the freshwater pond snail *Lymnaea stagnalis*).[85] The neuroscientist Lee Kavanau suggested that the evolution of vision led to an increase in learning and memory storage and that sleep evolved to avoid conflict between complex online activity and memory consolidation.[86] Kavanau's argument can be expanded to include species without sophisticated vision but with the ability to detect light, something that all animals seem able to

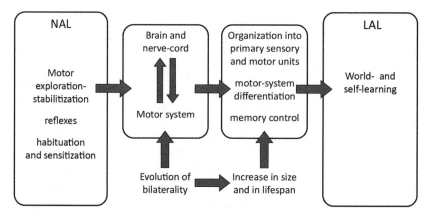

Figure 7.17

Factors in the evolution of LAL. Bilaterality in a moving animal drives the evolution of brains that coevolve with the motor (muscle) system. Increases in size and life span drive the organization of the brain into sensory and motor areas that can integrate sensory information and coordinate the activity of muscle sheets and of memory systems that enable associative learning—both world and self-learning. NAL: nonassociative learning; LAL: limited associative learning. Arrows denote evolutionary causation.

do.[87] We suggest that sleep acquired new functions and evolved with LAL as AL coevolved with sophisticated sensors and motor programs.

We summarize our suggestion for the evolution of LAL in figure 7.17. The figure illustrates the hypothetical relationship between the evolution of the brain and LAL, which led to the ability to memorize relations between different reflex circuits and to greater control of responses to value signals. The neural circuit underlying spontaneous fixed action patterns or simple reflexes became the neural unit of learning. Circuits that were formerly modulated through habituation and sensitization, or were only transiently linked through stimulus-dependent rewards and punishments, could now be combined to generate predictive behavior. Any weak or habituated US and any habituated innate action pattern could act as a CS through its association with reinforcers, as could any neutral (currently nonreinforced) stimulus. Many sensory inputs are likely to be neither noxious nor rewarding, and several stimuli, including many that have habituated effects, are likely to impinge on the animal simultaneously. The ability to use neutral, currently nonreinforced sensory stimuli to predict reward and punishment greatly increased the adaptive response repertoire of the animal, although at this evolutionary stage, we do not have to assume that distinct, semiautonomous reward/punishment systems were necessary.

This enhanced adaptability, based on the ability to reinforce the predictive relations between fixed action patterns, simple reflexes, and exploratory reactions, occurred, we suggest, in the context of the evolution of neural centralization and the increase in body size and life span that took place during the early Cambrian. Centralization facilitated making associations within and between different sensory modalities, and the ability to make these associations, which were at first crude, unstable, and insensitive to the sequence of stimulation, was modulated through selection to yield predictive, stimulation-sequence-dependent responses in a (relatively) long-lived species. This involved selection for a more sensitive long-term memory system, possibly based on the retrograde regulation of epigenetic nuclear memory in neurons and the transport of epigenetic factors to neighboring neural circuits. This, in turn, led to the retention of memory factors (protein complexes and small RNAs) in the synapse of the sensory cell and to the emergence of information-integrating interneurons and integration centers that could underlie processes such as lateral inhibition (e.g., Kamin blocking).

The transition to LAL therefore seems to have involved changes in the way neural information was encoded, stored, and retrieved. The elaboration of memory in a centralized nervous system can be seen as a progressive change in neural plasticity. If brains and nerve cords emerged several times during evolution, LAL is likely to have evolved in parallel, requiring only a few changes in the cellular networks underlying responsiveness to value signals and in the processes affecting cell memory.

Are Animals with LAL Minimally Conscious?

Animals with a capacity for LAL are not, according to our list of criteria, minimally conscious (table 7.2). They do not seem to show the fusion of sensory stimuli that enables discrimination among differently composed patterns of identical elements and therefore are not able to assign value to compound patterns of sensory signals or patterns of action. They also show no top-down attentional modulation, and the temporal thickness they exhibit, as manifest by trace conditioning, is very restricted.

Does this mean that the magnificent *Aplysia* is a living being with no subjective experiencing, more like a plant than like a dog? Does it not have even the fuzziest feeling of existence as it moves in its world, distinguishing

Table 7.2
LAL animals fulfill some but not all of the criteria for minimal consciousness.

Neurobiologists' criteria	Animal with the capacity for LAL (e.g., *Aplysia*, some nematodes, flatworms, annelids)
Global activity and differentiated states	Global activity is apparent in the centralized brain. There are far more differentiated states than in cnidarians because relations between innate circuits can be formed, stored, and reactivated.
Binding/fusion and unification	No indication of perceptual fusion; increased coordination is brought about by the centralization of the NS.
Intentionality	Neural structures mapping external relations onto internal relations, involving memory and exclusion of alternatives through blocking and integrating reafference, suggest that neural intentional structures (neural representations) are in place.
Selection, plasticity, learning and attentional modulation	Plasticity is substantially increased relative to cnidarians. Selection among action patterns suggests some attentional modulation based on the salience of stimuli; there is no evidence for top-down attention. Evidence for limited associative learning is rife.
Temporal thickness	Persistent temporal signatures accompany exploratory behavior and homeostatic states; there is some evidence for extremely short (limited in time) trace conditioning.
Attribution of values; emotions, goals	Values are still bound to stereotypic exploratory and reflex reactions, but the response to value signals is further modulated.
Embodiment and self	Agency is more complex and more subtle. There is an ability to discriminate between self-generated and world-generated stimuli, which is extended not only to reflexes but also to conditioned reflexes and positively and negatively reinforced exploratory actions; hence reafferent reactions are integrating.

itself from that world while coordinating the internal reactions and motor activity of its large body? And are animals that show (unlike *Aplysia*) uncontroversial, noncompound, "true" (beta) Pavlovian conditioning, such as some worms and large land snails, sentient?

As we discussed earlier, Hawkins and Kandel objected to the suggestion that there is a qualitative difference between alpha and beta conditioning, seeing the increase in learning complexity as an outcome of combinations of the basic and ubiquitous cellular mechanisms involved in habituation, sensitization, and simple conditioning. They called these simple processes "a cell-biological alphabet" from which more complex forms of AL, such as generalization, second-order conditioning, and blocking, can emerge.[88] This interpretation does indeed make sense for the types of learning processes that they studied and is dictated by the functional architecture of the system they described. Moreover, as we have argued, when considering the moving animal as a whole rather than an isolated preparation, even the simplest models of LAL entail sophisticated regulatory interactions between and within neuron circuits, which include integrating networks and excitatory and inhibitory feedback interactions, so there are many possibilities for functional combinations within the system. Yet, as table 7.1 suggests, the learning processes, even when they involve elemental (noncompound) beta conditioning, do not satisfy the criteria for minimal consciousness. So although it is tempting to assume that *Aplysia*, land snails, and some worms do have a measure of subjective experiencing, we have no *positive* reason to think so. We cannot rule out minimal consciousness in these animals, but we cannot rule it in. This, however, does not mean that alpha conditioning and compound forms of conditioning, such as compound beta conditioning, are qualitatively the same. As we proposed in chapter 5 and argue in the ensuing chapters, the emergence of learning about compound patterns of stimuli, as well as chaining of different trains of actions, required new memory systems and new integration and value systems that enable prior states of learned neural patterns to be compared with current stimuli. In other words, distinct, hierarchical, bidirectionally interacting neural circuits, which support motor-sensory-motor integration and new memory and value systems, must have evolved to enable open-ended AL. In the next chapter, we suggest that it is the interactions between these integrating neural circuits, which construct a dynamic model of the world, the body, and the animal's actions, that generate states that satisfy the conditions for minimal consciousness.

8   The Transition to Unlimited Associative Learning: How the Dice Became Loaded

The cognitive revolution that produced unlimited associative learning (UAL) led to the transition to subjective experiencing and to the emergence of sentience. This involved the evolution of new hierarchical levels in which neural information was stored; new types of functional units that integrated sensory, motor, and reinforcing signals; high-level sensorimotor association units; and dedicated memory systems. We present a toy model that suggests how these units interact during learning and show how the model can help flesh out the slippery notion of mental representation. We propose that what we call the dynamic "categorizing sensory states" (CSSs), which the embodied functional architecture of UAL implements, are as close as we can get to what one means by "mental representations." We describe the neural structures that underlie UAL in the three phyla manifesting it (vertebrates, arthropods, and cephalopods) and argue that the evolution of the functional architecture of UAL occurred in the context of selection for the flexible control of locomotion and may have involved novel usages of epigenetic mechanisms in nerve cells. Since the functional attributes of UAL, just like the behavioral ones we described in chapter 5, overlap those of minimal consciousness, they lead to informed suggestions about when, where, and why consciousness evolved.

> Mind is at every stage a theater of simultaneous possibilities. Consciousness consists in the comparison of these with each other, the selection of some, and the suppression of all the rest by reinforcing and inhibiting agency of attention. The highest and most elaborated mental products are filtered from the data chosen by the faculty next beneath, out of the mass offered by the faculty below that, which mass in turn was sifted from a still larger amount of yet simpler material and so on. (James 1890, p. 288)

UAL has, by definition, enormous generativity. The number of associations among stimuli and the number of possible reinforced actions that can be generated are practically limitless. The dramatic increase in adaptability that this enables has four major advantages: first, an animal with UAL, which can bind stimuli into compound percepts and integrate actions into compound activity patterns, can discriminate between combinations of stimuli or movements. Second, because local circuits, larger networks, and even larger maps are highly interconnected, a process of pattern completion that induces memory retrieval can occur even when only a subset of circuits is activated. UAL allows the recognition and reconstitution of a relevant compound pattern based on partial cues, so not all facets of a past experience are required for the retrieval of the learned response. Third, once reinforceable compound patterns become retrievable, the animal receives clues as to what to do, since some of the activated engrams are associated with successful exploratory activity toward the attractor-related stimuli (such as food, shelter, and their contexts) or away from repulsor states. The animal can now make an educated guess based on its past experience. Fourth, animals with UAL can make cumulative improvements to cope with novel conditions. They can learn, for instance, how to handle a new food source with increased efficiency, building on less-efficient past practices. Crucially, all these adaptations occur during the lifetime of the animal and are based on selection processes within the individual rather than (only) on natural selection among animals.

Yet the capacity for UAL, which depended on the evolution of sophisticated senses and locomotion, opened, like every revolutionary change, a Pandora's box of new problems. Even in the relatively simple animals described in the previous chapter, movement would have led to a maladaptive reflex reaction if the animal had not taken account of the sensory effects of its own actions and if it had not inhibited, through blocking, the overlearning of irrelevant, accidental stimuli. These two problems became enormously amplified with the evolution of animals with several differentiated sense organs and elaborate and unconnected muscle sheets that enable rapid and coordinated movement. The evolutionary differentiation of the sensory system, which involved the specialization of preexisting general-purpose receptors and neurons, expanded the repertoire of potential "neutral" stimuli.

Consider the evolution of sense organs and their neural structures, which include many differentiated sensory units, each specialized for receiving and processing different aspects of incoming information—for example, separate olfactory sensations or different features of a visual stimulus (angle, movement, brightness) that are processed by separate types of neurons. Learning that feature X predicts feature Y, even if the two features are not functionally preconnected, and responding to a compound rather than to the sum of its constituent parts enhance discrimination and predictive ability. Such inputs, which in the past could not have been connected into a fused pattern, became associable because the evolution of brains brought sensory circuits close together and led to the emergence of regions of association. Interestingly, with the evolution of increased sensory sensitivity, increasingly simple features of an input within a modality, such as an oblique line in the visual system, became the building blocks for perception. These features are in themselves inherently neutral, having no functional significance; only some conjunctions of the elements can have functional significance. In other words, greater synthesis entailed the evolution of greater analysis. For an animal having differentiated sense organs and a brain that processes and integrates inherently neutral sensory inputs, compound stimuli are interpreted as neutral unless they are ontogenetically reinforced.[1]

The greater the number of potentially learnable stimuli, the smaller the proportion of functionally relevant ones. To put it positively, as the number of possible unified associations grows, the "search space" expands, and the chances of dysfunctional responses are greatly increased. The same is true for the motor system: the huge number of reinforceable motor action patterns based on concurrent and sequential movements involving different muscle sheets grows exponentially, and there must be a way of selecting the right one. Neural chaos and behavioral dysfunction would prevail unless highly organized ways of structuring the nervous system were found. This point was eloquently made by James and was the basis of his argument that consciousness is a process of selection—of loading the dice:

> The dilemma in regard to the nervous system seems, in short, to be of the following kind. We may construct one which will react infallibly and certainly, but it will then be capable of reacting to very few changes in the environment—it will fail to be adapted to all the rest. We may, on the other hand, construct a nervous system potentially adapted to respond to an infinite variety of minute features in the situation; but its fallibility will then be as great as its elaboration. We can never be sure that its equilibrium will be upset in the appropriate direction. In short, a high brain may do many things, and may do each of them at a very slight hint. But its hair-trigger organization makes of it a happy-go-lucky, hit-or-miss affair. It is as likely to do the crazy as the sane thing at any given moment. A low brain does few things, and in doing them perfectly forfeits all other use. The performances of a high brain are like dice thrown forever on a table. Unless they be loaded, what chance is there that the highest number will turn up oftener than the lowest? (James 1890, pp. 139–140)

James's view of "lower brains" was somewhat naïve, and we now know that all animal brains, including "lower brains," are inherently and spontaneously active, enabling the organism to respond to some novel challenges.

Nevertheless, there is little doubt that a massive increase in neural connections in a closely knit brain creates the potential hazards that James alluded to. While in "lower brains"—in animals with only nonassociative learning or some form of LAL—the selection of perceptual stimuli and actions is largely based on the stabilization of preexisting innate pathways, this becomes impossible when the number of perceptual and motor stimuli grows. As we see it, the neural dynamics that enable the functioning of complex perception and action in what James called the "high brain" *is* loading the dice—it *is* minimal consciousness. It *is* what renders an animal sentient.

The big question, of course, is what "loading the dice" entails at the functional and neural level. James did not offer a functional-neural model, and he was unsure whether this was the right level of description for answering the consciousness question. We saw in chapters 3 and 5 that there are several suggestions on offer, but as we argued, none satisfies the criteria deemed necessary by neurobiologists and philosophers of mind in a way that can be applied to any animal taxon. UAL, we suggested, is a candidate evolutionary marker of minimal consciousness, which does satisfy these criteria and can shed light on its evolution and basic nature. In this chapter we flesh out this proposal by suggesting a toy model of this mode of learning.

To repeat: the core of our suggestion is that animals solved the loading-the-dice problem when they evolved neural integration, storage, and control systems that enabled them to learn about the relations between compound stimuli and actions. The driving force was strong selection for the ability to learn that certain compound patterns of novel stimuli and actions are reinforced, while other compound wholes made up of the same component stimuli and actions are not, and to use what has been learned as the basis for decision-making and second-order learning. The evolution of such learning (UAL) was therefore intertwined with several functional-computational (and structural) innovations: neural processes and structures that can integrate sensory stimuli into a compound and store the integrated percept as a memory trace were necessary. In other words, a way of retaining a compound engram or a representation (a troubling notion to which we come in a later section) had to evolve. These integration, exclusion, and memorization processes occur in a changing world, so mechanisms that enable the online updating and the off-line storage of updated neural (memorized) configurations had to be in place. The updating strategy is based on feedforward and feedback interactions between new signals received from the world or the body and preexisting connectivity patterns already stored in the system. As we discussed in chapter 7, such an updating strategy is already present in animals with LAL (seen in figure 7.12), where inhibitory signals are sent to sensory neural units from motor units (in the case of reafference) or from sensory units (in the case of blocking). However, with multifeature and multimodal integration, updating complex neural configurations must be much more complicated. We suggest that it is based on signals coming from an internal neural model, which embeds the learning history of the animal. Since multifeature integration is a multistep process, a bidirectional hierarchical updating strategy must have evolved. The neural-body dynamics of integration and exclusion, of learning and storing experienced compound stimuli and the actions that ensue are, we suggest, the dynamics of minimal consciousness.

In the following sections, we discuss in more detail the neural and functional processes required for UAL and their supporting neural structures in different taxa. We argue that online perception and action selection depend—evolutionarily and developmentally—on learning. On the basis of our functional analysis of UAL, as well as on what we know about its taxonomic distribution and supporting structures, we suggest that sentience/minimal consciousness can be inferred to exist in most vertebrates, many arthropods, some mollusks, and possibly some annelids. The ecological and selective context of their evolution is discussed in the next chapter, where we suggest that UAL and minimal consciousness first emerged in two groups in parallel during the Cambrian era and that an important driving force for the radiation of animals during this period was the outcome of the evolution of open-ended associative learning.

Learning through Models: The Patterns That Connect

> The really important first step in the evolution of advanced behaviour is the replacement of simple stimuli or simple patterns of stimulation for the genesis of behaviour, by an abstract model of objects in the real world—that same real world of objects with which our own naïve realism endows the world. The ant reacts to stone and so do we, rather than reacting to the very different initial sensory inputs by which these are detected by ants and men.
> —Pantin 1965, p. 588

How can an animal learn about complex spatial and temporal patterns and select the relevant stimuli and actions from the multitude of learning possibilities that are open to it? How can it discern the patterns of connections in the world,[2] and how are models of the causes of these patterns generated in the brain? What is a neural "model," and what is its relation to the notion of mental representation? To answer these questions, we return to learning and to current ideas about perception and action and tie them together within a learning-based functional framework by constructing a simple toy model of UAL.

The need to respond only to relevant, reinforced stimuli while (actively) ignoring all the rest is reflected in the prediction-error (PE) learning principle that we discussed in the previous chapter. The articulation of this principle stemmed from the observation that temporal contiguity in itself is insufficient to determine the strength of associative learning (as illustrated by the blocking effect described by Kamin) and that the PE—the difference between the actual and the expected reward and punishment—positively affects the formation of associations. In other words, the bigger the PE (the discrepancy between input effects and expectation), the greater the increase in learning success and vice versa. This principle is also at work when the animal moves: an efferent copy may also be seen as a PE signal, updating a "hypothesis" about the sensory effects of the animal's action. In both cases, updating using actual or expected discrepancies (PEs) modulates and restricts learning to the relevant relations only—something that is of great importance in a stimuli-rich world.

Reinforcement (or value) is inherent in the concept of learning. As we stressed in previous chapters, following Gerald Edelman and others, we use the term "value" to point to the salience of states of the world or body to the functioning of the animal. The term does not imply that mental states are involved, although, of course, mental states can have "value."

The molecular underpinnings of value signals are a focus of intense research, and all current theories of reward learning and conditioning, which are mostly based on mammalian learning but also apply to insect learning, propose a causal role for midbrain dopamine neurons in the signaling of reward PE. These studies show that dopamine neurons are excited by unexpected reward, unaltered by expected reward, and inhibited when an expected reward is omitted. In addition, the degree of their response is positively correlated with the magnitude of the PE. But although there is general agreement that reinforcement involves dopamine, the precise role this neurotransmitter plays in learning is still debated. While many believe that dopamine has both hedonic and motivational effects and directly codes for PE, the neurobiologist Ken Berridge, whose ideas were discussed in chapter 5, argues that dopamine plays a crucial role in motivation (wanting), but opioids are the neurotransmitters that code pleasure (liking), and it is liking that drives reinforcement.[3] It is now clear that several different neurotransmitters (e.g., glutamate, acetylcholine, opioids, and serotonin) are involved in hedonic learning. And although aversive learning is not as well understood as appetitive learning at the molecular level, we do know that steroids, serotonin, and dopamine (as well as other neurotransmitters such as noradrenaline, glutamate, and octopamine) are commonly involved, often through complex interactions among multiple types of secreting cells.[4] Studies based on the patterns of activation of the areas that participate in aversive learning (mediated through the amygdala and the periaqueductal gray, an area that in mammals is the primary control center for descending pain modulation) suggest that estimations of PE apply to aversive learning and involves the complex dopamine system as well as other neurotransmitter systems.[5]

The ideas about PE formulated by experimental psychologists and cognitive scientists have interacted with ideas in the field of machine learning, where scientists strive to develop algorithms that can identify and categorize complex perceptual patterns and use them as the basis for action and further learning. These studies are important for us because they shed light on the type of dynamics that constitute minimal consciousness or sentience in the embodied animal.

Integrating Information through Hierarchical Inference

Psychical activities are in general not conscious, but rather unconscious. In their outcomes they are like inferences insofar as we from the observed effect on our senses, arrive at an idea of the cause of this effect. This is so even though we always in fact only have direct access to the events at the nerves, that is, we sense the effects, never the external objects.
—Helmholtz 1867, p. 430

When we think about consciousness and sentience, we do not usually think about learning. What comes to mind are the immediate experiences of seeing a sunflower, of feeling a stab of pain or a flood of joy, of hearing the song of a blackbird, of smelling freshly baked bread, of tasting a ripe banana. These experiences can be fleeting, and we may soon forget them. But how did animals come, evolutionarily and developmentally, to have such experiences? As we have argued throughout this book, *evolutionarily*, learning was the basis on which this experiencing process was constructed. There is no point in sensory discrimination in a rapidly yet recurrently changing world if we are unable to remember it and use it to improve our lot. Moreover, we suggest that even fleeting experiences involve processes of stabilization and updating that are intimately related to learning, as is most obvious during early development. A young animal, like a baby, does not come equipped by evolution with ready-made percepts, although it does come with sensorimotor scaffolds on which percepts can be built. A baby *learns* to see distinct things, it *learns* to hear discernible voices, it *learns* to construct a model of its own body as it acts in the world. There are, indeed, no innate ideas, as John Locke told us in the seventeenth century. There are, however, many developmental constraints and affordances inherent in the morphology of the body and brain and in the structure of the developing sense organs that, when interacting with the regularities of the world, enable a young animal to construct its sensorimotor world through rapid and ongoing learning to make sense of the chaos of the world. All pattern-perception and compound-action selection depend on learning because all rely on the (learned) construction of updatable models of the world in which the animal acts. There must therefore be some common features in the mechanisms employed when we learn and update online (i.e., when we perceive while the stimuli persist).

Thinking about perception, action, and feelings in terms of learning processes that lead to the online updating of information leads us to the already familiar idea that both learning and online updating are selection processes that can be described as procedures for inferring what the world and the body are like. Importantly, the general notion of updating and learning through PE, which developed in cognitive science, incorporates aspects of cognition that are central to the understanding of minimal consciousness. Specifically, it highlights the entwined roles of memory-for-compound configurations and the layered, hierarchical architecture necessary for online perception and action processing. This more general notion of PE is, therefore, very much in line with our learning-centered framework and is based on the PE concept developed by behaviorists.

The basis of the general predictive coding or predictive processing perspective, which employs the PE concept, is a version of the idea elaborated by Helmholtz in the nineteenth century and presented in chapter 4 and in the quote beginning this section. Helmholtz suggested that the brain is an organ through which we infer the causal structure of the world (including the body and its actions) according to the signals ("evidence") that it receives. Today this idea is couched in terms of a Bayesian formalism (box 8.1), where evidence is used to update a prior guess or hypothesis (for example, a hypothesis that the cause of one's frequent thirst is type 2 diabetes) in the light of new evidence (e.g., new medical tests). This is done

Box 8.1

Bayes's theorem.

The theorem[6] has the following form:

$$P(H|E) = \frac{P(E|H) \cdot P(H)}{P(E)},$$

where

- | (*vertical line*) denotes conditional probability; it means "given."
- *H* stands for any hypothesis. Often there are competing hypotheses, from which one chooses the most probable according to new evidence *E* that corresponds to new data that were not used in computing the prior probability.
- *P(H)*, the prior probability, is the probability of *H* before *E* is observed. This indicates one's previous estimate of the probability that a hypothesis is true, before gaining the current evidence.
- *P(H|E)*, the posterior probability, is the probability of *H* given *E*; that is, *after E* is observed. *This tells us what we want to know: the probability of a hypothesis given the observed evidence.*
- *P(E|H)* is the probability of observing *E* given *H*. As a function of *H* with *E* fixed, this is the *likelihood*. It indicates the compatibility of the evidence with the prior hypothesis.
- *P(E)* is sometimes termed the "marginal likelihood" or "model evidence." The likelihood is the same for all possible hypotheses being considered.

According to Bayesian "unconscious inferences," psychologically updatable hypotheses or guesses drive perception and can explain diverse constancies and illusions in the perceptual system. A prevalent Bayesian-based notion of cognition suggests that the process of matching patterns of incoming sensory signals (*E*) leads to the updating, through ontogenetic learning, of hypotheses (*H*) about the causes of sensory evidence (sensory inputs). For example, the evidence (*E*) may correspond to sensory inputs received from a particular shape seen in the shade, and the "hypothesis" (*H*) is the (statistical) structure of this set of inputs; in this particular case, the hypothesis would be a model according to which light comes from above, and the effects of illumination depend on the angle. Different hypotheses are produced by a "generative model," a hypotheses-generating high-order representation.[7]

by taking into consideration the probability of the prior hypothesis (the likelihood that one has type 2 diabetes) and the probability of the evidence given the prior hypothesis (how good the match is), weighted by the probability of the evidence itself (the reliability of the medical tests).

How do brain computations lead to the updating of hypotheses? One Bayesian approach assumes that the neural mechanisms that underlie the process of inferring minimize the discrepancies between the sensory input that the brain receives from the world and the preexisting predictions about the nature of this world. These prior predictions are a result of the evolutionary history of learning (and of the general morphology and physiology of the body), combined with the memorized biases brought about by prior learning during development. Inference happens through reciprocal interactions between neural patterns—the specific "hypotheses" produced by a generative model (a high-level representation that affects the immediate lower level through a set of specific constraints) and the signals arriving from the world or the body. In the general scheme developed by Karl Friston and his collaborators, top-down predictions (hypotheses, incarnated as neural configurations or "representations") affect incoming information from neural patterns at lower levels ("lower-level representations"). Only information that is newsworthy—that is different from information already embedded in, and "predicted" by, the system—is transmitted from the bottom layers to the top layers of the brain to update the "hypothesis" (which means that a lot of information is excluded).[8] Perceptual constancies—for example, the constancy of shape and size in spite of changes of perspective and distance—indicate that top-down corrections, based on a model, must be operating.[9] As we already noted, the discrepancies between model predictions and incoming sensory signals are called, as in the learning literature, "prediction errors" (PEs), since in both cases the discrepancy is between the actual and the expected. The PEs update—or, as in the case of perceptual constancies, actively resist the updating of—prior hypotheses.

Updating through the reduction of discrepancies can occur in two ways: (1) by changing the hypothesis so that it fits the arriving perceptual signals, or (2) by changing the sensory input through actions—actively and selectively sampling the sensory inputs that fit the priors (the preexisting hypotheses). In addition, there is an evaluation of the precision (reliability) of the PE, which affects its contribution to the updating of priors (the more reliable the PE, the more updating will occur). The whole process of perception and action depends, according to this scheme, on a multi-level hierarchical structure: there are several levels of processing, with each intermediate level acting as a hypothesis for the level beneath it and as a PE generator for the level above it. Since the notion of hierarchical level is central to this framework, we need to say what we mean when using it.

A "hierarchical level" is not defined merely by the number of cells intervening between sensors and effectors, since there may be many intervening interneurons in LAL animals, although the structure of the circuit is essentially nonhierarchical. We suggest that one can talk about a hierarchical level when (1) there is a mapping-transforming relation between neural circuits so that relations in one circuit are mapped with transformations (e.g., condensed, not merely copied) into relations in another circuit, and (2) the "higher," "representing" level constrains and controls a "lower," "represented" level by providing "expectations" or a currently selective context that alters the functioning of the level below.[10] An example of hierarchical organization in part of the visual system is shown in figure 8.1. Such a hierarchical organization seems necessary for the perception of composite percepts and for composite action patterns. As machine-learning studies have shown, perception and action selection require mapping and recategorization at several hierarchical levels, over different time scales and levels of generality (the higher the levels, the more general and stable over time are the models). Since the neural patterns at the different levels of the hierarchy are the result of past learning and incorporate complex engrams, there must be memory mechanisms (and supporting structures) that allow their retention and recall. These supporting memory structures are part of the hierarchical framework and are dedicated to the storage of perceived compounds.[11]

What, on this hierarchical predictive processing (HPP) view, is experience—for example, the experience of seeing a drawing of a blue giraffe or the experience of feeling fear? We have a visual or emotional experience, the HPP proponents suggest, when the incoming sensory signals from the world or body (from the artist's picture or from the fiercely beating heart) are matched by a flow of top-down predictions constructed by hypotheses based on past experiences. Of course, the match need not be perfect; the clarity of the percept or the extent and sharpness of fear and their functional significance are determined by the degree of matching, which stabilizes the updated model. However distinct and clear the percept or the sensation, and however large or small the PE, the circular interacting hierarchical processes of predicting and readjusting lead to what we call "experiencing." Experiencing is therefore, as James suggested, a selection process involving a massive exclusion of possibilities. In addition, experiencing depends

A

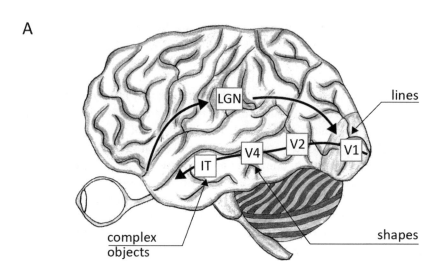

B

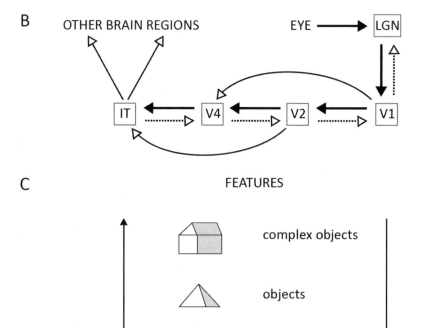

C

on preexisting neural patterns that are based on past evolutionary history and past ontogenetic learning—learning that is itself a selection process or, more precisely, a selective-stabilization process. Since UAL requires the kinds of cognitive processes that seem to be necessary for consciousness (perceptual fusion, memory for compound patterns, reinforcement that is highly sensitive to the animal's overall physiological state), can we construct a functional model of UAL that will help us think about minimal consciousness in a more structured way?

Figure 8.1

Interactions in the primate ventral visual system, which is engaged in processing the features of objects (but not their movement). *A*, the main ventral visual pathway in the brain. *B*, schematic diagram of bottom-up and top-down connections and interactions in the system. *C*, types of visual representations in the pathway. Full arrows denote feed-forward connections, and broken arrows denote feedback connections; thick arrows in (*B*) denote major connections. The lateral geniculate nucleus (LGN) in the thalamus acts as a relay center in the visual pathway; it receives direct sensory input from the retina and feedback from the primary visual cortex (V1); from the LGN onward, signals are integrated to form representations (maps) and then reintegrated and remapped into higher areas of the visual system.

V1, the primary visual cortex (striate cortex), the first station in the visual stream of processing, receives its main input from the LGN; it contains a map of the spatial information in the retina, and neurons within it can discriminate small changes in visual orientation, spatial frequencies (periodic patterns), and colors.

V2, the secondary visual cortex, is the first part of the visual association area; it receives input from V1 and sends output to V4 (and also to V3 and V5, not shown) and feedback to V1.

V4, the third cortical area in the ventral stream, is (like V2) part of the extrastriate cortex; it receives input from V2 and from V1 (directly), has feedback connections with V2, and sends output to IT; like V2, it is tuned for orientation, spatial frequency, and color, but it is also tuned to simple geometrical shapes, and exhibits strong attention modulation.

IT, the inferotemporal cortex, is considered to be the final stage in the ventral cortical visual system; it is associated with the representation of complex object features, such as houses and faces, and with their memory and recall. It receives input from V4 and also from V2, sends feedback to V2 and V4, and projects to many brain areas outside the visual system (such as the prefrontal cortex, basal ganglia, and amygdala). *A* and *C* are adapted by permission of The Association for Research in Vision and Ophthalmology.

A Toy Model of UAL: A Functional Characterization

The toy model of LAL presented in chapter 7 showed that the basic structure of the system implementing associative world and self-learning are the
same, although the temporal relations between perception, reinforcement,
and motor activity may differ in important ways. Based on the behavioral
analysis of UAL (discussed in chapter 5) and on the functional considerations
described in the previous section, we present a toy model of UAL (figure 8.2).[12] As with the LAL model on which it is based (figure 7.12), world
UAL (*A*) and self-UAL (*B*) have the same functional architecture. Of course,
as with LAL, the assumption that there is pure world or pure self-learning is
an idealization. Figure 8.2*C* shows both self- and world learning, in which
the animal learns about both its own actions and the object toward which
these actions are directed.

The UAL model includes the four central features that this type of learning demands: (1) hierarchical processing, which enables compound perception and compound action; (2) integration between perception and action
models; (3) memory for compound percepts and actions; and (4) a flexible
global reinforcement system. Three hierarchical neural levels are depicted
in figure 8.2, and this is the smallest possible number of levels in such a
system:

1. The level of the primary sensory units (Ss) that receive inputs from sensory receptors (Rs), and the primary neural motor unit (MU) that sends
   outputs to the effectors (e.g., muscles, M).

2. The level of the integrating structures MIU (motor integrating unit) and
   SIU (sensory integrating unit), which fuse sensory and motor inputs into
   compounds, and the REIU (the global reinforcing unit), which enables
   the reinforcement of any compound percept or action sequence; the SIU
   (which includes exteroceptive, proprioceptive, and interoceptive units)
   and the REIU are in constant reciprocal interactions via the integrating
   association unit (AU) at the next hierarchical level.

3. The control-modulatory level that includes the MEMU (integrating memory unit) and the AU (integrating association unit). The AU puts together
   information from various sensory and motor integrating units, sends predictions to the lower levels, and receives updating PE signals from them.
   These higher units control, constrain, and modulate the activity of the
   integrating sensory and motor units at the second level.

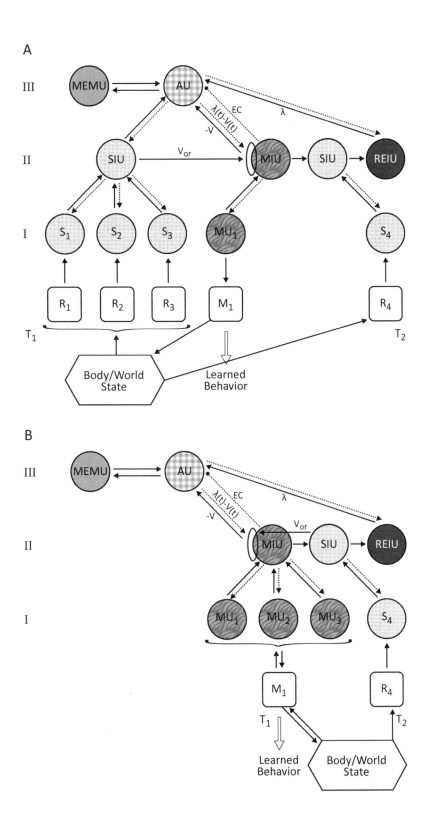

C

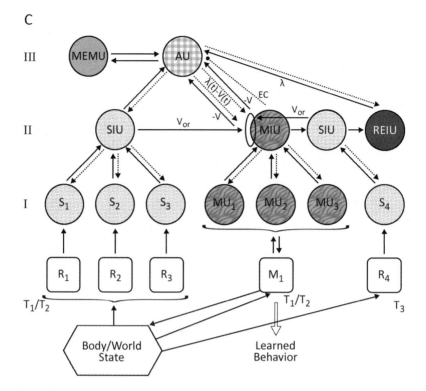

Figure 8.2

Toy models of UAL. *Circles*, functional units; *squares*, receptors and effectors; *solid arrows*, bottom-up and lateral interactions; *dashed arrows*, top-down interactions; *dashed line with a blob*, the efferent copy sent from the MIU to the AU; *white ellipse*, the local conjunction of circuits. *I*, *II*, and *III* are distinct hierarchical functional levels. R, receptor (e.g., retinal cell); M, effector (e.g., muscle); S, a primary unit processing sensory information; MU, a primary motor unit processing information sent to the muscles; SIU, sensory-integrating unit; MIU, motor-integrating unit (generates models of prospective action); REIU, reinforcing-integrating unit; AU, association unit (motor-sensory integration); MEMU, dedicated memory unit; EC, efference copy; PE, the prediction error, $\lambda(t) - V(t)$; ($\lambda(t)$, the expected reinforcement at time t; and V(t), the actual reinforcement at time t.

A, world learning. At time T_1, a compound CS (e.g., a black, round, and rigid ball) is processed in the sensory-integrating unit, SIU. The CS is composed of several sensory elements received by receptors $R_{1, 2, 3}$ and processed by sensory units $S_{1, 2, 3}$. The construction of the compound CS in SIU is produced through hierarchical predictive coding: the novel aspects of bottom-up information (*ascending solid black arrows*) are combined with top-down predictions (*dashed arrows*) based on available perceptual information in the SIU and on information in the memory unit, MEMU, transferred to the SIU via the association unit, AU. At T_2, a reinforcing stimulus (a US, such as food) activates the reinforcing unit, REIU, via receptor R_4 and sensory unit S_4. The SIU, the MIU, and the REIU systems send activating and EC signals to AU, which

Of course, for real, highly multifeatured patterns, there are several intervening layers between 1 and 3 in which the basic hierarchical structure depicted in the present figure is repeated. Crucially, the units in the model are described at the functional level and are not necessarily distinct anatomical units. For example, memory is distributed in many brain areas and, at least in some cases, several functions rely on partially, or even entirely, overlapping neural-anatomical structures so that one neural unit can implement more than one functional system, and different neuron types (sensory, motor, and

---

constructs the updated model of the active "self" in the world / body at T_2 and sends a PE to the local association circuit (*ellipse*). The PE is based on the discrepancy signal sent from the SIU (PE = $\lambda(t) - V(t)$; the received reward is λ, and the reward prediction at time T_1 is V). When the discrepancy is nonzero, the strength of the association between the SIU and the MIU will increase in proportion to the magnitude of the prediction error. This association circuit (*ellipse*) reconstructs engrams in the MEMU via the AU, which interacts directly with the MEMU. When the MIU is activated, it in turn activates a lower-level motor unit (MU_1), which leads to a response via the effector M_1. On later encounters with the stimulus, the SIU-MIU relation is primed for activity by the MEMU engrams. As a result, the compound CS will now elicit the adaptive M_1 activity. Direct interaction between the SIU and the MEMU (unmediated by the AU) are not included, although in real neural systems they are likely to occur; similarly, the EC signal can be sent directly to the SIU and the REIU as well as to the AU. Signals sent to the AU that are blocked are also not shown, nor are the targets to which the EC is sent from the AU.

B, self-learning. At time T_1, a compound action (e.g., sequentially pressing a button, turning a dial, and jumping) is triggered by the motor-integrating unit MIU, based on temporal and spatial combinations of various action patterns, which are implemented by MUs and are sequentially activated. The MIU sends a signal to the SIU that integrates proprioceptive and interoceptive stimuli, which send a signal (V_{or}) back to the MIU, which in turn sends a V signal to AU. At T_2, as a consequence of the actions of the animal, which are integrated in the MIU and are influenced by past behaviors (engrams in the MEMU), a reinforcing stimulus (such as food) activates the reinforcing unit, REIU, via receptor R_4 and sensory unit S_4. The reinforced action pattern is memorized both at the level of the local circuit (*ellipse*) via the AU and also in the dedicated memory unit MEMU. As in A, the relation between the motor pattern and the reinforcer is processed in the AU, which sends a PE to the association locus (*ellipse*). The possible direct interactions between the MIU and the MEMU are not shown—nor is the possibility that the EC signal can be sent directly to the SIU and the REIU as well as to the AU.

C, a combination of self and world learning (*A* and *B* combined), typical of most classical and operant learning.

interneurons) make up each unit. The model described is, we believe, the simplest model of UAL, which incorporates composite percepts, selection among composite percepts, and an implicit self-model.[13]

In the highly simplified functional model described in figure 8.2, the SIU can be thought of as a generator of models of the extrabrain environment and the MIU as a generator of models of prospective action. The SIU represents both the external world and the body, since BODY-WORLD is everything that is outside the brain. (Of course, the model of the body will have overlapping but separate topographical instantiation in the brain, just as different perceptual modalities have different brain localizations.) These units are influenced by the inputs from the dedicated memory unit (MEMU) that reflect the animal's history and by the current homeostatic state monitored by the reinforcement unit (REIU), and they send their outputs back to these units. Similarly, complex movements require a model of action that takes into account multiple levels of muscle coordination that interact with a model of the animal's body. Hence, proprioceptive, action, and body models must interact as the animal responds to the model of the external world that it has constructed. The REIU assigns value to percepts and actions according to the deviation of the system from a state of homeostasis, so water for a thirsty, dehydrated animal is a strong reinforcer, whereas for one that is fully hydrated it is not. The AU generates high-level models or predictions according to the reliability of the discrepancies between the expected effect of compound stimuli, actions, and reinforcements, changing their associability on the basis of an integration of lower-level models of the world, body, and action that represent these compounds. The AU is thus the "locus" in which the "self," as a dynamic process, is formed. All these dynamic models are modulated and, in turn, modulate the compound memory unit. As we described in chapter 7, blocking (as an instance of PE-based learning) and reafference can be implemented by a peripheral system, but in intact animals PE implementation involves central, integrating brain structures.[14]

According to our model, not only do the effects of binding in the SIU affect the construction of engrams in the MEMU, but the conjunction circuits, where the relationships between the SIU and the MIU are encoded, affect MEMU engrams (via the AU) through interactions with similar preexisting, overlapping MEMU engrams.[15] Such processes could occur in the context of engram reconstruction in the MEMU (which may itself represent a hierarchy of units). During these processes, multiple "engram drafts" would be formed, and the selective stabilization of engram ensembles would be based on reentrant interactions between the conjunction circuits and corresponding MEMU circuit engrams.

UAL entails the coupling of several generative models that are influenced by the animal's past history and are represented in one or more dedicated memory units. Both the generation of inputs that are transmitted from lower- to higher-level units and the generation of multiple models at high hierarchical levels that select and are selected by bottom-up inputs are, in our terms, exploration-stabilization processes. Hence, the UAL model can readily be reformulated in terms of the general predictive-processing framework suggested by Friston.[16] A "prior" is a top-down signal, shaped by the animal's evolution and past learning (e.g., descending signals from the SIU to the Ss), which constrains and modifies the ascending signals from a lower unit. The ascending signals—when different from the learned state—can be interpreted as "PEs" (e.g., ascending signals from the Ss to the SIU). The reliability of the PEs (termed "precision") is computed at each level through lateral inhibitions of alternative stimuli-response relations (these are not shown in the model). This prevents the blurring of the PEs by other competing signaling circuits. The exploratory processes as well as the "selecting" processes, such as blocking, are also not depicted in the model, although the AU (like all other units) receives and sends multiple signals that are not predictive and the effects of which are therefore suppressed and excluded.

The toy model can be applied to the processes of perceptual and motor online updating, processes that are not strongly reinforced and require ongoing stimulation for their persistence. Instead of the formation of lasting neural engrams at the junction of the MIU, SIU, and REIU and in the MEMU and AU, stabilization is transient, and no lasting engrams are formed. However, the same processes as those implementing UAL are involved in updating: hierarchical processing of information with its bottom-up and top-down interactions; the modulating effects of past memories; and reinforcements occurring during the formation of the transient motor-sensory-motor loops, which are based on the computation of discrepancies.[17] There are, however, important differences between learning and online updating: first, stimulus-independent endurance of the associations formed (which goes beyond working memory) is necessary for learning but not for updating, although because working memory is necessary for the generating compound percepts, weak engrams may be formed. Second, online updating overcomes some limitations of the memory system and is therefore richer than off-line recall. For example, one can distinguish online between very subtle shades of green, something that one cannot do off-line because the difference between the engrams is not large enough, and the two shades are stored as the same engram.[18] However, the basic architecture—the pattern of coupling of the units of the generative model and the memory and reinforcement units—is the same. As we suggested earlier, all perception, all experiencing, involves learning processes.

Our model is a functional, abstract toy model, not a neural-network computational model. It therefore does not include details such as the specification of the synapses formed at each level in the hierarchy and the fan-in (convergence of signals) and fan-out (divergence) architecture within and between different functional units;[19] nor does it show the inhibitory (e.g., blocking) and excitatory interactions among units within the same hierarchical level and between levels. Moreover, our model does not include some important features and therefore does not demonstrate some important ramifications of this type of AL. First, it does not include second-order conditioning and chaining, although it is not difficult to incorporate such conditioning: a second compound conditional stimulus (CS) can become primed for activity by the original, already reinforced compound CS.[20] We did not show this in order to avoid complicating further the already complex figure. Second, as already noted, only three levels of hierarchy are shown, while an actual UAL system is likely to have more levels. Third, we neither show the many processes of lateral inhibition that accompany the activation of each particular functional unit nor depict direct interactions between the SIU, MIU, and MEMU. Fourth, in the amalgamated world- and self-learning model (8.2C), we assumed that learning about self and world are distinctly encoded in the conjunction area (the ellipses on figure 8.2) as well as in the MEMU, but there may be several memory-storing levels in an MEMU. Fifth, the capacity for trace conditioning (when there is a temporal gap between perception and reinforcement) is not explicitly depicted in the model, although we see it as a necessary outcome of the reentrant processing that occurs during UAL. Sixth, many "background" processes are assumed but were not explicitly presented in the model, and lateral inhibition such as that involved in blocking is not shown. Seventh, in figure 8.2 memory is implemented by patterns of neural connections and dedicated structures, yet we know from the molecular studies of learning (discussed in the previous chapter) that memory also involves cytoplasmic and nuclear epigenetic factors. We will revisit these facets of memory later in this chapter and in the next chapters; here we wish only to point out that the epigenetic mechanisms involved need to be incorporated into a full picture of biological learning.[21] Eighth, we implicitly assumed, but did not show, that the inputs sent from the body are the outcome of multiple physiological processes, including hormonal and immunological activities, as well as the constraints imposed by the body's morphology. As we emphasized when discussing the ideas of Michael Levin and the notions of agency and self in chapters 4 and 5, the body's morphology is dynamically maintained (and changed) through bioelectric and biochemical gradients and fields, which both constrain and are constrained by the activity of the nervous system. Finally, although the reinforcement system plays a crucial role in the model, we did not describe how the reinforcements that accompany different types of world and body perceptions generate different types of emotions and drives.

The model in figure 8.2 is not a model of minimal consciousness. It is a schematic model of UAL showing, so to speak, the bare bones of the dynamic organization required for this kind of learning. But although the model is limited to a minimalistic depiction of UAL, the functional architecture of UAL (in an animal with a brain) requires that all the key elements of minimal consciousness are in place. In spite of its schematic nature, the model makes some predictions (which we discuss at the end of the chapter) and suggests plausible origins for the defects of learning apparent in some pathologies. These pathologies, for example, include associative agnosia (where association among different features occurs but their recognition and semantic meaning is gone) and apperceptive agnosia (failure to bond different percepts into a single, unified whole). Our model suggests that associative agnosia is the result of an impairment of the relation between the SIU (directly or via the AU) and the MEMU, while apperceptive agnosia is the result of a grouping deficit, the failure to integrate sensory information by the SIU. These interpretations are in line with what is known about the nature of these agnosias in humans.[22]

The normally coupled neural configurations, both fleeting and enduring, that are constructed by the functional architecture of UAL and the enabling system of which it is part, fit and entail all seven criteria in the list of characteristics of minimal consciousness. These seven characteristics appear together in some taxa belonging to three animal phyla: the vertebrates, the arthropods, and the mollusks. The interacting neural processes and structures in the UAL system depend on the *binding* of compound percepts of body, world, and action and their *global accessibility*. The neural states are motivating because they ascribe *value* to actions and percepts based on the overall (systemic) homeostatic state of the body in the world. *Selection* occurs through the exclusion of possibilities during the processes of integration, and attention (the active exclusion of most states and the amplification of a particular state) can be described in terms of PEs and the precision with which they are evaluated, as suggested by HPP theorists. UAL therefore leads to *intentional* states, in the sense that these states are "about" events and relations outside the brain, as well as in the sense that they express the individual's needs and its implicit and explicit goals.[23] Because UAL is based on interconnected activity within multiple networks and maps, it is time-consuming, allowing the representation to persist in short-term memory for a substantial time, which is the basis of *temporal thickness*. This seems to be the mechanistic foundation of working memory, which enables an animal to "hold on" to perceptions and to learn even when there is a gap of minutes between the CS and the reinforcer. Finally, the UAL architecture constructs specific configurations of coupled neural activities that are part of a globally active encompassing system (a mobile body in a world), with a distinct developmental and learning history; it constructs a differentiated yet global individual representation, a *self*. Its intrinsic and dynamic activity states are idiosyncratic and unique, depending on the reverberating neural, hormonal, and physiological activities within the whole animal, which are fully accessible only to itself. This characterization is the closest we come to a portrayal of a subjective mental state, a state with a quale. We therefore regard the dynamic neural configurations that are formed by an evolved and embodied UAL functional architecture as mental representations.

The term "mental representation" is one that we have studiously avoided throughout this book because it is used in confusing ways in both the neurobiological and philosophical literature. As we see it, mental states are by definition what a conscious animal generates. If UAL architecture is required for consciousness, we have to account for the generation of mental states within this framework. We need to start, however, by looking at what the general term "representation" (not just mental) actually means and why this loaded term is at all useful.[24] The large literature on representation does not make these questions easy to answer because the term is used either in a very liberal way or in an extremely limited context (usually, within the symbolic/linguistic framework). Since our focus throughout this book is on learning, we will characterize representations in the context of learning, which is better understood than online perception, although, as we shall argue, both are implemented by the same UAL architecture. We suggest that within the neurobiological framework, one can distinguish between three types of representations: neural, neural-mental, and neural-mental-symbolic (figure 8.3).

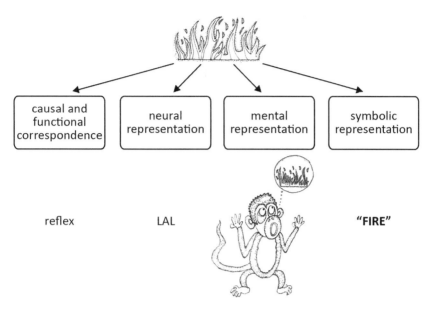

Figure 8.3

Functional correspondence including neural nonmental, neural-mental, and neural-mental symbolic representations. An example of cause-effect nonrepresentational mapping (*extreme left*) is a reflex arc where fire corresponds to an external cause, leading to a motor response. A neural (nonmental) representation is illustrated by an LAL reaction, where an animal relates a simple CS (a change in lighting) with a US, such as burning. Mental representation (fire in the head) is the main topic of the next section, where it is discussed at length. The word "fire" is a symbolic representation.

Mental Representations as Categorizing and Motivating Sensory States

> What is it that you see when you see? You see an object *as* a key, you see a man in a car *as* a passenger, you see some sheets of paper *as* a book.
> —Stan Ulam, as quoted in Rota 1986, p. 3

Mapping through structural or correlational correspondence, such as a cognitive map or a descriptive sentence, is the sort of thing that comes to mind when thinking of representation. Spencer suggested 150 years ago that correspondence between external and internal relations renders the internal relations representative of the external ones. This, however, seems too liberal a concept of representation, since any relation between causes and their effects can be so construed. For example, on this interpretation, a pattern of cracks on a thick glass windowpane would be a representation of the throwing of the stone that cracked it because this pattern bears some relation to the size and shape of the stone and can be said to "map" the action of throwing it. It therefore seems that for something to represent something else (for the mapping system itself, not for an external observer), more is required than a description of mapped cause-effect relations (although cause-effect mapping/correspondence is necessary). In a biological system, the mapped correspondence needs to make a functional difference to the mapping system itself, so in addition to a correspondence-based account, for a representation we may require an account in terms of teleofunctions and functional information. In the first chapter of this book, we defined functional information as "any difference that makes a systematic and causal difference to the goal-directed behavior of an encompassing system"; systematic implies that the effects are, on average, reliable and involve selection/sampling processes. Importantly, like all types of functional information, representations are the product of an evolved processing system that entails selective stabilization.[25] However, since all evolved biological systems can be said to embody functional information, this too is as yet an insufficient requirement.

Crucially, all representations are, by definition, *about* a pattern of relations, so representations can be correct or incorrect in the sense of accurately or inaccurately matching whatever they represent (in the symbolic case, they can also be true or false). Causal and teleofunctional correspondences do not have correctness conditions, although a functional system (e.g., a biochemical cycle) may be dysfunctional. Hence, aboutness cannot be described in functional terms alone. So what makes certain types of functional relations representational?

We suggest that representations are special types of teleofunctional mappings. For teleofunctions to be representational, three requirements are individually necessary and jointly sufficient: (1) there must be functional mapping between patterns or relations in two domains, so, for example, the patterns in the external world are mapped on the retina, and patterns in the primary visual cortex map downstream retinal maps; (2) "maps" should be understood broadly as the outcome of an ontogenetically and/ or phylogenetically selected pattern of correspondences. This implies that a mapped pattern has a value that satisfies a system-wide goal; and (3) the mapping enables one-to-many relations between inputs and outputs so that, for example, mapping the general type "cat" allows detection and discrimination among different cats. Hence, the map is a model of the referent. A token percept is categorized by the system: the token percept (Nudnik, the spotted, moving, green-eyed object) is an instantiation of the type CAT; it is seen *as* a cat. Perceiving "as" is based on capacities to map objects and their properties onto neural dynamic structures and to bind them and categorize them; this requires mapping to be both plastic and robust (usually, it requires the active exclusion of many patterns). Evolution must therefore have led to increased plasticity and canalization in the processing system so that different inputs can be categorized and processed by the representing system and lead to the same output (canalization), and the same input can be processed in different ways within the system to yield different outputs (plasticity). An additional feature, which is not necessary but occurs very frequently, is that the mapping relations can be dynamically retained in a latent state (i.e., there is memory), so a representation can be off-line as well as online.

Fulfilling these conditions as well as showing correspondence between cause and effect seems to constitute the concept of representation (neural and nonneural) as used in the biological literature. According to this proposal, the neural patterns that are observed in the simplest associative learning (LAL) are neural representations. Figure 7.12 in the previous chapter depicted this case: there is a pattern of neural relations that corresponds to, and can be said to model, the temporal relation between the CS and the US in the external world; many types of CS/US relations can be instantiated, and engrams of patterns of relations are retained locally in the circuits formed between sensory and motor units. In addition, the effects of uninformative stimuli and actions are actively excluded. (In figure 7.12 this exclusion is manifest by reafference and by the inhibition of one MU by another.)

The holy grail is, of course, the notion of *mental representations that are subjectively experienced*.[26] These are, on the naturalistic approach that we take, a special type of embodied and situated neural representation with a particular functional-dynamic architecture. The difference between mental and nonmental representation is not merely in the complexity of their architecture: mental (subjectively experienced) representations require *particular* functional-temporal relations between integrating neural units. Mental content arises out of this organizational-functional dynamics. We suggest that the dynamic architecture of UAL can provide us with some clues about the selective, value-laden generation of mental representations.

Our notion of subjectively experienced representation is influenced by, but also departs from, James's ideas about the relation between instincts (which are not necessarily consciously felt) and emotions (which are). When James discussed instincts, which he regarded as scaffolds enabling learning

A   B

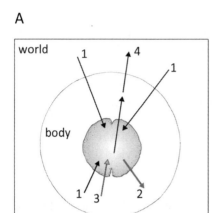   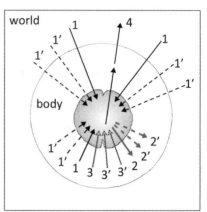

Figure 8.4

Feelings constituted during the first and "second" encounter with a stimulus. *A*, the first time: incoming stimuli (*solid arrows*) from the external world (such as day length) or from the body (such as maturational hormonal changes) are received in the brain (*shaded shape* within the body). The incoming stimuli (*1*) activate a reflex, resulting in motor and visceral responses (*gray arrow 2*) within the body; signals from sensors that are activated by motor acts and visceral activity are fed back to the brain (*gray arrow 3*), and activities in the brain lead to changes in the body, which in turn may change the world (*arrow 4*). Signals to the brain may include the sensory inputs from an elevated heartbeat and perspiration that accompany the flight response. These signals are integrated in the central nervous system and comprise the emotion. *B*, the "second" encounter: during subsequent encounters, in addition to activating the serial chain 1-2-3, representations that include traces of past encounters (*broken arrows 1′, 2′, and arrow 3′*) are also activated. These representations are memories of previous similar inputs, motor acts, contexts, and goals reached by the animal as it responded to the world and changed it (*arrow 4*).

and the formation of habits,[27] he distinguished between what he called the "first" and the "second" encounter of the animal with the world to which it reacts. The "first time" happens before learning occurs, when responses are constituted solely by innate ("instinctive") reactions, which then lead to physiological feedback, which is what he took emotions to be (figure 8.4A). When (using James's lovely example) a hen lays her egg for the first time, the sight of the egg triggers her instinct to sit upon it, leading, through complex sensorimotor feedback, to great tactile pleasure. Similarly, the first encounter with the mother's milk leads to the satisfaction of a baby's hunger, and the physiological feedback that ensues constructs the baby's emotion of pleasurable taste and tactile warmth. Neither the hen nor the baby has on the first encounter a motivation, a sense of a goal to be achieved. The round object is perceived *as* an egg, and milk is perceived *as* food, only following learning. When recall occurs on the basis of the first past experience/s, it includes the effects of the past state, and this is now part of the cause that leads to the construction of motivating emotion (figure 8.4*B*). The animal now *wants* to do certain things. During this "second time" (it need not be, of course, chronologically second; "second" denotes the postlearning state), the hen *wants* to sit on what she now perceives *as* eggs, and the baby *wants* the milk that is now perceived *as* food.

The idea that instinct is the scaffold on which habits and motivating emotions are built influenced Conway Lloyd Morgan, the psychologist and ethologist whose work we discussed in chapter 5. Morgan observed the behavior of chicks, ducklings, and moorhens youngsters that were raised in an incubator with no contact with adult birds. His observations of their initially unlearned instinctive behaviors, which following learning became adjusted to their specific environment and generated specific habits, led him to endorse and develop James's conclusions about the relation of instinct, learning, and emotions, which are the outcomes of visceral and motor feedback (which he called, very appropriately, "backstrokes"):

> Let us grant, in a word, that that professor James's theory [of emotions] holds good for the *first* experience of the emotions in question. What, however, about the *second* and all subsequent experiences? Association will have stepped in and exercised its modifying influence. We have seen that a chick which seizes a juicy worm gains thereby a bit of experience in the light of which worms become more alluring, since the sight of them at once calls up through associations a representation of this juiciness, a sort of anticipatory savour soon to be reinforced by the actual taste of the worm. In the case of an emotion, on the backstroke hypothesis, the same kind of effect will be produced by association. (Morgan 1896, p. 190; emphasis of *second* added)

Morgan thus added a subtle and important twist to James's view, which we are developing here within our UAL framework. At the Jamesian "second" time, the sum total of neural activity in the animal's nervous system, caused by the recurrence of stimuli plus the activation of memory traces, has a temporal, present-extending duration that produces an overall sensory state—a dynamic state with a specific sensory flavor or signature. An animal capable of learning compound perceptual patterns and actions recalls the past-encountered, learned stimulus-response relations. The neural consequences of learning are integrated with the original sensory state and are stored as engrams. During recall, in addition to the directly activating stimuli, the activated internal traces of the embodied triggers for action (James's egg in its embodied glory) and the activated internal traces of the action taken are reconstructed and guide behavior. The activated memory traces that are associated with the goal state and the ways of reaching it function as pointers to this goal. Hence, after learning, the animal has emotions *before* it overtly acts (for example, the hen has the excited expectation of the emotion of tactile pleasure when seeing the egg before she sits on it).

In our view, once UAL evolved the bound features that are perceived and recalled, such as those associated with food or with incubation, become inherently motivating. Consider the reaction of a starving young chick to a novel food item it encounters and consumes. The notion that this item is food does not at first exist. It is only after the animal has learned that this particular thing in the world is associated with the relief of hunger that it can identify and "embody" this thing as "food," and this percept can act as a mediating attractor, as an end. Once reached, such learning-based ends lead to the construction of corresponding internal attractors, which are stable neural representations of desired (homeostasis-promoting) states that are represented both locally and globally, at several levels. UAL also enables the animal to embody the far-from-homeostasis-triggering input, so an associatively learned cue, such as a certain pattern of vibrations, may become an indicator of a past-encountered dangerous predator. This is enormously beneficial, since the detection and response to stimuli associated with a far-from-homeostasis (repulsor) state can often be lifesaving.

The integration of multiple bodily states is required for such learning to occur. There are constant changes in the world of a mobile animal, and it needs to detect them with its sense organs and respond to them through the flexible coordination of its muscle sheets. The value-reinforcement system has to accommodate these world and action changes and deal with their potentially conflicting effects. (Should the animal first satisfy its need for food and approach object X or retreat because it smells the scent of a predator in the area?) The animal must therefore have a "common (value) currency" for consistently evaluating *types* of world and body states in spite of their inevitable variations and for preferring one type of state over another type according to its value, which is context-dependent. As several scholars have suggested, pleasures and displeasures can be seen as such overall and general currencies of value, which can evaluate *any* percept or action.[28]

After an animal has processed and evaluated the relevant sensory information, all neural connection patterns that were memorized and are reconstructed during the "second" encounter (whether it leads to the consolidation of past learning, to its modification, or just to a fleeting updating) become part of the ceaselessly active nervous system and part of the animal's overall, yet specific, sensory state. The entire coupled system of interactions depicted in figure 8.2C, with its distinct, history-dependent, persistent, overall dynamic sensory state (including the interactions with the body and the world) generates mental representations and mental states. These representational sensory states, which we refer to as CSSs (categorizing sensory states), categorize, through their dynamics, both input and output.[29] The inputs that elicit the mental representational capacities activate memory traces of other inputs of the same type (for example, inputs and memory traces related to hunger), and these inputs and traces also determine what type of response will occur as memory traces of the motor responses to the inputs and the stimuli associated with the attainment of the attractor (relief from hunger) become active. At the same time, particular CSSs are highly differentiated, since each CSS always maps a specific input-output relation.

We need to highlight one important aspect of a CSS—its dynamic nature. Suppose the animal is in a particular global sensory state, a state of food-deprivation that leads it to explore more vigorously. Among the activated parts of the network are memory traces of the past impact of food (which are related to the attractor state) and the route taken to reach it, but most dominant at first are activities related to the current exploration-enhancing repulsor state. The activated neural connections include innately given connections but also, and crucially, previously formed associations. Because initially the animal is hungry, this CSS is, in our terms, a repulsor CSS, and the animal will strive to change it. Exploration may be guided by the activation of memory traces that were formed during past encounters that led to a sensory state that corresponded to satiation. If exploration is thus guided by the memory of successful past motor actions, the chance that food will be found is increased.

According to this scheme, every CSS is built around basic, innately given exploratory and directed reflex trajectories. These reflexes are related to the fundamental physiological functions of reproduction and self-maintenance, such as acquiring and consuming food, and the maintenance of tissue integrity and salt balance—the kind of functions that Lamarck long ago suggested reflect the *besoins* of the animal. The internal state of homeostasis, or departure from it, is signaled by interoceptors, and exteroceptors are used to explore the external environment and find conditions that enable the restoration of homeostasis through the reinforcement of homeostasis-attaining states. However, once UAL evolved, the relatively fixed trajectories can be thought of as a neurophysiological scaffold, which can become redundant after its initial activation so that a CSS in the adult may be different from the original, innate response system. In the same individual at different stages of life, the CSS leading to food-seeking behavior, for example, may differ in detail because the learning history and memory of the encounters with food stimuli are different. The patterns of connections among neurons that are activated during a given type of response have family resemblances, but are not identical; they have many connections in common, but no single connection is obligatory. The effects of a new stimulus or response can become part of several different CSSs; they can belong to the CSS that elicits food seeking, or withdrawal, or mate seeking. But although CSSs have great flexibility, each is constrained to a certain domain of the animal's actions and is robust.

The number of CSSs that any one animal at any particular stage of life may have is very large, but the number of *potential*, newly learned associations within a CSS is far, far larger because through UAL many new links between stimuli and responses can belong to the same CSS. Once CSSs are formed, the neural effects of a single incoming stimulus are constructed in the context of other preexisting (innate and learned) neural trajectories. The effect of the integrated, hierarchically processed, and remembered learning-based internal states, rather than the effect of a single stimulus, becomes the cause of a behavior. The distinct yet unitary and global nature of the CSS, its particular "quintessence," guides the animal to the right realm of action, while its specific history-dependent features give the CSS an individual flavor and a more or less specific action direction. Hence, once a UAL architecture generating CSSs had evolved, CSSs became central organizing causes and navigators for the animal's goal-oriented actions. An overall sensory state became a state *for* the animal, not just a state *of* the animal. It became part of a new teleological system. Motives and emotions had come into being.

A particular CSS, then, with its activated traces and its idiosyncratic, yet global, sensory effects, informs the animal about its present deviation from the attractor state and guides it toward it by directing adaptive behavior based on past history. The CSS is therefore both a discriminator and a motivator: it evaluates inputs on the basis of the animal's present state and its previous learning history, which includes past triggers of exploration, the route taken, and the goal attained. Even in the absence of external attractor-related cues, the initial exploration is partially guided by the attractor-related internal memory traces. The animal is therefore endowed with a "remembered future," not just a remembered present. Thus, through the CSS, the animal "informs" itself about what to do, in what manner, and toward what end. Such an animal has an enormous adaptive advantage over one that does not have such future-oriented memories—such internal "guidance." According to this view, a mental representation is imbued with motivation.

The functional model presented in figure 8.2 highlights some of the essential processes and relations that go into the construction of CSSs. The integrated neural patterns that are generated are updated through top-down and bottom-up interactions between several successive levels within the central nervous system (CNS) hierarchy. As we stressed, the reason for this is that complex multifeature percepts require hierarchical predictive coding: economy of space and time requires hierarchical processing. Moreover, the more features a percept has, or the greater the number of conditions in which objects are perceived, the more a percept can be discriminated. Hence, in addition to a hierarchical system of representations that map and transform each other, there must be a dedicated memory (storage) system of updatable past experiences and processes that attribute an overall value to perception and action and exclude a vast number of alternatives. Percept selection and action selection are selective–stabilization processes and underlie what is described in cognitive terms as "attention." What is perceived and what action is taken are based, evolutionarily and developmentally, on learning, and hence on memory and past and present values. Hierarchical representation, memory for patterns, and overall value underlie the construction of CSSs.

The ontogenetic development of the relation between UAL and CSSs does not recapitulate phylogeny. According to our proposal, during phylogeny the evolution of UAL led to integration, memory, and HPP mechanisms, and these and their supporting neuroanatomical structures led to CSSs, so evolutionarily, UAL led to CSSs. In contrast, during the ontogeny of an animal that has the evolved potential to manifest UAL, CSSs form *before* UAL fully develops, guiding and facilitating it. An animal with the functional architecture for UAL has the system for constructing CSSs in place. It has acquired, during evolution, not only the relevant innate "scaffolds" but also the functional neural architecture that enables it to interpret its various states of activation as overall, integrated, and distinct sensory states—as experiences. Such an animal comes to the world with an architecture enabling multilevel integration, overall evaluation, and memory for compound patterns. This inborn functional architecture is the basis for the individual's responses and explains why a very young animal, such as a human neonate, who does not have UAL, can nevertheless feel; although some of its experiences (of visual objects, for example) may be more fuzzy than those of a child who has acquired UAL, others may be more intense (e.g., feelings that are not controlled and suppressed by mature higher cortical regions). In other words, when reflex-eliciting stimuli are processed by a brain that has an architecture that supports sentience/minimal consciousness, they become subjectively experienced because they are processed by high-level integrating units. Moreover, because such reflex-eliciting stimuli signal the widest departures from homeostasis, the experiences that they elicit are, as Derek Denton rightly noted, powerful and urgent.[30] Hence, while during phylogeny the functional architecture of UAL comes first and enables CSSs, during the ontogeny of an animal capable of UAL, CSSs come before an animal can display UAL. This means that the stereotypical flight reaction in an animal with UAL entails subjective experiencing, whereas in an LAL animal that lacks the functional architecture of UAL, such a reaction is nonconscious. Therefore, according to our reasoning, UAL is a sufficient (but not necessary) condition for minimal consciousness in present-day animals. It was, however, both sufficient and necessary when it first emerged during evolution.

How does our evolutionary view of UAL and of CSSs fit the current dominant theories of consciousness? The functional architecture of UAL is, in fact, either implicitly presupposed by the different models that we discussed in chapters 3 to 5 or is compatible with them. For example, the global neural workspace model (GNW, figure 3.11A) suggests that the perceptual, motor, memory, value, and attention systems come together to construct mental states. The same systems are part of our UAL architecture, although in our UAL model the coupling among them is explicit: we specify the temporal and functional relations between the relevant subsystems, and attention is instantiated by the precision of evaluating discrepancies through predictive coding rather than distinct attention networks. Gerald Edelman's dynamic core model (figure 3.7), which focuses on value systems and multilevel mapping and integration processes that relate compound inputs to past learning responses and future needs with states that have temporal thickness, is also compatible with our framework. Unlike Edelman, however, we are not committed to a specifically mammalian implementation, and our model uses the context of learning to spell out the specific relations among the different parts of the system, thereby allowing a more focused understanding of the interactions among neural maps. Tononi's integrated information theory (IIT) of consciousness posits that consciousness is based on composite, integrated (irreducible to noninterdependent subsets), intrinsic cause-effect processes, which exclude a lot of alternatives and can be described in informational terms.[31] Tononi therefore focuses on the unity and global accessibility aspects of consciousness; the other features that are considered central to consciousness are implicit in his theory or are seen as derivable from or implied by it. Like Edelman, he regards thalamocortical interactions as implementing consciousness, although his model is not committed to any specific neural process or structure. While our UAL architecture also requires the integration/exclusion of functional information and describes an intrinsic set of structures and processes, we explicitly include other aspects and relations. We specify which kinds of integration are necessary, point to temporal and modulating relations among different integrating units, highlight the deep hierarchical structure of the system, emphasize the importance of dedicated systems for memory for compound patterns and reinforcement-generating functional units, and stress the embodied nature of the process. These structures and processes, we argue, are central to both experiencing during learning and to online updating. Moreover, unlike the theories just outlined, our toy model includes an explicit behavioral (learning) facet, which can be tested experimentally.

The UAL model emphasizes the notion of an action-constructed bodily self and is similar in this general sense to other models of embodied cognition. The emotion-focused theory of Damasio, which was described in chapter 3, highlights the perceptions that the organism has of its own body through the continuous monitoring and updating of its internal state by the brain. Freeman's theory (also discussed in chapter 3) emphasizes the central role of exploratory actions in the construction of dynamic strange attractors—attractors that can be described, in our terms, as CSSs. The interactions of the REIU with the SIU and MIU systems (which are affected by the MEMU) and the AU system are compatible with Damasio's and Freeman's suggestions. However, in our UAL model, hierarchical representations are built into the levels between primary and secondary (and higher-level) integrating sensory and motor units, and we suggest *specific* patterns of interactions between the motor-integrating units and the sensory-integrating units. Merker's self-model, with its more explicit, though still overly abstract, functional architecture, is closest to our UAL model. His tripartite target selection, action selection, and motivation are instantiated by our integrating units and reinforcing unit. The motion-stabilized body-world interface organized around an ego center that Merker suggests is, in our model, the outcome of a temporally specific coupling among all the relevant high-order units that are integrated in the AU. The stable perspective that the animal adopts depends on the sensory world that it inhabits (which is not necessarily visual, as Merker's specific vertebrate model suggests) and on the anatomy of its sense organs.

Although compatible with Feinberg and Mallatt's focus on hierarchically organized neural maps, our account of minimal consciousness and mental states is incompatible with their distinction between affective, exteroceptive, and interoceptive types of consciousness.[32] According to our view, all conscious, subjectively experienced states are sensory, all involve motor-sensory-motor loops, all involve memory for compound patterns, and all are valued/stabilized. An animal may be conscious primarily of visual, or auditory, or tactile stimuli and may have experiences that stem from responses to the stimulation of receptors within the body and changed states of the CNS that result in internal pains, anxiety, imbalance, and fatigue, but these distinctions do not entail separate types of consciousness. All these experiences share the same basic patterns of interactions depicted in the UAL model, although, of course, the source of the sensory stimuli is different, and the relative contributions of the different integrating units that are involved are likely to be different too. Although there are clearly many different mental states, we believe that the idea that there are several types of consciousness is an error stemming from the attribution of consciousness to parts of the systems (to the activity of REIU, AU, MIU, and SIU) rather than to the activity of the system as a whole.

A related issue is the debate about which sensory experiences were evolutionarily primary: Were visual experiences primary? Or maybe olfactory experiences came first? Or were the first experiences interoceptive-affective? Or proprioceptive?[33] From our perspective, these debates miss the integrative nature of all subjective experiencing. Since movement and evaluated sensory inputs are always involved in the construction of consciousness, sensory signals from the surface of the moving body, proprioceptive signals, and interoceptors' signaling contribute to all mental states. The relative significance and richness of the exteroceptive signals in a lineage depend on the sensory evolution of that lineage, which was intimately related to its habitat. A sense of balance and the ability to detect vibrations were probably important in all moving animals, while in animals living close to the water surface, eyes and visual experiences were probably of special significance, and olfactory experiences may have played a larger role in animals that moved in the areas where little light penetrated. Whatever the sense/s most abundantly employed, it seems likely that all existing types of exteroception and interoception contributed to the ability of UAL animals to construct a multimodal representation of their "self" in the world.

In all these animals, the default, spontaneous, exploratory activity that led to distinction between the self and the world was rewarding (maybe even joyful, as Humphrey suggested), encouraging the animal to engage with its world. As Aristotle suggested, humans (and we extend this to other conscious animals) take delight in perception for its own sake because our senses and especially sight "makes us know and brings to light many differences between things."[34] Experience then brings joy because it leads to knowledge through learning, a high-level CSS. Exploratory activity was positively valenced because exploring animals had more learning opportunities and could adjust to the world by learning. We therefore see the SEEKING emotional system described by Panksepp (chapter 5) as reflecting animals' intrinsic motivation, exuberantly exploring their world. However, unlike Panskepp's view of the evolution of (affective) consciousness (figure 5.9), which suggests that primordial emotions preceded learning-related emotions, our model suggests that learning appeared long before basic emotions and drove their evolution.

UAL, therefore, involves intrinsic motivation and leads to the formation of CSSs that enable complex, flexible adjustments to the environment. This requires a complex set of temporal and spatial interactions among processes and structures, which are subjectively experienced. It is important, however, to emphasize what the UAL system does not require. While during the online encoding stage subjective experiencing is rich and the salient event is experienced in context, the recall stage need not be rich nor conscious. The example we gave in chapter 5 showed that the *encoding* during the learning of a compound novel stimulus (a composite picture of flowers or landscapes) requires consciousness, but the *recall* of composite percepts and action patterns is very often unconscious. An amnesiac patient, who experienced an unpleasant prick in her hand when shaking hands with her doctor, was reluctant to shake the doctor's hand again but could not remember why. This is true not only for amnesiac patients. Neurologically undamaged humans, too, often do not recall why they are afraid of X, or like Y. The reconstruction process that occurs during recall must be sufficient only for the activation of the MIU and reward systems; the exteroceptive, interoceptive, and proprioceptive states that encoded the engram do not have to be reexperienced. When a rat is presented with a compound stimulus—a white ball with dark stripes—that in the past was accompanied by a shock to the foot, it may show fear, but there need not be a process of reconstructing the actual past experience. In fact, the animal may express fear even when the previously learned compound stimulus is masked, and it is not aware of it. The most that we can say is that the animal has acquired *semantic* information—it knows facts about the compound objects (e.g., the striped ball) or the sequence of actions it performed and about their valences. However, it does not have to recall the patterned ball off-line, and certainly, it need not reexperience the rich spatial and temporal context in which it saw it and received a foot shock. For such *episodic* recall of the past, more is needed—the UAL system has to evolve further, something that seems to have happened in parallel in several animal groups (see chapter 9).

Our main message in this section is that the dynamic categorizing sensory states that are implemented by the functional architecture of UAL underlie the experiencing that occurs when an animal encodes or updates information. Since this UAL architecture is *about* patterns external to it and since its dynamics suggest that its enabling system is conscious, it is, by definition, a mental representation. To put it differently, we argue that the notion of mental representation can be operationalized by the functional model of UAL.

The Taxonomic Distribution of UAL and the Neural Structures Supporting It

> If you want to understand function, study structure. (I was supposed to have said in my molecular biology days.)
> —Crick 1988, p. 150

The existing information about the taxa that show UAL or that exhibit learned behaviors that can be used as proxies for UAL (e.g., an ability to learn rules, to estimate amounts and relations, to navigate in a novel terrain) is patchy. As indicated in previous chapters, there is no evidence of UAL in nematodes, flatworms, and aplysiid mollusks—nor is there evidence for this learning capacity in annelids.[35] UAL has been found in only three phyla: vertebrates, arthropods, and mollusks (see table 8.1).

What aspects of the anatomy of vertebrates, arthropods, and mollusks enable UAL? Species showing the capacity for UAL must have structures that support the integration of multimodal sensorimotor information, the memorizing of compound patterns, the assessment of their overall homeostatic state, and the assignment of value to the most salient stimuli. The task of describing these neuroanatomical structures is daunting, and we cannot do it in detail here, but we present the most important parts and connections of the MEMU, SIU, MIU, REIU, and AU systems in these three phyla in figures 8.5–8.8 and summarize the information in table 8.2.[36]

Table 8.1
Groups exhibiting UAL, or learned behaviors that can be seen as proxies for UAL.

Phylum	Type of learning
Mollusca[a]	Pavlovian conditioning (involving perceptual fusion)
	Operant conditioning (involving novel action patterns)
	Spatial learning
Arthropoda[b]	Pavlovian conditioning with compound CSs (involving non-elemental learning)
	Operant conditioning (involving novel action patterns and spatial learning)
	Conceptual learning[d]
	Number-based learning[e]
	Navigation learning[f]
Vertebrata[c]	Pavlovian conditioning with compound CSs (including non-elemental learning)
	Operant conditioning (involving novel action patterns and spatial learning)
	Conceptual learning[d]
	Number-based learning[e]
	Navigation learning[f]

Note: The literature on complex learning abilities in mollusks, arthropods, and vertebrates is vast, but in each phylum the number of species investigated for compound learning is small. Books and comprehensive reviews, as well as some of the original papers, that cover the literature on the subject are given in the next three notes. Torley (2007) provided a thoughtful and thorough analysis of the learning capacities of animals. He suggested that an animal can be called an "intentional agent" when it has a CNS, innate preferences and motor programs, sensory abilities, procedural memory, associative learning, cognitive mapping, action selection, and primitive concepts. His behavioral criteria partially overlap our UAL criteria.

[a] *Books*: Mather, Anderson, and Wood 2010 (octopus); Menzel and Benjamin 2013, chaps. 14–22 (gastropods), chaps. 23–25 (cephalopods). *Reviews*: Boal et al. 2000 (octopus); Hochner, Shomrat, and Fiorito 2006 (octopus); Watanabe, Kirino, and Gelperin 2008 (gastropods); Jozet-Alves, Bertin, and Clayton 2013 (cuttlefish). *General sources*: Corning, Dyal, and Willows 1973 (vol. 2, chap. 10; vol. 3, chap. 11); Perry, Barron, and Cheng 2013; Godfrey-Smith 2016c.

[b] *Book*: Galizia, Eisenhardt, and Giurfa 2012 (part VI, honeybee). *Reviews and papers*: Bhagavan and Smith 1997 (honeybee); Boisvert and Sherry 2006 (bumblebee); Collett and Collett 2009 (insects); Brembs and Heisenberg 2001 (*Drosophila*); Young et al. 2011 (*Drosophila*); Collett, Chittka, and Collett 2013 (mainly hymenopterans); Magee and Elwood 2013 (shore crabs); Mizunami et al. 2013 (cockroaches). *General sources*: Corning, Dyal, and Willows 1973 (vol. 2, chaps. 5–9); Chittka and Niven 2009; Giurfa, Devaud, and Sandoz 2011 (pp. 5–101, insects); Menzel and Benjamin 2013 (chap. 26, crustaceans; chaps. 27–42, insects); Perry, Barron, and Cheng 2013; Perry, Barron, and Chittka 2017.

Table 8.1 (continued)

---

[c] *General sources*: Razran 1971 (mainly mammals); Macphail 1982, 1987 (all vertebrates); B. R. Moore 2004 (mainly birds and mammals). *Papers*: Agrillo, Piffer, and Bisazza 2010 (fish); Newport, Wallis, and Siebeck 2014 (fish); Schumacher et al. 2016 (fish).

[d] Include: the transfer of a learned rule from one stimulus to stimuli that belong to the same category (e.g., matching a new large triangle with a previously-encountered small one), and comparing stimuli, even when involving a different sensory modality, and learning relations such as "same" and "different."

[e] Ability to distinguish between the number or amount of cues presented.

[f] A combination of both classical conditioning of landmark cues and operant conditioning of proprioceptive cues.

In vertebrates, the hippocampus is the part of the brain underlying memory for compound percepts and action patterns. The tragic case of Henry Molaison (the famous H. M. mentioned in chapter 5) shows that removing most of the hippocampus leads to the loss of the ability to retrieve recently presented perceptual patterns (long-term memory was not impaired). Subsequently, it was found that the human hippocampus has mechanisms that identify and complete (if the input is partial) compound (learned) patterns and that the neural representation of these patterns are then sent to the cortex, where perceptual predictions about the sensory inputs are reconstructed.[37] Midbrain and basal forebrain structures that integrate topographically arranged sensory input with memory and motivation and send commands to relay stations that control the motor output include the superior colliculus, which forms the roof of the midbrain and integrates mainly visual stimuli; the olfactory bulb, which integrates olfactory stimuli; the periaqueductal gray, which controls the modulation of pain and reciprocally interacts with the colliculus; the basal ganglia, which are involved in voluntary motor action, motivation, and emotions; and the hypothalamus, which integrates and regulates goal-directed, motivated, exploratory motor behaviors. Figure 8.5 (*A–E*) shows most of these structures, as well as the prominent regions in the human cortex that interact with them, and figure 8.6 (box 8.2) compares the organization of some of these regions in the mammalian (rat) and fish (zebrafish) brains.

Brain centers, which function and interact in a way similar to those found in vertebrates, are present in arthropods.[38] The insect brain is composed of fused ganglia and is divided into the protocerebrum, deuterocerebrum, and tritocerebrum. The protocerebrum is involved in processing

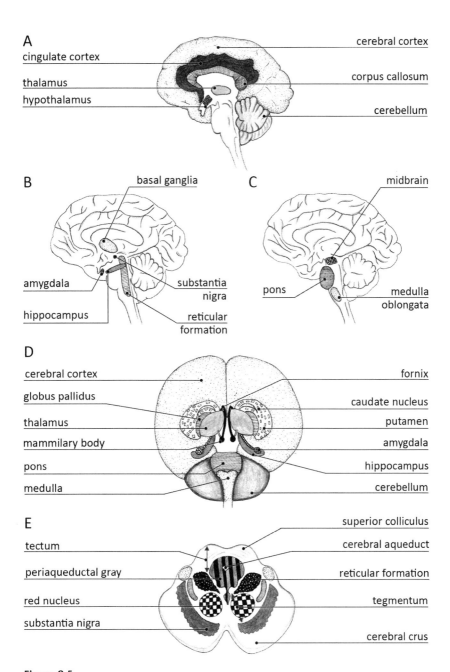

A

cingulate cortex

thalamus

hypothalamus

cerebral cortex

corpus callosum

cerebellum

B

basal ganglia

amygdala

hippocampus

substantia nigra

reticular formation

C

midbrain

pons

medulla oblongata

D

cerebral cortex

globus pallidus

thalamus

mammilary body

pons

medulla

fornix

caudate nucleus

putamen

amygdala

hippocampus

cerebellum

E

tectum

periaqueductal gray

red nucleus

substantia nigra

superior colliculus

cerebral aqueduct

reticular formation

tegmentum

cerebral crus

Figure 8.5

The human brain, highlighting areas involved in the generation of minimal consciousness. *A–C*, median sections showing different regions. *D*, view from the underside, with some structures outfolded. *E*, a cross section through the midbrain.

A, the cortex contains the animal's highest mental representations and is the highest controller of its actions. The cingulate cortex controls autonomic actions related to emotions and influences motivation. The corpus callosum is a neural band through which the two brain hemispheres communicate. The thalamus is a major relay station: it conveys to the cortex visual, tactile, auditory, and chemical sensory signals; it exchanges motor signals between the cortex, cerebellum, and basal ganglia, and it has a role in attention and pain perception. The hypothalamus controls internal homeostasis; affects hunger, thirst, sex drives, and emotions; and is involved in the sleep-wake cycle. The cerebellum controls body posture, coordinates and regulates sequences of motor actions, and has a role in motor memory (motor skills); in addition, it is involved in integrating cognitive functions.

B, the basal ganglia select, trigger, and control motor commands; the substantia nigra contain dopaminergic neurons that are essential for seeking rewards; the hippocampus is the center of long-term perceptual memory; and the amygdala controls emotions. The reticular formation (which is situated in the brain stem) plays a major role in alertness, wakefulness, and sleep.

C, the brain stem is composed of the midbrain, pons, and medulla oblongata. The midbrain and pons connect to structures that are involved in motor control. The medulla regulates basic functions (respiration, heart rate, and blood pressure) and controls basic reflexes (such as swallowing).

D, regions seen here, not depicted in *A–C*, are the mammillary bodies that function in memory recollection, along with the fornix, which is also involved in memory, sending signals to the hippocampus, mammillary bodies, and thalamus. The caudate nucleus, putamen, and globus pallidus are basal ganglia.

E, a cross section of the midbrain (top: posterior; bottom: anterior). The tectum forms the roof of the midbrain and contains the pair of superior colliculi, which are layered structures involved in preliminary visual processing and in the control of eye movements. In nonmammalian vertebrates they are called the "optic tectum," and their size, relative to other brain structures, is large. (See figure 5.5, chapter 5.) The tectum also contains the pair of inferior colliculi (not shown), which process auditory signals. The tectum as a whole functions in directing behavioral responses to specific areas in space. The tegmentum forms the floor of the midbrain and contains several subcortical nuclei, among them the red nucleus, which has a role in motor coordination, and another basal nucleus, the substantia nigra. The most anterior section of the midbrain consists of the cerebral crus (also called "cerebral penduncles"), which link the rest of the brain stem to the thalami, and through these, to the cerebrum. In the posterior center of the midbrain, surrounding the hollow cerebral aqueduct, is a mass of gray matter, the periaqueductal gray, which is involved in generating drives and emotions, in responding to pain, and in coordinating body posture (for discussions of the contribution of midbrain structures to basic consciousness, see Merker 2007; Woodruff 2017).

Box 8.2
Vertebrate brains.

The organization of the telencephalon, the embryonic structure from which the cerebrum develops, has been conserved in vertebrates, from fish to primates. In mammals, the dorsal telencephalon, the pallium, develops to become the cerebral cortex, and the ventral telencephalon, the subpallium, becomes the basal ganglia. Figure 8.6 compares the four subdivisions of the zebrafish pallium with those of the rat; it shows the localization of zebrafish regions that corresponds to brain centers in the rat.

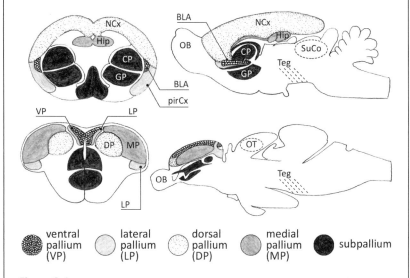

Figure 8.6
A comparison between some midbrain and higher brain structures that are relevant to mental representation in a rat (*top*) and a zebrafish (*bottom*). Two sections are shown for each brain—coronal (*left*) and sagittal (*right*). The zebrafish pallium, like that in the rat, is divided into four main sections: the dorsal pallium (*DP*), which corresponds to the mammalian neocortex (*NCx*), which includes most of the cerebral cortex; the medial pallium (*MP*), homologous to the mammalian hippocampus (*Hip*); the ventral pallium (*VP*) homologous to the pallial basolateral part of the amygdala (*BLA*); and the lateral pallium (*LP*), corresponding to the piriform cortex (*pirCx*) that processes olfactory signals. *CP* and *GP* are two subcortical basal ganglia (caudate putamen, globus pallidus) located in the subpallium (*SB*). In the sagittal sections on the right, the midbrain tegmentum is marked, as well as the superior colliculus (*SuCo*) of the rat and its homologue in the fish, the optical tectum (*OT*). The olfactory bulb (*OB*), which transmits olfactory stimuli, is in the most forward part of the brain. Adapted by permission of Thomas Mueller.

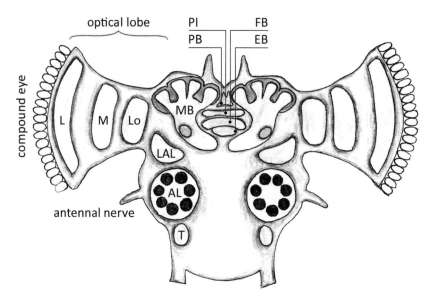

Figure 8.7

A section through the insect brain showing the mushroom bodies and the central complex. Structures within the protocerebrum include the mushroom bodies (*MB*), the lateral accessory lobes (*LAL*), and the central complex substructures: the ellipsoid body (*EB*, also called the "lower central body"), the fan-shaped body (*FB*, also called the "upper central body"), and the procerebral bridge (*PB*). The supporting structure at the midline of the brain, the pars intercerebralis (*PI*), contains neurosecretory cells. On either side of the central body lie the optic lobe structures—the lamina (*L*), lobula (*Lo*), and medulla (*M*)—that process visual signals received from the retina of the compound eye and hold visual retinotopic maps. The antennal lobes (*AL*), which process the incoming information from the antennae, are part of the deutero-cerebrum. The small tritocerebrum is marked *T*. Reproduced by permission of www .cronodon.com.

vision, combining information from different sensory modalities, controlling movements, and memory. It consists of the ganglia located in the uppermost part of the head and contains the mushroom bodies, the central complex, and the lateral accessory lobes (figure 8.7). The mushroom bodies, which receive inputs from the lobula of the optic lobe and from the antennal lobes, play a prominent role in insect memory and learning.[39] The central complex receives inputs from the mushroom bodies and consists of the ellipsoid body, the fan-shaped body, and the procerebral bridge; these three regions contain visual, tactile, and additional sensory representations; they are also centers of motor command and action selection and in addition are involved in motivation.[40] In some insects, such as the monarch butterfly, the procerebral bridge contains neurons that may form a direction-specific polarization map for navigation.[41] Two prominent optical lobes, responsible for processing the inputs from the compound eyes, are attached to the protocerebrum but are not part of it. The deuterocerebrum contains the antennal lobes, which process the information coming from the antennae and from the body wall. The tritocerebrum is the smallest part of the brain and though little studied, probably functions in tasting food.

There is remarkable overall similarity between the neural architecture and the functional organization of the mammalian and arthropod brains. The mushroom bodies and the central complex, situated in the insect protocerebrum, are believed to be either homologous or analogous to the hippocampus and basal ganglia, respectively. There is also striking similarity between the cerebellum and the insect mushroom body and the central complex in motor learning and between the vertebrate tectum and the insect fan-shaped body. Similarly, the hemiellipsoid bodies of crustaceans exhibit

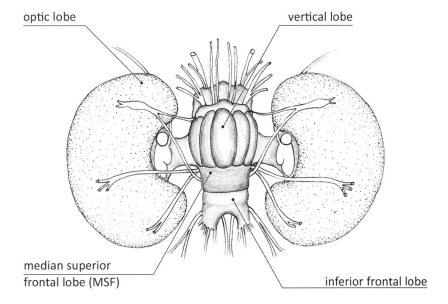

optic lobe   vertical lobe

median superior
frontal lobe (MSF)   inferior frontal lobe

Figure 8.8
The octopus brain. The central brain consists of dozens of lobes and 50 million neurons. It integrates processed information from the enormous optic lobes (with their 80 million neurons) and coordinates the peripheral nervous system of the eight arms (320 million neurons), which are endowed with a high degree of autonomy. The vertical lobe, median superior frontal lobe, superior frontal lobe (not shown), and inferior frontal lobe are thought to play a role in memory and learning. Reproduced by permission of Oxford University Press.

functional homologies to the hippocampus, and these brain structures can be found in arthropod and vertebrate fossils from the Cambrian era.[42]

In cephalopod mollusks such as the octopus and cuttlefish, two main structures of the brain, the vertical lobe and the superior frontal lobe, form complex networks that together are analogs of the vertebrate hippocampus. These integrating brain structures underlie cephalopod learning and memory (figure 8.8).[43] Cephalopods are clearly able to learn in an open-ended manner, and the behavioral flexibility and learning ability of the octopus are legendary. Octopuses have even been observed to use tools: some individuals of the veined octopus retrieved discarded coconut shells, manipulated them, transported them, and reassembled them to use as a shelter.[44] Since their arms contain huge ganglia and many types of chemoreceptors and mechanoreceptors, one may wonder whether the octopus arm, even when detached from the brain, can learn—and if it can, what kind of learning it is (box 8.3).

Box 8.3

What is it like to be an octopus arm?

The octopus arm is one of the marvels of nature, something that roboticists view with envious wonder and try to imitate. Each of the eight extremely flexible arms is covered with suckers (each having its own ganglion), and crammed with chemoreceptors and mechanoreceptors. The arms can move in all directions, allowing the animal to walk on them, catch and reject objects, and expertly reach an obstructed object of interest (figure 8.9). The arms are controlled by a huge number of neurons—two thirds of all neurons are in the arms' ganglia. We know that a severed octopus arm can show reflex reactions (e.g., withdraw following noxious stimuli),[45] and that the basic motor program for voluntary movement is embedded within the neural circuitry of the arm itself,[46] but, as Benny Hochner and his colleagues have shown, octopuses do not seem to have skill-learning at the level of the arm (although the central brain can direct the arms to engage in ever more successful strategies).[47] Yet, since according to J. Z. Young, to whom we owe much of our basic knowledge about the neurobiology of the octopus, the nervous system in the octopus arms is functionally equivalent to the spinal cord in a vertebrate, we wonder if ganglionic learning by an octopus arm is a possibility.

As we noted in chapters 1 and 5, very limited ganglionic learning is possible in a rat or human whose spinal cord has been severed from its brain: a limb of a spinal animal can be conditioned to withdraw following a light touch if

Box 8.3 (continued)

the light touch is paired with a strong one. Can the octopus arm similarly learn, or perform even more amazing learning feats, such as guide the learning of *new* things? Can inputs from several different types of chemoreceptors in the arm become integrated into a compound percept? Can a brain-less arm perceive taste? Can the arm ganglia store a memory of past brain-connected events? What is the nature of such memory, if it exists at all? Unfortunately, as yet there are no answers to these questions, although curious biologists do ask them.[48] It is very probable that the arm is capable only of autonomous reflex reactions, or of nonassociative learning, or, at best, LAL; this would mean that an arm cannot experience, and cannot be said to have mental representations. Hence, it is very likely that it is like nothing at all to be an octopus arm—it is probably as nonexperiencing as a quivering, detached, lizard's tail. Yet, given the great sensory and motor complexity of the arms, their relative autonomy, and their probable capacity for LAL, we should keep an open mind about the complexity of their neural representations. Nevertheless, even if it is absolutely nothing to-be-like an octopus arm, we must still ask what effect the arms' autonomy has on the subjective experiences of the octopus. In other words, what is it like to be a sentient animal with semi-independent (sometimes "rebellious") parts? Godfrey-Smith, who has been observing and writing about octopuses' behaviors for many years, suggests that "In the octopus's case there is a conductor, the central brain. But the players it conducts are jazz players, inclined to improvisation, who will accept only so much direction. Or perhaps they are players who only receive rough, general instructions from the conductor, who trusts them to play something that works."[49]

Figure 8.9
An octopus's idea of reaching out.

Thanks to behavioral, learning, and neuroanatomical studies, there is a growing consensus that all vertebrates, many arthropods, and some cephalopod mollusks have neural structures that can support mental representations.[50] They also have the neural structures supporting UAL. For example, the insect mushroom bodies and central complex implement two of the hierarchical levels depicted in our UAL model: the second level of integration (MIU and SIU) and global reinforcement (REIU) and the third level of memory for compound patterns (MEMU). The main brain structures and centers that subserve UAL and mental representation in vertebrates, arthropods, and mollusks are summarized in table 8.2.

At the other end of the spectrum, there is general (albeit usually implicit) consensus that animals that have limited learning—and a shallow hierarchical neural architecture—do not have mental representations. The neural representations of animals like *Aplysia* seem to rely on interactions within a restricted hierarchical system that nevertheless contains complex feedforward and feedback loops that render it robust. There are, for example, interactions between motor networks that are responsible for competing actions (such as feeding and escaping) and between different sensory units that receive inputs at different times. Very simple action selection and predictive coding (Kamin blocking) are therefore possible. As one expert put it when we asked him if the sea hare has something like an integrating brain with specialized internal regions for processing and associating sensory and motor information: "The closest thing in *Aplysia* would be the cerebral ganglia, which do not appear to be specialized for any sensory, motor, or physiological function and contain some 'higher level' interneurons."[51] Similarly, the sea slug *Pleurobranchaea californica* lacks complex hierarchical neural organization and a dedicated memory system.[52] The organization of the CNS in these sea slugs is therefore compatible with what we know about their limited learning abilities.

Deducing the mental status of other mollusks and nonarthropod invertebrates from their brain anatomy is difficult. In annelids that have mushroom bodies and complex movements and in gastropod mollusks that have more complex brain ganglia than the sea slugs, the presence of elementary mental representation is an open question. Although at present there is no evidence for any unambiguous associative learning in annelids (there are only a few behavioral investigations of their learning capacities), the presence of mushroom bodies in these groups should encourage more work on this issue. In the terrestrial gastropods *Helix* and *Limax*, the degree of open-ended learning (UAL) is unknown. Their protocerebral lobes, which are the most differentiated part of their brains, consist of a pair of head ganglia that

Table 8.2
Brain structures implementing the functions of minimal consciousness in vertebrates, arthropods, and mollusks.

	Integrating into compound patterns (corresponds to the SIU, MIU, and AU)		Globally acting value mechanisms and factors (correspond to the REIU)	Memory for compound patterns (corresponds to the MEMU)
	Exteroceptive and interoceptive (perception of the external world and of body parts)	Proprioceptive (movement of the body in space)		
Vertebrates	Cortex/pallium, superior and inferior colliculi, periaqueductal gray, cerebellum	Superior colliculus, periaqueductal gray, cerebellum	Cortex/pallium, basal ganglia (nucleus accumbens and ventral striatum), cingulate cortex, amygdala, reticular formation, substantia nigra, thalamus, periaqueductal gray, hypothalamus, cerebellum, mammillary bodies, pituitary. Prominent transmitters systems participating in the value systems are: cholinergic, dopaminergic, GABAergic, serotonergic, and noradrenergic	Cortex/pallium, hippocampus, basal ganglia, cingulate cortex, fornix, mammillary bodies, cerebellum
Arthropods **Insects**	Mushroom body, central complex	Central complex, lateral accessory lobe	Lateral accessory lobe; the mushroom body and the central complex (fan shaped body and ellipsoid body) have specific neurons with dopamine receptors; dopaminergic and octopaminergic systems are prominent	Mushroom body; central complex
Crustaceans	Hemiellipsoid body	Central complex	Central complex; octopaminergic and serotoninergic systems	Hemiellipsoid body, central complex

Table 8.2 (continued)

		Integrating into compound patterns (corresponds to the SIU, MIU, and AU)		Globally acting value mechanisms and factors (correspond to the REIU)	Memory for compound patterns (corresponds to the MEMU)
		Exteroceptive and interoceptive (perception of the external world and of body parts)	Proprioceptive (movement of the body in space)		
Mollusks	**Cephalopods (e.g., octopus)**	Superior frontal lobe, vertical lobe, and peduncle	Brain and peripheral nervous system	Vertical lobe; octopaminergic and serotoninergic systems are prominent	Superior frontal lobe, vertical lobe

Sources: The information in this table relies, primarily, on Strausfeld and Hirth 2013; Barron and Klein 2016; Feinberg and Mallatt 2016; and Wolff and Strausfeld 2016. The role of the cerebellum in conveying information about reward (and expectation) is described in Wagner et al. 2017. For more detailed information on the arthropod central complex, see Loesel, Nässel, and Strausfeld 2002; Homberg 2008; Pfeiffer and Homberg 2014; and Stegner, Fritsch, and Richter 2014. For detailed information about proprioceptive integration, see Tuthill and Azim 2018. The information on value mechanisms and factors in invertebrates relates to reinforcement and reward learning in these animals. It is based on Barron, Søvik, and Cornish 2010, and additional sources for individual species. For a specific insect neuron (VUMmx1) that serves the function of a value system, see Menzel and Giurfa 2001. For studies showing the presence of specific dopamine neurons in *Drosophila*, see Hige et al. 2015; Poo et al. 2016, p. 16. For biogenic amines and behavioral reinforcement in crustaceans, see Kravitz and Huber 2003; Kaczer and Maldonado 2009. For biogenic amines involved in reinforcement learning in the octopus, see Shomrat et al. 2015. Benny Hochner (private communication, August 20, 2016) provided the information regarding the value system of the octopus.

receive and process mainly olfactory input and have additional inputs from the statocyst, the internal organs, and the posterior body wall.[53] These lobes are involved in relatively advanced learning (enabling Kamin blocking and second order AL), and the gastropods' olfactory memory resembles mammalian mechanisms at several processing levels.[54] It is possible that this neural architecture can support very low-level consciousness, but this is an example of the large gray area one encounters when studying evolutionary transformation, and there is an urgent need for further neurobiological and learning studies in this group. Neuroanatomical information about integrating control regions should direct the choice of taxa in which to search for UAL, and conversely, when UAL is detected in an animal, the study of brain anatomy and physiology is likely to uncover an interesting functional architecture.

Another way of approaching the distribution question is to look at the organization of the *bodies* of animals that have brains with UAL architecture. An animal must have not only distal senses (for smelling, seeing, and hearing) but also the ability to control its movements in a flexible way if it is to learn about compound objects that guide spatial orientation and enable the learning of compound action patterns. To perceive and respond to compound objects, the animal's whole-body movements and the movements of its appendages must have many degrees of freedom, so studying bodily architecture can inform us about cognitive capacities. Taking a comparative approach, the neurobiologist Ilan Golani and his colleagues found that in both an arthropod (the fruit fly) and in vertebrates, drug-induced transitions into immobility begin with the animal moving along straight lines and then along increasingly more curved lines that narrow down the spatial spread of the animal's path; drug-induced transitions out of immobility generate patterns of movement in the opposite order. They suggest that the shared generative models of locomotion in the two groups might point to the common descent of arthropod and vertebrate body plans.[55]

Michael Trestman has taken an embodied cognition approach and has argued that the bodies of arthropods, vertebrates, and the cephalopod mollusks are the only ones that satisfy the design requirements necessary for implementing object-oriented spatial cognition. The "cognitive embodiment" of animals in these groups, he suggests, can complement our neurocentric view of animals, contribute to our understanding of the Cambrian explosion, and help us construct a fuller picture of the evolution of the mind.[56] Although Trestman does not discuss the sentience of animals with this type of embodied cognition, his approach provides another, complementary, way of investigating UAL and, by implication, sentience or minimal consciousness.

Based on the anatomical, taxonomic, behavioral-functional, and theoretical considerations that we have presented in this chapter, we conclude that the first arthropods, vertebrates, and mollusks that could learn about their world through UAL were minimally conscious, sentient beings.[57] They had what Lamarck called "a feeling of existence"—they felt themselves distinct from the environment and actively explored and categorized their world through their tactile, visual, olfactory, and other integrated CSSs. These states have all the attributes of mental states and entail a "self" that "owns" them. With these animals, a new mode of living emerged: the sensitive soul had come into being.

Our approach not only suggests fairly detailed predictions about the distribution of consciousness in the animal kingdom but also has testable predictions about which modes of learning and reasoning require consciousness. First, as already noted in chapter 5, our model predicts that only animals that are conscious can engage in the open-ended discrimination learning of novel compound sensory stimuli and of novel patterns of actions; for example, consciousness is required for second-order learning about compound novel stimuli and for the performance of successive operations (do novel acts A and B, and only then do novel act C). Second, only conscious animals will be able to recognize new goals by assigning priority to different motivating states (X is positive in context A, negative in context B; X has priority in context D and is of secondary importance in context E), and only conscious animals will be able to engage in causal reasoning.[58] Third, the interacting structural, neural, and molecular systems that underlie UAL are predicted to be strongly correlated with those that underlie consciousness. Additional predictions concern the coevolutionary relations of UAL (and therefore minimal consciousness) and other systems in animals, such as the stress response; these will be discussed in the next chapter.

From LAL to UAL: The Multiplication of Developmental Neural Modules and a Novel Use of Engrams

If, as we have argued, one can approach the transition to minimal consciousness by following the transition from LAL to UAL, it should be possible to follow the evolution of the critical structural and functional attributes that were involved in the transition. Two major and related evolutionary developmental changes in brain organization must have been involved in the generation of representations of compound stimuli and actions that could drive UAL, and both are apparent when we compare the functional LAL and UAL models (figures 7.12C and 8.2C, respectively). The first is the addition of hierarchical levels within and between sensory modalities (the evolution of the SIU) and motor neuronal systems (the evolution of the MIU), and the second is the addition of more general-purpose, high-level integrating and value units (the evolution of the AU and the REIU) and memory centers (the evolution of the MEMU). When following the evolution of these functional units, we need to know what developmental hereditary changes were involved. This question, however, cannot be answered without addressing questions about the innervation and morphology of the rest of the body because the brain is not something that evolved independently of other neural and nonneural structures.

Intriguingly, and in line with Trestman's suggestion about the "cognitive embodiment" of arthropods, vertebrates, and cephalopods, animals that exhibit UAL also seem to have, in addition to complex brains with complex integrating areas, large ganglia distributed along the body that control motor activity and are necessary for object-oriented and self-oriented spatial learning. In arthropods there are large, paired, segmental ganglia along the nerve cord and in vertebrates in the spinal cord; the coleoid cephalopods (octopuses, squid, and cuttlefish) have large distributed ganglia in the arms. In all these cases, the ganglia enable flexible movements based on combinations of different action patterns, which means that self-learning of compound patterns can occur. Of course, moving animals with distributed ganglia-containing parts require a brain that can model the moving body (an MIU), and the evolution of the MIU may have been an important factor in selection for the perception of compounds that depend on the SIUs. At the same time, we must remember that increased distal perception through the evolution of sense organs such as eyes and noses may also have driven the evolution of compound actions. The relation between compound perception and action is mutually reinforcing: compound perception (target/object selection) is particularly advantageous for a moving animal with many degrees of freedom of movement, which can control its actions, and the selection of compound actions is particularly advantageous for an animal with compound perception that can discriminate among multiple facets of the world. Small biases in the developmental strategies of ancient bilateral animals as increased size evolved—biases toward primitive modular organization (e.g., segmentation) and the perception of distant objects—may have led to the repetition of neural circuits in the brain and in potentially moving segments/parts during development. The repetition of

parts required the repeated activity of the gene networks underlying ontogenetic development at different locations of the body, as well as at different
times. In animals with modular somatic organization based on repeated
moving parts, spatial cognition could have driven both self-learning and
world learning (and their inevitable coordination), leading to UAL.

Is neurogenesis in animals with the capacity for UAL different from that
seen in taxa that learn nonassociatively or through LAL? We regard this as
an important question that may shed light on the enabling conditions for
UAL. However, at present there is no evidence for a single developmental
neurogenetic strategy that is shared only among animals with UAL. The limited work on neurogenesis in bilaterians suggests that a modular construction and the versatility of the genetic networks that increase the number of
neuronal functions and their regulation underlie the increased complexity
of the CNS in groups that show complex behavior.[59]

Figure 8.10 summarizes the various processes that may have contributed to the evolutionary transition from LAL to UAL. The evolution of the
action-modeling MIU was driven by the development of body parts with
large ganglia that could control the flexible movement of these parts, something that required a high level of coordination. The increase in the size of
sense organs (e.g., eyes and olfactory and auditory organs) required sensory
integration, which drove the evolution of SIUs that model the world and
the body. Learning in such animals depended on integrating regions (AU),
which could associate SIU and MIU models and enable spatial cognition
that took past experience into consideration. If this suggestion is valid, UAL
brain architecture would not have emerged without the evolution of flexible movement, based on semiautonomous large ganglia. Thus, movement,
which drove the evolution of the nervous system and of LAL, also drove
the evolution of spatial and self-oriented cognition, which is a facet of UAL.

It was in the context of the evolution of associative learning, we believe,
that a new function for epigenetic memory evolved. As we argued in the last
chapter, memory traces (engrams) are stored not only in patterns of synaptic
connections but also as epigenetic marks of neurons (e.g., in DNA methylation and histone modifications in chromatin), in the three-dimensional
structures of self-templating proteins, and in migrating RNA molecules
(which we call "migRNAs") packed and transported in exosomes.[60] Epigenetic, intracellular engrams (EEs) can be far more stable than synaptic
engrams, and the two types of memory can be decoupled. Two networks
might be identical with regard to the architecture of their synaptic connections and their connection weights but, because of their learning history, differ with respect to their EEs. Since EEs can affect future learning,

A

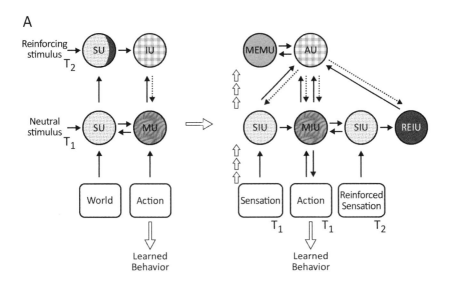

B

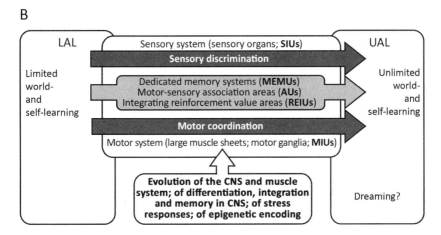

Figure 8.10

The evolutionary transition from LAL to UAL. *A,* a simplified scheme based on the models described in figures 7.12 (LAL, *left*) and 8.2 (UAL, *right*). Note the new hierarchical levels and new dedicated integrating and memory structures that are necessary for UAL. Vertical empty arrows indicate that there are additional intervening hierarchical levels between the integrating sensory and motor units and the higher-level units. *B,* factors and processes involved in the transition from LAL to UAL. The evolution of stress responses and the control of memory are discussed in chapter 9.

knowledge about synaptic connections might be insufficient to predict the learning potentials and biases of the animal. For example, the EEs of a neural network may lead to a rapid response to stress-associated stimuli because the relevant chromatin configuration is primed by past experience to respond to the signals, even when the synaptic connections show no traces of past learning. Memory is stored in both potentially highly stable intracellular epigenetic patterns and in more labile intercellular patterns of synaptic connections. Since more information can be stored in such a dual system, the network can learn more efficiently. Moreover, because engrams can be transferred among cells (through regulatory proteins or migRNAs) to higher (or lower) levels of the neural hierarchy, they could be the basis of very precise copying and updating of neural information, underlying processes such as Kamin blocking. Although cellular memory based on epigenetic mechanisms has very ancient origins, occurring in all eukaryotes (including unicellular ones), it could have been the recruitment of this system in neurons that enabled the integration of past memory traces and their long-term storage.

A potentially important quality of epigenetic engrams, which may have been of significance for the evolution of UAL, is that epigenetic encoding might be "symbolic." A pattern of epigenetic engrams in a neural network could encode several different properties of a percept. For example, DNA methylation and histone modification patterns in several regulatory loci within the neurons of a network could encode relations such as the CS-US interval, the rate of firing, and durations and distances between perceived features. Through migRNAs, such relations could be communicated to a higher level of the learning/cognitive hierarchy and reencoded. Epigenetically marked DNA or histones that share a certain property (e.g., encode a particular relation in a face, such as the distance between the eyes) could transfer this information through migRNAs to higher levels of the brain hierarchy, where it could be reencoded in chromatin. Each EE at this level can be said to represent an individual relational property that makes up the composite percept. The rebinding of such relational facets could construct a corresponding type of percept in a new context (e.g., the same face in different visual conditions) through progressive top-down processes, mediated by migRNA copied from the marked chromatin regions in the high-level regions, which are sent downstream. The predictive processing perspective can be applied here: the highest top-down encoded information can be seen as a high-level "hypothesis" (e.g., one that provides the engrams for a generic face) and the effect of the bottom-up processes (via the upward transport of the specific migRNAs that correspond to the particular face) can be seen as updating PEs. What we suggest is, of course, totally speculative, but we believe, as does the cognitive psychologist Charles Ransom Gallistel, that memory is encoded in intracellular engrams and not just in synaptic-weight patterns and bioelectric fields, and that such encoding may be the basis of generating something like symbols in the brain.[61] The similarity of this route of information storage and transmission to that of the genetic information system is striking: both involve stable chromosomally stored information, signaling RNA molecules, and proteins. However, while in the genetic system information flows unidirectionally from DNA/RNA through messenger RNA to proteins (there is no reverse translation), in the neural memory system, information flow is bidirectional. In the CNS, information flows from the synaptic signaling factors (proteins, neurotransmitters, and RNAs), through migRNAs to chromatin (DNA-methylation and histone acetylation patterns), and back again. The computational possibilities that such organization enables during development, especially in long-living animals with a capacity for UAL, are expected to be as vast as those afforded by the genetic system during evolution.

In whatever way it is neurally instantiated, the ability to encode relations among facets of a percept and among percepts conferred a huge advantage on animals having this capacity, but what drove its evolution? There is no doubt that a quantum leap in the ability to discriminate and select among actions has great advantages, but we believe that in addition to the obvious advantages, the transition from LAL to UAL—and the evolution of increasingly more efficient UAL—was partially driven by the problems of overlearning. For an animal that learns by association, a particular cue may be associated with stimuli with different valences: a pattern of vibrations may sometimes be associated with a predator and sometimes with a nonthreatening passing animal, but since flight is less costly than injury, overreaction to the cue is inevitable. In such conditions it would be beneficial to remember and recall a compound CS because a compound stimulus is a better predictor of the reinforcement. The evolution of binding may therefore have been driven by overlearning. Similarly, it is beneficial to inhibit any inappropriate response to irrelevant stimuli; the evolution of Kamin blocking and more complex lateral inhibitions can be seen as consequences of the evolution of associative learning in the growing chaos of a stimulus-rich world. Remembering compound percepts and action patterns was, however, a partial and transient evolutionary solution to the problem of overlearning. As the world of UAL animals expanded dramatically because of their new cognitive architecture, so sifting between the relevant and the irrelevant information became urgent once again. Learning about

Table 8.3
Vertebrates, arthropods, and cephalopods fulfill all seven criteria for minimal consciousness.

Neurobiologists' criteria	Vertebrates, arthropods, cephalopods
Global activity and differentiated states	Widespread brain activity during the performance of complex cognitive tasks, such as discrimination among compound stimuli, navigation in new terrains, and decision-making, is apparent in animals belonging to the three groups.
Binding/fusion and unification	Perceptual fusion enabling discrimination among compound patterns of stimuli is observed in animals of the three groups.
Intentionality	Flexible goal-directed behavior, requiring cognitive maps and involving dedicated integration and memory areas, is observed in all three groups.
Selection, plasticity, learning, and attentional modulation	In all three groups, behavioral plasticity is far greater than that found in animals of other phyla; all show evidence for top-down modulation of attention and the ability for serial operations; all show complex forms of learning.
Temporal thickness	Working memory lasting minutes and prolonged trace conditioning are displayed in all three groups.
Attribution of values; emotions, goals	Animals in the three taxa show an ability for priority scheduling; all display emotional reactions and complex action-selection, pointing to goal-oriented behavior.
Embodiment and self	There is integration of bodily responses in a manner suggesting a self-model (although this integration has been less studied, and seems to be of a different nature, in cephalopods).

Sources: Table 8.3 is based on data reviewed in chapter 3 (which was centered on mammals, from which the seven criteria were derived), in chapter 5 (which provided some of the evidence for homologies in both neuromorphological and cognitive traits in birds, reptiles, and other vertebrates), and on evidence reviewed in this chapter and summarized in tables 8.1 and 8.2. Merker (2007) and Feinberg and Mallatt (2016) have presented detailed evidence for cognition and brain organization in vertebrates that explicitly describes or implicitly points to all the listed features. The studies of arthropod cognition, brain morphology, and neurobiology summarized in this chapter, as well as reviews by Chittka and Niven (2009), Barron and Klein (2016), and Perry, Barron, and Chittka (2017), show all these features; Hochner and colleagues (Hochner, Shomrat, and Fiorito 2006; Hochner 2013; Hochner and Shomrat 2013; Hochner and Glanzman 2016) and Godfrey-Smith (2016a, 2016c) provide the evidence for cephalopods, which is less extensive than that for vertebrates and arthropods. Tye (2017) reviews and discusses much of the evidence for cognition, with a special stress on affect in vertebrates and arthropods.

compound percepts and action patterns could lead to the formation of an unmanageably vast number of memory traces. Under these conditions, selective forgetting and selective memory consolidation became crucial, and this may have enhanced the role of sleep as a controller of memory (and forgetting). We come back to these ideas in the next chapter, when we look at the coevolution of the learning and stress responses.

We now want to return to the relation between UAL and minimal consciousness. We argued that UAL animals display all seven characteristics in the neurobiologists' list. If we look at the phylogenetic distribution of the seven capacities without relating them specifically to learning, do they actually cluster together in vertebrates, arthropods, and cephalopods? The answer, based on the evidence we have assembled so far, is yes. When taken together, studies on vertebrate, arthropod, and cephalopod cognition, even when *not* based on tests of their learning abilities, show that the seven criteria are all clustered in these groups (table 8.3). This lends independent support to our suggestion that UAL, which leads to the same grouping, is an appropriate transition marker of minimal consciousness.

The observation that UAL and minimal consciousness seem to have evolved in arthropods, vertebrates, and cephalopods does not resolve all questions about the evolution of UAL: Did UAL evolve independently in the three groups or did it emerge twice? The time of origin and the marked difference in the brain organization of cephalopods as compared to vertebrates and arthropods, which evolved in the Cambrian, suggests that UAL may have evolved independently in coleoid cephalopods about 250 million years later.[62] The similarity of the time of emergence of the neural architectures supporting minimal consciousness in vertebrates and arthropods allows for two interpretations: (1) the remarkable functional homology between insect and vertebrate brains and their shared molecular kit mean that UAL in arthropods and vertebrates had a common origin, and (2) the UAL-supporting brain structures evolved independently in all three groups. If, as we argued in the previous chapter, several groups, including arthropods and primitive vertebrates, independently evolved LAL architecture, this may have been the foundation for the parallel evolution of UAL in these two groups, provided that the sensorimotor biases enabled by their modular ganglia and brain organization were also in place in both. In the next chapter, we suggest that during the early mid-Cambrian, UAL and sentience rapidly evolved in both arthropods and vertebrates (with the arthropods leading the way).

9 The Cambrian Explosion and Its Soulful Ramifications

What led to the emergence during the Cambrian of unlimited associative learning (UAL) and sentience? We argue that it was initiated by the evolution of a central nervous system (CNS) that enabled associative learning (AL) and generated positive feedback loops between learning-driven adaptations. This produced evolutionary arms races, which accelerated the rate of metazoan radiation and resulted in UAL in arthropods and vertebrates and in diverse behavioral and morphological adaptations in the species that interacted with them. But with UAL, as with every great innovation, there was a price to pay. UAL could lead to overlearning, and overlearning would have led to stress, neurosis, and illness. There was therefore strong selection for active forgetting in animals manifesting UAL, which eventually contributed to the reduction in the high rates of evolution by the end of the Cambrian. It was thus the coevolution of learning, the neurohormonal stress response, and the immune system that drove the evolutionary dynamics of the diversification of Cambrian arthropods and vertebrates. However, the evolution of cognition and consciousness did not stop when the Cambrian explosion ended. Building on the functional architecture of UAL, millions of years later vertebrates, cephalopods, and possibly even some insects evolved imagination, dreaming, and the capacity to plan and flexibly choose among alternative future actions. These animals, too, had to "pay" for their ability to select internally, which, as Popper put it, "permits their hypotheses to die in their stead." In these "Popperian organisms," selection for ameliorating the potentially confusing effects of imagination and dreaming led to self-monitoring and to the further control of memory, forgetting, and emotions.

There is something seductive about the proposal that the great transition to Aristotle's sensitive soul first occurred during the Cambrian, the most spectacular era of animal diversification. We first recognized this possibility in the early 2000s while working on the evolution of AL and its relation to minimal consciousness.[1] The additional taxonomic, anatomical, and functional information presented in the previous chapter lends further support to the idea. The first part of this chapter develops and extends our proposal that not only did minimal consciousness emerge during the Cambrian but that it also drove the explosion of new taxa that occurred during this era. In the second part of the chapter, we look at the further evolution of learning based on UAL—at the evolution of Popperian, imaginative, dreaming animals.

From Ediacaran Gardens to Cambrian Arms Races

> To the question why we do not find rich fossiliferous deposits belonging to these assumed earliest periods prior to the Cambrian system, I can give no satisfactory answer. ... The case at present must remain inexplicable; and may be truly urged as a valid argument against the views here entertained.
> —Darwin 1872b, pp. 286–287

Before the Cambrian, during the Ediacaran period (635–542 million years ago), life on Earth seems to have been very odd from today's perspective. Large and thick microbial mats covered the seafloor, and multicellular sessile organisms with bodies resembling fronds (*Dickinisonia*) and quilted air mattresses (*Swartpuntia, Charnia*) inhabited the oceans. Some Ediacaran creatures displayed symmetries unlike anything we know in current living forms, and their classification is, unsurprisingly, controversial, with some biologists suggesting that they represent a now-extinct kingdom of life, while others insist that they are stem group metazoans.[2] A few more familiar forms of life—some that are undisputedly sponges and others that are possibly cnidarians—lived toward the end of this era, as did, arguably, some small bilaterian-like animals (figure 9.1). Among them we find a 3-cm-long creature similar to a slug (*Kimberella*) and another similar to an annelid worm or arthropod (*Spriggina*), which may have crawled on the seafloor, grazing the microbial mats. It was, so the fossil evidence suggests, a world with few active predators, although if cnidarians with sting cells inhabited it, they may have preyed upon each other and on some of the larger bilaterians, and there is little doubt that parasites were abundant and host-parasite interactions proliferated. Hence, calling it, as the paleontologist Mark McMenamin did, "The Garden of Ediacara," is probably a little misleading.[3] Yet there was little "neural buzz" in this world. Only cnidarians and ctenophores and the few relatively large, mobile slug-like or worm-like creatures, which must have had a nervous system to coordinate their crawling, and some microscopic bilaterians with very simple nervous systems may have modestly "buzzed" as they went about their lives.

Figure 9.1
The "Garden" of Ediacara.

This bizarre world came to an end with the Cambrian explosion, an era beginning 542 million years ago and ending 485 million years ago, with a peak of animal diversification between 530–520 million years ago. Almost all animal phyla appeared during the Cambrian (figure 9.2). A whole orchestra of neural circuits started the great symphony of intensely active, highly interactive, and competitive animal life of the type we know so well. The complex tripartite brains of arthropods and vertebrates, which we described in the previous chapter, made their unmistakable appearance, as did learning-based predation and escape from predation, and one can infer that social (e.g., sexual) interactions among conspecifics began to dominate animal relations. These neural and locomotory changes were accompanied by various types of rigid skeletons, protective shells, armored plates and spines, claws, and striking and efficient mouthparts. Remarkably, many of the Cambrian animals, however strange, can be classified in currently existing phyla, and no animal phylum has evolved since. Arthropods of all sizes and shapes, including some fantastic creatures that challenge the imagination—*Anomolocaris*, the most ferocious meter-long predator of the era, and *Opabinia*, with its five eyes and a trunk-like proboscis ending in a grasping claw—thrived during the Cambrian. Rapidly swimming

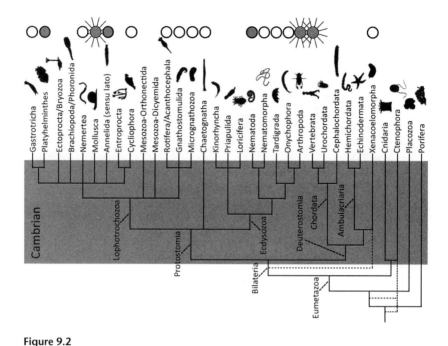

Figure 9.2

Phylogenetic relationships (not to scale) among major animal phyla (alternative possible positions are indicated by dashed lines) and their relation to the Cambrian explosion and the evolution of brains and AL (based on figure 7.13). *Empty circles,* animals with brains but no AL; *filled circles,* animals with brains and LAL; *radiating filled circles,* animals with brains and UAL. Modified by permission of Elsevier.

vertebrates, such as *Haikouichthys* and *Metaspriggina*, 3-cm-long fish-like filter feeders with gills, eyes, and nostrils, swam hurriedly in the Cambrian seas to escape the arthropod predators, and the members of other phyla evolved various ways of camouflaging and protecting themselves. The only Ediacaran animals that persisted through and after the Cambrian and did not become extinct were the sponges, the cnidarians, the ctenophores, and the indistinct placozoans. The Cambrian world, where intra- and interspecific competition abounded, is depicted in figure 9.3.

Two big questions come to the fore when we consider the Cambrian explosion: First, what spurred it and why was diversification so rapid? Second, why was the evolution of new animal body plans ("Bauplans") confined to the Cambrian? We believe that the evolution of AL and, in two groups (vertebrates and arthropods), of UAL, were factors in the explosive diversification during the Cambrian (see figure 9.2). But before

Figure 9.3

An artist's impression of the Cambrian battlefield. The Cambrian had various magnificent arthropods, such as the *Anomalocaris* (A), the biggest known arthropod of the era, with large eyes and claws, which became extinct at the end of the Devonian; the versatile trilobites (B); and *Opabinia* (C), with five eyes and elongated claws with grasping spines. Among them swam some small, inconspicuous fish (D) with a great evolutionary future. One of these fish is shown above *Opabinia* and another above the multilegged worm (E).

discussing our proposal, we must briefly look at the suggestions of other scientists, especially the argument that the Cambrian explosion is an artifact.

As the quotation with which we began this section shows, the sudden appearance of animal-rich strata in the Cambrian troubled Darwin, for it suggested that evolution may not have been as gradual as he proposed. Because he could not offer a good biological explanation for the lack of animal fossils prior to the Cambrian, he concluded it was an artifact, the consequence of the poverty of the fossil record.[4] This, we know today, cannot be the whole story, for there is sufficient fossil evidence from the preceding Ediacaran period to contradict the "poverty of data" argument. Nevertheless, there are still some people who adhere to Darwin's gradualistic view, mainly because comparative molecular data can be interpreted as suggesting that Cambrian animal phyla had pre-Cambrian, small, soft-bodied bilaterian ancestors whose fossilized remains have not yet been found.[5] This view is based on two assumptions: that the rates of molecular and morphological evolution are linearly correlated and that the rate of molecular evolution during the Cambrian was similar to that before and after it. Both assumptions are questionable. First, there is evidence that molecular and morphological evolution are often decoupled so that molecular evolution can be much slower or much faster than morphological evolution.[6] Second, as Mark Pagel and his colleagues have shown, an increased rate of molecular evolution is correlated with periods of speciation: the rate is greater in taxa with phylogenetic trees with multiple nodes than in the trees of related taxa of the same chronological age but having fewer nodes.[7] Since the Cambrian was undeniably a period of intense, unprecedented diversification (including speciation), the effects of a high speciation rate must be factored into the dating estimates, and when this is done, the explosion hypothesis is reinforced. Hence, molecular evolution could have been much faster than traditionally believed.[8]

It is, however, the morphological (and by implication the physiological-neural) diversification that is of the most direct interest to us, and here paleontologists are the authorities we must consult. After more than 150 years of digging, many new discoveries, and much novel and sophisticated analysis, the fossil evidence still stubbornly suggests that in a geologically very short period, most extant metazoan phyla made their first appearance in the fossil record and diversified.[9] So although the first (molecular) roots of this evolutionary explosion are still disputed, and some are undoubtedly deep, the occurrence of a morphological big bang is not in doubt. Many paleontologists believe that minute bilaterian animals were present during the preceding Ediacaran era, possibly 570 million years ago, before the larger benthic (bottom-dwelling) Cambrian bilaterians appeared, yet argue that extensive morphological and some genomic repatterning occurred during the early Cambrian.[10] Others contend that a full-blown explosion, both molecular and morphological, occurred during the Cambrian and that the absence of earlier ancestors in pre-Cambrian rocks, when fossilization conditions were excellent, does indicate the actual absence of pre-Cambrian bilaterian ancestors.[11] Whatever their assumption about the existence or absence of tiny ancestral bilaterians before the Cambrian, all those who study this era agree that a remarkable ecological and morphological diversification occurred during the Cambrian and that the nature and causes of this diversification are pertinent and important evolutionary questions.[12]

So what scenarios are suggested? Can we justify our suggestion that the appearance of AL and its underlying CNSs was not a mere by-product of the rapid evolution of animals but, rather, was one of the drivers of the Cambrian explosion?

The Many Causes of the Cambrian Explosion

The causes of the Cambrian explosion have been sought in tectonic, geo-chemical, climatic, and biological processes, and the results of studies from all these perspectives paint a picture of the Cambrian as a unique junction period, with different and interacting factors contributing to the explosion. The climatic and geochemical factors include biologically significant increases in oxygen concentration, beginning approximately 635 million years ago, which led to the diversification of the Ediacaran fauna and the appearance of the first calcifying metazoans approximately 548 million years ago;[13] pulses of global warming, the result of methane release associated with polar movements, which led to increased nutrient cycles and productivity;[14] and changes in sea level that led to the flooding of continental margins, which greatly increased the range of habitable shallow-water areas and led to the rapid input of erosional by-products that brought about changes in the chemical constitution of the oceans, including an increase in calcium and phosphate concentrations (the permissive conditions for biomineralization, which animals exploited).[15] All these factors interacted and had effects that were crucial for the occurrence of the Cambrian explosion. However, these global processes do not explain the special features of the dramatic morphological diversification within the *animal* kingdom.[16] An explanation of this aspect of the Cambrian explosion must incorporate molecular, developmental, and ecological factors that are specific to animals.

Given the importance of the question, it is not surprising that many hypotheses attempt to explain the biological changes underlying the explosion of animal morphological forms during the Cambrian.[17] Table 9.1 is a nonexhaustive list of different hypotheses, ranging from those that have some empirical support to those that are, at present, entirely speculative.

We believe that just as there is no single geochemical factor that spurred the Cambrian explosion, no single biological factor can account for it. It is far more plausible that interactions among several biotic factors created a positive evolutionary feedback loop that culminated in the explosion of bilaterians, which, through their activities, altered the face of the planet and their own evolution. As Butterfield puts it: "Early animals did not simply fill up previously existing but unoccupied niche space; they created

Table 9.1
Biological causes and prerequisites suggested as explanations of the Cambrian explosion.

Innovations, prerequisites, and causes	Contribution	References
Developmental/ecological innovations or prerequisites		
Macroscopic predation	Arms races among predators and prey, leading, among other things, to burrowing and the evolution of external and internal hard parts.	Bengtson 2002; Dzik 2005; Wray 2015
Social selection (e.g., sexual selection, selection of nesting sites, and brood caring)	Diversification driven by social selection related to the evolution of the senses.	West-Eberhard 1983, 2003; Miller 2012
Anal breakthrough: evolution of anus, gut, and directed movement	Improved sorting of food and waste, anterior-posterior-directed movement, leading to diversification.	Cavalier-Smith 2006, 2017; Sperling and Vinter 2010
Set-aside cells	Complex morphologies arose through the evolution of specifications for adult body plans within set-aside cells (sequestered in simple ancestral lineages). Set-aside cells are assumed to have been homologous among some phyla, so the common ancestors of these phyla had body plans similar to an ancestor of their larvae.	Davidson, Peterson, and Cameron 1995; Peterson, Cameron, and Davidson 1997
Spiking neurons	Spiking neurons may have emerged once animals began to eat each other. The first sensory neurons could have spiked in response to other animals in their proximity, alerting them to perform precisely timed actions, such as fleeing or attacking.	Monk and Paulin 2014
Central nervous system	CNS evolution led to improved motor coordination and sensory discrimination; learning led to developmental and ecological complexity.	Stanley 1992; Cabej 2008, 2013; Wray 2015
Consciousness	Emergent processes at the quantum level involving microtubules, which led to increased behavioral and ecological complexity.	Hameroff 1998

Table 9.1 (continued)

Innovations, prerequisites, and causes	Contribution	References
Visual perception	Arms races between predator and prey species; invasion of new niches.	A. Parker 2003
Burrowing (associated with an anterior-posterior axis)	Burrowing animals changed the ocean soil by mixing it, leading to the recycling of organic matter. This enabled rapid animal evolution (and altered the preservation conditions, preventing the ready fossilization of fragile cells and tissues).	Brasier 2009
Beginning of deeper burrowing	From an anoxic seabed with a plastic-like coating, off-limits to animals, a change in grazing style led to a transformation of the seafloor. Animals began to burrow several centimeters into the sediments beneath the mat, accessing nutrients and at the same time hiding from predators.	Carbone and Narbonne 2014; Fox 2016
Unlimited associative learning (UAL)	Expansion and formation of new niches; arms races and cooperative alliances.	Ginsburg and Jablonka 2007b, 2010b; this chapter
Basic cognitive embodiment (BCE)	Distal senses guided mobility, involving complex control of body and object manipulation; led to predation and initiated predator-prey interactions.	Trestman 2013a, 2013b
Genomic causes or prerequisites		
Dynamical patterning modules	Generation of morphological novelties by core developmental genes mobilizing physical processes (cohesion, viscoelasticity, diffusion, and more) in cell aggregates.	Newman and Bhat 2009
Epigenetic signaling systems	Reversible deposition of methyl groups depends on oxygenases, and hence epigenetic signaling systems could evolve only after an increase in atmospheric oxygen. Epigenetic systems are essential to cellular differentiation in animals.	Jeltsch 2013

(continued)

Table 9.1 (continued)

Innovations, prerequisites, and causes	Contribution	References
Changes in the genetic regulatory networks (GRNs)	Differences in GRNs among phyla; new organization in bilaterians.	Knoll and Carroll 1999; Arthur 2000; Erwin and David-son 2002, 2009; Davidson and Erwin 2006
Establishment of body patterning involving the CRFC-Hox kernel	CTFC (CCCTC-binding factor), a master regulator, first appeared in bilaterians. The Hox-CTCF kernel led to diverse body patterns.	Heger et al. 2012
Recruitment of patterning gene clusters of *Hox, NK,* and *ParaHox* genes	ANTP-class homeobox genes, which encode transcription factors involved in body patterning; NK (patterning mainly mesoderm); *Hox* (patterning CNS); and *ParaHox* gene clusters (mouth, midgut, and anus) were recruited and duplicated.	Holland 2015
Role for conserved noncoding genetic elements	Cisregulatory enhancers located near major developmental genes, resulting from duplications and transpositions, were associated with the rewiring of the genetic circuits involved in morphological divergence.	Vavouri and Lehner 2009
Mobilization of transposable elements	Increased genetic variability; affected regulation of gene networks.	Oliver and Greene 2009; Zeh, Zeh, and Ishida 2009
Epigenomic stress responses	Stress led to epigenomic changes, which increased selectable variation.	Zeh, Zeh, and Ishida 2009
Temperature stress that affected chaperone proteins such as Hsp90	Relaxation of canalization and exposure of cryptic variation to selection.	Baker 2006
Neurohormonal stress leading to epigenomic destabilization	Increased genetic variability, especially of genes expressed in the brain.	Ginsburg and Jablonka 2010a; this chapter

Table 9.1 (continued)

Innovations, prerequisites, and causes	Contribution	References
Accumulation of miRNA genes	Accumulated miRNAs led to increased precision of gene regulation and increased evolvability.	Hornstein and Shomron 2006; Heimberg et al. 2008; Peterson, Dietrich, and McPeek 2009; Erwin et al. 2011
Horizontal gene transfer	For example, transfer by virus-like particles of skeletonization-related genes that allowed the establishment and spread of animals with hard parts and led to arms races and ecological complexity.	Mourant 1971; Jackson et al. 2011; Ettensohn 2014
Extensive hybridization	Interspecies hybridizations led to immediate expansion of the genome and expanded morphological possibilities; this led to the origin of larvae.	Williamson 2006
Duplication of genes/ genomes	Duplicated genes/genomes diverged and new functions evolved.	Lundin 1999; but see Miyata and Suga 2001; Holland 2015; Grant 2016
Duplication of introns	Increased protein diversity.	Babenko et al. 2004

the space itself."[18] The relationship among the suggested developmental-ecological hypotheses is depicted in figure 9.4. The ecological factors include the intriguing possibility that some Ediacaran taxa, which coexisted for a while with the Cambrian bilaterians, scaffolded the bilaterian's ecological niche by providing organic carbon resource heterogeneity that rendered increased motility and burrowing beneficial.[19]

The transition from the nearly two-dimensional, mostly benthic existence of the pre-Cambrian animals to the three-dimensional world of burrowing and swimming creatures opened up enormous ecological opportunities. This transition would not have been possible without a throughput gut, muscles, an internal or external skeleton, and a CNS that could coordinate internal movements and locomotion. Although all these aspects of animal morphology are crucial, we would like to highlight the role of the CNS. We follow here the viewpoint of Nelson Cabej, who argues that the evolution of the CNS drove the Cambrian explosion: "My hypothesis is that

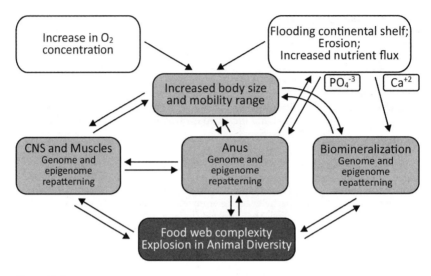

Figure 9.4
Interactions among geological, geochemical, and biological factors that have been suggested as major contributors to the Cambrian explosion. Behaviors such as burrowing, predation, fight and flight, mate choice, and other social interactions guided and accompanied the changes in morphology and physiology.

the centralization of the nervous system, which represents evolution of the full-fledged ICS [integrated control system] in animals, was the driving force behind this momentous landmark of animal evolution."[20] Cabej points to the CNS's control of adaptive behavior, such as predatory or reproductive behavior, and dwells on a crucial and little-appreciated factor: the effect of the CNS on morphology and cell differentiation. He points out that once the maternal resources that are deposited in the egg and control the first stages of development are exhausted, ontogeny in neural animals comes under the control of the nervous system. The development of muscles, the circulatory system, and the gastrointestinal tract, as well as the processes of regeneration and the switches between discrete developmental stages (e.g., switches between larval and mature morphologies) are all regulated by the CNS. The regulation of development and morphology by the CNS is therefore a key to understanding the evolution of morphological diversity.[21]

We agree with Cabej that the evolution of the CNS was an important factor driving the Cambrian explosion, but we add to his proposal the more specific suggestion that it was the evolution of AL (which was dependent on the evolution of the CNS) and, in the arthropods and vertebrates, of UAL, that spurred the evolution of animals during the Cambrian. This hypothesis, we will argue, incorporates, as special cases and preconditions, several of the previous explanations listed in table 9.1.

An evolutionary explanation requires not only an account of preconditions, evolutionary stages, and selection regime/s but also an account of the type of dynamics involved. Since the evolution of learning is an instance of the evolution of plasticity, we focus on explanations that emphasize the evolutionary dynamics of behavioral plasticity. We base our reasoning on Mary Jane West-Eberhard's argument that the evolution of organisms is shaped by the modulations and effects of morphological, physiological, and behavioral plasticity. Although plasticity can slow down evolution because fluctuating conditions make some hereditary variations invisible to natural selection, plastic responses to these fluctuations can also drive evolutionary change.[22] Hence, we expect that lineage-specific and social-ecology-specific morphological and behavioral modifications will result from selection for learning about particular regularities in a new environment, as well as selection for more domain-general learning based on exploration-stabilization strategies, such as trial-and-error reinforcement learning. Another factor that has bearing on the evolution of learning is related to our growing understanding of the role of stress in the generation of epigenomic variations. As we argue in what follows, the breadth of learning and active forgetting evolved partly as a response to learning-induced stress and involved the coevolution of neurohormonal and immune responses that ameliorated the stress induced by aversive AL.

The Evolution of Behavioral, Learning-Driven Plasticity

The plasticity afforded by the evolution of learning was an important topic of discussion at the end of the nineteenth century. Several leading biologists and psychologists, notably Fairfield Osborne and James Mark Baldwin in the United States and Lloyd Morgan in England, suggested that if learning enables individuals to adapt ontogenetically to new conditions, those individuals that can most effectively adapt through learning—for example, those that learn to dig burrows in the ground to hide from a newly introduced predator—will be the most likely to survive and reproduce. Hence, through natural selection, the capacity to deal with the new conditions (e.g., how to dig, where to dig, when to inhabit the burrow) will improve, and eventually, through the selection of chance genetic changes in the lineage's hereditary constitution, the behavioral adaptation that originally required much learning would occur without, or with only a little, learning. Such processes, they suggested, can explain the evolution of instinctive, innate behaviors like the fear reaction to snakes or the smell of a predator, which require very little learning—sometimes just a single exposure to the stimulus is sufficient to elicit the behavior. This evolutionary process, through which selection for the buffering or canalization of specific perceptual, motivational, and cognitive capacities against genetic and environmental perturbations leads to less reliance on learning, is known as "the Baldwin effect."[23] In the mid-twentieth century, it was generalized by Waddington and Schmalhausen, who applied it to *any* adaptive environmentally induced character modification, not just to behavior.[24] Waddington called the canalizing-selection process leading from a stimulus-dependent response to a response that is independent (or less dependent) of that stimulus "genetic assimilation," and Schmalhausen called it "stabilizing selection." Later, Avital and Jablonka suggested that when considering the evolution of behavior, which involves a sequence of actions, genetic assimilation can lead not only to specialization and less reliance on learning but also to an increased sophistication of behavioral repertoires because genetic assimilation "releases" cognitive-learning resources, thus allowing additional learning that can lead to the lengthening of behavioral chains of action. They called the process leading to the sophistication of behavior through the assimilation of one of the parts of the behavioral sequence the "assimilate-stretch principle."[25]

The idea that selection for responsiveness can modulate plasticity was expanded by West-Eberhard. Central to her argument is the observation that any change in genotype or in environmental conditions leads to phenotypic accommodation—to the reorganization of development, through "adaptive mutual adjustment, without genetic change, among variable aspects of the phenotype, following a novel or unusual input during development."[26] Any heritable variation that improves phenotypic accommodation, regardless of whether it increases or decreases developmental canalization, will be selected. Hence, genetic changes supporting phenotypic accommodation can lead to increased genetic control, to more complex genetic regulation, or to an extended, general, exploration-stabilization strategy. In other words, within the context of learning, it is not just behavioral specializations (e.g., "instincts") that can be explained by canalizing selection but also domain-specific increased plasticity and, importantly for our argument, increased domain-general learning ability. West-Eberhard called the process leading to the genetic modulations of plasticity that begins with phenotypic adjustment "genetic accommodation" and gave many plausible examples of the ways in which learning can lead to such accommodations, increase diversity, and even result in adaptive radiations.[27] For example, she suggested that the divergence of African lake cichlids, a famous example of a recent adaptive radiation, was driven by a positive feedback in conditions of food scarcity between learned food preferences and variations in developmentally plastic jaw morphology. While all cichlid fish prefer soft food, when food is scarce, those fish with jaws that can more easily crack mollusks take a greater proportion of mollusks, and those with a jaw morphology that makes them better at handling soft food specialize in soft food. In these conditions, preexisting variation in flexible morphology affects diet, and the resultant divergent behavior reinforces selection for a specialized (and in this case, less plastic) jaw morphology that matches the preferred diet and feeding behaviors.

The role of behavior as a driver of adaptive radiations has surfaced repeatedly during the last sixty years. In the 1960s, the zoologist Alister Hardy proposed that behavior can drive morphological evolution and underlie diversification; he suggested that the dramatic radiations of reptiles, mammals, and birds were all driven by their enhanced learning abilities. The extent and nature of learning in these groups, he argued, led to the invasion of new niches and thus to new selection regimes.[28] Later, the biochemist Alan Wilson and his coworkers called the guiding effect of behavior on evolution and adaptive radiations "behavioral drive,"[29] and behavioral ecologist Patrick Bateson called behavior "the adaptability driver."[30]

Our suggestion that AL drove the Cambrian explosion is based on similar reasoning: we believe that the plasticity afforded by nonassociative learning was the major enabling condition for the evolution of the far more extensive plasticity and opportunities for phenotypic accommodation that were conferred by AL. In chapter 7 we suggested that AL emerged as mobile, macroscopic bilateral animals with CNSs evolved. The evolutionary increase in size, which began in the late Ediacaran, involved larger muscle sheets and larger sense organs with many more cells. Developmental adjustments had to be made among the tissues involved in the control and coordination of motor behavior and in the integration of signals arriving from the multiple sensors in the larger sense organs. The continuous adjustments that were necessary throughout the animal's growth and development led to the evolution of the neural control of early ontogeny. In bilaterians, the closely packed, specialized, and integrating areas in the anterior part of the animal facilitated associations within and between different sensory modalities and motor inputs, enabling better coordination. Initially, the sensory and motor associations formed in the brain were presumably insensitive to the sequence of stimulation, and learning involved global and local sensitization and habituation to external stimuli and actions. However, in large, mobile, relatively long-lived bilaterians, with their increased scope for motor exploration, there was selection for more efficient and rapid sensorimotor coordination. This, we suggest, drove the evolution of more elaborate anatomical organization and control in the anterior "center of communication." Since the molecular tool kit that underlies sensitization and habituation and their simple modulations was already in place, selection for predictive responses (which involved associations between specialized, closely packaged neural networks) occurred in animals that encountered recurrent conditions. At the anatomical level, this required selection for information-integrating circuits and for distinct, but closely interacting, sensory and motor integration centers. At the molecular level, the modulation of adenylyl cyclase and N-methyl D-aspartate (NMDA) receptors and the selection for long-term association between particular synapses, possibly based on retrograde signaling and epigenetic changes in neurons, led to the evolution of the control of memory and sophistication of learning.

We believe that comparative molecular analyses may help both to pinpoint the molecular tools involved in AL and to unravel the evolutionary relations among different animals that manifest AL. However, comparisons based on a large number of protein-coding genes in many representative taxa can provide only initial, general information about the evolution of neural plasticity and learning because protein-coding genes are highly conserved.[31] Studying gene networks and the chromatin conformation of regions involved in the transcriptional regulation of memory-associated genes, the miRNA that evolved in protostomes and deuterostomes, the targets of these miRNAs in the nervous system, and the genomic targets of epigenetic mechanisms involved in the establishment and maintenance of long-term memory are needed to shed light on the key processes that were modified during the evolution of AL.[32]

As we discussed in chapter 7, animals with long-term memory and limited associative learning had an enormous selective advantage compared to their non–associatively learning competitors: they were able to adapt ontogenetically to a variety of biotic and abiotic environments and to use new resources. Although still limited by their inability to learn about compound stimuli and actions, their learned behaviors could guide them to where and how to look for food, mates, and protection and enabled them to locate and react to cnidarian predators and to competitors. These learned behaviors were fundamental to the construction of the niches that they and their offspring inhabited and to the evolution of social and ecology-specific adaptations. For example, if an animal learned to exploit a novel rich and safe food source and consequently tended to stay and reproduce in areas where this resource was abundant, its offspring would have the same learning environment and learning opportunities and would seek a similar niche; this would lead to habitat-specific habits and styles of parental behavior. Any behavioral, physiological, or morphological feature that improved adaptation to this specific learning environment would be selected. This includes learning specializations: we see Garcia conditioning—the specialized rapid conditioning to particular salient and reliably recurrent aspects of the niche—as such an ecological or social learning-specific specialization. We therefore disagree with Moore, who suggested that Garcia conditioning preceded more typical forms of conditioning (chapter 7), and propose that this type of conditioning was a derived, ecological learning specialization based on standard forms of conditioning. The new behaviors and new learning biases in novel ecological conditions would have been accompanied by new, matching morphological adaptations.

Morphological diversification is a hallmark of the Cambrian explosion, but how, according to our scenario, can one explain the lack of subsequent morphological evolution at the Bauplan level? Why did the morphological big bang stop? Stuart Newman and his colleagues have emphasized the importance of the constraints imposed by the physical nature of multicellular animals: viscoelasticity and chemical excitability that leads to self-organizing and interacting processes of free diffusion; immiscible liquid behavior; the oscillation and multistability of chemical states; reaction-diffusion coupling; and mechanochemical responsivity. These physical-chemical processes lead to a restricted number of morphologies that correspond to the hollow, multilayered, and segmented morphotypes seen in the gastrulation-stage embryos of modern-day metazoans, which are the basis of all subsequent evolutionary variations.[33] We believe that the evolution of the CNS and AL imposed additional developmental constraints as well as affordances on animal forms. First, as discussed in chapter 6, following the evolution of the CNS, distant tissues could interact and coordinate their activities, enabling the development of complex, distinct body plans, which relied on increasingly precise and flexible CNS control. In other words, we propose that the evolution of the CNS and its control of morphological development entered into a positive feedback relation, leading to further improvement of both. Second, since, as the hierarchical predictive-processing (HPP) framework suggests, the brain generates models of the body in context,[34] a relatively stable generative model of adult form would be an advantage in an animal that has to update its world-body relations through learning. We therefore suggest that the evolution of a sophisticated CNS, with multiple hierarchical levels and top-down control, selected for distinct and stable adult morphologies that can be accurately "modeled" and updated. This new constraint may have contributed to the morphological stasis (at the Bauplan level) in animal evolution while at the same time contributing to the further diversification and sophistication of learning.

According to our scenario, the arthropods were the first group of early Cambrian bilaterians to take the evolution of learning one crucial step forward.[35] We assume that tiny pre-Cambrian protoarthropods already had a body plan with a modular somatic organization based on repeated moving parts controlled by semiautonomous ganglia. As protoarthropods evolved and increased in size, this organization, which was particularly amenable to segment differentiation, enabled flexible movements and, crucially, rendered self-learning particularly advantageous. This in turn reinforced the use of distal senses and led to the evolution of spatial cognition. Hence, as we suggested in the previous chapter, it was principally the evolution of locomotion and motor control that drove the further improvement in sensorimotor coordination and complex learning in macroscopic animals. However, learning based on a simple adding up of multiple sensory or motor inputs has its dangers because individual stimuli (e.g., persistent vibration) may be part of several different inputs with different valence and lead to inadequate or conflicting responses (for example, persistent vibration may be linked to both predators and food). The ability to *fuse* inputs into composite percepts and to generate and distinguish between different action patterns drove the evolution of UAL, producing a neural architecture that supports UAL, which is based on the early developmental multiplication of brain modules with some degree of specialization (e.g., modules "for" memory or "for" sensorimotor integration). This functional architecture made the arthropods the cognitively most sophisticated animals of the early Cambrian. (Some of them, the Cambrian trilobites, are depicted in figure 9.5.) They became the most efficient predators of that era and the most important drivers of the evolution of all other groups, both in their own phylum and in other phyla. Their superior learning abilities, we suggest, led to coevolutionary arms races in interacting species and to intraspecific diversification that drove speciation.[36]

An important factor in these coevolutionary dynamics was the evolution of hard parts, including external skeletons. This was made possible by the calcium release that followed the lengthy pre-Cambrian erosion of rocks and soils, which led to a threefold increase in the concentration of calcium in the seas. The arms race is also likely to have led to the evolution of camouflage and toxicity in prey and competing predators and to improved

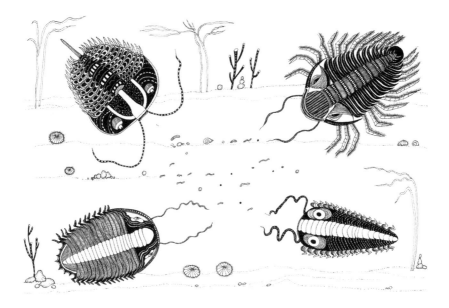

Figure 9.5
Cambrian trilobites. Early trilobites are thought to have been predators of benthic invertebrates.

locomotion and more sophisticated sensors in species that interacted with conspecifics or with individuals from other species. The evolution of the eye, which Parker believes drove the Cambrian explosion,[37] was, according to our scenario, just one of the several consequences of the evolution of AL, each of which had strong evolutionary domino effects. As the learning ability of a predator improved, there would have been strong selection for morphological and physiological adaptations in its prey species; for stress responses associated with flight and fight; and for the prey's ability to learn to avoid predators, attack, or escape.

We believe that sexual selection based on compound signals, as well as mimicry of other species, made their appearance at the Cambrian, since AL enabled animals to perceive and discriminate between complex percepts. As Darwin noted, sexual selection implies the existence of complex cognition.[38] If, for example, male trilobites or male fish were able to display complex visual patterns of behavior to attract the opposite sex, this would indicate that females are able to perceive and appreciate these patterns, pointing to the existence of visual experiencing. Unfortunately, direct

Figure 9.6
The artistic male puffer fish (*Torquigener* sp.) constructs large geometric circular structures on the seabed to attract females.

evidence for sexual selection and social coevolutionary dynamics from this era is lacking, but we know that sexual selection has led to some remarkable behaviors in present-day animals. Figure 9.6 depicts the complex artistic patterns created by a 12-cm-long male Japanese puffer fish to attract females.[39]

Many Cambrian species were probably driven to extinction by associatively learning arthropods, and over time some minimally conscious invertebrates may have lost their complex and expensive brains (and the accompanying consciousness). For example, fossil evidence suggests that penis worms (priapulids) may have lost their originally tripartite, segmented CNS, which was similar to that of arthropods, leaving them with a highly simplified CNS and no consciousness (if their ancestors had any).[40] Other taxa evolved learning-related countermeasures to competition and predation, including, in some cases, long-term memory and limited associative learning and in vertebrates, UAL. Hence, the evolution of AL (in many taxa) and of UAL (in arthropods) led to the convergent evolution of AL capacities in other groups. The parallel evolution of UAL and minimal consciousness in the grazing vertebrates of the Cambrian was an evolutionary response to the strong selection imposed by their UAL arthropod predators and social competitors. The enabling conditions in vertebrates that allowed this convergent evolutionary response were presumably similar to those of arthropods, including segmented organization of the body and good control of motion by semiautonomous ganglia. In addition to the genetic accommodation of morphological and behavioral adjustments, the learning-driven occupation of new niches may have resulted in frequent hybridizations and speciation via hybridization, which is beginning to be recognized as an important process in macroevolution.[41] The occupation of large arrays of new niches may have exposed the animals to different types of bacteria and viruses and forged new types of parasitic and symbiotic relations, all of which could have contributed to their rapid evolutionary diversification.

As we noted in chapters 5 and 8, the idea that vertebrate consciousness has Cambrian origins has been explored by Feinberg and Mallatt. They argued that in ancient Cambrian fossil fish, the brain structures in the tectum that enabled these animals to form the multisensory maps of the world, which in their view are the hallmarks of consciousness, were already present. They suggested that vertebrates evolved these brain structures, which allowed superior perception, to escape arthropod predators.[42] Their observations support the conclusions we reached when we first noted that associatively learning animals seem to have evolved during the Cambrian.[43]

Framed in the terminology of phenotypic and genetic accommodation, we suggest that mobile arthropods with segmented (and at first poorly differentiated) body parts were able to explore their environment far more than other, nonsegmented mobile animals and that their explorations led to more flexible locomotion and handling of prey. This behavioral plasticity enabled self-learning, as well as combined world and self- (e.g., spatial) learning, behavioral accommodations that were at first not very reliable and required many repetitions but nevertheless had significant benefits for animals who were able to learn in this way. Genetic accommodation led to more reliable learning—to the evolution of the functional architecture of UAL—first in the arthropods and then in vertebrates. The evolution of UAL capitalized on the multiplication of modular neural units and on the already existing gene networks and epigenetic mechanisms that controlled and coordinated the development of the nervous system and body patterning. However, UAL was not just a huge jump in the ability to discriminate and regulate actions—it also opened a new Pandora's box of problems.

Stress, Learning, and Their Coevolution

> For in much wisdom is much grief: and he that increaseth knowledge increaseth
> sorrow.
> —Ecclesiastes 1:18 (King James Version)

The great advantages of AL in general, and UAL in particular, are obvious: such learning leads to increased discrimination and action selection, which are important for survival. It enables, for example, the more frequent satisfaction of appetitive needs because the detection and guidance of associated cues make it easier to find food or mates. Similarly, animals are better able to cope with antagonistic interactions because they can learn to avoid or counter them. However, as already noted in chapters 7 and 8, AL can lead to overreaction, and inhibitory effects (Kamin blocking in animals that manifest LAL and the more complex and extensive types of inhibition of the effects of irrelevant stimuli during learning in animals with UAL) are necessary evolutionary modulations of learning in the chaos of a stimuli-rich world. The evolution of these mechanisms did not, however, solve the problems of overlearning because with UAL, partial compound cues may serve (through pattern completion) as predictors of several complex stimuli with different valences, and many partial cues are therefore likely to be false positives. As long as the benefits of avoiding actual danger exceed the costs of responding erroneously to possible danger, an animal that overreacts will have an advantage. For example, flight-or-fight reactions that prepare the animal to either flee or fight an adversary, though costly, are less expensive than injury or death. We therefore expect, and indeed observe in extant animals, flight and a readiness to fight when neither predator nor foe are around. Randolph Nesse has called the principle underlying these types of overreactions to adversity (he focused on flight and anxiety reactions) the "smoke detector principle."[44] Since the cost of encountering "fire" (predator, foe, or any other type of potentially grave danger) is very high and the cost of flight from it is lower, it is advantageous to flee upon any cue associated with grave danger ("smoke"). Responses to false alarms are therefore inevitable. However, since these stressful false alarms trigger the mobilization of organismal resources, there are considerable costs to overlearning, and countermeasures must have evolved to cope with the problem. We would like to look at these countermeasures, specifically, active forgetting, more closely now; we suggest that they are crucial for effective learning and that they evolved in the context of the overlearning that was an inevitable facet of all forms of AL, especially UAL. Since we believe that the control of the stress response, which includes immunological responses, is deeply entangled with the evolution of learning and its inevitable perils, a short digression about stress responses is warranted.

The Neurally Mediated Stress Response in Vertebrates and Invertebrates

The life of all organisms is a precarious affair—they are always coping with threats and perturbations, often on the verge of annihilation. All living organisms have therefore developed ways of coping with their risky existence. Since perturbations leading to a far from equilibrium state (defined as stressful or leading to stress[45]) are unavoidable and since a return to homeostasis as soon as possible is mandatory, all living organisms have evolved stress response systems—complex sets of reactions that manage out-of-equilibrium states and facilitate a return to homeostasis. Given the universality and inevitability of stress, it is not surprising that the stress response is based on a highly conserved network of cellular reactions and factors that can be found across all kingdoms of life. These include MAP kinases (a highly conserved family of serine/threonine protein kinases that are also involved in fundamental cellular processes such as proliferation, differentiation, and motility) and the highly conserved heat-shock proteins (HSPs), which are upregulated following stress by the heat-activated HSF-1 transcription factor and act as molecular chaperones, maintaining protein conformation under stress, refolding misfolded proteins, and targeting irreversibly damaged proteins for degradation; in addition, there are various cellular immune mechanisms for distinguishing between self and unwelcome nonself. Like the other stress-managing systems, immune systems, which cope with pathogens and other damaging factors, are present in all living organisms and include both innate responses that evolve on the phylogenetic timescale and acquired responses that cope with specific challenges and evolve at the ontogenetic timescale.[46] Crucially, all these factors and systems are intimately involved in learning and memory.

In all neural animals, the stress responses involve an interaction between the neurohormonal system and the immune system, two systems that show amazing similarities and interact with each other. Enzo Ottaviani and his colleagues point to four major types of interactions between the two systems:

> (1) molecules usually described as hormones and neurotransmitters also bind to specific receptors on immunocytes and modulate their activity; (2) soluble

products of the immune system (i.e., cytokines) can act on cells of the neuroendocrine system, modifying the latter's functions; (3) immune stimuli and hypothalamic releasing hormones both induce lymphoid cells to synthesize neuropeptides that, in turn, might influence the activity of the neuroendocrine system; (4) cytokines and cytokine-like peptides that are potentially able to modulate immune cell activity are produced by cells of the nervous system.[47]

The similarities and the interactions between the immune and the neurohormonal systems are so close that the distinctions between hormones, neurotransmitters, and cytokines are open to debate, as are the distinctions between immunocytes and neuroendocrine cells. The close connections between the systems have led to the repeated suggestion (already mentioned in chapter 6) that neurons and immune cells evolved from a single multifunctional cell type and that their communalities are the foundation for their close interactions during neural development and learning.[48]

In animals, the vertebrate stress response is the best understood. The main parts of the CNS involved in the response are the hypothalamus and locus ceruleus, and peripheral contributions come from sections of the hypothalamic-pituitary-adrenal axis and the sympathetic and parasympathetic systems. As they respond to stress, a sequence of events and processes, which have been divided into four phases, occur in mammals and, with some variations, in other vertebrates:[49]

1. CRH (corticotropin-releasing hormone) and AVP (arginine-vasopressin) are released by the hypothalamus into blood in the brain.

2. Noradrenergic cells in the locus ceruleus and adrenal medulla are activated and trigger an alarm reaction; they produce norepinephrine that stimulates autonomic and neuroendocrine responses.

3. Triggering the rapid fight-or-flight response involves the sympathetic nervous system (with a concomitant silencing of the parasympathetic nervous system). The response includes physiological changes in the endocrine, cardiovascular, respiratory, gastrointestinal, and renal systems.

4. The release of CRH and AVP activates the rest of the hypothalamic-pituitary-adrenal axis. As a result, the adrenocorticotropic hormone is released from the pituitary into the general bloodstream; this in turn results in the secretion of glucocorticoids (GCs; e.g., cortisol) from the adrenal cortex. GCs produce an array of effects in response to stress, including cytokine inhibition, which accounts for many of the inhibitory effects on the immune response during stress. However, GCs eventually contribute to ending the response via inhibitory feedback.

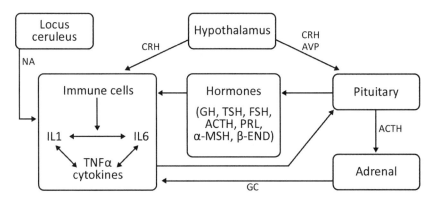

Figure 9.7

Stress axis and main neuro-immuno-endocrine interactions. CRH, corticotropin-releasing hormone; AVP, arginine vasopressin; ACTH, adrenocorticotropic hormone; GC, glucocorticoids; NA, noradrenaline; IL1 and IL6, interleukins (cytokines, small proteins involved in signaling); TNFα, tumor necrosis factor α (a cytokine involved in systemic inflammation); GH, growth hormone; TSH, thyroid-stimulating hormone; FSH, follicle-stimulating hormone; PRL, prolactin (luteotropic hormone); α-MSH, alpha melanocyte—stimulating hormone; β-END, beta endorphin.

Figure 9.7 summarizes some of these processes. Not surprisingly, given the many similarities in the functions of their nervous systems and immune systems, invertebrates such as arthropods and mollusks, which do not possess a hypothalamus or pituitary or adrenal glands, nevertheless show parallel neuroendocrine activities that participate in the response to stress and have similarly organized dopamine-glutamate circuits that are implicated in reward and aversion.[50] Like vertebrates' nervous systems, the invertebrates' nervous and endocrine systems interact intimately with one another, and the immune cells of invertebrates (circulating hemocytes) contain not only ACTH- and CRH-like materials but also biogenic amine neurotransmitters, including dopamine, noradrenaline, adrenaline, serotonin, and histamine.[51]

Neurohormonal Stress and the Evolution of Forgetting

> To think is to forget a difference, to generalize, to abstract. In the overly replete world of Funes there were nothing but details, almost contiguous details.
>
> —Jorge Luis Borges 1942/1962, p. 115

The Cambrian animals, with their enormously increased exposure to challenging stimuli, needed a flexible memory/forgetting system: as the Argentinean writer Jorge Luis Borges so eloquently noted and as the Russian neurobiologist Alexander Luria demonstrated with the case of his mnemonist patient, perfect memory can be a curse.[52] Forgetting is clearly not just passive decay, although decay happens.[53] Forgetting is a basic design feature required for flexible learning because brain space, especially in Cambrian animals, was limited and needed to be recycled as new learning occurred. But there is, as mnemonic pathologies and experimental interventions suggest, more to forgetting than this design constraint. A flexible nervous system requires that both memory and forgetting are regulated because an animal must choose among courses of action based on past experiences. This means that irrelevant information must be suppressed. The active regulation and restriction of the storage and the retrieval of engrams is therefore to be expected. The initial evidence from both behavioral and neurological studies showed active, retrieval-based forgetting resulting from interference during recall: percepts and actions presented to an animal immediately before or after a learning session trigger the forgetting of what was learned during that session—and so do traumatic experiences.[54] More recently, the active erosion of engrams, not just the failure to retrieve intact ones, is emerging as an additional facet of forgetting. In fact, it seems that when animals learn, both active memory and active forgetting mechanisms are simultaneously recruited,[55] reflecting the intense selection/exclusion processes required for effective learning to occur.

We believe that the evolution of AL, forgetting, and the stress response are linked. It is likely that AL, which evolved in Cambrian metazoans, induced chronic stress due to overlearning and led to the coevolution with AL of the stress response and forgetting. The neuroendocrinologist Robert Sapolsky described the well-managed stress response in mammals as the reason "why zebras don't get ulcers" (the title of his book on the subject), in spite of the presence of menacing predators. He showed that if stress cannot be managed efficiently, it results in immunosuppression, sickness, and weakness.[56] Since extreme or prolonged stress is also related to memory and forgetting impairments, it is plausible that before the neurohormonal-immunological stress-managing systems evolved, stress-induced responses were less regulated, and stress had greater deleterious effects (figure 9.8). The Cambrian probably witnessed the first nervous breakdowns: the first Cambrian subjectively feeling animals would have been nervous wrecks, with the equivalents of paranoia and posttraumatic stress disorder (PTSD) dominating their anxious lives. Mechanisms at the cellular, neurohormonal, and immunological level that restricted the duration and extent of memory, that promoted active forgetting, and that controlled and limited

Figure 9.8
Living in terror. A small fish living in a world inhabited by predators (a young *Anomalocaris* surveys the area) suffers the debilitating effects of extreme and chronic fear, having an as-yet-imperfect control system for stress management.

arousal must have been selectively very advantageous for both the mental and physical health of the animals.

Affective states with different levels of arousal, such as fear, pain, hunger, excitement, and satisfaction, were part of what minimally conscious animals experienced as they evaluated the relation between the state of their body and the world. As we argued in the previous chapter, departures from homeostatic states were categorized and used as proxies or "rules of thumb" for directing the type of action necessary (search for food, hide, attack, mate), and the action-promoting mismatch between the out-of-homeostasis state and the homeostatic state felt like "wanting." These affective categories were central components of what we have called CSSs—categorizing sensory states that are based on value-laden, compound, interoceptive, proprioceptive, and exteroceptive signals. These CSSs—including the basic emotions of fear and rage; the intrinsic positive valence produced by motor exploration; and positive and negative visceral affects such as varieties of pain, deprivation, and satisfaction or pleasure—were constructed during the evolution of UAL Cambrian animals. Such affective states were (and remain) highly idiosyncratic, depending on the developmental history of the individual, but they were categorized into types of valenced systemic states (primordial emotions) that directed types of actions. They were the basis on which in some groups additional emotional and motivational states such as care, playfulness, and various social emotions were constructed during subsequent evolution.

Neurohormonal stresses induced by AL had additional effects that would have been subject to modification by selection. An important one was the epigenomic destabilization brought about by induced changes in gene expression. Widespread modifications in chromatin and protein structure and alterations in small RNA profiles would have occurred in all cells and tissues, including the nervous system and gametes.[57] Some of these epigenetic changes would have had persistent, and sometimes heritable, phenotypic effects.

There are good reasons for thinking that epigenetic inheritance had ancient origins, because epigenetic mechanisms are found in all living organisms. It has probably played a crucial role throughout evolutionary history because whenever ancestral conditions predict present conditions (a situation that is quite common), epigenetic inheritance can be advantageous. For example, in randomly and regularly fluctuating conditions, if the cycle of changes is longer than the generation time of the individual, offspring benefit if they inherit the parental epigenetic state.[58] Moreover, through epigenetic inheritance, ancestral environments can prime gene expression so that offspring can respond more effectively (e.g., more rapidly) than their ancestors to the same type of inducing conditions. The advantages of epigenetic priming in this case are similar to those of long-term memory following strong sensitization. Not surprisingly, the epigenetic mechanisms of ontogenetic neural memory and of transgenerational memory are similar.

Through the effects on the epigenome, AL may have had a direct effect on variation in Cambrian animals: neurohormonal stress may have increased the generation of new, heritable, epigenetic variations and hence increased evolvability. The initially unmanaged stress response could have contributed to the very rapid phylogenetic diversification through the generation of stress-induced heritable, selectable variation. Once mechanisms that modulated and restricted the scope and extent of the stress response evolved, the transmission of stress-induced variants (e.g., miRNAs) in the germ line would have been reduced, which may have been part of the reason for the post-Cambrian slowdown of diversification.[59]

As table 9.1 shows, there is no shortage of ideas about the nature and causes of the epigenomic and genomic changes involved in the morphological evolution of bilaterians during the Cambrian. If we look at these in the light of neural learning and neurally mediated stress responses, two features stand out. The first is that all the genomic networks and factors that have been suggested as contributors to the Cambrian explosion have a role in neural development and/or learning. For example, Hox gene networks, which appear to be important molecular players in the Cambrian explosion, are also key players in neural development;[60] heat-shock proteins, which cells produce in response to exposure to heat and other stressful conditions, are expressed in the brain and are involved in neural development and cognition;[61] miRNAs that are expressed in the brain play a role in AL, and their dysregulation following trauma leads to pathologies such as PTSD or chronic stress disorder (as do all the epigenetic processes and factors discussed in chapter 7, such as those involved in DNA methylation and histone modifications). There is also increasing evidence that in present-day animals (e.g., mammals and insects), stress-associated learning, such as aversion learning following a painful encounter, is remembered for a very long time and involves epigenetic repatterning in the cells of the nervous system.[62] Some observations even suggest that there are changes in the DNA sequence in neurons as a result of stress-related learning.[63]

The second observation that table 9.1 highlights is that gene duplications, modifications of gene regulatory networks and cisregulatory elements, new miRNA sequences, and the other factors that form part of the genomic changes associated with the Cambrian can all be consequences of responses to stress. One of the first people to recognize the relevance of stress to evolutionary change was Barbara McClintock. The lecture she delivered when she received the Nobel Prize in Physiology or Medicine in 1983 was titled "The Significance of Responses of the Genome to Challenge."[64] From her work, much of which was carried out before the chemical nature of the gene and its activities was known, she concluded that cells can actively sense and adjust to internal and external threats. They have programs that deal with inevitable, predictable problems such as heat shocks and chromosome breaks, but they also have ways of coping with unexpected and unfamiliar mishaps such as unrepaired chromosome damage, viral invasions, hybridization, and totally new environmental conditions. One of McClintock's many discoveries about the genome was that chromosomes contained transposons, genetic elements that are capable of moving from one site to another, especially in stressful conditions. We now know that the rate of transposition is increased under stress by dysregulating epigenetic mechanisms such as DNA methylation and RNA-mediated silencing, which usually keep transposition under control. The stress-induced movement of transposable elements alters genome organization in ways that can generate new genes and new regulatory systems. Bursts of transposition can therefore lead to the rapid diversification of lineages and adaptive radiations. According to McClintock, "stress, and the genome's reaction to it, may underlie many formations of new species,"[65] and as table 9.1 shows, it has been suggested as one of the factors that drove the Cambrian explosion.[66] So, too, have chance events such as hybridization, chromosome doubling, and the transfer of genes between species by parasites. In the distant past, just as today, events like these probably led to major genetic and epigenetic repatterning because the regulatory activity of the RNAi machinery and chromatin-marking systems were compromised.[67]

Even evolved systems, such as the heat-shock response, which deal with the more predictable problems cells face, may have contributed to the genomic reorganizations that took place in the Cambrian. The heat-shock response is ancient and universal; it is triggered by various stresses, such as high or low temperatures, exposure to toxins, salinity, or hypoxia, all of which upset protein folding. Misfolded proteins cannot carry out their usual functions and often form aggregates in the cell. However, their presence induces a rapid, transient increase in the production of HSPs, some of which are chaperones that refold misfolded proteins while others clear away irreversibly damaged molecules and aggregates. Jeff Schwartz and Bruno Maresca have suggested that when stress exceeds the capacity of HSPs to effectively refold misfolded proteins and disaggregate abnormal protein complexes, proteins involved in DNA replication and repair may retain their abnormal conformations and cause mutations in both soma and germ cells, leading to a greatly enhanced mutation rate.[68] Michael Baker took a somewhat different view of the significance of HSPs in the Cambrian explosion.[69] He pointed to the evidence that in normal conditions the presence in a population of genetic variants of proteins is often masked because the chaperone HSP90 enables them to fold in a way that allows them to do their job. In times of stress, however, HSP90 is diverted into refolding stress-damaged proteins, so the variant proteins may adopt abnormal conformations that affect phenotypes. Formerly cryptic genetic variation is thus exposed to selection. According to Baker, through their effect on the role of HSP90, the earth's thermal fluctuations before the Cambrian generated new, selectable phenotypic variation by revealing genetic diversity that already existed.

In multicellular organisms, the selection of stress-induced ontogenetic variations may have affected selection at the population level indirectly (through genetic accommodation processes) and also directly. The direct relation between developmental selection and population level natural selection and its role in the generation of evolutionary novelties in response to new stressful challenges has been explored in yeast by Erez Braun and his colleagues. They argued, and showed experimentally, that novel stressful conditions can launch an intraorganism exploration process in which many gene expression patterns are tried out, but only the functional ones eventually become stabilized and inherited between generations.[70] This type of intraorganism exploration and selective-stabilization process, which has been also found in fruit flies, may be common during the initial stages of adaptation to new environments, where it would occur alongside conventional interorganism selection.

In neural animals, stress is always mediated by the neurohormonal system and has effects on both somatic and germ cells.[71] Whatever the causes of the stress and by whatever mechanisms it affects the genome and the epigenome, these effects will be evolutionarily significant only if they are transmitted to later generations. As yet, information about the directly inherited transgenerational effects of stress is limited to well-studied animals like nematodes, fruit flies, and mammals. In mice, chronic maternal separation, extreme fear (exposure to the smell and sight of an interested cat), or a painful shock to the foot (associated with the neutral smell of cherry blossoms) all lead to the transgenerational transmission of the effect of stress to unexposed offspring or grand-offspring.[72] Similar inherited effects have been found in other animals, including humans.[73] If neurohormonal, as yet poorly regulated stress was rife during the Cambrian, aversive learning may have triggered not only persistent genetic and epigenetic changes in nerve cells but also, as a by-product and in parallel, common and extensive changes in the germ line too, rendering the offspring of stressed UAL ancestors anxious and neurotic. Regulating such transgenerational effects by limiting the duration of stress through active forgetting or by preferentially transmitting compensatory effects that lead to stress reduction would clearly have been advantageous. The latter effect has been demonstrated in psychologically stressed mice. Isabelle Mansuy and her colleagues showed that the offspring of stressed male mice were more stress-resistant than their fathers when challenged by an acute cold-swim stress: the concentration of N-acetylaspartate, a putative marker of neuronal health, which was reduced in their fathers, had returned to normal in their cortex and hippocampus. Moreover, the level of this marker was not reduced (as expected in control mice) by acute cold stress.[74]

What, then, was targeted and honed by selection for regulating and reducing neurohormonal stress? The various neurotransmitters and regulatory interactions described in figure 9.7 (which all have equivalents in arthropods) affect memory and learning, so they could clearly have been targets of selection. For example, the negative feedback that GCs exert on their own production is expected to have been under positive selection during the Cambrian. There are, of course, many other potential targets of selection. For example, in the crab *Chasmagnathus*, octopamine administration impairs aversive memory reconsolidation,[75] so we would expect that the duration and extent of octopamine secretion under stress would have been targets of selection for the not-too-long retention of aversive memory. Similarly, since serotonin is instrumental in reducing poststress anxiety in mammals (and presumably in other animal taxa as well),[76] selection for an increase in serotonin secretion following stress probably occurred during the Cambrian. Our more specific suggestion is that selection for active-forgetting mechanisms operated during this period and that the stress-response system was intimately involved in this process.

There is some experimental support for the proposal that the stress response and memory/forgetting are related. In mice, the c-Jun N-terminal kinases (JNKs; proteins that belong to the conserved MAPK family involved in the stress response) that are expressed in the hippocampus are activated after extreme stress and lead to impaired fear memory; memory is enhanced—in both extreme and mild stress conditions—by JNK inhibition. In nematode worms, when the gene for JNK-1 is mutated, the worm retains its memories, so it is clear that the normal state of the worm is one in which memories are (eventually) forgotten. Active forgetting in nematodes is also under the control of factors that are involved in translational repression and the regulation of the actin cytoskeleton structure in neurons, so this regulatory system may also have been a target of selection for forgetting during the Cambrian.[77] Other factors affecting active forgetting are two signaling proteins: Rac, which is involved in actin remodeling in the mushroom body, and Cdc42, which affects filopodia extension in developing neurons and dendritic spine number in adults. In fruit flies, the rapid forgetting that follows stressful aversive learning requires the activation of the Rac gene, and its inactivation leads to memory retention. A decrease in *Cdc42* gene expression prolongs memory.[78]

Another target of selection for optimal forgetting may have been the regulatory factors that control heat-shock proteins' activity—the heat-shock chaperone proteins that are involved in the folding of proteins and the disaggregation of protein aggregates. Kyle Patton suggested that heat-shock proteins were recruited during the evolution of learning and became part of the control of memory and forgetting at the synaptic level.[79] In previous chapters we described how cytoplasmic polyadenylation element-binding

(CPEB) protein, a protein essential for long-term memory in many taxa, may be involved in the retention of memory because when in its aggregated, self-templating, prionic conformation, it activates the translation of synaptic proteins and leads to the persistence of synaptic activity. Patton suggested that the extent of disaggregation of the prionic form by chaperon proteins such as the HSPs may control the level of retention and "forgetting" at the cell level, and this cellular remembering/forgetting is reflected in these functions at the behavioral level. It is telling that neurons secrete serotonin in response to heat in the nematode worm *Caenorhabditis elegans* (*C. elegans*), and this affects both remote somatic and germ line cells. Serotonin, as we indicated, is implicated in the stress response of animals and is intimately involved in aversive learning, with its secretion increased when animals expect adverse conditions. The release of serotonin is enhanced after a heat shock through the activity of thermosensory neurons, and the secreted serotonin activates HSF1 (heat shock factor 1) in remote tissues, including gametes. The neurobiologists who made this discovery propose that this is likely to be a general mechanism triggering the activity of HSF1 in many noxious conditions, not just during heat stress.[80] We have argued that a major recurring condition was neurohormonal stress resulting from AL. We find the protective effect of HSF triggering in gametes particularly interesting, because if stress leads to protein misfolding that renders gametes defective, the germ line would be a prime target for protection by this mechanism.

In chapters 7 and 8, we discussed some of the evidence showing that small RNAs are involved in the formation of memories. Small RNAs also ameliorate the effects of stress: they play a role in the canalization of development and protect the germ line from transposition (which is increased under stress).[81] Controlling the biogenesis of small RNAs in neurons and their transport via exosomes may therefore have been an important target of selection in the stressful conditions of the Cambrian. Similarly, since chromatin-modifying factors such as DNA methylases and histone acetylases and deacetylases are involved in different stages of memory formation under stressful learning conditions, the regulation of their activity was probably another target of selection during the Cambrian.[82]

Sleep, as we argued in chapter 7, may have been particularly important in animals that manifest AL. Although sleep is multifunctional, there is growing evidence from studies in both vertebrates and invertebrates that it plays an important role in memory consolidation and forgetting irrelevant information, that it reduces stress, and that sleep impairments lead to learning impairments.[83] We know that animals that have highly centralized nervous systems and can sleep are able to learn by association, although some animals—schooling fish that allegedly do not sleep—have alternative mechanisms that enable memory consolidation.[84] Furthermore, animals with UAL not only sleep but have dedicated brain areas (the mushroom body in insects and the hippocampus in vertebrates) that regulate both sleep and memory.[85] We therefore suggest that mechanisms linking sleep with selective forgetting were important targets of selection when AL evolved. Many of the molecular factors that we have discussed have probably been involved in the evolutionary elaboration of sleep in animals that manifest AL, including the factors and mechanisms underlying both retrieval-based and storage-based forgetting.

Our suggestion that the evolution of AL during the Cambrian led to cognitively handicapping overlearning, to emotional stress, and to epigenomic destabilization, and that these side effects of AL not only promoted selection for perceptual fusion but also contributed to selection for active forgetting, has some testable predictions. First, we predict that changes in DNA sequences involved in the regulation of the neurohormonal stress responses, in particular in the regulation of active forgetting, will be found to date to the Cambrian in all groups of associatively learning animals. Second, since the immune and neurohormonal systems are intimately linked, new immune mechanisms are expected to have coevolved with AL and brain centralization and may have been lost in animals, such as penis worms, which seem to have lost their brain during evolution—something that can be tested through comparative molecular analysis. Third, the function of sleep as a regulator of memory and learning is assumed to have evolved during the Cambrian, and we expect variations in the anatomical and molecular factors underlying this function to be traced back to this era. Fourth, we expect that interfering with the mechanisms regulating the stress response will lead to a greatly reduced ability to control memory and active forgetting and that this will be reflected not only in altered gene expression patterns but also in the disordered formation of chromatin marks and unregulated transposition. Fifth, neurohormonal destabilization following extreme trauma or persistent stress is expected to have transgenerational effects at the epigenetic level.[86]

The evolution of AL, and especially of UAL, had many consequences, driving ecological specialization, arms races, and coevolutionary correspondences among different systems. If it indeed was, as we have suggested here, one of the factors driving the Cambrian explosion, then it has shaped the grand pattern of animal phylogeny. But the story of the evolution of complex cognition only starts in the Cambrian—it is the first step in the exuberant diversification of minds and of modes of being. The next major transition in the evolution of learning, cognition, and consciousness, which was built on the foundation of UAL and minimal consciousness and was continuous with it, was the evolution of Popperian animals—of animals endowed with imagination.

The Evolution of Popperian Animals

> Think left and think right and think low and think high,
> Oh, the thinks you can think up if only you try!
> —Dr. Seuss 1975/2003

The evolution of animal imagination requires a book-length treatise, going far beyond the scope of this already lengthy book. Here, in our very short excursion into the topic, we explore the idea that the evolution of new learning and memory systems, based on the UAL functional architecture but extending it further through the construction of additional levels of representation, brought forth imaginative souls with richer consciousness and cognition.

Before we start, we need to clarify how we use the term "imagining." By imagining we mean the off-line (currently stimulus-independent) recombination and transformation of facets of episodes that were experienced in the past and that lead to the construction of new virtual episodes. Animals are engaged in imagining when they plan (imagining is a necessary, although not sufficient, condition for planning) and when they dream (where it seems to be sufficient). Although we usually do not think that our recall of the past involves imagining, the reconstruction processes occurring during the recollection of past episodes and when imagining future ones are, in fact, very similar.[87] As we have done throughout this book, we will focus on the evolution of learning and memory systems—in this case on the evolution of processes that led to the ability to recall episodes, which is required for imagining. We suggest that the evolution of this memory system coevolved (but does not necessarily codevelop) with other cognitive capacities, such as dreaming and new emotions. The evolution of such systems of learning and memory, and the enriched consciousness that accompanies them, took place in several taxa at different periods following the Cambrian.

The Episodic-Like Memory System

Nearly fifty years ago, Endel Tulving suggested that there are two distinct, although intimately related, types of declarative ("knowing what") memory systems. The first is semantic memory, the memory of the facts of the matter ("Nuts are in the hollow of the oak tree."); the second is episodic memory, the memory of events and the spatial and temporal context in which they happened ("As I arrived to this wood early in the morning today, I saw X hiding nuts in the hollow in the oak tree.").[88] Such episodic memory, which has been delineated and characterized in humans by a combination of behavioral and neuroimaging studies and through studies of patients with brain damage, is accompanied by what Tulving calls "autonoetic consciousness"—the conscious experience of recollecting. Tulving believes that this phenomenological aspect of episodic memory is unique to humans. In his own words:

> Episodic memory is a recently evolved, late-developing, and early-deteriorating past-oriented memory system, more vulnerable than other memory systems to neuronal dysfunction, and probably unique to humans. It makes possible mental time travel through subjective time, from the present to the past, thus allowing one to re-experience, through autonoetic awareness, one's own previous experiences. Its operations require, but go beyond, the semantic memory system. Retrieving information from episodic memory (remembering or conscious recollection) is contingent on the establishment of a special mental set, dubbed episodic "retrieval mode." Episodic memory is subserved by a widely distributed network of cortical and subcortical regions that overlaps with but also extends beyond the networks subserving other memory systems. The essence of episodic memory lies in the conjunction of three concepts—self, autonoetic awareness, and subjectively sensed time. (Tulving 2002, p. 5)

The study of the autonoetic aspect of episodic memory in nonhuman animals, which cannot linguistically report about their mental states during recollection, is clearly very difficult. However, when many aspects of a past episode are recalled, the richness of the recollection can give researchers clues about the memory and learning system of the animal and possibly about the likely first-person mental state involved in the process. This type of multifactor, narrative-like recollection in nonhuman animals has been studied extensively in scrub jays by Clayton, Dickinson, and their colleagues and is referred to as "episodic-like memory," which for brevity we shall call ELM. As Clayton, Bussey, and Dickinson pointed out, for ELM to be inferred, three behavioral criteria have to be satisfied: (1) the recollection has rich *content*; it has information about *what* happened, *where*, and *when* (on the basis of a specific past experience), often referred to as the

"*WWW* requirement"; (2) an *integrated representation* of these component parts of the past experience is formed; this means that the different *W* facets are bound together into a single representation and that any one facet of the episode will automatically retrieve the others (completion), allowing discrimination between episodes that share some of the same features; and (3) there is *flexibility*—that is, a flexible deployment of the information acquired so that a component part (e.g., a representation of "when") will be differently deployed in different contexts.[89]

The design of the experiments with scrub jays that the Dickinson group carried out and we briefly describe here has subsequently been used (with appropriate modifications) to study episodic-like memory in other species. Western scrub jays, which are food-caching birds, cached perishable worms (which they prefer) and nonperishable peanuts (second-best) in distinct sites in sand-filled caching trays and were allowed to recover their caches later (figure 9.9). If a short time had elapsed between caching and recovery, the jays were apparently aware that their preferred worms were still fresh because they tended to recover the worms. But if a long time had elapsed between caching and recovery, the birds apparently knew the worms had decayed, since they preferred to recover the peanuts. The jays could also update their knowledge about the decay rate of worms if that information was changed (by the manipulating experimenter). Furthermore, the birds also showed evidence of future planning based on experience: scrub jays that stole another bird's caches (thieving jays) subsequently recached food in new sites but only when they had been observed caching by other jays. Nonthieving jays did not; jays apparently extrapolate about the future actions of others based on their own practices. Clearly, scrub jays know what, where, and when something is happening; can bind the different facets of the episode into one recollectable representation; and can deploy each facet flexibly. Hence, they satisfy all three behavioral criteria for ELM.

Scrub jays are not alone in having ELM, although for most other species we have a far less rich and detailed picture. Other species of bird (e.g., magpies, black-capped chickadees, and hummingbirds), fish (e.g., wrasse and zebrafish), mammals (e.g., voles, rats, mice, pigs, monkeys, and apes) and invertebrates such as cuttlefish and bees show many aspects of ELM, and we suspect that this is the tip of a large iceberg.[90] It is interesting to note that spatial learning, which these ELM-exhibiting species seem to excel at, especially learning-based navigation using external and self-learned actions as cues, not only requires something like a cognitive map but also enhances the sense of time emanating from the self-learning of action sequences.

A   B

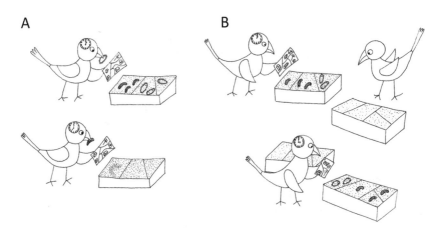

Figure 9.9
Western scrub jays in action. *A*, a jay remembers where and when it stores nuts and worms (*top*), which decay at different rates, and because a long time had elapsed since hiding, retrieves only the durable nuts when it gets back to the tray (*bottom*). *B*, the jay remembers where it hid the nuts and the worms (*top*) and changes the hiding place after being observed by a potentially thieving conspecific (*bottom*). *Source*: Based on the cartoons in Clayton, Bussey, and Dickinson 2003, p. 688, p. 690.

From our point of view, the relation between episodic-like recollection and future planning is of special interest. This is not just because future planning and recollection are closely related in humans.[91] It is because we are interested here in the evolution of a new type of consciousness that builds on, but goes beyond, the minimal consciousness that we have described up to this point. With ELM and future planning, imagination can be inferred, and with its emergence a more advanced kind of consciousness, that of Popperian animals, has made its evolutionary debut.[92]

Building on the UAL Model: The Functional Organization and the Evolution of ELM

ELM has been reported in some but not all animals with UAL: in the vertebrates it is found in mammals, birds, and some fish; in arthropods it is found in some social insects like the honeybee; and in the mollusks it was demonstrated in cuttlefish (but is probably also present in squid and octopuses). Although the evolutionary continuity between UAL and ELM is obvious, there is a big difference between the two. With UAL, committing compound percepts or action programs to memory always requires reward or punishment, whereas the commitment of episodes to memory does not always require such direct reinforcement. However, as we noted in the previous chapter, online updating, which is the basis of all perception, requires working memory, which may generate ephemeral engrams even for non–learning reinforced compounds, especially when the percept is attended to. The perception of episodes—complex changing scenes—requires substantially greater integration, so information needs to persist for a longer time for the representation to be formed. Past episodes are therefore likely to leave memory traces even when the episode is not directly reinforced by learning, but is attended to. Yet although a UAL model cannot account for ELM, it is, we argue, the basis on which an ELM architecture has been evolutionarily constructed.

The three behavioral criteria of full-blown ELM—the *WWW* content; the binding of the different *W* facets; and their flexible, context-dependent deployment—suggest how the UAL functional architecture needed to be extended to accommodate ELM. The first thing to notice is that the different *WWW* facets are types of compounds: compound objects (e.g., nuts) are *What* facets, while the compound *Where* facet is the spatial context (e.g., the hollow of the tree on the left side). The *When* facet requires either the representation of relations of before and after (e.g., X happened/was perceived before Y) or contextual temporal cues (e.g., the sun was setting as Z happened).[93] For ELM all of these compound representations of percepts and actions and their specific relations need to become bound together in a single metarepresentation, which can be reconstructed on the basis of partial (compound) input. In terms of our UAL model, several sensory-integrating units (SIUs), several motor-integrating units (MIUs), and the relations between them need to be represented in the integrating-memory unit (MEMU). The integration of the relations among these representations requires a new hierarchical level of mapping in the MEMU, so the integration that occurs during encoding in the association unit (AU) is mapped back to a new hierarchical level in the MEMU. The AU is differentiated into two areas: an AU where the episode is constructed through interaction with the higher level MEMU and an executive AU, where the planned action is constructed and also interacts with the MEMU (figure 9.10). This last interaction has an interesting and important consequence because the AU is the neural locus in which the representation of self is formed during encoding and action planning. The representation of the self at a higher MEMU hierarchical level means that the self is not only experienced but also memorized and may be recalled. Animals with ELM have a metarepresentation of the self and therefore of the body and the world in which this self is embodied, so they are

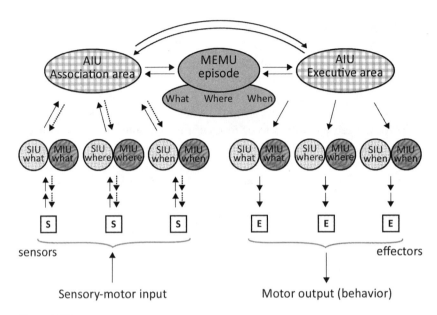

Figure 9.10

A schematic illustration of episodic-like memory (ELM), which adds another hierarchical level of integration and representation to the functional model of UAL presented in figure 8.2. Each of the sensory- and memory-integrating units (*SIU and MIU, twin circles*) corresponds to the primary association area (AU) in the UAL model and includes both sensory and action aspects of *WWW* (*What* one perceived and *What* one did; *Where* one perceived and *Where* one acted; *When* one perceived and *When* one acted). The additional hierarchical level includes the second-order sensory-motor integrating association area (*AIU, left*), the second-order executive associative area (*AIU, right*), and the additional (*uppermost level*) memory integration area in the MEMU. Note that the episode is constructed through interactions between the second-order association areas and the hierarchically organized memory system. Within the MEMU, there is both integration and differentiation of the three *W* facets of experience. As recall occurs, and as action is planned in the executive area, memory is reconstructed. As with UAL, we are not committed to the anatomical identity of the functional units, which are distributed. Top-down dashed arrows transmit priors, or PEs.

able to engage in self-monitoring during planning, as well as online. ELM is reinforced whenever the animal is attending to the episode, so reinforcement accompanies attended episodes rather than follows them: attending to episodes is inherently reinforcing. And since in animals with ELM temporal thickness is extended, these animals have a working memory that can sustain information for longer periods; trace conditioning in them should enable a longer gap between the episodic CS and its reinforcement, and the "specious present" of the encoded episode is greater than that of animals with elementary UAL.

The type of consciousness that animals with ELM display is therefore richer than that of animals with UAL: the complexity and range of representations are greater, as are the time depth and the sense of self. All these aspects of episodic representation require consciousness during the encoding phase, and in this sense it is similar to the kind of consciousness required during encoding a unit compound in UAL. The main difference, however, is not only the richness of the experience but also the fact that, in contrast to UAL, consciousness is also required during the recall phase. When there is episodic recall, the representation is reconstructed (reencoded), and this time-consuming reencoding process is experienced. Hence, we expect that when animals are engaged in a memory task requiring complex episodic reconstruction, an experimental masking of the reconstruction-triggering stimulus would fail to lead to adaptive behavior (there will be no successful retrieval of the nonperishable nuts that were cached a week ago in tray 3). We are not implying that the consciously recalling animal necessarily relives its past caching experience: it may well be that the recall is only richly semantic (the jay has semantic knowledge of where and when and what it cached but does not relive the past experience). We do believe, however, that there may be selective benefit to the animal if episodic recall involves reliving past experiences. When comparing rhesus monkeys who had witnessed one of a number of tubes being filled with food with monkeys who did not know which tube contained food, it was found that the monkeys who had not seen the food placed into the tube—and therefore had no memory of it—were more likely to look into the tubes first. Such awareness of what one remembers can plausibly be explained in terms of the conscious recollection of past experiences.[94] However, episodic recall may have a cost, because it involves inferential processes, relying on generally (but not always) reliable contextual cues, and can therefore occasionally lead to false memories.[95] We therefore also believe that when the episode is rich and details such as "X cached the worms in tray A earlier today, and Y was in the area; the nuts were cached a few days ago in tray D and no one was watching" are necessary to direct successful action choice, it is useful to reexperience the series of actions and factors that led to the caching—it is a way of keeping this information tractable and more trustworthy. With such episodes, semantic recall (even probabilistic semantic knowledge, incorporating uncertainty) may not be sufficient. However, perhaps counterintuitively, reliving the past is even more necessary when one plans complex future events, even though the emotional aspects of past experiencing, which may be overwhelming, are likely to be partially sup-pressed during planning. With future planning, the flexibly bound *WWW* facets of the virtual episode require off-line manipulation and (sometimes) self-monitoring,[96] and it is difficult to see how episodic counterfactuals can be generated without a measure of consciously experienced reliving of past episodes from which only some facets of *WWW* can be extracted.

The Evolution of ELM

The extension of the UAL functional architecture and the emergence of mental metarepresentations were driven in different taxa by social- and ecological-specific demands that led to a buildup of bound, possibly domain-specific constellations of different aspects of *WWW*. It is also likely that the integration of *What* and *Where* aspects of an experience preceded the inte-gration of the *When* facet into the episode. Hence, like others,[97] we expect that partial or proto-ELM will be found in different species, although we are not aware of comparative studies that have systematically looked for partial ELM. The list of species in which ELM has been found suggests that it evolved in parallel in some groups of vertebrates (birds, mammals, some bony fish), arthropods (honeybees), and cephalopods (cuttlefish). Whether or not it evolved in parallel in birds, mammals, and other vertebrates (some fish and maybe some reptiles), or whether it evolved in the common ances-tors of birds and mammals and independently in some fish, is at present not clear.[98] In mammals, ELM is subserved by the prefrontal cortex, cortical association areas, and the parahippocampal and hippocampal areas. There is evidence for *What-Where* coding in the CA3 area and for *What-When* cod-ing in the CA1 area of the hippocampus.[99] In birds, parallel structures play a role in ELM, including the hyperpallium, the dorsal ventricular ridge (DVR) association areas, and specific parahippocampal and hippocampal areas.[100]

It is difficult to produce a single list of obligatory conditions for the evolution of the ELM system. Sociality seems to be related to ELM in both vertebrates and arthropods: in arthropods ELM has been reported for the honeybee, all mammals are social in the fundamental sense that a milk-providing mother is the definition of the taxon, and there is no bird hatchling that does not require parental care. Similarly, the cleaner wrasse engages in providing social services to needy clients, and zebrafish aggre-gate in shoals. It is therefore plausible that social learning and the increased discrimination of self from social others is conducive to the evolution of ELM and leads to an enhanced representation of self as an agent in the world. However, the supposedly asocial cephalopods seem to break the rules, although their multiple, complex relations with their many predators and prey may be construed as social and require the acquisition of flexible, context-specific information.[101] Planning for the future and inferring relations between different *WWW* aspects of an episode on the basis of past experience would be useful to all animals living in complex ecological conditions that involve interactions with other animals. These capacities, which require discriminating between past (recalled) and planned (future-oriented) episodes, may have spurred the evolution of self-monitoring. We assume that learning under safe, parentally controlled conditions allowed young, neurally plastic animals opportunities to learn from knowledgeable adults and siblings.[102] Such repeated individual and social-learning opportunities gave the young better chances of learning about the different *W* aspects of their social and ecological environment. There would have been strong selection for early learning processes that allowed the encoding and remembering of episodes. Once again, the evolutionary dynamics that led to progressive changes in ELM were probably initiated by phenotype-first developmental changes, based on early behavioral explorations and selective stabilization.

Our hypothesis also leads us to expect coevolutionary relations between episodic-like memory, dreaming, and playing. As we see it, dreaming (figure 9.11) is an inevitable by-product of ELM that was honed by natural selection to control memory and forgetting. Dreams are semirandom combinations of facets of *WWW* that were experienced in the past, and although animals cannot tell us what they dream about (though sometimes we may guess), we suggest that all ELM animals dream, including those without rapid-eye movement (REM) sleep. (REM sleep characterizes the homeothermic dreaming birds and mammals, but it is not obligatory for dreaming to occur.)[103] In humans, dreams are typically forgotten, pointing to sleep's role in ensuring the forgetting of episodes that did not receive high value during waking time. Not remembering dreams also ensures that the virtual episodes experienced during dreaming will not be confused with waking experiences, so the forgetting of dreams was presumably enhanced by positive selection for forgetting. In other words, since imagining during sleep might lead to actions that are guided by scrambled representations, the evolution of imagination may have led to positive selection for forgetting dreams. If so, we expect to find dream-forgetting (with the exception of strong, emotionally reinforced dream episodes) in nonhuman animals too. On the other hand, remembering an episode that was strongly reinforced when the animal experienced it, such as an episode that led to strong pain, is further reinforced, since the already strengthened connectivity patterns are more likely to be reactivated by stochastic neural activity during sleep.

Figure 9.11
A dreaming elephant.

In humans, rats, birds, and bees there is a replay of learned behaviors during sleep,[104] suggesting that dreams play a role in memory consolidation through reexperiencing.

Play in humans and other animals is an inherently rewarding, spontaneous, repetitive, and often extravagant pattern of behavior that occurs when the animal is healthy and unstressed. It is very often accompanied by positive affect—by the emotion of joyful exploration, or playfulness. Patrick Bateson and Paul Martin suggested that "playful play" leads to new forms of behavior in both humans and other species, generating novel ways of dealing with the environment, a few of which may turn out to be useful.[105] Interestingly, as pointed out by psychologist Kelly Bulkeley, playing and dreaming share many important features: both occur in a special environment, may involve strong emotion, and take their contents from the survival concerns of the animal yet tend to be exaggerated, suspending the rules of ordinary life.[106] Playing, like dreaming, is in this sense imaginative, and it is interesting that all taxa that display ELM, including fish, have also been reported to play.[107] As we noted in chapter 5, Cabanac suggested that play behavior is one of the characteristics of basic consciousness and that it emerged in the early amniotes. Although we have argued against the amniote origins of consciousness, we think that play (especially playful play) is one of the markers of Popperian consciousness and that the emotion of playfulness, which was highlighted by Panksepp as one of the basic emotions, is a new type of affect that appeared when imagination evolved. This is just one example of the coevolution of emotions and imagination: we believe that the control of emotions was enhanced in Popperian animals and that all basic emotions (caring may be a particularly cogent example) have been expanded in scope and intensity in imaginative animals through the greater crystallization of the sense of self and the distinction between self and others.

Play, dreaming, and new and more controlled and nuanced emotions (such as disappointment) are only part of the cognitive suite of characters of imaginative conscious animals. The ability to imitate both motor and vocal patterns of behavior and to developmentally construct a representation of the actions of others through a mirror neuron system are probably necessary aspects of ELM:[108] one needs to represent and recall not only one's own actions to plan future activity but also represent and recall the observed and imagined past actions of others. The ability to read other minds and to infer causal relations and monitor one's state of knowledge have all been influenced by ELM and may have coevolved with it in some groups, although they usually do not codevelop with them: with highly plastic, learning-dependent traits, ontogeny rarely recapitulates phylogeny. Hence, dissociation studies based on brain-damaged adults cannot be used (on their own) to infer phylogenetic relations.[109]

Does ELM require not only an enriched hierarchical processing architecture but also a new type of neural evolutionary principle that goes beyond the selective stabilization of neural patterns described by Edelman and Changeux? The theoretical biologist Eörs Szathmáry and his colleagues have suggested (and illustrated with computer simulations) that a solution to a problem that requires novel insight and imagination becomes possible when there is replication, multiplication, and selection of newly formed, variant neural patterns within the brain, a process they call "Darwinian neurodynamics."[110] Since ELM animals seem to be able to solve challenging problems by employing imagination, it would be interesting to see if they, unlike animals that manifest UAL but lack ELM, actually employ such neurodynamics when faced with insight-requiring problems. We add another speculation: the intracellular epigenetic marking that we suggested was recruited for encoding high-level features in animals that have UAL may have been used at higher-level association areas in ELM animals, forming long-lasting epigenetic engrams representing more abstract concepts and relations.

To do justice to what we know about ELM and its relations with other aspects of sophisticated cognition and consciousness, a far more detailed evolutionary account is needed than the sketchy outline just given. What we wanted to highlight here is that this richer form of consciousness is built on the minimal consciousness architecture that evolved through the selection for UAL. Using the same reasoning and the assumption of evolutionary continuity, we believe that the next major transition in the evolution of consciousness was an extension of the ELM architecture, when animals could not only imagine but could also communicate to others what they had imagined and instruct the imagination of one another. This occurred in our hominin lineage and led to the evolution of the rational soul. Like the evolution of Popperian organisms, the evolution of rational ones also deserves a separate book because it is one of the most intriguing and intensely researched topics in evolutionary biology. As we show in the next and final chapter, where we summarize our arguments and conclusions, the heuristics used for studying the sensitive soul can also be fruitfully employed to study the rational soul.

Darwin wrote in the last paragraph of *The Origin of Species*: "Thus, from the war of nature, from famine and death, the most exalted object of which we are capable of conceiving, namely, the production of the higher animals, directly follows."[111] It was in the Cambrian, we suggest, that the great war of nature began to involve suffering—the subjective experiencing of anguish and pain. It eventually led, somewhat paradoxically, not only to dreaming and imaginative animals but even to animals that collectively build utopias of a life without suffering.

10 The Golem's Predicament

Though I think not
To think about it,
I do think about it,
And shed tears
Thinking about it.

—Ryokan Taigu[1]

The last chapter of a book is an opportunity to scrutinize the approach taken and to reflect on how it illuminates related topics. Since writing this book took a very long time, during which we presented our arguments to colleagues, students, friends, and other patient souls, we have heard recurring questions and objections, as well as received many useful suggestions, some of which have been left unaddressed in the previous chapters. Philosophical discussions since Plato have often taken the form of a dialogue between intellectual adversaries or colleagues, and this was used by one of us (E.J.) in a book, *Evolution in Four Dimensions*, written with Marion Lamb. The adversary to the arguments in that book was a character called Ipcha Mistabra (I.M.), which is Aramaic for "the opposite conjecture," a term that embodies the argumentative dialogue style used in the Talmud and encourages critical thinking. I.M., like many imaginary figures, in time acquired a life of his own, and as we were writing this book, we had many imaginary discussions with this quarrelsome, highly educated, somewhat grouchy, and infinitely curious character (who also began, not surprisingly, to share some characteristics with the late Professor Leibowitz). We (WE) therefore decided to write the last chapter of this book in a dialogue style, with I.M. as our partner, in an attempt to address some of the many topics that were left out or that remain troublesome and unclear.

Back to Basic Assumptions

I.M.: I have some general questions concerning your evolutionary hypothesis, and I also want to more fully comprehend what follows from the seemingly simple and sensible assumption that is the basis of the whole argument. You assume that there is continuity between life and consciousness and that this observed continuity is the result of evolutionary processes. You have by no means exhausted the corollaries of this continuity, although you made some suggestive hints in your first chapter. I therefore would like you to expand on those. But first things first. I start with what I see as a fundamental question about your core assumption. I can accept that by characterizing unlimited associative learning (UAL), you have identified a transition to a new level of cognition. This is indeed convincing, but I find the link to consciousness an encumbrance. There will always be those who argue that consciousness requires more—or fewer—cognitive capacities, and of course there are those who maintain that cognition and consciousness belong to different categories, and no measure of cognitive research can explain consciousness. I know that your motivation was to unravel the dynamics of consciousness through an evolutionary approach and not to identify a transition in the evolution of cognition, but maybe, to turn a biblical allusion on its head, you went in search of kingship and found some nice, healthy donkeys.[2] That is not a bad achievement.

WE: The continuity between life and consciousness and between cognition and consciousness are topics we too would like to explore, but let us start with your first question. We are glad that you see UAL as a distinct level of cognition, but we insist that it is also an evolutionary marker of minimal consciousness. This is not vague intuition—we do argue the case. In fact, we had to change our own initial intuitions about the matter—not an easy thing, as you may appreciate. Our main argument is that one can reverse engineer an embodied, conscious, biological system from the functional architecture of UAL. True, we have not accomplished this—we do not have a Gantiesque model of minimal consciousness—but we have explained why UAL is a good candidate for an evolutionary transition marker, similar to the unlimited heredity that is necessary for the reconstruction of sustainable life. We argued that a functional, evolved, biological system that realizes UAL is both part of an enabling system that has all the characteristic properties of a system that can be said to be conscious and is a point of departure toward the reconstruction of such a system. Hence, we believe that with our UAL model we have made a first step toward a model of consciousness.

I.M.: But why should it be such an elaborate learning system? Why is limited associative learning (LAL) not enough for sentience? I thought you subscribed to Jacques Loeb's opinion that "consciousness is only a metaphysical term for the phenomena which are determined by associative memory."[3] Why do you deny minimal consciousness to the charming though somewhat cognitively limited *Aplysia*? Why does minimal consciousness require such heavy-duty functional architecture?

WE: We like Loeb's emphasis on the involvement of associative learning (AL) in consciousness, but we do not subscribe to his unqualified position. Unlike Loeb, we do not see consciousness as determined by or synonymous with AL. We see AL (UAL, to be precise) as an evolutionary marker for consciousness, and this is a very different thing. LAL seems to be insufficient because there is evidence that it can occur without consciousness—for example, in spinal animals and humans. Moreover, LAL does not satisfy the seven criteria for a minimally conscious system. In animals that show only LAL, there is no evidence of perceptual fusion, there is very limited temporal depth, the representational system is shallow, and the value system is inflexible. We cannot, as we stressed, be sure that *Aplysia* is devoid of minimal consciousness, but we have no good reason to attribute it to this animal. As we emphasized several times, having the capacity for UAL is a positive indicator of consciousness; its absence is no proof for the absence of consciousness. As to the simplicity or complexity of the functional architecture of consciousness, we do not expect a conscious system to have a simple functional cognitive architecture. Remember that a minimal living system, such as Gánti's chemoton, is not simple (although the toy model is, of course). The chemoton requires coupling and feedback among three chemically quite elaborate systems, each of which, like every complex chemical system, is stochastic and variable. Stochasticity, as we have stressed throughout this book, is a feature of living systems, not a bug.[4] The fundamental processes of exploration stabilization would be impossible without the variability generated by stochastic processes. The enabling system for living, which can be reverse engineered from the unlimited heredity subsystem, certainly has a demanding functional architecture and a very complex biochemical one. It is as simple as it can get, but not simpler.[5] The same holds, writ much, much larger, for UAL and its enabling system. Chalmers's claim that consciousness needs to be explained in terms of a physical primitive, like mass or electric charge, rendering a theory of consciousness more like "elegant physics" than "messy biology" is, we think, profoundly wrong.[6] Consciousness is not only based on "messy biology," it is superbly and uniquely organized messy biology.

I.M.: Any biological system is "messy" in this sense, I agree. However, your arguments are convincing only if I accept the basic premise that UAL is the evolutionary marker of minimal consciousness. Yet this is just what I question. Why is learning of compounds necessary? Why are compound percepts and action patterns that are not encoded in memory insufficient for generating consciousness? Surely one can experience such percepts without committing them to memory!

WE: Indeed one can. We do it all the time, and the example of those unfortunate victims of brain injury who have lost their memory but not their online experiencing testifies to it most dramatically. But wouldn't you agree that it is difficult to see how such a system could have evolved without assuming that it provides learning-related advantages? How can ontogenetically constructed, compound percepts and action patterns, which are practically unlimited in number, be assigned a value? If the environment was very stable, the ability to discriminate among compound stimuli online, without an off-line memory system, might indeed be an asset, but this is not the environment that mobile animals inhabit. Moreover, since learning appeared very early in evolution, the assumption that the evolution of learning drove compound perception and action is reasonable. Once in place, the enabling system of UAL allows conscious perception even without the animal actually exhibiting UAL, but this is not how it appeared in evolution. As we stressed, during early ontogeny conscious perception precedes UAL, and the baby learns about already consciously experienced and valued percepts. But it took a lot of evolution to get there. The conscious percepts are developmentally constructed through Bayesian learning processes based on the evolved enabling system of the baby. We still know little about the processes involved in this early developmental construction, and we think this is an important direction for research.

I.M.: Let me then ask you the complementary question. If I accept that the enabling system provides so much, then once there is an enabling system in place, why can't the animal learn about novel compound percepts and action patterns unconsciously? After all, a lot of learning goes on unconsciously.

WE: And a lot does not. The difference between conscious and unconscious learning is important. We assume that when there is a novel composite percept with high salience, consciousness is involved. Let's take two examples: a person seeing an elephant marching on Fifth Avenue, and a hungry cat stalking a nearby bird. In the elephant case, the novelty leads to a large prediction error in the amazed observer, which is reliably evaluated through the repeated and active sampling of this surprising percept, a process that is a feature of focused attention. In the case of the cat stalking the bird, the context-dependent behavior of the bird (which is never identical to past behaviors of stalked birds, though it may be similar) must be precisely evaluated to update the priors in order for it to lead to a successful capture and satisfy the cat's need for food. The cat is attentive and alert and monitors every move of the bird. The whole system is aroused through the generous release of neurotransmitters, such as dopamine. This arousal and the strong and time-consuming top-down and bottom-up activities of the network of relations among the high-level processing subunits (the sensory-integrating unit, the memory-integrating unit, the integrating association unit, and the dedicated memory unit subsystems) render this predictive processing (PP) conscious. It is a system-wide process, based on the animal's actions (repeated sampling of the percept), leading to integration, evaluation, and action selection from the point of view of an active, aroused, perceiving individual. Similarly, the initial construction of a complex train of actions (riding a bike), before it becomes a habit, requires alertness and consciousness. We expect that many compound percepts and actions can be learned unconsciously, but these will usually not include novelties that require discrimination. As we noted in chapter 8, this can be tested experimentally: we predict that UAL will not be observed under conditions of unconscious processing. We also predict that the neural processes and structures that are essential for consciousness in humans will also be necessary for their UAL and that the homologs or analogs of these structures and processes will be necessary for UAL in other animals. UAL is a *sufficient* (but not necessary) condition for minimal consciousness in *evolved extant* animals. It was, however, both necessary and sufficient when it first emerged during the Cambrian in vertebrates and arthropods and 250 million years later in coleoid cephalopods.

I.M.: I got the point about UAL being a positive marker of minimal consciousness in present-day animals. But can you point to a negative marker, one whose absence would suggest that an animal cannot be conscious? And a related question: Is there a gradation with respect to consciousness or would you say that there is a switch, a metaphorical lightbulb that is turned on or off?

WE: In extant animals there are many necessary conditions, specified in our list, that are individually necessarily and jointly sufficient for minimal consciousness and UAL. What prevents consciousness has been empirically studied only in mammals, especially adult humans. As we stressed, not all the conditions for UAL will be present in the conscious human neonate, who tries to make sense of the "blooming, buzzing confusion" of its world, to use James's phrase. We think that it is likely—not only for humans and other vertebrates but also for arthropods and cephalopods—that if a percept is presented very rapidly or if it is masked, there will be no conscious perception; if wide-spreading of activation and reentrant interactions are prevented as a percept is presented, there will be no consciousness of perceiving it; and if there is no hierarchical organization and PP, subjective experiencing will not occur. If this turns out to be the case—and this is testable not just in humans but in other animals too—one may be tempted to say that the absence of these processes marks the lack of minimal consciousness in present-day animals.

This, however, is not a satisfying answer because one may ask: How widespread must the activation be? How long must the neural processes persist? How deep must the hierarchy be? An answer to the negative marker question depends on how we answer your second question about the all-or-none nature of consciousness. It seems to most people that an animal is either conscious or not conscious (whatever the range and scope of conscious experience), just as an animal is either alive or not alive. It seems that there is a sharp threshold. But this intuition, as we argued in the first chapter, is misleading when we think about life. When considering the evolution of living, there is a gray area where the decision about the living or nonliving status of a complex chemical system is moot and is defined a priori by one's theory of life. If we consider the evolution of consciousness, we encounter a similar gray area. And the same is true for embryonic development. Gray areas are something we expect when taking an evolutionary or developmental perspective. The metaphor of the familiar lightbulb leads to an all-or-none position about the gradation question because the familiar lightbulb is either on or off. However, if you think about an exceedingly sensitive dimmer instead, we would not be so sure. Our answer then is that gradations are evolutionarily very likely, although threshold effects during development may occur. If a sharp threshold were to be found, it would be easier to find the negative marker for consciousness, though it might be different in different taxa. We think, however, that this is unlikely. If we are right and there is no sharp threshold, we shall not be able to describe a single negative marker simply because there is no such marker.

I.M.: Since you have come back to the PP framework that is central to your account, there is something that bothered me. On the one hand, the PP idea—indeed, all Bayesian learning—is representational. There is a complex, dynamic, neural representation of the world, of the body, and of the perspective-dependent relation between them, and the processing of this neural, brain-based model is, you tell us, experiencing. In fact, you tell me that your account fleshes out mental representation. On the other hand, you are very committed to the embodied approach to consciousness, and you emphasize time and again the role of mobility—of the sensorimotor coordination of moving about and the activity of the internal organs of the body, which, to use Maturana and Varela's apt phrase, "bring forth a world."[7] This, in my understanding, is a nonrepresentational approach. How can you have your cake and eat it too?

WE: Your question about representation and embodiment worries many of the people with whom we discussed our ideas. We suggested that the evolution of the nervous system was driven by the increased size of mobile animals. They required novel ways of coordinating the movements of tissues separated by long distances (accomplished by nerve-muscle interactions), getting enough food for such a big body (for which the motions of a throughput gut are required), and coordinating its development and growth. When viewed in this way, we hope you see that the PP model and the commitment to embodied and situated cognition, with a focus on motor exploration, are not contradictory. A crucial part of the PP framework and of our UAL architecture is the modeling of action and body. As we have argued throughout this book, if there is no body and no action, there is no consciousness because the active exploration and selection of actions and perceptual targets, and their stabilization through reinforcement, are central to an account of UAL. We do indeed assume that there must be neural representations, but these feed off the inputs from the body as it perceives and acts. There is no UAL without sensory integrating units that process and update and evaluate not just exteroceptive signals but also proprioceptive and interoceptive signals; nor is there learning without a motor system that provides some of the crucial inputs to other units. Inference, in the language of the PP modelers, is typically active. The philosopher of mind Andy Clark, who interprets all cognition in terms of PP, says that active inference "names the combined mechanism by which perceptual and motor systems conspire to reduce prediction error using the twin strategies of altering predictions to fit the world, and altering the world to fit the predictions. This general schema may also—perhaps more transparently—be labelled 'action-oriented predictive processing.'"[8] This is well put, and like Clark, we see a necessary complementarity between this form of representation and embodiment, rather than a contradiction. We think that without UAL-type neural modeling by the CNS (the central nervous system, which is, one should not forget, part of the body) and the strange attractors (or strange loops, as Hofstadter would put it) thus generated, there would be no agency and no consciousness. And without the additional interactions between an active body-brain and the world, the world cannot be made to appear: there would be nothing to update and to model, and no consciousness would be possible. That is why we regard the idea of the "brain in a vat" as an empty fantasy.

Cognitive Cells and Conscious Living

> Life and mind share a common pattern of organisation, and the organisational properties characteristic of mind are an enriched version of those fundamental to life. Mind is life-like. But a simpler and more provocative formulation is this one: Living is cognition.
>
> —Thompson 2009, p. 81

I.M.: The constitutive role of cognition in the construction of consciousness is central to your view. Hence, understanding the role of cognition in nonneural organisms and grasping why it cannot constitute consciousness in such creatures seem important. Your list of criteria for minimal consciousness was compiled on the basis of human and mammalian studies, and these criteria may be irrelevant for living organisms like microbes, for example. Remember that for a long time Maturana and Varela's idea that "living systems are cognitive systems and living as a process is a process of cognition" was seen as a category mistake, an annoying misuse of language.[9] Their way of thinking has started to penetrate only recently, after half a century. Our intuitions have been changing, as Patricia Churchland suggested some time ago.[10] Karl Friston's work, for example, provides a general, computational interpretation of living processes in terms of cognitive-like PP; sensorimotor coordination was identified with cognition by Keijzer and colleagues; and philosopher Pamela Lyon interprets bacterial behavior in cognitive terms.[11] She pointed out that bacteria have cellular networks of incredible complexity: they transduce signals, have sensors and effectors, move, select their environment and the targets of their sensations, collectively construct their world, integrate and coordinate sensorimotor information, engage in social interactions, and exhibit limited learning. A dramatic study, which may be a bit unsettling from your point of view, is that of the "eye" of the mind-boggling, warnowiid dinoflagellates, which are free-swimming protists that cannot, at present, be cultivated and that are rarely found. In these protists—single-celled organisms, I want to stress—there is a subcellular organ that has been found to be amazingly similar to a camera eye. This is not your commonly found eyespot that directs photons onto photoreceptors. It is a full-blown, complex organ, called the "ocelloid," that is so like an eye that scientists at first thought it to be a dead jellyfish eye that had been scavenged by the creature. The ocelloid of these protists is made up of subcellular components that resemble a lens, a cornea, an iris, and a pigmented retina, the retinal body. The cornea-like layer is made of mitochondria, and the retinal body is made of anastomosing plastids that originated through an ancient endosymbiosis with a red algae.[12] Have a look at this picture (figure 10.1) and be amazed! Do these protists *see* with these eyes? How would their networks need to operate for you to grant them a spark of consciousness? Are your seven criteria for minimal consciousness to be adhered to at all costs, like holy commandments, at all levels of organization?

WE: We have no problem ascribing minimal cognition to flexibly sensing and responding bacteria, archaea, plants, fungi, and sponges, although distinctions between levels of cognition need to be made.[13] We believe that thinking about biological organization and function in terms of cognition provides a good framework, unifying theoretical and empirical studies in biology and the cognitive sciences. The ideas of Maturana and Varela have been of great importance. In fact, we think that what we are learning about the nervous systems of brainy animals has implications for research on

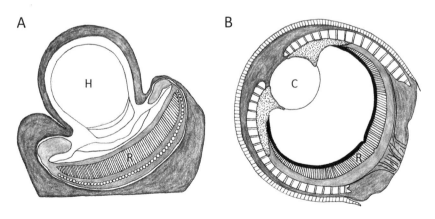

Figure 10.1
Comparison between the structures of the ocelloid (*A*) and the vertebrate eye (*B*). *H*, hyalosome, a structure that serves as the refractive lens of the ocelloid; *R*, retinal body/retina; *C*, crystallin lens. Adapted by permission of *PLoS One*.

unicellular organisms. For example, understanding how neural networks operate directs attention to intracellular networks and their sophisticated computational capacities, alerting us to the possibility that nerve cells may turn out to be more cognitively complex than we have ever dreamed of. Your example of the ocelloid is a glimpse into uncharted biological wonderlands. As we pointed out, a sensory organ enabling the formation of a picture, such as those comprising the visual system in the box medusa, is a necessary but not sufficient condition for discriminating between images made up of individually neutral components that are presented in a different arrangement. Complex learning is needed for such discrimination to occur. The warnowiid dinoflagellate and the box medusa do not exhibit such learning and do not challenge our distinction between conscious and nonconscious organisms.

As this is not a positive answer to your question, however, let us go back to what we said in the first chapter of this book: we claimed that for something to be deemed conscious, consciousness has to be losable. A living dog can be conscious or nonconscious (in the intransitive sense, of course). A cell can be alive, potentially alive (e.g., frozen), or dead. When a dog loses consciousness, its cells, which are alive in the full sense of the word, are not losing consciousness. The effect of anesthetics on plants, or on neurons in culture, which reversibly alters their membrane and electrical properties, does not mean that the plants or the neurons are conscious.[14] Your neurons, when you are aware and thinking, are extremely sophisticated cells sending endless signals, but it is a real category mistake to say that they are conscious.

We think that the identification of life and cognition with sentience and consciousness is a big part of the problem. You may argue that only when cells are autonomous organisms, like the dinoflagellates, or like a bacterium, can one assign consciousness to them. That would imply that cell consciousness is lost when cells join together to form a multicellular body. This is absurd and has no biological meaning. But some biological predictions about the cognitive abilities of the organism can be made if you adopt our approach: going back to the dinoflagellates and their eyes, we predict that they may learn nonassociatively and even display some LAL in the visual domain. The list of seven characteristics we extracted from the neurobiological literature is not a mammalian or human-biased list, although it was compiled mostly on the basis of studies of mammals. It can be applied, as we have shown in this book, to arthropods and mollusks. It is a list that tries to capture properties that we would ascribe to any conscious being, but of course it will be updated and possibly modified as we learn more about the system. What is encouraging and supports our position is that these seven characteristics, which were suggested on the basis of cognitive capacities and neurobiological processes that did not center on learning, cluster together and form a natural grouping in the same taxa that exhibit UAL.

I.M.: Yet you maintain that your UAL marker is only a positive marker of consciousness. Maybe there is a gray area between the undeniably conscious UAL animals (if one accepts UAL as the positive evolutionary marker of consciousness) and other creatures with different sensorimotor networks of interactions? These creatures too may be conscious, although we have as yet no criterion other than their embodied cognitive complexity that could clinch the matter. Maybe multidimensional strange attractor dynamics, such as that you described in chapter 3, will turn out to be a good criterion. I do agree with you that identifying consciousness with life and cognition is a problem. In fact, this vague word, "consciousness," is the root of a lot of evil and confused thinking. It is identified with human reflective consciousness as well as with the elementary process of living, which is, in turn, described in terms of sentience; this all occurs at the same time without any awareness of the different usages. I am not sure that your readers will find your "minimal-consciousness equals sentience" easy to absorb. Words are such little dictators! Maybe we can go back to a more semantically friendly question. You said that distinctions among levels of cognition are needed. Which distinctions? Jacques Monod famously said, "Anything found to be true of *E. coli* must also be true of elephants,"[15] but you seem to disagree. Maybe you can spell out why. And a related question: How does the growing understanding of the neurobiology of undeniably conscious animals like elephants illuminate our understanding of simpler organisms?

WE: As to your comment about the gray area, all we can say is—maybe. As long as we lack a good alternative marker for UAL, this will remain very much in the inarticulate terrain of intuition, and we want to go beyond this. In fact, we do agree with Monod. Monod's phrase, pointing to the unity in the living world, was not a claim about identity. He did not say that anything true of elephants is true of *E. coli*, which would be ludicrous. There is indeed great unity and conservatism at basic levels of molecular organization, and the more we learn about the molecular aspects of the living organism—any organism—the more communalities we find. Monod was pointing to this basic biochemical unity and agency, and we agree with him. But it is clear that multicellularity and the evolution of the nervous system have altered, in profound ways, the cognition and the evolution of living organisms. New types of individuals have appeared. We think that neural organization has opened up new evolutionary trajectories. Behavioral plasticity, especially the plasticity afforded by learning, has become central to the evolution of animals. This kind of plasticity, we have argued, was an important factor in the Cambrian explosion, and it has been guiding the evolutionary trajectories of animals ever since. We therefore agree with Lamarck's early idea that the emergence of the nervous system shaped the evolutionary histories of animals. Neural organisms have a new type of cognition. We identify two levels of cognition in preconscious neural animals: nonassociative learning–based cognition and LAL. UAL, with its entailment of minimal consciousness, is yet another level of cognition, and episodic-like memory and symbolic language–based cognition are two more levels. Altogether, we recognize five levels of cognition in neural animals and at least two more in nonneural living beings: minimal cognition at the single-cell level and cognition at the level of nonneural integrated multicellular organisms, such as trees or mushrooms.

Although we see a qualitative difference between neural and nonneural animals in terms of cognition, we think that studies of neural animals like elephants and rats do provide some insights into biological questions pertaining to microbes. One example is our knowledge of the epigenetics of learning and memory, which comes from studies that were conducted mostly with mammals. Neurobiologists have realized that they need to consider not only synaptic memory but also the epigenetic memory embedded in the cell nucleus and the transmission of "memory molecules" between neurons (and other cells); these memory molecules include small RNAs, which can alter both the nuclear epigenetic memory within neurons and the synaptic connections between neurons. Thus, there are additional biochemical memory systems in the nervous system, not just the well-studied synaptic one, and although these mechanisms interact, they can be semi-independent, with nonlinear interactions. If we want to understand learning and memory in neural organisms, we need to consider all these mechanisms. Crucially, the epigenetic mechanisms of cell memory are very ancient; they preceded the evolution of neurons and are found in all living organisms. And so are simple forms of learning. Nonassociative epigenetic learning is the likely explanation of many behaviors of protists such as paramecia. There is also a theoretical possibility for LAL in these protists, although here the behavioral data are less convincing.[16] Furthermore, even pea plants can learn by association! They learn to grow toward the direction of a breeze (a conditional stimulus, CS) that in the past was associated with a source of light (an unconditional stimulus, US).[17] Clearly, the next step is to discover the cellular epigenetic memory mechanisms that underlie learning in microbes and plants. The studies of the epigenetic changes that underlie and reflect learning in rats and mice—and presumably in elephants and humans too— direct us to such research.

I.M.: This focus on cell memory as a starting point for thinking about learning more generally brings to mind Richard Semon's ideas. In previous chapters you mentioned him as the originator of the term "engram" and of an insightful notion of neural memory, as well as the first to suggest something like the epigenetic cell engrams. Are there conceptual and possibly mechanismic links between his ideas and present-day notions?

WE: There are such links, and he would have felt vindicated by present-day thinking. We feel we should say a little more about this original zoologist and theoretician, who is still largely unknown despite Daniel Schacter's inspiring book about him.[18] As the brilliant and promising student of Ernst Haeckel, the great German evolutionary biologist, Semon (figure 10.2) was a rising star in the German biological firmament of the late nineteenth century. When forced to retire from institutional academic life because of a passionate and scandalous love affair, Semon became a private scholar, a theoretician studying the interlocking problems of memory and heredity. He thought about engrams very generally, both as something inscribed in interactions among cells and as an intracellular memory trace. As we noted in chapter 7, Semon thought that the processes that lead to the development of new behaviors and other characteristics acquired by an individual through learning or the direct effects of the environment leave engrams in the individual's biological organization, some of which are transmitted to descendants. He suggested that engrams are reconstructed and retrieved when similar or past-associated conditions occur during the development of an individual, thus envisaging developmental induction during embryogenesis as a recall process. After a long period of neglect, there is now intense interest and research into cell memory, epigenetic marks, and the transgenerational inheritance of some cellular marks, although the link between Semon's early ideas and this research remains largely unrecognized. The study of cell memory and learning in nonneural organisms such as protists and plants and the intimate relation between the epigenetic mechanisms of cellular memory and neural memory would certainly have delighted him. There is, as he suspected, not only a conceptual link but also a subtle mechanism connecting cell memory and neural memory, and hence the cognition of neural animals is continuous, at this level, with the cognition of cells. In chapter 8 we speculated that epigenetic engrams (EEs) are necessary for the rich memory displayed by animals with the capacity for UAL and

Figure 10.2
Richard Semon (1859–1918). Reproduced by permission of University of Zürich.

that the evolution of the propagation of EEs in the nervous system led to a quasisymbolic mode of encoding memories, allowing complex cognition.

I.M.: Do you think that the new type of integration and individuality that the nervous system imparts to animals depends on cell memory mechanisms? Is it not first and foremost the consequence of the dynamic neural activity-connectivity patterns in the body-brain? Are you not in danger of trying to read animal cognition from cell cognition? You may end up deciding that neurons, too, are conscious after all!

WE: Cells are not conscious for the same reason that molecules are not alive. But yes, we do think that the kind of PP that we described in our UAL functional model is, in evolved animals, the result of CNS activities that would not be possible without the molecular memory and learning processes occurring at lower levels, both in neurons and between neurons. The CNS is not made up of "stupid" components. Neurons are highly sophisticated cells—more sophisticated than any computer that a human has ever yet constructed. They are capable of cognitive and computation feats that we do not begin to fathom. And we think that some effects of the higher-level neural activities are reflected at the level of the cell—this is what the epigenetic learning-related patterns in neurons suggests—and that we may be able to relate behavioral and epigenetic learning. However, this does not mean that we need to start, bottom up, from cell epigenetic learning profiles to behavior. The correspondence is more likely to be the outcome of the behavioral learning affecting, in a top-down direction, the epigenetic profile of the neurons, which may then lead to the stabilization of behavior. The reasoning is the same as that we used when explaining genetic accommodation: just as genes are followers in evolution, low-level intracellular neural features, such as neuronal epigenetic profiles, are "followers" of stabilization at the behavioral-developmental level.

I.M.: Do you think, then, that in principle learning and memory cannot lead to the generation of consciousness unless the epigenetic levels of memory are included? Do you think that the existence of multiple memory levels—in cells' nuclei and cytoplasm, between cells, in bioelectric fields, in dedicated brain areas—which seems to be the case in evolved animals, requires that all these levels of memory will need to be simulated in robots if robots are ever to become conscious? Will robots need to have not only a highly sophisticated central processor but also multiple "processor ganglia" and a linked network of billions of tiny, clever processors spread throughout their bodies, as well as millions of sensors, effectors, and EEs to give them a new kind of sensitive soul? Or will robots, or golems, as the early versions of such fictitious creatures were called in the Jewish tradition, remain non-conscious, albeit highly interesting, artifacts?

The Golem's Predicament

> God, whose law it is
> that he who learns must suffer.
> And even in our sleep, pain that cannot forget
> falls drop by drop upon the heart,
> and in our own despite, against our will,
> comes wisdom to us by the awful grace of God.
> —Aeschylus, *Agamemnon*, 179–184, quoted in Hamilton 1930

WE: We don't know what shortcuts are technologically possible to render a robot or a golem conscious. It is interesting that in the early golem stories, the golem, usually made of clay by a great rabbi, is endowed with life and sometimes with consciousness, but often not with language and reason. And yet it is through the use of language that it is both created and destroyed. The Talmud tell us that the Babylonian Jewish sage Rava created a man and sent it to his colleague Rabbi Zeira, who spoke to the created being, but it did not answer him (because artificially created men in this version of the story do not have the ability to speak). Rabbi Zeira then told it: "You have been created by one of my colleagues.... Return to dust."[19] In this tradition, words, and specifically the name of God, have a mystical creative power that has to be used to generate the golem—a golem can be brought to life by the rabbi reciting the proper name of God, or by putting God's written name in its mouth, or by affixing it to its forehead. A great deal of symbolic power is required.

You are right in supposing that we do not believe that a central processor, however sophisticated, will suffice to animate a golem. We think that conscious golems will end up having complex movable bodies with communicating parts, many sensors and many effectors, and a massive network of tiny powerful computers as well as a hierarchy of higher-level processors. We are assuming that there will have to be ongoing processes of integration and predictive coding based on the inputs from this complex, flexible, ever-changing body that will enable values to be ascribed to the constantly changing states of the system as a whole. Because we know that conscious animals are organized in this way, we are assuming that something like a golem/robot categorizing sensory state (CSS) can be constructed in this way too (unless a detailed virtual reality of body and world is provided to a general processor). In whatever way such robots are constructed, their creators will benefit from the insights of biologists about the affordances and constraints imposed by the biology of actual living animals.

I.M.: But you are not questioning the assumption that this is possible, are you? And are you suggesting that the kind of cognition that you have described, the UAL cognition, with all its many affordances, will facilitate a robot's transition to consciousness?

WE: We believe that conscious robots/golems will be constructed one day, although we are not sure that their consciousness will be comparable to that of animals or humans. We think that if conscious robots are to be constructed, they will have to be immensely complex systems. Their enabling system will have to instantiate something like CSSs, which depend on the systemic valuation of body-in-world states, and that requires a fine-grained, flexible, and sensitive embodiment. We think that it is likely that a capacity for UAL will be a facet of their cognition. There may be more than one way of constructing conscious robots, but we suspect there will be rather few, and they will be very demanding—the fact that a robot would need to have a sensitive, active, physical body and require a complex cognitive architecture to support consciousness are strong constraints. Importantly, we are not assuming that a UAL cognitive architecture is a cultural evolutionary marker of robot consciousness. We argued that UAL is a cognitive capacity that is an evolutionary marker of minimal consciousness in living animals, not that it is a marker of consciousness in the general sense in any entity that may implement it. As you realize, and as we have repeatedly stressed, ours is a biological account, based on an analysis of the biological properties that must be in place to implement minimal consciousness in animals. In the same way that life as we know it depends on complex chemistry, consciousness as we know it depends on complex biology. We gave the example of a genetic algorithm that is an unlimited heredity system, yet it does not indicate that the computer chip in which it is installed is alive, since the computer chip is not an autopoietic system that re-generates itself and that staves death; similarly, implementing UAL in a machine need not indicate that the machine is conscious. Consciousness entails cognitive autopoiesis—incessant, embodied, re-generation of conscious states. We stress: we have been discussing consciousness as we know it, trying to better understand it by tracing its evolutionary history.

I.M.: Consciousness as we know it is a product of the evolution of living organisms, and biological UAL as we know it is a product of cognitive development in animals with the appropriate cognitive-neural infrastructure. However, animals, including humans, do not come into the world with a mature UAL. They develop this capacity during ontogeny, as they learn to perceive and control their bodies and make sense of their world. And even though animals are born—as you argued for babies—minimally conscious, you suggested that it was variations in cognitive development that were targeted during the evolution of UAL. Do you think that, unlike animals, robots will be able to become minimally conscious without evolutionary and developmental processes of construction? And if you think that development is necessary for the generation of conscious robots, and that the construction of molecular epigenetic engrams is part of this developmental-cognitive process, then the equivalent processes at the micro level may need to be included in the construction of future robots. There is also the vexed question of the materials used to build them. Can there be a metallic conscious robot, or will they have to be made of "soft matter"—condensed viscoelastic materials such as colloids, liquid crystals, foams, and gels? I am wondering how deep the analogy with developing biological organisms must go.

WE: We don't really know, but we think the analogy needs to go deeper than the popular imagination suggests. What we do know is that biological materials are necessary for consciousness as we know it. They may turn out to be necessary for any conscious system. Soft biological matter and the processes it makes possible provide great affordances: epigenetic memory mechanisms are just one class of a vivid example. In fact, some roboticists believe that robots made of soft materials will have much more sophisticated capacities than metallic ones. Moreover, the rich connectivity between the nervous system and the other body tissues is crucial—one's sense of self depends on the extensive connectivity between the brain and the body, which is incessantly re-generated; changes in body ownership too are based on this richly constructed self-model. Soft matter provides great affordances, and some roboticists believe that robots made of such material will have much more sophisticated capacities than metallic ones. It is therefore likely that conscious robots will be made from such materials, at least partially.[20] We also think that developmental processes will be necessary for conscious robot construction because the cognitive complexity underlying consciousness cannot be formed in a single stroke: it requires cumulative, context-sensitive, ongoing, developmental selection and, of course, evolution as well. There is a very thriving discipline called "developmental robotics" or "epigenetic robotics," which implements developmental processes in robots (figure 10.3 depicts a robot child). Developmental roboticists are trying to understand the principles that underlie the development of natural cognition in human babies and apply this knowledge to design the cognitive capabilities of robots. Incorporating intrinsic motivation and implementing mechanisms that allow autonomous learning in developing robots can lead to increasingly sophisticated sensorimotor, cognitive, and reasoning capabilities that may result in robots who make their own decisions and do not need ongoing human control.[21] Another well-established discipline, evolutionary robotics, is based on evolutionary computation, employing, for example, a genetic algorithm with variations and selection occurring in a population of candidate controllers. However, evolutionary and developmental robotics do not seem (as yet) to communicate with each other very much. All in all, our impression is that many roboticists take very seriously the possibility that further developments will eventually lead to conscious robots. This may mean that robots will become conscious as they develop, just as animal embryos do. The possibility of conscious robots is

Figure 10.3
A robot child.

exciting, but it has horrifying ethical implications. The philosopher Thomas Metzinger is one of the people who has explored some of the dangers. He argued that constructing such robots will lead to suffering because the first conscious robots are likely to be only partially functional and therefore handicapped; furthermore, they may also suffer because of their enslavement to humans, if they become self-aware in the reflective sense. Humans' ongoing record of horrendous and shameless cruelty toward other humans and toward conscious animals does not bode well for future conscious robots.[22]

I.M.: I am glad that this ethical issue is beginning to be discussed. I have little faith in the benevolence of humans. Suffering seems to me, more generally and more fundamentally, to be the conscious golem's inevitable predicament. Once negative valence is subjectively felt (and "owned") by a conscious being, be it a robot or an animal, it will suffer. This was a major point in your discussion of the neurotic, minimally conscious Cambrian animals. The richer the learning, the more extensive are the joys and, unfortunately, even more so, the pains of living. This brings me to a troubling point. Constructing conscious robots will invert the evolutionary life-to-consciousness continuity. Once a robot is conscious, it seems that it will also be alive. Consciousness may therefore render a nonliving entity alive. This has a strong whiff of vitalism—it is somewhat like animating the golem by putting the written name of God in its mouth. This consciousness-to-life relation problematizes our notion of life if we insist that the conscious robot is a living being.

WE: Yes, it does. It may not be life as we know it, but conscious robots will have an embodied cognitive organization that generates a self-maintaining, autopoietic "self," which, we predict, will depend on a very complex "soft" body. A robot may well "die" not only when its "central processor" is badly and irreversibly damaged but also when its body is beyond repair and therefore essential feedback from and to the "brain" is impaired. Since the sense of self the robot will develop will have to be dynamically maintained, it follows that processes of selective inclusions as well as selective exclusions of representations need to be in place to incessantly reconstruct the image of the self. How this neural-cognitive autopoietic process relates to the reproduction and reconstruction of other aspects of organization, such as morphology, is not clear, but as we noted in previous chapters, morphological standardization in animals seems to be controlled by the CNS. We must wait and see whether or not the organization and behavior of conscious robots entail morphological self-production and self-maintenance. We suspect they will because the self-model depends on the gradual construction of a unified, stable, situated, body-in-the-world model. We therefore doubt that mass produced, factory-manufactured, conscious robots will become a reality. To end up with the capacity for experiencing, robots may need to develop, to be treated like growing children, in ways that allow for many developmental changes, many social interactions, and very possibly—for care and love. They will "die" because serious damage will lead to the collapse of their unique, developmentally constructed internal states, which will be practically impossible to reconstruct.

I.M.: I am very skeptical about the feasibility of this science fiction world, although I have little doubt that experiences based on weird and fantastic virtual realities, and enhancement of sensory and motor perceptions in humans and possibly other animals, are all around the corner. I do hope that if robot consciousness is to be, it will come slowly and gradually. However this turns out, it seems to me that the relation between consciousness and life opens up philosophically interesting and less fictitious questions: Is a beehive (or a human tribe) conscious? Put differently, is the beehive or the human tribe, rather than the individual bees or humans, a living conscious organism? There is no doubt that along some trajectories, a social group can be a systematic target of selection—what philosophers call a "Darwinian individual."[23] It can also be a highly functional unit because there are processes of collective decision-making based on multiple criteria and feedback relations among interacting individuals, and there is no doubt that a beehive and a tribe display collective emergent properties. Would you attribute consciousness and life to such groups?

WE: In our opinion the beehive or the human tribe are not living or conscious entities. Groups are not hungry and thirsty and do not experience pain, whereas conscious individuals most certainly do. We think that phrases such as "a group is alive" or "a group is conscious" are metaphors, stemming from social groups' collective, emergent cognitive properties. Let us clarify this point by looking at the more manageable case of life, because life can be unambiguously attributed to some multicellular units, not just to the cells of which they are composed. Clearly, at some point in the evolution of multicellularity, a group that started as a loose aggregate of living cells did become a single unified living entity—a multicellular living organism that lives and dies. But at what stage? Cells of an organism must share a common fate and be strongly constrained by organism-level morphology and physiology; the interests of the components must be aligned (they must cooperate rather than compete), and there must be processes that ensure this. These include powerful homeostatic mechanisms, division of labor, a single-cell bottleneck, soma-germ line segregation, chromatin resetting, strong top-down control, and the delineation of the spatial boundaries of the individual by something like an immune system.[24] Such a "group" is far more autonomous in its control of internal and external conditions than are its component parts, and only when the autonomy of the parts is small relative to that of the group can we say that the "group" (e.g., a mushroom, not just its cells) is alive. This is not the case with social groups. Although there is plenty of communication in complex social groups and a great deal of functional integration based on communication and groups may be good Darwinian individuals, the autonomy of an individual animal is far greater in both scope and extent than that of the collective. This may be in part the result of the lack of ongoing close physical contact among individuals and the resultant absence of topological constraints. In social groups that consist of conscious animals, the relative autonomy of the group is decreased because the individuals' autonomy is greatly increased—and that increase is the result of sociality. The social individual's consciousness develops and becomes enriched through social interactions. As we argued when we discussed animals with ELM, the self becomes defined vis á vis other group members, not just vis á vis the world. Sociality in conscious animals increases the autonomy of the individual—its distinct, personal, unique sense of self. The social conscious individual is far more "individuated" than an asocial conscious one. This is most dramatically obvious in the case of humans, and it is what sets us apart from the other beasts of the earth. On the one hand, human communication, especially through language, is open-ended to an unprecedented extent, enabling feats of social coordination that are unthinkable for other animals; on the other hand, this symbolic communication system with self and others adds layers of uniqueness to individual, highly personal human consciousness.

I.M.: At long last, we have reached the rational soul! In the previous chapter, you implied that the evolution of the linguistic communication system is the key to the human rational soul. What exactly do you mean by "rational soul," and why is language your chosen Archimedes lever to human nature? Why not the feeling of shame and guilt, as the first chapter in the Old Testament suggests? Or the ability to read other minds?

The Evolution of the Rational Soul

What is the mind?
It is the sound of the breeze
That passes through the pines
In the Indian-ink picture.
—Ikkyū Sōjun[25]

WE: What we mean by "rational soul" is not just the capacity for logical thought, for engagement in syllogisms, although this is an important part of it. It is the conception of an objective world but also, at the same time, the emergence of new abstract concepts and societal-cultural goals, such as the desire for justice, for beauty, for good. Our notion of "rational" is very similar to what the developmental and comparative psychologist Michael Tomasello calls "objective-reflective-normative thinking," but we put special emphasis on the human imaginative ability that generates publicly sharable, reified, meta-metarepresentations, as expressed most cogently in human language. Of course, hominins needed to have the appropriate cognitive and emotional infrastructure for linguistic evolution to take off. We have suggested, in previous work, that social emotions such as shame, guilt, embarrassment, and pride were a necessary condition for the evolution of language and that language and emotional control coevolved.[26] Humans are not just very clever animals with the emotional profile of crocodiles, though sometimes one may be fooled. We are emotionally very different from other animals, including our apish ancestors. For example, neuroscientist Barbara Finlay convincingly argued that the threshold for pain decreased during hominid evolution due to hominids' greatly enhanced reliance on the help of others.[27] In our own work, we suggested several requirements for the evolution of the human linguistic capacity, which coevolved and codeveloped with it. These include the ability to read other minds, the emotions and cognition that accompany alloparenting (caring for the young by individuals other than the mother), the domination of causal reasoning and learning to follow rules, the ability to teach, and the ability to exercise fine motor control and make tools.[28] Language is built on and is entangled with these cognitive and emotional capacities. However, before we can justify the claim that language, more than any other capacity, is a good evolutionary transition marker for the rational soul, we need to characterize it.

We adopt the view of language developed by our colleague Daniel Dor, a linguist and communications scientist. It is a view that takes imagination one crucial step forward from what we saw in Popperian animals. According to Dor, language is a communication technology for the instruction of imagination. Dor starts his account with the fact that each and every actual or imagined individual experience is always private and idiosyncratic. As he puts it, we (and all other conscious animals) are separated from each other by experiential gaps. This fundamental experiential gap is only partially bridged by communication. Communication, according to Dor, creates common ground either by enabling the direct sharing of experiences, which is what nonlinguistic communication systems do, or through the translation of private experiences into a socially constructed code that can be decoded back into private experience, which is what human language does. Language is a cultural communication system, based on a jointly identified set of conventional signs and norms. In Dor's words: "The communicator provides the receiver with a code, a plan, a skeletal list of the basic coordination of the experience—which the receiver is then expected to use as a scaffold for experiential imagination. Following the code, the receiver raises past experiences from memory and then reconstructs and recombines them to produce a novel, imagined experience."[29] Through social negotiations and cultural evolution, a web of socially agreed-upon signs (the "semantic landscape") and a set of prescriptive rules for the regulation of the process of instructive communication (the "protocol") are formed gradually, through joint identification and joint agreement. There is, according to Dor, an intervening level of meaning—semantic and social—between the private representations of the communicators and the signifiers (e.g., words) they use.

I.M.: Surely, language has some genetic basis. Although the famous prophet Balaam's she-ass spoke biblical Hebrew, this required the intervention of a special power.[30] How, then, did language evolve genetically? Do you think that like UAL and episodic-like memory (ELM), language evolution was driven by plasticity and genetic accommodation? Can you build a model of language on the basis of your UAL and ELM models?

WE: We start with your second question, about the modeling. Yes, we think that by adding further systems of representation to the ELM model, a functional toy model of language can be built. Following Dor, these additional high-level representation systems will include the learned representation of the semantic landscape and the learned representation of the language-specific signifiers, the words of the language. Clearly, all the systems of representation—that of episodic experiences, that of the semantic landscape, and that of the signifiers—will influence each other. Constructing such a system is, of course, easier said than done. It is a very complex hierarchical system, but we believe that it should be possible to construct a model once the "simpler" models of UAL and ELM are worked out in sufficient detail and different types of engrams, which operate at different timescales and enable quasisymbolic encoding, are considered. A neural-function model of language also requires a new type of neural selection dynamics. If Szathmáry is right about the need for Darwinian neurodynamics to solve problems that require insight, such dynamics would be most evident in language-endowed humans. We also expect that intracellular epigenetic engrams would be employed to encode linguistic symbols. It is therefore possible that human neurocognition will provide us with the clearest and most elegant implementations of the cognitive encoding of rules, of concepts, and of quantities.

When considering language evolution, we believe that one must start with more than the developmental plasticity of the individual. With UAL and ELM, the behavioral plasticity of the individual guides the evolutionary process; it is an individual-phenotype-first evolutionary scenario. In the case of language, it is a culture-first process. Dor and Jablonka have argued that language began to evolve in the way that all technologies do—as a long succession of gradual modifications, reversions, and occasional techno-logical revolutions.[31] Throughout the process, language gradually spread and diversified, grew in size and efficiency, demarcated itself from mimesis-based communication, and asserted itself as an autonomous tool of com-munication with a unique internal structure. And unlike many human technologies, it was partially genetically assimilated through culturally driven genetic accommodation. Technologies and capacities fashioned by culture are the result of collective activities over generations; these activities con-struct a social and epistemological niche that selects genetic variations that fit the persistent features of this niche. Thus, selection for the consistent aspects of communicative practices that allow flexibility increases plasticity further, opens up more learning opportunities, generates new problems, and so on; it leads to positive feedback, which accelerates the tempo and scope of language evolution. Going back to where we started, this view of language implies that the evolved biological processes and structures that support human language involve cognitive and emotional preadaptations that facilitated the integration, articulation, and decoding of instructive signs—signs that are acquired within the group and are the shared communi-cative norms of the group.

I.M.: I am afraid we lost the rational soul on the way. How exactly does language render the experienced world "objective" and at the same time inhabited by abstract values?

WE: It is the normative, collective aspects of language that make the world objective. Through the sharing of conventions about the use of lexicon, a form of normative, transindividual communication logic emerged, embod-ied in the syntax or the protocol. The common ground of the shared words led to the reification of what they signified—the signifieds became "objec-tive" because they were shared through collective identification and went beyond the individual's idiosyncratic experience. Michael Tomasello ties this objectification to the growth of human cultural groups and to the need to coordinate activities and intentions with anyone from the group. He argues that "modern human individuals came to imagine the world in order to manipulate it in thought via 'objective' representations (anyone's perspective), reflective inferences connected by reasons (compelling to any-one), and normative self-governance so as to coordinate with the group's (anyone's) normative expectations."[32] This objectification is related to the evolution of the human social emotions of guilt, pride, shame, and self-deception, which were expressed when the individual became exposed to the actual or internalized public "gaze" that was objectified as a binding societal and moral law. It led to a new type of self-monitoring that reinforced the feeling of having a self, a self that internalized the norms of the society in dialogue with itself, having what we call "inner language." This self-monitoring enabled the development of autobiographical memory, which partially solved one of the new problems that language introduced (yet another Pandora's box of troubles): the problem of distinguishing between what one was told about and what one imagined on the basis of personal past experience.[33]

I.M.: You reverse engineer from language to normativity to objectivity and to self-reflectivity, to the reified notion of an individual self inhabiting the poor golem. Does this guilt-ridden, self-deceiving, rational soul of yours acquire a free will? I am not referring just to the ability to make flexible choices and to the lack of brute determinism. Our understanding of quantum mechanics and of the chaotic and practically indeterminate effects of some types of nonlinear interactions, which can be extremely sensitive to small perturbations, have taken care of that. Moreover, as Björn Brembs has rightly stressed, predictable behaviors are disadvantageous in a world that is constantly changing, where novel resources need to be actively explored, and which is inhabited by adversaries who can exploit the predictability of their victims' behavior.[34] Not only do humans and most animals spontaneously increase their behavioral variability when they face new and challenging conditions, but unpredictable behavior has actually evolved. Behaving unpredictably—"protean behavior"—is a well-known adaptive strategy in the animal world. Brembs argued that animals make flexible choices based on exploratory actions that occur spontaneously and cannot be explained solely as slavish responses to external stimuli. This is obviously necessary but not sufficient for free will, which is about conscious decisions. However, some animals—the conscious ones, who can be said to have a self in the Merkerian sense that you described—may well make conscious, flexible decisions. They can decide to behave against certain motivations—for example, not to jump on the table and consume the sausage, something we can recognize as choice. But it seems that there is more to human free will than that. Do you think that Brembs's account of free will is sufficient for the human case?

WE: No, but it is a basis on which we can build. We think, as you clearly do, that the question about whether or not there is free will is misleading. Free will is not a metaphysical entity. It is a feeling, a feeling of having a free will, that is as real and compelling as any other conscious feeling. Animals that are conscious have a feeling of willing—of wanting this or that—and this wanting is the feeling that accompanies the process of attempting to reach a goal. And they, too, as you pointed out, can have conflicting wants and hesitate between actions. This is not yet a feeling of free will, but it is the precursor of the feeling of free will in humans. We think that in humans the feeling that comes with choosing one alternative over another is very different from that found in other animals. The human feeling of free will is a consequence of represented and imagined collective norms (a kind of superego, which provides the "ought") interacting with the individual's feeling of agency, which is linked to people's ability to imagine themselves in many situations, acting in many different ways and toward many different goals. In other words, the high-level, imagined system of symbolic-social values can constrain and bias the imagined choices of humans. It is the interaction between these products of the imagination that generates a feeling of free will. On the substrates of such subjective agency feeling and collective norm–related feelings, human morality emerged. Our view is that free will exists as a powerful feeling in self-reflecting humans, and morality is forced upon us because of human rationality and human values and therefore has an imperative, logical-normative validity. Free will has huge social and individual consequences, but it is the feeling that exists, rather than something that is beyond and above it.

I.M.: The sensitive and rational soul of humans, with its feeling of free will, its imperative moral sense, its excessive joys and pains, its guilts and shames, seems very different from the sensitive soul of other animals. You imply a great difference in kind. Yet surely this rational soul has functions. You argued in chapters 1 and 4 that consciousness, like life, is not a property nor a process, that it does not have a function—that it is a mode of being imbued with teloi, with felt needs. Yet it seems to me that what you call "rationality" is advantageous for the survival and reproduction of the animals so endowed and therefore that it has functions. Similarly, one may argue that consciousness has multiple functions, allowing the systemic yet precise categorizing of inputs and outputs, instantiated by your CSSs, which have causal efficacy. You said yourself that with CSSs an overall sensory state became a state *for* the animal, not just a state *of* the animal. Survival and reproduction can also be said to keep the living system going. If you insist that consciousness does not have functions, does it mean that in a sense you see it as a by-product of cognition, which obviously has many functions? In other words, do you see it as an epiphenomenon?

WE: We think that human consciousness is profoundly different from the consciousness of other animals, although there is, of course, an evolutionary gray area here too. The consciousness of prelinguistic humans, of *Homo erectus* hominins who lived one million years ago and had many of the cognitive and emotional building blocks necessary for language, is of enormous interest. As to your function question: you can, of course, argue that within the larger system of life, consciousness must contribute to survival and reproduction. This is trivially true in the sense that the sensitive soul must also be what Aristotle called "nutritive" and "reproductive," so the system as a whole must be self-maintaining. But you would not say that a living system, conscious or not, has a function. Self-maintenance and reproduction define living, they are not functions that contribute to the goal-directed behavior of the whole. They are the goals of the living system and are instantiated by processes and parts that do have functions, such as metabolic activities, membranes, and heredity mechanisms, all of which contribute to these goals. Having a function is not a precondition for living; it is, rather, a necessary consequence of any teleological organization. Similarly, the notion of qualia, instantiated as we suggested by CSSs, is a necessary outcome of the process of experiencing, and the notion of symbolic values is a necessary outcome of the dynamics of the rational soul. Of course, mental states have causal efficacy and are functional, but mental states are not conscious, in spite of the sloppy language of philosophers and biologists. It is the animal that is conscious, not its states. Mental states can be said to be the organizing attributes of the conscious mode of being, just as nonmental functions are the organizing attributes of the nonconscious organism's being. We agree with James that consciousness is *"a fighter for ends,* of which many, but for its presence, would not be ends at all."[35] Hence, consciousness, which according to James is a process involving selection, opens up a new realm of goals, which are realized by various functions, including UAL and other more or less related cognitive and emotional subsystems. If you insist, you can say that opening a new realm of goals is the function of consciousness, but we think that this is misleading, and it is simpler and more accurate to say that consciousness is a goal-directed and goal-directing system. It is interesting that the idea of consciousness having a function has also been criticized from another, related, direction by Hobson and Friston, who draw an analogy between consciousness and genetic evolution. They argue that, like the process of genetic evolution, the process we call "consciousness" requires a selection principle and has no special physical "locus" or function.[36] Our own analogy—between living and experiencing—is different from Friston and Hobson's suggestion, although the broad notion of selection is central to living, experiencing, and reasoning, and as table 10.1 shows, we see an analogy between the open-ended evolution that is necessary for sustainable life and the open-ended learning that is necessary for consciousness. Our conclusion, however, is the same: attributing function to consciousness is a category mistake. We think that it is the result of the conflation of the concepts of telos and function.

What, you may ask, about the goals or functions of the rational soul? The goals of the rational soul, we suggested, are abstract values like the beautiful, the good, and the just, toward which this soul strives. These values are humans' goals, not functions, although they are realized by cognitive processes that have been, on the average, teleofunctional. The claim that experiencing is a by-product or an epiphenomenon of the evolution of complex cognition, or that rationality is the by-product of metacognition, makes as much sense as saying that life is the by-product of complex chemistry. It is nonsense to make such a claim because in all cases, the new mode of being (living, experiencing, rationality) is constituted by the biochemical, representational, and metarepresentational processes. It is not a side effect. To use a different type of example, an animal is not a "by-product" or an epiphenomenon of its cells. We think the wrong type of causal language is leading to this conceptual confusion.

I.M.: I think that the idea that consciousness and rationality have no function will strike people as rather nonevolutionary, which is strange given your evolutionary approach.

WE: A teleological naturalistic approach requires an evolutionary framework writ large. To make this explicit, we have expanded the table we presented in the first chapter (table 1.2), when we first laid out the conceptual analogies among the three levels of the soul. We find these analogies (table 10.1) intriguing, and we hope that what we understand or have become accustomed to considering reasonable about one system may illuminate less intuitive notions about another system. Highlighting the parallels between the familiar nutritive/reproductive soul (column two) and the sensitive soul (column three) is also a way of summarizing and putting together many of the arguments in this book. To fully explain the fourth column, which points to the analogous processes and concepts with respect to the rational soul, really requires (as we have already said) another book or books, starting with the evolution of ELM. Selection and evolution are crucial aspects of all systems. We think that employing the heuristics we have suggested, which involves tracing the evolution of the transition marker for the rational soul, can shed light on many of the strange antics of the human soul. But first, of course, the case for language being the evolutionary transition marker of the human rational soul must be made convincingly.

Table 10.1
A comparison of the three teleological transitions.

Transition	To life (nutritive/ reproductive soul)	To consciousness (experiencing, sensitive soul)	To symbol-based, reflective experiencing (rational soul)
Telos	Self-maintenance, reproduction; value (fitness) ascribed to differential survival and reproduction	Felt needs; values are ascribed to newly learned complex stimuli and actions	Abstract regulative ideas (e.g., justice, beauty); values ascribed to symbolic concepts
Organizing autopoietic principle	Functional (functions emerge)	Mental; mental representations emerge (CCSs, qualia)	Symbolic; metamental representations emerge (abstract values and concepts)
Closure	External/internal distinction	Self/world/neural-mental models; closed, strange, neural body-world loops	Collectively shared metamodels forming a coherent system
Heredity	Unlimited (e.g., genetic)	Unlimited memory/learning	Unlimited cultural variation
Developmental construction	Induction	Recall	Social-symbolic construction
Selection principle	Natural selection, open-ended	Exploration and selective stabilization— ontogenetic (UAL)	Cultural selection; replicative selection in CNS?
Material preconditions	Complex polymers; autocatalytic cycles; operational closure	Central nervous system; associative memory for compound percepts and action patterns	Social conditions for the communication of subjective representations
Evolution	Open ended, phylogenetic; new niches and selective regimes are constructed	Open-ended, ontogenetic; cumulative experiencing is possible; new niches and selective regimes are constructed	Open-ended, symbolic; cumulative cultural evolution is possible; history is open-ended; new niches and selective regimes are constructed
Evolutionary transition marker	Unlimited heredity	UAL	Unlimited symbolizing

I.M.: If we, rational linguistic souls imbued with values, adopt your view of animal consciousness, how should this affect our practices? Certain conclusions seem to be forced upon us if we accept, as you argued, that many animals are sentient. This is particularly pertinent with regard to the animal food industry and animal experimentation, which cause much suffering. And what about the big debates on abortion? Do you accept the far-from-simple societal implications of your view?

WE: As far as we know, there is no definitive argument or finding showing that sentient animals do not feel pain and other forms of suffering, and there are good arguments to the contrary. Our reasoning ability and what we have learned about animals' sentience and its evolution therefore inflict this moral responsibility upon us—we must spare other animals from suffering. It is quite likely that in a hundred years' time the barbaric way we treat animals and our pathetic justifications for our practices will be looked upon with disbelief and disgust by our descendants, just as we regard with incredulity and revulsion our ancestors' practices and justifications for slavery and the subjugation of women. Social practices will have to change, as they have throughout history, in those places that we regard as enlightened. We have, fortunately, some examples of cultural practices, such as those of some Buddhists, on which we can draw with regard to our treatment of animals, and this is therefore not such an insurmountable problem. We sympathize with the great British biologist J. B. S Haldane, who was reported to have said: "My claim to a more immortal soul than a goat is not strong enough to justify me in eating the goat" and who argued that we should not do an experiment on an animal that we are not ready to do on ourselves.[37] We do not see this common-sense implication of our view of animal consciousness as a profound metaphysical issue, although it is obviously a hugely important ethical and practical one. Our growing understanding of animals' evolutionary biology, especially the biological evolution of their sentience, makes this conclusion difficult to escape. As to abortion: this is a complex question, because it involves not only or mainly embryological considerations about sentience. It is cultural values that are pivotal, and the embryological considerations are often used as legitimation of preexisting cultural positions. If you ignore specific cultural values and focus on the question when the embryo is beginning to be sentient— sentient like most vertebrates, some arthropods and cephalopods—then according to our arguments, this is when the relevant brain structures develop during embryogenesis; if the question is when the rational human specific soul can be first said to emerge, then, again, the first appearance of the enabling brain structures in the embryo may be indicative. But this is, we stress, just a small facet of the debate: as we clearly see when we study how people from different societies think and feel about abortion, the feelings of the parents and the related values of the community are central. The symbolic-cultural considerations are dominant.

I.M.: Indeed, actual biology is not apparent in your "rational" column of table 10.1. It all seems to be about symbols and abstract values. Yet "messy biology" is a major thread of your evolutionary tale and must also be part of the human evolutionary path. You seem to be moving with great lightness between different levels and types of description and abstraction and are in danger of forgetting the messy biological aspects on which it is all built.

WE: It is all messy biology. The second part of this book, which traces the evolution of UAL and minimal consciousness, starts with the "white noise" of overall sensation, the sensorimotor neural buzz that is the hallmark of every nervous system, including the most primitive. Spontaneous activity and stochasticity of interactions in single cells and in multicellular nonneural, neural, and "rational" organisms are necessary for the exploration part of the exploration-stabilization general principle, without which there is no life-as-we-know-it. We have stressed this principle throughout this book, and we hope that it is firmly entrenched in the minds of our readers. You are right, however, about our moving between levels, scales, and types of description. We tried to give evolution-based "material," "efficient," "formal," and "teleological" accounts of minimal consciousness and some of its ramifications, and we kept moving, as the subject matter dictated, from an emphasis on one Aristotelian "cause" to another. But there is a firm basis to it all in the inherently active actions and reactions of biological entities.

I.M.: So we have come a full circle now, back to Aristotle. It was, indeed, a long odyssey you took us through, from Lamarck's inner feeling and the white noise of preconscious nervous systems to the symphonies and operas of the sensitive and rational souls. One last question then: How do you see your contribution to the solution of the "hard problem"? Have you solved it?

WE: Since we argued that there is no unbridgeable gap to start with, it is a problem that has to be dissolved, not solved. We hope we have contributed toward its dissolution.

Notes

Chapter 1

1. An important exception is Emil Du Bois-Reymond's (1874) lecture *On the Limits of Our Knowledge of Nature* (*Über die Grenzen des Naturerkennens*). He declared that the origin of life is subject to physical explanation, but consciousness will always be beyond the limits of science.

2. P. S. Churchland 2002, p. 133.

3. "Phenomenal consciousness" and "prereflective consciousness" describe the experiences involved in feeling, perceiving, and thinking—experiences associated with nonreflective thought. The terms usually used with reference to nonhuman animals or to infants are "primary consciousness," "primordial consciousness," "core consciousness," and "basic consciousness." The cognitive scientist Stevan Harnad (2011) prefers the term "feeling," or sentience, which is equivalent to him to basic consciousness ("sentience" refers to feeling and perceiving, not merely responding to sensory stimuli). A similar term, "Feeling" (always with a capital *F*), was also the preferred term of Herbert Spencer. In both cases, the term denotes what we call "subjective experiencing." Terms such as "second-order consciousness," "access consciousness," and "reflective consciousness" usually refer to forms of (conscious) self-awareness, although these terms are not fully overlapping. Allen (2010) provides a useful and clear analysis of some of the different terms, and the multiple uses of the term "consciousness" and of related terms are summarized and discussed in Revonsuo 2010, chap. 3. We return to some of the distinctions in later chapters.

4. For usage of the term "cognition" in this general sense, see Thompson 2007. For a discussion of the controversies over what counts as cognitive, see Akagi 2016.

5. Aristotle 1984d, 415b, 9–10.

6. "This [self-nutrition] is the originative power the possession of which leads us to speak of things as living at all, but it is the possession of sensation that leads us for the first time to speak of living things as animals; for even those beings which

possess no power of local movement but do possess the power of sensation we call animals and not merely living things" (Aristotle 1984d, 413b, 1–5). The basic categories of soul belong to "plant, beast, and man" (414b, 34). Within the sensitive-soul category of animals, Aristotle discussed different gradations (those endowed with movements and those endowed with imagination). We discuss the evolution of imagination in chapter 9.

7. Aristotle 1984d. Our reading of Aristotle is influenced by Nussbaum 1986; Nussbaum and Rorty 1992; Juarrero 1999; Lennox 2000; Thompson 2007.

8. As Jessica Riskin (personal communication) has noted, many eighteenth- and nineteenth-century writers considered a rudimentary sort of consciousness to extend all the way down to the simplest organisms and even to plants (and, in extreme cases—for example, Diderot and La Mettrie—to inanimate things such as rocks). In the twenty-first century, the biologist Lynn Margulis (2001) suggested that any living entity should be considered as conscious. However, her notion of consciousness seems to be synonymous with cognition—an "awareness" of or selective responsiveness to the surrounding environment. Chalmers (1995, 1996) maintained that subjective experiencing is an aspect of information processing and can be seen as a primitive property of any material entity (including a sugar cube).

9. Yokawa et al. (2018) showed that anesthetics may exert similar effects on plant and animal membranes (including, perhaps, on their electrical properties), but from this it does not follow that plants have consciousness, which is lost upon use of these agents; anesthetics have similar effects on neurons in culture (Platholi et al. 2014).

10. We list these characteristics in chapter 3.

11. Learning by neurons in petri dishes is described in Shahaf and Marom 2001. Spinal learning is reviewed in Allen, Grau, and Meagher 2009.

12. See Lennox 2017 for a discussion of Aristotle's views on the functional unity of biological entities; he shows how recent discussions in the philosophy of biology and theoretical biology that stress developmental systems biology views can be construed as neo-Aristotelian. Alicia Juarrero (1999) has shown that Aristotle's notion of causality, which included teleological and formal causes in addition to material and mechanismic causes, is fully compatible with the modern notion of causality used in systems biology, which takes into consideration the complex relations between wholes and parts. For example, the heart is more than the mere sum of the cells that constitute it; once cells begin to interact to form a heart, the properties of the component cells change, and the cells' morphology and physiology change as the organ begin to shape up and function. We use the term "intrinsic teleology" as shorthand for describing goal-directed systems in which the goal is instantiated by the dynamics of the system. The chemoton, which is described in a later section in this chapter, is a model of such a system.

13. See Fry 2000 for a philosophical analysis.

14. For heuristic purposes, in the sections that follow we discuss lists of characteristics and criteria for life, the organizational principles, the evolutionary scenarios, and the transition marker separately. In the work of the scientists who developed this research program, they were all interwoven.

15. For an excellent discussion of old and new theories of the origin of life, see Fry 2000; a good, readily accessible review of the subject can also be found at http://en .wikipedia.org/wiki/Abiogenesis. A multiperspective analysis on the definition of life can be found at http://www.springerlink.com/content/y854g2x3055p5g78.

16. Eduard Trifonov (2011) analyzed 123 different definitions and characterizations of life and concluded that the consensus is that life is "self-reproduction with variation." However, without a lot of unpacking and clarification, this consensus is not informative.

17. Boden 2009.

18. See Maturana and Varela 1980; Varela 1997. In fact, there was a whole family of systems approaches to life around the same period (Kauffman 1969; Rosen 1970, 1991; Eigen and Schuster 1979; Dyson 1984), but we focus here on Maturana and Varela's and Gánti's models because of their emphasis on both cybernetic and structural (cellular-like) organizational principles.

19. Maturana and Varela's major book was titled *Autopoiesis and Cognition*. For a discussion of autopoiesis and its relation to evolutionary biology, see Etxeberria 2004. Bourgine and Stewart (2004) discuss autopoietic systems as cognitive systems. For a detailed account of autopoiesis and its philosophical implications, see Thompson 2007.

20. See Di Paolo 2009.

21. For a discussion of Gánti's ideas since his 1971 book, computer simulations based on them, and the various approaches they encourage, see Gánti 2003. It is important to emphasize that his model is idealized: a realistic chemoton has to incorporate chemical side reactions, ion fluxes, and a more heterogeneous membrane structure, which may be absolutely necessary for the compartment to work effectively. However, the model provides a specific enough framework for asking concrete questions and devising research agendas.

22. This far-from-self-evident property is due to Gánti's wish to describe a minimal living system that will have all the properties we recognize in living systems on Earth. Since the role of such informational subsystems in life today (in cells) is accomplished by DNA or RNA, Gánti built them into his chemoton.

23. A similar, more recent approach has been taken by Mark Bedau (2012), who has explored what he calls CMP models. The *C* in CMP stands for the "Container" that

keeps the system together (Gánti's membrane system), the *M* for the "Metabolism" that extracts usable resources and energy from the environment (Gánti's metabolic cycle), and the *P* for the "Program" that controls the protocell's processes and carries replicable and inheritable combinatorial information (the informational polymer). As in Gánti's chemoton, the three component systems are coupled so that each supports the operation of the other components. Bedau has analyzed different representations of such a system, which makes it easy to envisage different evolutionary routes leading to a cohesive and fully coupled CMP.

24. The chemoton is a concrete example of such unity, but the unity of formal and teleological accounts can be even more readily understood when we consider an abstract and formal description of an autopoietic system (as shown in figure 1.3). A very general computational model that illustrates the unity of formal and teleological causes and points to the cognitive behavior of the system (minimization of free energy) vis-à-vis fluctuations in the external world has been suggested by the cognitive scientist Karl Friston (2013a). The system he modeled behaves as if it engages in active inference about the state of the world to preserve its internal state.

25. Very good discussions of the chances of the emergence of living organisms, as well as other central problems related to the origin of life, can be found in Ruiz-Mirazo and Luisi 2010.

26. In volume 18, p.18 of *Life and Letters of Charles Darwin* (1887), Francis Darwin included this part of Darwin's 1871 letter to Hookeras as a footnote to an earlier (1863) letter.

27. The discussion in this section is based on Fry 2000.

28. Bernal 1949.

29. Fox and Dose 1977.

30. The different scenarios suggested for the origin of life have been accompanied by thorough theoretical analyses (e.g., by Eigen, Dyson, and Kauffman, whose work is discussed in Fry 2000). The analyses point to the problems inherent in the chemical evolution of such complex autocatalytic systems.

31. For an account of the complexity of the chemical conditions in living and protoliving systems, see Hoffman 2012; this complexity is discussed in relation to Gánti's chemoton in papers in a special issue of the *Journal of Theoretical Biology*, volume 381. For an overview, see Szathmáry 2015b.

32. Maynard Smith and Szathmáry 1995.

33. Function is a central concept in biology, and it has been thoroughly analyzed in the philosophy of biology literature. We use "function" in a more general sense than the original usage of L. Wright (1973), who defined a part's function as the part's

effects, which have evolved though natural selection. We combine Wright's notion with the sense of function used by Cummins (1975), who defined functions in terms of their causal roles in the encompassing system's behavior, rather than with respect to evolutionary theory. Our notion of function is close to that discussed by Kitcher (1993), who sees function as a *general* design-based notion, although we believe that it may be possible to ascribe function to processes and parts operating within the framework of a self-sustaining protochemoton—for example, a chemoton that does not have an informational polymer. For a more detailed discussion of function and of the notion of biological (functional) information, see Jablonka 2002.

34. In preliving systems that are already in the gray zone, "function" is the contribution of parts, mechanisms, or structures to the self-stabilizing behavior of the system (i.e., Keller 2011). However, attributing telos to a self-sustaining chemical loop/cycle and function to the components of such a cycle does not add anything to the description of a system in terms of simple autocatalysis.

35. Bateson (1979) defined information as "any difference that makes a difference."

36. When a chemical system becomes organized in a way that leads to self-maintenance, its parts can be said to have functions. By saying that function cannot be reduced to a chemical description, we mean that an understanding of the system's goal-directed behavior is necessary for a chemical analysis in terms of function to make sense.

37. Kant 1790/1987, part II. For excellent analyses of Kant's critique of teleological judgment, see McLaughlin 1990; Thompson 2007, chap. 6. Interestingly, it was Kant who introduced the notion of self-organization into biology. See Kant 1790/1987, p. 253.

38. Wittgenstein 1922, 6.521.

39. Nagel 1974. Joseph Levine (1983), who accepted Nagel's point, expressed the same idea by suggesting that there is an explanatory (epistemological) gap between mechanisms and functions, on the one hand, and experience, on the other.

40. Chalmers 1995, 1996.

41. Chalmers's position is not to be confused with that of several other scholars who suggest that already existing physical principles (e.g., quantum coherence in the microtubules of neurons), which are not used by most biologists today, have to be incorporated and given center stage to solve the "hard problem." These scholars include Hameroff (1994, 2007) and Cairns-Smith (1996).

42. Nagel 2012. Complex adaptations like the genetic code are also supposed to be illuminated by the addition of the mental facet and the teleological drive.

43. Dennett 1996.

44. Kant (1790/1987, p. 311) discusses the eminent German anatomist Johann Friedrich Blumenbach's vitalist position in the second part of his *Critique of Judgment* (called *Critique of Teleological Judgment*), criticizing Blumenbach's "formative impulse" because it seems to him inscrutable and unsatisfactory for solving the "hard problem" of living. Garrett (2006) discusses the position of an earlier vitalist, Nehemiah Grew, the seventeenth-century English botanist. Grew just could not see how living organization can be reduced to the mechanical reactions of inert matter. He too saw living organization as a "hard problem" (and his solution was to invoke God). It is interesting that a change in the conception of matter—the new conception of matter as dynamic—was necessary for an evolutionary approach to the origin of life. Chalmers could have used this change, in the physicists' views, to support his own contention that an understanding of experiencing will require an analogous change in the view of matter. However, he does not make this comparison, presumably because it would make the understanding of experiencing too close, conceptually, to understanding life. Such an argument is, according to our approach, unnecessary; our current understanding of biology is enabling us to fathom the underlying dynamics.

45. "About the weak points I agree. The eye to this day gives me a cold shudder, but when I think of the fine known gradations, my reason tells me I ought to conquer the cold shudder" (Darwin, in a letter to Asa Gray, February 1860; see F. Darwin 1887, 2:273). For a full discussion, including a discussion about the limitations of the imagination, see Darwin 1872b, chap. 6. In the same chapter (pp. 164–165), Darwin quotes Helmholtz (the great nineteenth-century vision physiologist), who showed that the eye is not the perfect optical organ that Paley and others took it to be and described in detail its many imperfections, for which active behavior must compensate.

46. Thompson 2007. The body-body problem is central to our discussions in the second part of the book.

47. At the conceptual level, we see that people like Alicia Juarrero (1999), Evans Thompson (2007), and, more recently, Terrence Deacon (2011), conduct a systems-biology analysis of life and mind and see them as fundamentally goal-directed and evolved systems. We started publishing about the topic in 2007, and other evolutionary biologists are also developing evolutionary-based accounts of consciousness. Chapter 5 documents existing approaches and part II (chapters 6–10) is devoted to our evolutionary proposal.

48. Dennett 1997. Like Dennett, we do not subscribe to the "ladder" view implied by the generate-and-test tower metaphor; the tower is a diverging tree, not a ladder.

49. We do not endorse Dennett's neo-Darwinian approach to evolution, which, for example, ignores the genomic plasticity of living organisms that enables them to

generate nonrandom mutations (Shapiro 2011). Our view of evolution is presented in Jablonka and Lamb 2005/2014.

50. Skinner 1981 (quoted in Dennett 1995, p. 374).

Chapter 2

1. For an excellent study of theories about the evolution of the mind, see Richards 1987.

2. For a review of associationism focused on the eighteenth century, see R. M. Young 1968.

3. Aristotle (1984c, 451b, 17–24) suggested that several principles of association govern recall: similarity, contrast, and adjacency (in time and space) can all reconstruct past experiences or ideas. Aristotle believed that association takes place in the "common sense," where the look, the feel, the smell, and the taste of an apple, for example, join together to become the idea of an apple. Aristotle, however, did not apply these principles to other faculties of the mind, as did later associationists.

4. Locke called these innate principles "abstract ideas." See Locke 1690, vol. 2, chap. 1. Ideas were explained in terms of sensations alone (without abstract ideas) by later thinkers.

5. Hume added contrast to his three basic principles. However, he suggested that contrast is a mixture of causation and resemblance (Hume 1748/1993, in the section "On the Association of Ideas," p. 15n14).

6. Hartley 1749. For a useful article about Hartley's life and ideas, see plato.stanford .edu/entries/hartley/.

7. Richards 1993. According to Staum (1980, p. 372) and Head (1980, p. 260), the ideologues were members of the Institut National and supported the *écoles centrales*. They were staff members or contributors to the journals *La décade philosophique* and *Le conservateur*. The term "ideology" in the sense of a science of ideas seems to have been first used by Telleyrand in 1797 (Pietro Corsi, personal communication, October, 2013).

8. Much of this discussion of Cabanis is based on Cabanis (1802/1981) and on Staum (1980), who shows the similarities and differences between Cabanis's ideas and those of other scholars.

9. Cabanis 1789–1824/1956, 1:142.

10. Diderot, Erasmus Darwin, and Buffon had somewhat different ideas about this capacity for adaptation and change. For a discussion of their ideas and the effect of sensationalist tradition of evolutionary theories, see Richards 1979; Riskin 2002,

2016. Corsi (1988, 2011) documents the prevalence of evolutionary-like theories during the late eighteenth century.

11. Our account of Lamarck's theory is based on primary sources, as well as Burkhardt 1977 and Corsi 1988. For discussions of the transformation of Lamarckism from the nineteenth century to the present day, see Gissis and Jablonka 2011, chaps. 1–9.

12. Lamarck believed that although "affinities" between different components of matter allow some substances to harmoniously join others, there is still a need for some external binding forces to hold them together.

13. Corsi 2011. Corsi is convinced that Lamarck was an atheist (personal communication, October, 2013).

14. The first principle is described by the famous two "laws" enunciated in Lamarck 1809/1914, p. 113.

15. Lamarck 1809/1914, p. 10.

16. Lamarck's notion of sensibility is different from that of Cabanis and other natural philosophers. For Lamarck, sensibility is a faculty of animals with a nervous system capable of sensation—that is, animals that have a brain. Another faculty, irritability, was for Lamarck a type of reactivity that is specific to animals; because of irritability, stimulation can lead to movement and local contraction. It is, he believed, a property of the particular type of material that animals are made of. For an analysis of Lamarck's approach to the nature of feelings, see Gissis 2010.

17. Although many eighteenth-century scholars thought and wrote about the self and the feeling associated with selfhood, they did not use the term in the physiological sense used by Lamarck. The term "feeling of existence" was introduced in 1794 by the French philosopher and ideologue Maine de Biran (1766–1824), who stressed that the actions of the organism are crucial for the construction of its consciousness.

18. Lamarck 1809/1914, pp. 380–381.

19. Lamarck 1809/1914, p. 386.

20. Todes (2014, p. 82) describes Spencer's influence on young Pavlov.

21. Putting it in modern terms, we would say that breaks in symmetry lead to increased heterogeneity in a system open to the flow of energy and material.

22. Spencer 1862, p. 291.

23. Spencer 1857. Although Spencer (1864/1867) argued that there is an (inevitable) increase in organizational complexity in many lineages, he *did not* think that organic evolution was always progressive and ladder-like, and he discussed this issue many times in his *Principles of Biology* (see, for example, the illustration in vol. 1, p. 303).

The only exception was human evolution, which Spencer believed would prove, in the long run and following many erratic fluctuations, to be progressive, leading to innately cooperative and altruistic humans.

24. Spencer was influenced by and responded to the ideas of the Scottish intuitionist Thomas Reid and his followers (Francis 2007, chaps. 10 and 11). Reid asserted that a "sure mark of a first principle [is] that a belief of it is absolutely necessary in the ordinary affairs of life" (see Francis 2007, p. 173). Spencer "evolutionized" this idea.

25. Spencer 1855, p. 548.

26. See, for example, Spencer 1855, pp. 586, 597.

27. Spencer 1855, p. 530.

28. Spencer 1870, p. 545.

29. For a discussion, see Boakes 1984.

30. Spencer 1870, p. 192. These ideas are further developed in Spencer 1890.

31. Spencer 1855, p. 567.

32. Mayr 1982, p. 386.

33. Maynard Smith 1986, chap. 1.

34. Darwin 1871, 1:39.

35. Darwin 1871, 1:48–49.

36. Darwin 1871, 1:35–36.

37. Darwin's principles are very similar to those enunciated by Spencer a year earlier. Not insensitive to priority claims, Darwin stressed the independence of his work and conclusions, pointing out in note 11 in the introduction to *The Expression of Emotion* (first edition) that he had started writing notes on the subject in 1838.

38. Darwin 1872a, p. 28.

39. Darwin 1872a, p. 28.

40. Ekman, in his introduction to Darwin 1890/1998.

41. We discuss this and Romanes's studies in chapter 5.

42. Richardson 2006, p. 238. For the biography, see Perry 1935.

43. Richardson 2006, p. 293.

44. Richardson 2006, p. 521.

45. James 1890, 1:224.

46. James 1890, 1:137.

47. Richardson 2006.

48. Weismann 1889.

49. Mill 1843, vol. 3, chap. 7, quoted in James 1890, 2:634.

50. C. Wright 1873.

51. James 1890, 2:637.

52. Mill's mental chemistry, which we discussed earlier in this chapter, acknowledges the nonadditive effects of simple associations.

53. It is surprising that James did not refer here to a paper by his friend and mentor Chauncey Wright, who suggested many years earlier (in 1873) that human consciousness, which enables us to become rational, abstracting, scientific, and reflectively self-conscious creatures, is the by-product of an improved memory that enabled the representation of signs, mainly communicative, to be memorized and spontaneously recalled. Even more surprising is the fact that James's own favored cognitive candidate for explaining human consciousness, the ability to reason by analogy, is not discussed within an evolutionary framework. See James 1890, 2:348–360.

54. James 1890, 1:281.

55. James 1890, 1:234.

56. The term was coined in 1882 by E. Robert Kelly, who is known under the pseudonym E. R. Clay, https://en.wikipedia.org/wiki/Specious_present.

57. Unfortunately, James contradicted himself on this point. While in chapter 24 (on instinct) he repeatedly stressed how many impulses and instincts humans have and how these are the basis for their elaborate learning (e.g., James 1890, 2:390), in chapter 22 (on reasoning) he explains that humans are special because they inherit no settled instinctive tendencies (James 1890, 2:368).

58. Lang's theory is focused more on the physiology of emotions than on their subjective facet (which was central to James). Lang also focused specifically on cardiovascular interactions, suggesting that the emotion mediator is located in the vasomotor centers of the brain stem.

59. This point was made strongly in the chapter on movement (chapter 26) in James 1890, vol. 2.

60. For an early and detailed critique, see Cannon 1927. See Lang 1994 and Friedman 2010 for a review of alternative theories.

61. Wittgenstein 1988. It is worth noting, however, that chemicals like fluoxetine (brand name Prozac) that act on the brain (though not the stomach) can relieve some of the suffering of the soul.

62. Theories of emotions based on the James/Lang theory, especially Damasio's theory, are discussed in the next chapter. In chapter 8 we make use of James's framework to explain the nature of mental representations.

63. See James 1890, 1:274.

64. For an excellent analysis and discussion of the history of behaviorism until 1930, see Boakes 1984; for a history of behaviorism and the cognitive revolution, see Baars 1986; see also http://plato.stanford.edu/entries/behaviorism/.

65. See Macphail and Bolhuis 2001 for a comprehensive review. It is important to recognize that the conditional notion of associative learning used by the behaviorists is different from the broader notion used by the early associationists (which we described earlier) and from the narrower notion used by the computational learning theorists of the twenty-first century. The latter assume that when there is concurrent activity of linked elements in a neural network these links are necessarily reinforced. This assumption is based on Hebb's law ("neurons that fire together wire together"). We describe Hebb's law and its connection to conditioning in chapter 7, and in chapter 8 we suggest that the formation of compound percepts is based on dynamic hierarchical models (themselves products of phylogenetic or early ontogenetic learning), which involve both top down and bottom up interactions.

66. Macphail 1982, 1987. See chapters 7 and 8 for discussion of associative learning in all animal taxa.

67. Here we use the terms "conditional" and "unconditional" because they are the correct translations of the original Russian terms; however, since "conditioned" and "unconditioned" are now widely employed, we will use them interchangeably.

68. See Razran 1971.

69. The law of effect was published by the American psychologist Edward Thorndike. It states that activities that produce a satisfying effect in a particular situation become more likely to occur again in the same or a similar situation, while activities that produce a discomforting effect become less likely to occur again.

70. Following Skinner, B. R. Moore (2004, p. 319) argues for a clear distinction between instrumental and operant conditioning.

71. For an account of cognitivism, see Boden 2006.

72. For a good review of the various cognitive studies, including those of Miller, Chomsky, and the gestalt psychologists, see Baars 1986. With regard to Piaget, we are not implying that he was a cognitivist. With his emphasis on the centrality of action

in psychological development, Piaget, a constructivist, fits much better within a pragmatist framework. However, the different European traditions became more widely known and appreciated as behaviorism began declining.

73. As Heyes (2012) puts it, behaviorism was never proven to be false, although it was overambitious, conceptually limited, and jargon-ridden. She rightly argues that behaviorism is a specific interpretation of the much broader study of associative learning (p. 2699).

74. Dennett 1978; Hofstadter 1979; Hofstadter and Dennett 1981.

75. It is important to introduce a terminological clarification here because the term "function" as used by philosophers of mind is not what biologists and most other people mean by "function." For biologists, function is the specific and causal contribution of a part or a process to the goal-directed behavior of an encompassing whole (think of the function of the heart in the body, of chlorophyll in a plant, of a brake in a car, or of rational reflection in a human). For cognitivists, a mental function is the causal relation of a particular mental state to other mental states and to sensory inputs and motor outputs (cognitivists call the biological function "teleofunction"). Functionalists are interested in these relations however they are instantiated. For connectionists, on the other hand, neural-like instantiation is crucial.

76. Searle (1980) imagines himself in a room following instructions for responding to questions in Chinese characters slipped under the door. He does not understand Chinese, yet by following an instruction manual, he produces strings of Chinese characters that are meaningful answers to the questions, making those who send him the Chinese text believe an intelligent Chinese speaker is in the room. Since Searle is imitating a digital computer, this thought experiment shows that although a digital computer may appear to understand the meaning of messages, it does not.

77. P. S. Churchland 1986; P. M. Churchland 1988.

78. Humphrey 1992.

79. See Griffin 1976. We discuss recent studies on animal consciousness in chapter 5.

80. Varela, Thompson, and Rosch 1991; see also Thompson 2007, chap. 1.

Chapter 3

1. For descriptions of these and other techniques used to follow and measure brain activities, see http://thebrain.mcgill.ca/flash/capsules/outil_bleu13.html.

2. NCC is sometimes used for "neural correlate (singular) of consciousness" and sometimes for "neural correlates (plural) of consciousness." We shall use NCC as a singular and NCCs to indicate the plural form.

3. For a discussion of the distinctions among different types of correlates, see P. S. Churchland 2002. The causal mechanisms explaining fundamental properties of consciousness, such as its unity and diversity; its subjective first-person nature; and its emotional, motivational, and intentional aspects, are what Seth (2009a) calls "explanatory correlates of consciousness" (ECC). Stanislas Dehaene (2014) calls consciousness-specific causal neural correlates "consciousness signatures."

4. For attempts to distinguish between the NCC and its prerequisites and consequences, see, for example, Aru et al. 2012; de Graaf, Hsieh, and Sack 2012; Sergent and Naccache 2012.

5. Every consensus has it dissenters. The dissenter of the emergentist consensus is the neuroscientist Samir Zeki (2003), who claims that consciousness is basically disunified, and there are brief microconscious states. For example, in the visual domain, there are microconscious states of location and color that occur at slightly different times and can be decoupled in various pathologies. The microstates bind to form macrostates, which last longer, but the microstates (atomic quale) are nevertheless conscious, though one becomes aware of them only under special conditions.

6. We discuss this issue further in the next chapter. A summary about the relation between consciousness and attention may be found in http://www.scholarpedia.org/article/Attention_and_consciousness.

7. We use the term "temporal thickness" here in a neural sense. While James and the phenomenologists who followed him used the notion of temporal thickness to describe the temporal structure of subjective experience, here the term is used to refer to the temporal persistence of neural reverberations that is a precondition for any experiencing.

8. The ability to learn through positive or negative reinforcement is based on the existence and operation of "value" (reward/punishment) systems and was thought by late nineteenth-century experimental psychologists to be an indicator of phenomenal experiencing.

9. Bjoern Merker's (2007, 2012, 2013) approach, which involves the dynamic construction of a self or an "ego center," is described in chapters 4 and 5 and applied to our own framework in chapter 8. For accounts of hierarchical predictive processing (HPP), see Clark 2013; Seth 2015; Seth and Friston 2016. We describe and discuss this framework in chapter 8, although some aspects of it are anticipated by our earlier discussion of associative learning in chapter 7.

10. Searle 2004.

11. Crick and Koch 2003, p. 119.

12. Logothetis 1998.

13. Crick and Koch paid special attention to the claustrum, a thin sheet of subcortical gray matter lying below part of the cortex. Although little studied, the claustrum is known to receive inputs from virtually all parts of the cortex and also projects back to them; in addition, two-way connections between the claustrum and subcortical structures are involved in emotion. Crick and Koch thought of the claustrum, because of its widespread connections, as the conductor of an orchestra who creates a symphony (an integrated whole) from the individual performances of the players—the activities of the various cortical regions. However, recent studies show that the claustrum is not essential for maintaining consciousness (although it may be important for regaining it after it is lost; see Chau et al. 2015).

14. Crick and Koch 2003; Koch 2004.

15. A few years later, Koch changed his mind about the relation between attention and consciousness (see Koch and Tsuchiya 2007).

16. Crick and Koch 1990. For an overview of binding, see Singer 2007. For early suggestions that binding might come about through the synchronized activity of the neurons that process different features of an object, see Milner 1974; von der Malsburg 1981. Pioneering work by Singer and collaborators (Gray et al. 1989; Singer and Gray 1995) has lent support to the hypothesis by showing that assemblies of neurons in the visual cortex that fire simultaneously represent seen objects.

17. For a discussion of these reservations, see Shadlen and Movshon 1999.

18. Crick and Koch 2003.

19. Koch 2012.

20. For reviews and critiques of these studies, see, for example, Rees and Frith 2007; Weil and Rees 2010; Aru et al. 2012; Dehaene 2014. A review by Koch et al. (2016), based on data from imaging neuroanatomical and neurophysiological studies of human and mammalian brains, locates the NCC of both global and content-specific consciousness in a posterior cortical "hot zone" that includes sensory areas.

21. Llinás 2002.

22. Llinás and Ribary 1993, p. 2078.

23. Ribary 2005; Ward 2011.

24. Not to be confused with his father, Walter Jackson Freeman II, the infamous lobotomist.

25. Axel 2004.

26. Freeman 2000.

27. Since the two orbits in this strange attractor are finite and hence cannot diverge forever, they "fold over" themselves. This process of "stretching" and "folding"

occurs again and again, generating folds within folds and pleats within pleats. A strange attractor is therefore a fractal—a mathematical figure or object, such as a snowflake, that reveals increasingly more detail as it is magnified. Fractals show geometrical regularity, called "self-similarity" or "scale invariance," meaning that when fractal structures are examined at different scales, one meets very similar elements again and again. Chaotic events, displayed by dynamic systems, exhibit similar patterns of variation on different time scales, and in this sense they resemble the geometric scale invariance of fractal objects that show similar patterns on different spatial scales.

28. Freeman 2000.

29. The idea of consciousness as a strange attractor is described by Humphrey (2011); the idea that consciousness can be conceptualized as a strange loop was developed by Hofstadter (2007).

30. Walling and Hicks 2003, 2009.

31. This can be seen in the titles of articles by well-known scholars that began to appear from the mid-twentieth century—for example, Campbell 1965, "Variation and selective retention in socio-cultural evolution"; J. Z. Young 1979, "Learning as a process of selection and amplification"; Popper 1978, "Natural selection and the emergence of mind"; Skinner 1981, "Selection by consequences."

32. See Calvin 1996. The book is available online at http://williamcalvin.com/bk9/bk9ch1.htm.

33. Dennett refers to his early work, which became the basis of his multiple draft model, in Dennett 1991.

34. Changeux, Courrége, and Danchin 1973; Changeux and Danchin 1976.

35. Monod 1971.

36. Changeux 2010.

37. Edelman 1987.

38. Since the "Darwinian" nature of neural evolution is not straightforward because there is no multiplication and replication, the term was strongly criticized by Francis Crick (1989), who sarcastically suggested the term "neural Edelmanism."

39. Edelman and Tononi 2000, p. 85.

40. For the cognitive neuroscientist Victor Lamme (2006, 2010), recurrent processing (reentry), when it involves binding and integrating *corticocortical* interactions (rather than thalamocortical interactions), is the hallmark of consciousness. He claims that this is an objective neural instantiation of a conscious process and that consciousness *should not* be identified with higher (e.g., linguistic) cognitive functions.

41. For many years, "degenerate" has been part of the jargon biologists use when talking about the immune system and the genetic code. It simply means that the same output may be produced from different inputs in many ways.

42. Edelman and Tononi 2000, pp. 87–88.

43. Tononi's IIT theory and the nature of phi are summarized in Tononi 2012, 2015. For an accessible review of the theory (and references to earlier versions), see Tononi and Koch 2015.

44. See Seth 2009a; Seth, Barrett, and Barnet 2011. A causal interaction between elements is said to occur when the past state of signal X can predict the present state of signal Y better than the past state of signal Y alone can predict its present state: this description of causality is known as "Granger causality."

45. Summarized in Baars 1997. In its original form, Baars's model can be seen as a late twentieth-century cognitivist rendition of James's model of consciousness.

46. Prinz 2012.

47. Dehaene and Changeux 2011.

48. Simons and Chabris 1999. A video based on the experiment first performed by Simons and Chabris can be found at http://www.youtube.com/watch?v=v JG698U2Mvo.

49. Fisch et al. 2009; Malach 2012. Malach interprets this single-neuron signature as the result of neural assembly dynamics.

50. Lamme 2006; van Gaal and Lamme 2012. Experiments by Bronfman et al. (2014) show that people could estimate the color diversity of an array of letters even though they were not attending to the whole array.

51. LeDoux 2012. See also the interview in *Edge*, http://www.edge.org/3rd_culture/ ledoux/ledoux_p2.html.

52. Rolls 1999, 2005, 2013.

53. For a short analysis of the history of emotions, see Dror 2001. For evidence of emotions in decerberated humans, see Merker 2007.

54. LeDoux 1996; for discussion of the problems of unifying the notion of "emotions," see Griffiths 2004.

55. Damasio 2010, chap. 5.

56. The account of Damasio's ideas in this section is based on Damasio 1994, 1999, 2010; Damasio and Carvalho 2013.

57. Damasio 2010, pp. 165–166.

58. Hobson 2009.

59. Varela, Thompson, and Rosch 1991; Noë 2009; Northoff 2012.

60. Tononi and Koch 2008, p. 240.

61. Even if maturation of some brain structures can occur without back and forth interactions with the surrounding body, these structures are the products of evolutionarily stabilized changes in neural development that probably did involve such interactions.

62. McGinn 1999.

Chapter 4

1. Bennett and Hacker 2003, p. 405.

2. Searle 2004, p. 131.

3. The "field of lived experience" is a term used by Gallagher and Zahavi (2012, p. 87). Varela's neurophenomenological approach is described, for example, in Varela 1999. For attempts to form bridges between phenomenological and neurobiological approaches, see Thompson 2007; Gallagher and Zahavi 2012.

4. McGinn 1989, pp. 364–365; Chalmers 1995.

5. These characteristics of consciousness (or constraints, as Metzinger calls them) are presented in Searle 2004 and Metzinger 2003, chap. 3.

6. A good and accessible introduction to the philosophy of mind and a critical discussion of the approaches to the mind-body problem, including the various versions of dualism, epiphenomenalism, and physicalism, can be found in Searle 2004. An excellent book about neurophilosophy is P. S. Churchland's *Brain-Wise* (2002) and Metzinger's 2003 book provides an in-depth analysis of consciousness, which is focused on the organizing process of "self" construction. Revonsuo 2010 gives accessible summaries of many views of philosophers of mind; Prinz 2012 analyzes, on the basis of extensive empirical studies, the source of consciousness in humans and argues that consciousness arises when intermediate-level representations are modulated by attention.

7. For one of the first twentieth-century discussions of this distinction, see Maxwell 1968, p. 127.

8. Jackson 1982, p. 130.

9. Animals that are tetrachromatic have four different types of cone cells in their eyes. Tetrachromacy is widespread in birds and exists in some fish, reptiles, and insects. Interestingly, there is some evidence that certain women are endowed with

tetrachromacy, a condition that enables them to distinguish many more shades of color than the rest of us, who are trichromatic (see Jameson and Highnote 2001; Jordan et al. 2010).

10. Dennett 2005.

11. P. M. Churchland 1985, pp. 25–28; 1989, pp. 64–66.

12. Beaton 2005, p. 23.

13. See Avital and Jablonka 2000, chap. 8, for more details and references.

14. Rizzolatti and Craighero 2004; Jabbi, Swart, and Keysers 2007. We discuss mirror neurons and their evolutionary relation with social cognition and the evolution of the social self in chapter 9. The fact that mirror neurons seem to be the outcome of associative learning, as argued by Heyes 2010, is not relevant to this particular argument.

15. Varela 1997; Petitot et al. 1999; Thompson 2007; Noë 2009; Gallagher and Zahavi 2012.

16. Hatfield, Cacioppo, and Rapson 1994; Gallotti and Frith 2013.

17. Nagel 1974, p. 324.

18. Teng and Whitney 2011.

19. Thaler, Arnott, and Goodale 2011.

20. Some philosophers will argue that this is begging the question and that Mary will not, and cannot, have a new experience in this way. Our point is that there is an empirical way of finding out.

21. P. S. Churchland 2002, pp. 192–193. This point was also made by Maxwell (1968).

22. Even if Mary cannot imagine because of inherent limitations to her imagination, she may still be able to distinguish (without *experiencing* color) between yellow, red, and blue (Dennett 2005).

23. Searle 2004.

24. Lamme 2006.

25. Edelman 2005, chap. 7.

26. Aristotle 1984a, 715a, 4–7.

27. For Hilary Putnam's argument, see Putnam 1975. Papineau (2013) adds that causes have to be proportional to the effect. If you think about the precise neuronal arrangement that underlies your wish to drink a cup of coffee and compels you to get up and brew one, this unique arrangement is too specific (and therefore not

proportional) as a cause relative to its effect because even if this neural arrangement were excluded, many other similar arrangements would have made you get up and make the coffee.

28. This definition of information was given by G. Bateson (1972), p. 315.

29. For a discussion of the implications of the nonconservation of information, see Jablonka 2002.

30. A discussion of emergence, which is especially relevant to the approach we present in this section, can be found in Moreno and Mossio 2015, chap. 2. Denis Noble's (2017) discussion about the constraints imposed by the organized whole on the parts has many illuminating examples.

31. Ruby and McFall-Ngai 1992; Jones and Nishiguchi 2004.

32. This type of camouflage by counterillumination employs ventral light from the animal's body to match light coming from the sea surface (see, for example, Haddock, Moline, and Case 2010). To witness the phenomenon of counterillumination, see http://biolum.eemb.ucsb.edu/organism/squid.html.

33. See Szabo-Gendler and Hawthorn 2002 for discussions of the conceivability arguments. For a good argument against Chalmers's position about the conceivable existence of zombies, see Malec 2015.

34. Harnad 2011.

35. Seth 2011, commenting on Harnad 2011. We return to the robot question in chapter 10.

36. For an exposition of these ideas, see Feinberg 2012.

37. This point was also made by Hofstadter (2007). Feinberg believes that the properties and constraints he lists can provide a neurobiological basis for Searle's claim that consciousness is causally but not ontologically reducible. We think that the affordances and inherent limitations of neural organization explain why we *feel* that consciousness is ontologically irreducible and not why *it is* (ontologically) irreducible. We therefore see the resolution as epistemological rather than ontological.

38. Fodor (1981) entertained and expressed such ideas, arguing that the distinction computer science draws between hardware and software makes sense of both the causal and the relational character of the mental claimed by the functionalists.

39. See the discussion on Freeman's model in chapter 3; Varela 1999; Heisenberg 2013. A helpful discussion of embodied and situated cognition can be found in Thompson 2007, chap. 11.

40. Clark 2008. A good critical exposition is provided by Shapiro (2011), who argues that the cognitivists' and dynamicists' views described in chapter 2 are compatible.

41. For effects of actions and sensory-motor feedback, see Noë 2009. For the effects of the immune system of learning, see Ziv and Schwartz 2008. Shapiro (2011, chap. 5) summarizes neurodynamic models of embodied cognition.

42. This discussion is based mainly on Levin 2012, 2013 and Tseng and Levin 2013. For additional publications by the Levin group, see http://ase.tufts.edu/biology/labs/levin/publications/.

43. James 1890; G. E. Moore 1903; Metzinger 2003, 2007, 2009; Revonsuo 2005.

44. The poem is by Ikkyu Sojun (1394–1481), a Zen Buddhist monk. See Berg 2000.

45. Gabriel 2015; this argument was made with regard to visual color perception by Chirimuuta (2015).

46. For a discussion of the need to have reflective awareness or higher order thoughts (HOTs), see chapter 5.

47. Zahavi 2003; Nelson et al. 2009.

48. Merker 2005.

49. Von Holst and Mittelstaedt (1950/1973) based their model on experiments performed with hover flies. Sperry (1950), who worked on fish, used the term "corollary discharge" to denote the motor commands that influence sensory processing. Poulet and Hedwig (2007) and Crapse and Sommer (2008) provide good reviews of reafference. We will return to this topic when we discuss the evolution of the nervous system in chapter 6. The work on the complex system underlying reflexes and reflex control through motor-sensory feedback is described by Edwards, Heitler, and Krasne (1999).

50. Merker 2005, 2007. Morsella (2005) emphasizes the central role of action selection, which relies on potentially conflicting skeletal motor plans that require supramodular integration. He suggests that consciousness is necessary for suppressing action tendencies, such as pain withdrawal, blinking, or reaction to muscle fatigue, in order to reach a goal.

51. For a full account of his views, see Metzinger 2003, 2009. For a concise exposition of his self-model, see Metzinger 2007.

52. The ego tunnel is defined in Metzinger 2009, p. 6. Metzinger emphasizes the complex dynamics of the construction of self by discussing robots that are built with the ability to form an internal and updatable model of their body, the world, and the relation between their body and the world yet cannot be said to be conscious (Blanke and Metzinger 2009; Metzinger 2009; Seth 2009a).

53. Blanke and Metzinger 2009, p. 8; Metzinger 2009.

54. For examples of pathologies resulting from the breakdown of the minimal self-model, see Metzinger 2003, 2009; Blanke and Metzinger 2009.

55. Seth 2013; Seth and Friston 2016.

56. Hofstadter 2007, p. 203.

57. Hofstadter 2007, p. 205.

58. Dennett 1991.

59. More recently, Humphrey (2015) suggested another metaphor, that of consciousness as a work of art, to highlight the artistic genius of the virtual reality generated by the brain and to avoid the unsettling implication of it being a mere illusion. Humphrey is not suggesting an implementation of a Cartesian theater but rather is attempting, like Dennett, Hofstadter, and Metzinger, to stress the ongoing generation of a feeling of self and distinguish it from the prevalent reified notion of the self.

60. Ramachandran 2005; Ramachandran and Rogers-Ramachandran 1996. In Michael Levin's terms, such body images can be thought of as CNS "target morphologies" that got locked into vicious circles.

61. Metzinger 2009, pp. 98–101. The virtual body illusion experiment is one in which Metzinger took an active part (Lenggenhager et al. 2007), exemplifying the intimate relation between conceptual analysis and scientific work.

62. See, for example, Slater and Sanchez-Vives 2016.

63. Dennett 2005.

64. Romanes 1883; Baldwin 1896b. We return to these ideas in chapters 6–9.

65. Seth 2009b, p. 286. Note that integration of complex visual stimuli can occur subconsciously, without awareness (Mudrik et al. 2011). It is only when the integration is shown to be capable of leading to flexible, goal-oriented motor behavior that awareness is required (Morsella 2005).

66. Block (2007, 2008) bases his claims on experiments carried out in the 1960s, which show that report to be limited to a certain number of items: when subjects were presented with an array of many letters, they claimed to have seen all the letters but could report on only four or five. However, when given certain cues, they could recall more, thus showing that all letters were indeed experienced and registered in working memory. Lamme (2006) claims on the basis of many similar experiments and various other types of neurobiological data that failures of reportability and failures of consciousness are not identical and that to make sense of the mismatches between them, the notion of consciousness must include certain neurobiological processes. Lamme suggests that recurring processing is a defining feature of consciousness, even when reportability is absent.

67. For a good discussion of this aspect of experiencing, see Gallagher and Zahavi 2012.

68. Searle 1999.

69. Searle (2013) makes a similar point, emphasizing the multiple functions that are entailed by consciousness. However, Searle believes that thinking in terms of mechanical causation and function over evolutionary time (Darwinian evolution) has *replaced* teleology. As we and other scholars (mentioned in chapter 1) believe, intrinsic teleology has not been replaced; it has been *explained* by evolutionarily informed systems biology. We return to the subject of the telos of consciousness in chapter 10.

70. Tinbergen did not assign a "cause" status to subjective states because he believed that they cannot be scientifically studied and observed: "The ethologist does not want to deny the possible existence of subjective phenomena in animals, he claims it is futile to present them as causes, since they cannot be observed by scientific methods" (Tinbergen 1951, p. 5; see also Tinbergen 1963).

Chapter 5

1. For the full declaration, see http://fcmconference.org/img/CambridgeDeclaration OnConsciousness.pdf.

2. See Liljenström and Århem 2008; Allen and Trestman 2015. The phrase "who problem" is used by Prinz (2005).

3. G. J. Romanes 1896, p. 14.

4. Burghardt (1985) presents a summary and discussion of G. J. Romanes's views.

5. Boakes (1984) provides a clear and balanced discussion of G. J. Romanes's work on mental evolution and its critical reception.

6. G. J. Romanes 1883, p. 20.

7. See G. J. Romanes 1883, p. 62.

8. G. J. Romanes 1883, p. 61.

9. Morgan 1903.

10. Liljenström and Århem 2008.

11. http://fcmconference.org/img/CambridgeDeclarationOnConsciousness.pdf.

12. People with blindsight can process visual cues but do not have subjective visual experiences because of lesions to the primary visual cortex. See Rosenthal 2005 for a discussion; for a critique of the blindsight case, see Merker 2007. For a clear philo-sophical analysis of the problems inherent in HOT and similar views, see Tye 2017, chaps. 2 and 3.

13. Carruthers 2000.

14. Rolls 1999, 2005, 2013.

15. Macphail 1998, 2008. Descartes and his followers held this opinion, which is described and discussed in Tye 2017, chap. 3.

16. See Allen, Grau, and Meager 2009. Ganglionic learning has been documented in insects. Leg withdrawal can be conditioned in insects—for example, in headless cockroaches or even in isolated leg and thoracic ganglion preparations (discussed in Torley 2007, p. 242).

17. Rose's influential article, published in 2002 in *Reviews in Fisheries Science*, was written at the request of recreational fishing societies in the United States to counter the arguments of pain researchers that angling causes fish pain and therefore should be stopped. Lynn Sneddon (2011), who argued that fish do feel pain, attempted to respond to a later paper by Rose et al. (2014) submitted to the journal *Fish and Fisheries* in 2012, but her response was not accepted for publication; it was eventually published elsewhere (Sneddon et al. 2014). See Tye 2017, chap. 6, for a detailed discussion of fish pain and fear. Rose's argument has been repeated by Brian Key (2016), and detailed answers and discussion of his claims (including their frail factual basis) were published in *Animal Sentience* 1(3), http://animalstudiesrepository.org/animsent/vol1/iss3/.

18. For the suggestion that crabs and insects feel pain, see Elwood and Adams 2015; Tye 2017, chap. 8. The idea that bees may not feel pain is based on their failure to consume a readily available morphine solution when clips were affixed to their legs to create the sensation of a continuous pinch (Groening, Venini, and Srinivasan 2017). However, as the scientists suggest, this means that pain in bees, which they regard as subjectively experiencing animals, may be different from pain in vertebrates and must be further studied.

19. For a summary of these arguments, see Macphail 2008, pp. 115–116.

20. Recent experiments suggest that the cause of this amnesia may be totally different: neurogenesis in the brains of young children may interfere with the storage of long-term memories. See Akers et al. 2014.

21. Seth, Baars, and Edelman 2005.

22. Edelman, Baars, and Seth 2005; Edelman and Seth 2009. These authors also suggest that cephalopods may be conscious.

23. Butler et al. (2005) show that the critical structures (according to the theories of Crick and Koch and Edelman and Tononi) that are assumed to be necessary for consciousness in mammalian brains (see chapter 4) have homologous counterparts in avian brains. Rial et al. (2008) and Århem and Liljenström (2008) discuss the scenario in which consciousness originated independently in birds and mammals and compare it to other scenarios such as it emerging in the reptilian lineage.

24. See Pepperberg 2002.

25. A comprehensive comparison of mammalian, avian, and reptilian brain characteristics that are linked to consciousness is given in Århem and Liljenström 2008, table 4.2, p. 85.

26. See Northcutt 2002 for a discussion of bony-fish brains, the size of which overlaps with that of birds and mammals.

27. Cabanac 1992; Cabanac, Cabanac, and Parent 2009. See also Rial et al. 2008; Århem and Liljenström 2008.

28. Mikulka, Vaughan, and Hughes (1981) discuss learned aversion in toads. Behavioral sleep-like states in tree frogs were reported by Hobson, Goin, and Goin (1968). Volkoff (2012) reviews data showing that orexins (hypothalamic neuropeptides that regulate sleep in other vertebrates) are present in amphibians and seem to control arousal. According to Burghardt (2005), poison frogs have been observed to engage in what looks like play.

29. See Allen 2012.

30. Olds and Milner 1954; see also Shizgal 1999.

31. Berridge 2003; Berridge and Kringelbach 2013.

32. Craig 2002, 2009.

33. Rolls 1999, pp. 60–61. LeDoux (2012) thinks that fear behavior and physiology (and other "emotions") in nonhuman animals are not indicators of subjective experiencing.

34. Panksepp 1998, 2005, 2011a, 2011b; Panksepp and Biven 2011.

35. This alternative "dimensional" view is described and reviewed by J. A. Russell 2003; Scherer 2009; Mendl, Burman, and Paul 2010; James, Gross, and Barrett 2011.

36. Denton 2006, p. 205.

37. Denton 2006, p. 7.

38. As Merker (2007) makes clear, this proposal was discussed in the 1950s but later dismissed.

39. These structures include the hypothalamus, which forms the floor of the diencephalon; the superior colliculus that forms the roof of the midbrain; and the ventral thalamus, the periaqueductal gray, the ventral tegmental/substantia nigra, and the centers serving navigation between the midbrain reticular formation.

40. Alkire, Hudetz, and Tononi 2008; Långsjö et al. 2012; Gili et al. 2013; Mashour and Alkireb 2013.

41. Merker 2007, 2016. See Feinberg and Mallatt 2016 for arguments on the central role of the interaction between the tectum and the pallium and Woodruff 2017 for an emphasis on the role of the pallium. Balcombe (2016) describes the sophisticated behavior of fish, and Braithwaite (2010) discusses the behavioral and neural evidence for the proposition that fish feel pain.

42. See Griffin 1976, 2001; Griffin and Speck 2004.

43. Trestman (2013a, 2013b) argues for the presence of highly complex embodied cognition in arthropods and suggests that it evolved during the Cambrian era.

44. Gilissen 2007.

45. Barron and Klein 2016.

46. Fiorito and Scotto (1992) discuss observational learning in octopuses; Mather (2008) reviews behavioral, neural, and developmental evidence suggesting that octopuses may have a global neural workspace (GNW) and primary consciousness. Montgomery (2015) and Godfrey-Smith (2016a, 2016b) describe octopuses' cognition and behavior in ways that allow the reader to imagine their world.

47. The Cambrian era is named after Cambria, the Latinized form of Cymru, the Welsh name for Wales, where extremely well-preserved fossilized deposits were found in exposed rocks.

48. Our argument that associative learning and minimal consciousness emerged during the Cambrian explosion and drove it was presented in Ginsburg and Jablonka 2007a, 2007b, 2010a, 2010b.

49. Feinberg and Mallatt 2013, 2016.

50. In addition to Barron and Klein (2016) and Feinberg and Mallatt (2013, 2016), Fabbro et al. (2015) suggested that many vertebrates have early evolved limbic structures underlying primary consciousness, and Godfrey-Smith (2016a) presented a general picture of the evolution of consciousness that is very similar to our early proposals.

51. Dehaene 2014.

52. Some specific suggestions limit consciousness to explicit goal-directed behavior (Dickinson 2008) or to actions that involve conflicts with skeletal muscle plans (Morsella 2005).

53. The analogy with life and a more detailed discussion of the idea that UAL is the evolutionary transition marker of consciousness are given in Ginsburg and Jablonka 2015. The idea that open-ended associative learning is an evolutionary marker of consciousness was first described in Ginsburg and Jablonka 2007b.

54. Proposals that flexible behavior is a mark of consciousness include Dickinson and Balleine 1994; Dretske 1999; Lovibond and Shanks 2002; Ginsburg and Jablonka 2007a, 2007b, 2010a; Cleeremans 2008, 2011.

55. See Allen, Grau, and Meagher 2009 for a discussion of spinal learning.

56. Several of the various definitions of learning are presented in Roediger, Dudai, and Fitzpatrick 2007.

57. Bergson (1896/1991) presented his highly influential view of memory and the distinctions between different forms of memory in *Matter and Memory*, first published in 1896.

58. Danziger (2008) provides a historical and conceptual discussion of the concept of memory. For a shorter review of the conceptual intricacies involved, see also http://plato.stanford.edu/entries/memory/; for a developmental approach to autobiographical memory, see Nelson and Fivush 2004.

59. Many scholars have found fault with the old literacy metaphor that suggests that past events are imprinted on the mind, like marks on malleable wax or as words on paper, and then "read." Metaphors of storage—of a container (a house, a cage, an attic) in which memories are stored and then found or the more recent computer-based "memory storage"—all imply that "storage" and retrieval are passive and therefore miss the active, constructive role of the individual in the process. The metaphors used to describe memory and the problems these metaphors cause are discussed in Danziger 2008. The forgotten essay by Robert Hooke about memory, published in 1705, describes a dynamic model of memory encoding, storage, and retrieval that can be interpreted as being in line with current memory research (Hintzman 2003).

60. Semon's work on memory is discussed, within the context of his time, in Schacter 2001. Semon's own term for processes of encoding and retention is "engraphy," and he terms retrieval "ecphory" (Semon 1904/1921). Poo et al. (2016) review the different molecular and structural instantiations of engrams among and between cells, and Josselyn et al. (2017) provide an overview of the term "engram" and the main scholars who contributed to its study since Semon introduced it. Tonegawa et al. (2015) expand Semon's framework by introducing concepts such as "engram cells" (neurons in which a retrievable engram, an "engram component," is stored) and "engram pathways" (connections among engram cells within a network, generating a given memory); such cells and relations comprise an "engram complex." We return to Semon's ideas about memory, linking them to epigenetic memory within cells, in later chapters. As Semon stressed, in many cases encoding and retention cannot be decoupled—for example, when the process of encoding involves positive feedback, leading to self-sustaining processes. Moreover, retrieval always involves encoding and may include the destabilization and restabilization (with modifications) of retained internal states. Nevertheless, the distinction between

encoding, storage, and retrieval remains important. As we indicated when discussing Macphail's view of consciousness, unfortunate confusion arises because the fact that the retrieval of associatively learned information can be nonconscious and implicit was seen as proof that the encoding process, too, does not involve consciousness.

61. The notion of such an external memory trace was suggested by Donald (1991) in the context of his discussion of human evolution.

62. Semon's major book, *The Mneme* (translated into English in 1921), was first published in 1904 and ran into three editions. It generated some critical interest, but his views were criticized and eventually discarded and forgotten, although, as Schacter (2001) documents, in the 1970s, after decades of neglect, his conception of the multiple distribution of engrams and of retrieval as an active constructive neural process became today's common wisdom. His notion of resonance among engrams describes reentrant interactions among different engrams. The evolutionary sophistication of the three facets of memory that Semon highlighted, and their manifestations at different levels of biological organization, can account for the progressive (and branching) evolution of associative learning.

63. We developed the operational definition of UAL with Zohar Bronfman (Bronfman, Ginsburg, and Jablonka 2016b), so throughout this subsection "we" is triadic rather than dyadic.

64. Discussions of "perceptual fusion" or "spontaneous configuration" can be found in Razran 1965; Konorski 1967; and Bellingham and Gillette 1981. See Bronfman, Ginsburg, and Jablonka 2016b for further discussion.

65. Young et al. (2011) examine some forms of nonelemental learning in *Drosophila*. Honey, Iordanova, and Good (2014) summarize a series of sophisticated experiments showing the obligatory role of the hippocampus in nonelemental learning in vertebrates and discuss the relations between elemental and nonelemental learning.

66. For experiments and discussion of second-order conditioning, see Pavlov 1927; R. A. Rescorla 1976.

67. Enquist, Lind, and Ghirlanda (2016) show (through modeling) that chaining simple reinforced behaviors into a long sequence can explain complex behaviors like tool use, social learning, self-control, or expectations of the future.

68. Global accessibility is central to the GNW model presented in chapter 4.

69. Crick and Koch (2003) and Edelman (2003) stress the importance of binding.

70. James (1890) discussed the central role of discrimination in cognition and consciousness. Razran (1971) and Wells (1968) highlighted the role of discrimination in the evolution of complex forms of learning.

71. The hot plate example is described and discussed in Morsella 2005.

72. The experiment is described in Reddy, Reddy, and Koch 2006.

73. This topic is discussed in Edelman 2003. Dudai, Roediger, and Tulving (2007) offer a discussion of complex forms of learning that involve the formation of representational maps.

74. For an example of the late stage of visual processing of a multifeature percept, see Bodelón, Fallah, and Reynolds 2007.

75. See, for example, Seth 2013; Tsakiris and Haggard 2005. Additional references can be found in Bronfman, Ginsburg, and Jablonka 2016b. We discuss the construction of perception, affect, and action that UAL requires in chapter 8.

76. See Öhman and Soares 1993 and Soares and Öhman 1993 for evidence that the masking of a compound neutral stimuli does not support learning, while the masking of stimuli eliciting innate responses does. Note that the experiment was less demanding than what we demand as a proof of UAL: no discrimination between pictures with the same but differently arranged elements was tested, nor was second-order learning investigated (although humans are known to form discriminable compound patterns that can be the basis of cumulative learning).

77. The neurologist and psychologist was Edouard Claparède. See Claparède 1911.

78. We are aware of only one study that found unconscious instrumental learning using a nonreflex CS: people who were presented with masked symbols (made of lines and circles) were asked to press a response key (if they thought they might be rewarded). Each masked symbol was associated with either a monetary reward or with punishment. The researchers found that the participants' tendency to act (press the key) on the unseen cue increased for positively reinforced masked cues and decreased for negatively reinforced ones. For details, see Pessiglione et al. 2008. A more recent study, which does not, however, involve compound stimuli nor reinforcement, was performed by Scott et al. (2018). They showed that an association can be formed, without reinforcement, between two noncompound, masked stimuli across two modalities. Clearly, more studies of the limitations of unconscious learning needs to be performed to test the predictions of our model.

79. See Perry, Barron, and Cheng 2013.We summarize the information in chapter 8. Based on our suggestion that associative learning drove the Cambrian explosion, Trestman (2013b) proposed that object-oriented spatial cognition (a complex type of AL) required correspondingly complex bodies, with distal senses and articulated appendages that could move flexibly and manipulate objects, and that the emergence of animals with such bodies—the arthropods—drove the Cambrian explosion. We return to this idea in chapters 8 and 9.

Introduction to Part II

1. Dennett 1995.

2. This type of approach *is not* progressionist: it does not assume that an increase in complexity is inevitable; it accepts that evolutionary regressions occur quite frequently and that evolution is often mosaic and parallel. However, no one would deny that one consequence of evolution is an increase in complexity in some lineages.

3. Maynard Smith and Szathmáry 1995.

4. Szathmáry (2015a) summarized the progress made since his 1995 book with Maynard Smith was first published and suggested several amendments to the original classification. He merged transitions 1 and 2 into a single transition (the transition to protocells); omitted the transition to sexuality from the list, regarding meiosis and syngamy (the fusion of two cells) as a coevolving facet of the transformation to a higher evolutionary unit (the eukaryotic cell); and added a new transition, that to plastids incorporated within the eukaryotic cell, to the list.

5. For a discussion, see Jablonka and Lamb 2006. Szathmáry (2015a) regards the origin and evolution of new Darwinian systems within an existing hierarchy, such as the nervous system in most metazoans and the adaptive immune system in vertebrates, as "filial transitions," which contribute to the complexity and richness of the evolutionary transition-based view.

6. Gilbert 2003.

7. The price equation is $w_{av} \bullet \Delta z_{av} = Cov(w,z) + E(w \bullet \Delta z)$. The individual's character value is z, its number of offspring is w (or more generally, its fitness), and the difference between the character values of itself and its offspring is Δz. The subscript av denotes average; Cov and E denote covariance and expectation. The equation has two terms: the first describes the change in the frequency of a trait following selection in this broad sense—a statistical covariance between individuals' character values and their relative abundance. The second term describes changes in the trait value due to the process of transformation from one "generation" (understood in a very wide sense) to another, which in the genetical case would be the result of changes due to recombination or mutation. Because of its generality, Price's equation has to be qualified when applied to specific cases, and this has been productively done. It was used, for example, to derive formulations that can be applied to several levels of selection (e.g., individuals and groups). See Gardner 2008 for a primer on the Price equation.

8. At the most general level, Price was interested in the relation between selection and information, and hoped that a link will be forged between information theory and selection theory. Such a link is now actively explored by theoretical evolutionary

biologists. See, for example, Bergstrom and Lachmann 2004; Donaldson-Matasci, Bergstrom, and Lachmann 2010; Frank 2012; Lachmann 2013. Price's letter is cited by Gardner 2008.

9. See, for example, Stolovitzki et al. 2006; Stern et al. 2007. The work of this group is reviewed in Braun and David 2011.

10. West-Eberhard 2003.

11. Deneubourg et al. 1989.

12. Popper (1972) and Campbell (1990) discuss cultural processes in these terms, and Plotkin (1997) discusses learning in animals in terms of selective processes. For selective stabilization during spindle formation, see Gerhart and Kirschner 1997; West-Eberhard 2003; Kirschner and Gerhart 2005; and Tavory, Jablonka, and Ginsburg 2012. Tavory, Ginsburg, and Jablonka (2014) provide many additional examples.

13. Gerhart and Kirschner 2007.

14. Soen, Knafo, and Elgart 2015.

15. Levi-Montalcini et al. 1996; Buss, Sun, and Oppenheim 2006.

16. Reviewed by Munno, Naweed, and Syed 2003; Colón-Ramos 2009; Tripodi and Arber 2012.

17. Fernando, Szathmáry, and Husbands (2012) challenge the idea that S-selection can account for complex neural learning. They argue that replication-based evolutionary change can lead more readily to complex adaptations than sample selection and suggest that evolution occurs in complex brains through replication-like processes (of synapses or patterns of neural activity).

18. See chapter 2 for a discussion of Bain's, Spencer's, and Skinner's views. For an accessible book about learning by consequences, see Schneider 2012.

19. Friston 2010, 2012.

20. Free energy is expressed as energy minus entropy. The estimation of variational free energy is based on methods that allow an approximate Bayesian inference, given some data, and a model for the generation of those data. As Friston (2010, 2012) has shown, neural processes, such as those involved in perception and learning, can be measured in terms of variational free energy, which is minimized as the animal adaptively changes its state (a change that may be mediated by neuromodulators, such as dopamine and acetylcholine). The value of the decrease in free energy can be calculated given information about the animal's sensory state and about the causes of data distribution in the environment, which are inferred from the internal states that represent them. However, the number of internal states a nervous system can be in is not known, and the measure depends on the theory one adopts about the

relation between the microstates measured and a macrostate the system is in. A proxy for a decrease in variational free energy must be used instead—for the case of learning, the proxy can be the decrease in prediction error.

Chapter 6

1. IZQuotes, http://izquotes.com/quote/55876.

2. Motile sponge larvae are 50 µm to 5 mm in size, and placozoans range from less than 0.5 mm to 3 mm.

3. The significance of the movement of muscle sheets for understanding the organization and the evolutionary origin of neural animals was emphasized by Pantin (1956) and Passano (1963). Keijzer, van Duijn, and Lyon (2013) developed these ideas in light of recent information about animal evolution. Keijzer and Arnellos (2017) focused on the epithelial features of premuscle contractile tissues and the importance of mechanotransduction.

4. The importance of coordinated movements in internal organs was highlighted by Jékely, Keijzer, and Godfrey-Smith 2015.

5. See Pantin 1956; Brembs 2011b; and Heisenberg 2013.

6. The information metaphor was attributed to Adrian by Garson (2003).

7. In their discussion of the requirement for complex behavioral agency, Moreno and Mossio (2015, chap. 7) stress what they call the partial "decoupling" of neural signaling from specific metabolic organization. Since the circulatory systems and the immune systems also display this type of behavior, we do not see decoupling as the best way of describing the uniqueness of the nervous system. We stress instead the "common language" of electrical signaling that allows cells with different metabolic states and biochemical profiles to communicate in highly specific ways (somewhat like language in a human linguistic community).

8. In his Nobel lecture, Ramón y Cajal (1906) explained at length why he adopted the cellular view.

9. There are some exceptions to this general rule about the one-way transmission of nerve impulses, notably in certain neurons that make up nerve nets (e.g., in cnidarians).

10. The invention of the term "synapse" is credited to the British classics scholar Arthur Verrall. The word first appeared in 1897 in part 3 of the seventh edition of Michael Foster's *A Textbook in Physiology*; Sherrington assisted in writing this part ("the nervous system"). For a short historical account of the choice of the word "synapse," see Tansey 1997. For a fuller account, see Shepherd and Erulkar 1997.

11. There are also electrical synapses between neurons ("gap junctions") that pass action potentials without transducing the electrical signal into a chemical signal; the gaps across such synapses are extremely narrow. Electrical synapses do not seem to contribute to neural plasticity, and we do not discuss them here.

12. Hartline 2011.

13. Santello and Volterra 2009; Tremblay 2011.

14. Gehrmann, Matsumoto, and Kreutzberg 1995.

15. G. P. Morris et al. 2013.

16. If there are three types of memory, neural network models based on synaptic memory alone do not do justice to real biological neural networks and cannot account for their performances. We say more about the cell memory systems and their relation to learning in the next chapters.

17. Maynard Smith and Szathmáry did not include the neural transition among those described in their 1995 book *The Major Transitions in Evolution*, although they later realized that this was an omission and intended to add it to a later edition. For a discussion of this omission, see Szathmáry and Fernando 2011, p. 301. The justification for recognizing the transition to neural animals as a major evolutionary transition is given by Jablonka and Lamb (2006).

18. For the documentation of homologies between channel proteins in prokaryotes and different eukaryotes, see Liebeskind, Hillis, and Zakon 2011; Cai 2012; Kristan 2016.

19. Originally, the PSD was identified by electron microscopy as an electron-dense region at the membrane of a postsynaptic neuron, in close apposition to the presynaptic active zone, the area from which transmitter molecules are released into the synaptic gap that separates the presynaptic and postsynaptic cells.

20. Sakarya et al. 2007.

21. Kleinenberg 1872; his contributions are described by Anctil 2015, *Dawn of the Neuron*, which presents the history of the nervous system.

22. Mackie 1970.

23. Jékely 2011.

24. Moroz 2009. Brunet and Arendt (2016) take the damage hypothesis further. They suggest that the depolarization-contraction-secretion coupling seen in neurons evolved from an emergency response in unicellular stem eukaryotes to uncontrolled and toxic calcium influx following membrane rupture.

25. Keijzer, van Duijn, and Lyon 2013. For a discussion of the epithelial features of cell sheets from which muscles and nerves may have evolved, see Keijzer and Arnellos 2017.

26. For a discussion of the role of the nervous system in development, see Cabej 2013. We return to this idea in later chapters.

27. Rusten et al. 2002; Denes et al. 2007; Arendt 2008; Martín-Durán et al. 2018.

28. For a discussion of Xenacoelomorpha and their taxonomic status, see Philippe et al. 2011; Telford, Budd, and Philippe 2015; Rouse et al. 2016.

29. Maxmen 2016.

30. Erwin 2015.

31. For different views about basal animal phylogeny, see Pang et al. 2010; Ryan et al. 2010; Edgecombe et al. 2011; Simmons, Pang, and Martindale 2012; Ryan et al. 2013; Cavalier-Smith 2017; Shen, Hittinger, and Rokas 2017.

32. Leys 2007. For recent claims about the status of sponges as the sister group to all animal phyla, see Pisani et al. 2015; Cavalier-Smith 2017. There is a dispute about the timing of the first appearance of the group. Love et al. (2009) claim that traces of sponges were already evident 635 million years ago. Antcliffe, Callow, and Brasier (2014) argue that confirmed sponge fossils are only evident at the beginning of the Cambrian; Yin et al. (2015) present evidence of a 600 million-year-old sponge fossil.

33. Leys, Mackie, and Meech 1999. Much of the body of glass sponges is composed of syncytial tissue. Electrical signaling does not seem to occur in cellular sponges (Leys 2015).

34. Renard et al. 2009.

35. Sakarya et al. 2007; Kosik 2009; Nickel 2010.

36. Emes and Grant 2012.

37. See Newman and Bhat 2009 for a description of the physical, chemical, and cybernetic constraints and affordances that shaped the first morphological forms in pre-Cambrian, nonneural small animals. In chapter 9 we return to the idea that the nervous system, particularly the CNS, played an important role in the evolution of standardized morphologies. We do not assume that morphological specificity *depends* on a nervous system (flowers can have a highly standardized morphology) but believe that once a nervous system had evolved, it led to the standardization of morphology in animals.

38. Dunn et al. 2008; Ryan and Chiodin 2015. For powerful arguments against this suggestion, see Leys 2015.

39. Some researchers argue that jellyfish and comb jellies have ganglia-like concentrations of neurons. There is also dispute over the claim that ctenophores have only two germ layers, as discussed, for example, in Röttinger, Dahlin, and Martindale 2012.

40. Shu et al. 2006.

41. Moroz et al. (2014) base their conclusions on the systematic analysis of the *Pleurobrachia* genome and ten other ctenophore transcriptomes. Ctenophores have unique protein components in their gap junctions, and their synapse has an unusual structure. They also lack many other molecular characteristics that most animals share; for example, they do not have microRNAs (major regulators of gene expression in other organisms), they do not possess *Hox* genes that encode transcription factors necessary for forming the body axis, and they do not have several classes of genes that in other animals are essential for immunity (Maxwell et al. 2012; Ryan et al. 2013). Marlow and Arendt (2014) argue, however, that it is premature to accept the idea that neurons appeared separately in ctenophores and other animals. For a debate about this issue, see Telford, Moroz, and Halanych 2016.

42. Albert (2011) describes the varied behaviors of *Aurelia*.

43. For a historical account of Romanes's contributions, see Anctil 2015, chap. 5.

44. Cnidarian nervous systems are described in Mackie and Meech 1995a, 1995b, 2000; Spafford, Spencer, and Gallin 1998; Anderson 2004; Meech and Mackie 2007.

45. McFarlane 1982; the three conducting systems have been found in other sea anemones and even in corals, although their functions are not entirely clear. McFarlane, Graff, and Grimmelikhuijzen (1989) have suggested that they act as command fibers to control tentacle and sphincter muscles.

46. The nerve ring connects the rhopalia and the pedalia (points of attachment of the tentacles), and several sets of clusters, or "ganglia," are in the path of the nerve ring: four radial clusters situated where the nerve ring branches to connect to the rhopalia and four pedal clusters near the points from which nerves set off from the ring to the tentacles. The rhopalia are filled with nerve fibers and cell bodies, referred to as a "rhopalial ganglion" (Coates 2003). The optics of cubozoan eyes are described by Nilsson et al. (2005).

47. Garm, Oskarsson, and Nilsson 2011.

48. Examples are Satterlie (2011) and Koizumi et al. (2015), who consider nerve rings in jellyfish as CNS-like structures; Albert (2011) goes even further, claiming that the nervous system of *Aurelia* includes a brain.

49. See Perry, Barron, and Cheng 2013 for a discussion of the importance of knowing what animals *cannot* do when using a comparative approach to understanding learning and cognition.

50. We first used this metaphor to describe the neural buzz of preconscious overall sensation in Ginsburg and Jablonka 2007a.

51. "Simple reflex" refers to a response system with only two or possibly three neurons interposed between receptor and effector. For more complex nervous systems, the notion of simple reflex is inappropriate.

52. See Red'ko, Prokhorov, and Burtsev 2004 for a short description of functional system theory and its application to the construction of an animat (a simulated animal). The architecture of a functional system requires thinking in terms of anticipatory models in the nervous system.

53. Grosvenor, Rhoads, and Kass-Simon (1996) describe the feeding behavior of *Hydra*.

54. Hooper 2001; for CPGs in the box medusa, see Stöckl, Petie, and Nilsson 2011.

55. Albert (2011) provides a detailed description of the behavior of *Aurelia*.

56. Razran 1971; Dyal and Corning 1973. Marinesco, Duran, and Wright (2003) point out that general sensitization can be more or less general, depending on the proximity of the site that elicits the reflex to the sensitizing site where the noxious stimulus is applied. We describe the molecular mechanisms underlying neural sensitization in the next chapter.

57. The RNA synthesis that occurs during long-term memory involves epigenetic changes—alterations in chromatin, such as DNA methylation and histone modifications, and changes in the profile of small regulatory RNAs (see chapter 7).

58. The molecular and anatomical features underlying long-term cellular and behavioral learning in invertebrates have been the focus of an enormous amount of work. For reviews, see Ardiel and Rankin 2010; Glanzman 2010; Kandel 2012; Mayford, Siegelbaum, and Kandel 2012; Rahn, Guzman-Karlsson, and Sweatt 2013.

59. For details of learning in different cnidarians, see Rushforth 1973; Logan and Beck 1977, 1978; Johnson and Wuensch 1994; Kass-Simon and Scappaticci 2002.

60. To the best of our knowledge, backed up by discussion with the cnidarian experts Brigitte Galliot (personal communication, May 21, 2014) and Osamu Koizumi (personal communication, June 6, 2014), nothing has been published on the molecular basis of sensitization and habituation in Cnidaria. It is not even clear if habituation and sensitization in Cnidaria involve chemical or only electrical synapses. Answering such questions may be of great importance for understanding the evolution of learning.

61. The evolutionary relationship between habituation and sensitization is unclear. Since habituation seems to be more common than sensitization (Miller and Grace 2003, p. 360) and since Rankin and Carew (1988) showed that in *Aplysia* habituation emerges in the juvenile developmental stage, while sensitization emerges at least sixty days later, it is sometimes suggested that habituation appeared before

sensitization in evolution. However, both habituation and sensitization have been reported and are apparent at the cell level in ciliates (reviewed in Ginsburg and Jablonka 2009) and can be implemented by different (more or less complex) mechanisms. Hence, we believe that a generalization is not possible.

62. Summation in several different cnidarians is described by Spencer (1989) and by Meech and Mackie (2007).

63. See Kass-Simon 1988 for evidence of synaptic inhibition in *Hydra*; see also Meech and Mackie 2007, according to whom synaptic inhibition is absent in Cnidaria.

64. This experiment is described in Haralson, Groff, and Haralson 1975.

65. For a failed attempt to reproduce the results, see Rushforth 1973. Torley (2007) conducted a literature search, supplemented by personal inquiries from leading scientists working on learning and/or on cnidarians, and was unable to find studies that demonstrated classical conditioning in Cnidaria.

66. If different types of eyes are specialized for specific tasks, then as Nilsson et al. (2005) suggest, the complex filtering of visual information in cubozoans may occur much earlier in the signal transduction chain than in other visual systems. Whereas in vertebrates the registration of specific complex visual features (like edges and angles) occurs only after many other processing steps, in cubozoans, because they have several different types of eyes, complex features may be processed without multiple earlier steps. Each type of eye, it is suggested, has fewer, more specific visual features to process (say, edge for one type of eye, angle for another). Perhaps the early evolution of the visual system started with "eyes" that were involved only in single visual tasks.

67. For a general review of the various forms of sleep in animals, see Lesku, Martinez-Gonzalez, and Rattenborg 2009; Siegel 2009; Aulsebrook et al. 2016. Sleep in cubozoans is described by Kavanau (2006). He suggests that nonurgent processing is deferred to periods of sleep. Without sleep, selection would favor the evolution of circuitry "dedicated" to a single task or just a few tasks, with lower efficiency. The close link between detailed focal vision and sleep is supported by the observation that in several marine mammals and in many birds, half of the brain instantly falls asleep or awakens when one set of eyelids closes or opens (Kavanau 2006). However, since sleep has also been found in other cnidarian species (Nath et al. 2017), the origins of sleep may not be related to vision but may have a more general function related to the regulation of homeostasis, especially in the nervous system.

68. If cubozoans can learn by associating visual stimuli and this ability dates to the Cambrian or post-Cambrian era, the genetic profile of these animals is likely to reflect this.

69. We asked several expert fieldworkers about the learning abilities of cubozoans. Unfortunately, the learning abilities of these animals are unknown (personal

communications with George Mackie, November 1, 2006; George Matsumoto, November 1, 2006, and July 2, 2012; and Dan Nilsson, April 29, 2014).

70. Thompson and McConnell 1955, p. 65.

71. The mechanism enabling an animal to distinguish sensory inputs that result from its own action from those that do not was suggested by von Holst and Mittelstaedt (1950) on the basis of their experiments on flies; see discussion in chapter 4 and Brembs 2011a, 2011b.

72. Denton 2006. See chapter 5.

73. Gagliano et al. 2016. We return to this intriguing finding in the last chapter of the book.

Chapter 7

1. Papini (2002) provides a summary of the different approaches to the evolution of learning. For more details about the prevalence of AL, see Macphail 1982, 1987; Menzel and Benjamin 2013. The significance of species-specific adaptation is emphasized, for example, by Shettleworth (1998), and Macphail (1982, 1987) expounds the centrality of invariant mechanisms of general intelligence.

2. See Thorndike 1898; Rescorla and Wagner 1972.

3. Rescorla and Solomon (1967) review the two-process view of conditioning.

4. See Brembs and Heisenberg 2000.

5. For studies of action-based perception in the visual and vibrissal system, see Ahissar and Knutsen 2008; Ahissar and Arieli 2012; Ahissar and Assa 2016.

6. Ahissar and Assa 2016.

7. Anokhin's system view was described in the previous chapter. We discuss the interactions between different neural functional systems in chapter 8, where we present "toy models" of UAL. The view that operant and classical conditioning are evolutionarily entangled is described in Bronfman, Ginsburg, and Jablonka 2018.

8. See Wells 1968.

9. Wells (1968) assumed that long-term memory, which is often associated with Pavlovian conditioning, became advantageous only when discrimination was increased and sensitizing stimuli could be specifically predictive. He constructed toy models of neural circuits illustrating different stages in the transition from global sensitizing stimuli to Pavlovian sensitization.

10. See Razran 1971.

11. A second process, which is superficially similar, is the modulation of habituation known as "protection from habituation," which occurs when the linking of an

additional cue with the habituating one decreases habituation. This process occurs regularly when a potentially habituating stimulus, caused by the animal's own movements, is inhibited (Bryan and Krasne 1977). However, in this case it seems that the inhibiting stimulus is *any* animal movement, so we doubt that this process is directly related to the evolution of Pavlovian conditioning.

12. See B. R. Moore 2004 for a review of different types of learning and their evolution. Garcia's experiments are described in Garcia, Hankins, and Rusiniak 1974.

13. See B. R. Moore 2004, p. 317.

14. Kandel (2000) summarizes his contributions and the personal background that led to his work in his Nobel Prize lecture at http://www.nobelprize.org/nobel_prizes/medicine/laureates/2000/kandel-bio.html.

15. An important part of the circuitry of the gill withdrawal reflex resides in its abdominal ganglion. The complete circuit has over one hundred neurons, including many interneurons (Frost and Kandel 1995).

16. When the facilitatory interneurons (FAC INTs) are active, they enhance the release of transmitters from the terminals of the siphon sensory neurons (SNs). Although not shown in the figure, the tail sensory neurons also make direct contact with the motor neurons.

17. Kandel 2001; Hawkins, Kandel, and Bailey 2006.

18. The NMDA receptor (named after its strongest ligand, N-methyl D-aspartate) contains an ionic channel through which calcium may enter the cell, but it is normally blocked by magnesium ions, and its activation requires unblocking. The removal of magnesium is achieved by depolarizing the postsynaptic neuron, which occurs as a result of binding glutamate by AMPA receptors (named after their strongest ligand, the artificial glutamate analog α-amino-3-hydroxy-5-methyl-4-isoxazole-propionic acid) and the consequent flow of sodium ions into the cell through the AMPA receptor channels. During long-term memory, AMPA receptors are trafficked into the postsynaptic membrane from pools in nonsynaptic zones of the cell.

19. For a review of nonassociative learning in *Aplysia* and other invertebrates, see Byrne and Hawkins 2015.

20. Hawkins et al. 1983; Walters and Byrne 1983; reviewed by Hawkins and Byrne 2015.

21. As with long-term sensitization, the persistent activation of PKA leads to the mobilization of transmitter vesicles from the storage pool to the release sites, and this facilitates the release of transmitter independent of the calcium influx. In this case, cAMP acts in parallel with PKC, which is also activated by longer exposure

to serotonin (PKC can both prolong action potentials and bring about vesicle mobilization).

22. Later it was discovered that another cellular mechanism (in addition to presynaptic facilitation), involving the postsynaptic cell, contributes to conditioning in *Aplysia*. See Murphy and Glanzman 1997.

23. Strengthening a synapse when the action potentials in the presynaptic cell coincide with action potentials in modulatory neurons was proposed as an "associative rule" many years earlier by Kandel and Tauc (1964).

24. Donald Hebb's (1949) learning rule states: "When an axon of cell A is near enough to excite a cell B and repeatedly or persistently takes a part in firing it, some growth process or metabolic change takes place in one or both cells such that A's efficiency, as one of the cells firing B, is increased" (p. 62).

25. In mammals, the equivalent to long-term facilitation (LTF) is long-term potentiation (LTP).

26. In mammals, application of the gas nitric oxide (NO) to hippocampal slices can produce LTP but only when coupled with presynaptic activity. The advantage of this combination may be to ensure that LTP initiated in one postsynaptic cell does not spread to neighboring cells (by diffusion of the retrograde signal; only the active presynaptic pathways are potentiated; inactive presynaptic terminals are not affected).

27. The activity-dependent amplification of the cAMP pathway is not unique to the conditioning in the gill withdrawal reflex of *Aplysia*. Mutants in *C. elegans* that could not form long-term memory due to the absence of AMPA receptors have also been found (McEwan and Rankin 2013), and in *Drosophila*, the gene that encodes a calcium/calmodulin-dependent adenylyl cyclase, causes the cyclase to lose its ability to be stimulated by calcium/calmodulin (Livingstone, Sziber, and Quinn 1984).

28. In *C. elegans*, for example, CHR1, the analog of CREB1, is required for long-term olfactory learning (Alcedo and Zhang 2013).

29. See Wolff and Strausfeld 2016 for a discussion of this molecular homology.

30. See Hawkins and Kandel 1984b; Kandel 2006.

31. Reviewed by Hawkins and Byrne 2015.

32. Brembs et al. 2002; Baxter and Byrne 2006; Lorenzetti et al. 2006; Lorenzetti, Baxter, and Byrne 2011. Research on the molecular differences between operant and classical conditioning (in feeding) in *Aplysia*, reviewed in Hawkins and Byrne 2015, is still in its infancy.

33. For a review of mechanisms that may underlie memory, from epigenetic alterations to synaptic strength and dendritic spine modifications, and from synaptic

assemblies and "engram cells" to synchronous wave bursts in the hippocampus, see Poo et al. 2016.

34. Semon's view and subsequent research on cell memory are discussed in Ginsburg and Jablonka 2009; Bronfman, Ginsburg, and Jablonka 2014, 2016a.

35. For more extended discussions, see Bronfman, Ginsburg, and Jablonka 2014, 2016a.

36. S. Chen et al. 2014; Bédécarrats et al. 2018.

37. See studies by Johansson et al. 2015; Jirenhed et al. 2017.

38. Reviewed in Jablonka and Lamb 2005/2014, chaps. 4 and 11.

39. Colbran and Brown 2004.

40. Si et al. 2010.

41. For reviews, see Blaze and Roth 2013; Zovkic, Guzman-Karlsson, and Sweatt 2013; Bronfman, Ginsburg, and Jablonka 2014, 2016a; Yang et al. 2016.

42. For a gene-specific effect see, for example, Miller and Sweatt 2007; Miller et al. 2010; for a general review, including a review of genome-wide effects, see Bronfman, Ginsburg, and Jablonka 2014, 2016a.

43. Reviewed in Kandel, Dudai, and Mayford 2014.

44. For a model of the relation between the behavioral and epigenetic learning curves, see Bronfman, Ginsburg, and Jablonka 2014. Studying epigenetic learning curves may open up potential therapeutic avenues for treating neuropsychological pathologies that involve defects in learning and memory by detecting differences in the epigenetic dynamics that correlate with them.

45. Bédécarrats et al. 2018.

46. Kamin 1969.

47. This is the original formulation by Rescorla and Wagner (1972). For discussions of PE, see Schultz, Dayan, and Montague 1997; Schultz 1998; Schultz and Dickinson 2000. Niv (2009) presents an elaboration of the basic PE model, which takes into consideration second-order learning and predicts future outcomes. Schultz (2015) presents a comprehensive review of the many aspects of reward-based learning and decision making, and Holland and Schiffino (2016) review the interpretation of PE in terms of the associability of the CS and the US, taking into account attentional modulations. An experimental prediction of the PE principle is that the rate of learning acquisition (i.e., the learning curve) will not be linear but rather concave (since the PE progressively diminishes, rendering the learning rate progressively slower). Rao and Ballard (1999), Friston (2005), and Clark (2013) present more general discussions of predictive coding, a subject to which we return in the next chapter.

Schultz (2015) reviews the evidence for the involvement of dopamine in the computation of PE.

48. For blocking in *Aplysia*, see Colwill, Absher, and Robert 1988; for planarians, see Prados et al. 2013; for *Drosophila* larvae, see Schleyer et al. 2015. As we argue in the next chapter, such effects in UAL animals, which can learn about compound stimuli and actions, depend on *models* of the world, body, and action and therefore involve integrating brain areas.

49. These processes could be implemented by epigenetic mechanisms. For example, assume that the association between the conditioned stimuli (CS) and the reward response (R) is formed and strengthened through the mediation of dopamine neurons and that DNA methylation marks are gradually removed from biogenesis sites of specific small noncoding RNAs (ncRNAs) in the CS neurons as learning proceeds. When the association between the CS and the reward response is strong, the activation of CS neurons is associated with small prediction error and with the expression and export of small ncRNAs, secreted in vesicles (see Hoy and Buck 2012). These ncRNAs are then taken up specifically by dopamine neurons and inhibit their activity, thus preventing the formation of new or additional associations. Such epigenetic processes among interacting neurons could thus lead to blocking. (A model describing this process is presented in Bronfman 2017.)

50. The experiment is described in Hawkins, Greene, and Kandel 1998. Hawkins and Kandel (1984a, 1984b) argue that combinations of the cellular processes that underlie habituation, sensitization, and classical conditioning in *Aplysia* can lead to complex forms of learning.

51. Trace conditioning was shown in *Aplysia* by Glanzman (1995) and in *Drosophila* by Galili et al. (2011), Shuai et al. (2011), and Dylla et al. (2013). The unsuccessful attempt to find trace conditioning in *C. elegans* was described by Bhatla (2014), in his PhD thesis.

52. Koch 2012.

53. Edwards, Heitler, and Krasne (1999) summarize information about the giant axons of the crayfish nerve cord that drive tail-flip escape responses. Figure 1 of this paper gives an idea of the complexity involved.

54. For a discussion of what distinguishes a brain from other neural structures, see Matheson 2002. Arendt et al. 2008 provide a characterization of a brain.

55. In chapter 6 we predicted that box medusae may exhibit limited AL in the visual modality because of the neural centralization and the bilateral symmetry in these headless animals, but it appears that their learning abilities have never been studied. Indeed, there are no recent studies on learning in any cnidarian (George Mackie, personal communication, November 1, 2006).

56. For studies on learning in these invertebrate groups, see Razran 1971; Corning, Dyal, and Willows 1973; Rushforth 1973. For starfish studies, see Valentinčič 1978, 1983; Migita, Mizukami, and Gunji 2005. If associative learning is possible in brainless animals, we predict that it will be extremely constrained.

57. Marinecsco, Duran, and Wright (2003) describe variations in nonassociative heterosynaptic modulation in three species, and Hoover et al. (2006) show differences in the duration and extent of conditioning.

58. The origins of bilaterality are discussed, for example, by Martindale, Finnerty, and Henry 2002.

59. See Holló and Novák 2012; Holló 2015.

60. The definition of CNS is based on Arendt et al. 2008, p. 1523.

61. See Bullmore and Sporns 2012.

62. Watts and Strogatz 1998. The "small-world" topology has been found to characterize many biological systems (such as molecular biochemical networks) as well as nonliving infrastructural networks (such as transportation systems).

63. Fazelpour and Thompson (2015) called the intrinsically active brain a "Kantian brain" because Kant argued that perception is based on intrinsic, spontaneous, self-organizing brain activities. We prefer to use the term "Lamarckian brain" because Lamarck employed the same notion of the brain more broadly, in the context of the evolution of all psychological faculties.

64. Moroz 2009, 2012.

65. The chimeric brain hypothesis is discussed in detail in Tosches and Arendt 2013; Arendt, Tosches, and Marlow 2016.

66. Described by Tomer et al. 2010.

67. For a study comparing brain organization through memory-related gene expression patterns in insects, vertebrates, annelids, and acoelomorphs and pointing to the similar regenerative capacities of the mushroom body and the hippocampus, see Wolff and Strausfeld 2016.

68. See, for example, Kristan 2016.

69. See Kocot et al. 2011; Moroz 2012; Shomrat et al. 2015.

70. See Martín-Durán et al. 2018 for a detailed study of the anatomy and the expression patterns of transcription factors during CNS development in different bilaterian taxa.

71. Arendt (2018) summarizes the available gene expression data in the developing nervous systems of different invertebrates and vertebrates and concludes: "No firm

statement can yet be made about the cellular neural architecture of ancient ecdyso-zoans, lophotrochozoans, or bilaterians" (p. R226).

72. See Tosches and Arendt 2013; Wolff and Strausfeld 2016.

73. The hemiellipsoid body in crustaceans, the simple mushroom body–like struc-tures in acoels, the mushroom body–like structures in annelids, the procerebral lobe in gastropods, and the vertical lobe in cephalopods are all integration centers with varying degrees of structural and functional complexity, and we discuss their func-tional significance in the next chapter. For a review of the evolutionary elaboration of mushroom bodies in different groups of invertebrates, see Farris 2015.

74. Nargeot and Simmers 2012.

75. Hawkins (personal communication, December 31, 2015).

76. The learning abilities of Planaria are described in Jacobson, Horwitz, and Fried 1967; Nicolas, Abramson, and Levin 2008; Prados et al. 2013.

77. The biology of Planaria is described in Agata 2008.

78. Bhatla (2014) documented the absence of delay and trace conditioning in *C. elegans*.

79. Ardiel and Rankin (2010) reviewed the flexibility of contextual AL learning in *C. elegans*; Lüersen et al. (2014) described the limitations of the worm's spatial learning.

80. For the role of oxygen in the evolution of life, see Reinhard et al. 2016.

81. See Ginsburg and Jablonka 2010a, 2010b.

82. Velvet worms of the species *Euperipatoides rowelli* may be good candidates. As shown by Reinhard and Rowell (2005), these worms have complex social behaviors, but their learning ability has not been investigated. It is possible they also have the capacity for very minimal UAL.

83. The analysis can be found in Rapoport's (2016) master's thesis (supervised by Eva Jablonka). See also Rapoport and Neuhof 2017.

84. Sleep in the upside-down, largely sessile jellyfish *Cassiopea* is described by Nath et al. 2017.

85. For sleep and sleep-like states in invertebrates, see Michel and Lyons 2014. Sleep in insects is reviewed by Eban-Rothschild and Bloch 2012; Potdar and Sheeba 2013. For sleep in *C. elegans*, see Raizen et al. 2008. For sleep in the mollusks *Aplysia* and *Lymnaea stagnalis*, see Vorster et al. 2014 and Stephenson and Lewis 2011, respec-tively. Kavanau (1998) explains the reported lack of sleep in schooling fish as a con-sequence of their life in featureless open waters with meager visual stimuli and little requirement for learning.

86. See chapter 6; Kavanau (2006) explains the occurrence of sleep in cubozoans in these terms.

87. Tosches et al. (2014) showed that in the larvae of the marine annelid *Platynereis dumerilii*, melatonin, the "hormone of darkness," is produced in brain photorecep-tors. The release of melatonin at night induces the rhythmic burst firing of cholin-ergic neurons that innervate locomotor-ciliated cells, and this modulates the length and the frequency of ciliary arrests and leads to the vertical descent of the larvae during the night. This melatonin-based nocturnal behavior has some interesting parallels with vertebrate sleep, and it has been suggested that it is an ancient precur-sor of sleep.

88. Hawkins and Kandel 1984a, 1984b. An "alphabet" on which complex learning is based was suggested earlier by Clark C. Hull, one of the leading learning research-ers in the United States, who used the behavioral principles of simple forms of learn-ing as building blocks for constructing more complex forms (Hull 1943).

Chapter 8

1. Whether or not a stimulus is reinforced, and to what extent, can be influenced by evolutionary history, rendering some association or conjunction of features more readily learnable than others. For example, for humans, learning that an angry face is threatening is easier than learning that a blue flower is threatening.

2. The phrase "patterns that connect" is taken from Gregory Bateson (1979), who, in a very different context, asked: "What pattern connects the crab to the lobster and the orchid to the primrose and all the four of them to me? And me to you? And all the six of us to the amoeba in one direction and the back-ward schizophrenic in another?" (p. 6).

3. For the role of different neurotransmitters in learning in the rat, see Myhrer 2003. For a discussion of Berridge's work and the role of dopamine, see chapter 5.

4. For reviews of both reward and aversive learning, see Hu 2016; Scaplen and Kaun 2016.

5. For the role of PE in aversive learning, see, for example, Dayan and Huys 2009; McHugh et al. 2014; for the involvement of the periaqueductal gray in pain, see Behbehani 1995. The involvement of both metabotropic glutamate and NMDA receptors in the consolidation of conditioned fear has been shown by Handford et al. 2014.

6. Bayes's theorem is named after Thomas Bayes (1701–1761), the British statisti-cian who formulated a specific case of the general theorem. See https://en.wikipedia .org/wiki/Bayesian_inference for a clear exposition.

7. See M. Rescorla 2016 for a detailed analysis.

8. Friston's framework is not the only one that attempts to account for cognition. There are competing Bayesian accounts: see M. Rescorla 2016 for a discussion of different Bayesian approaches to goal-directed bodily action.

9. Gillam (2000) describes perceptual constancies and their cognitive basis.

10. Our definition of hierarchy is different from that suggested by Feinberg and Mallatt (2016, pp. 25–26). In our opinion hierarchical levels are not increased by just introducing an intervening neuron or circuit. We base our definition on Engel, Fries, and Singer 2001. For a useful discussion of levels, see Rauss and Pourtois 2013.

11. Schapiro, Kustner, and Turk-Browne 2012.

12. The models of UAL presented in this section are versions of those described in Bronfman, Ginsburg, and Jablonka 2016c.

13. Recognizing and memorizing compound smells in the mushroom bodies of insects (described by Stopfer [2014]) seem to require a simpler model than that we suggest. Peng and Chittka (2017) constructed a neural network model based on the special architecture of mushroom bodies (which includes converging and diverging signaling, two loci of association, and negative feedback) and showed that this network can lead to nonelemental learning of odorants. However, this network does not include integrated action, a self-model, and compound multimodal interactions. Agency, which involves action-selection processes, is lacking. The Peng and Chittka model can be considered a network partially implementing the first two levels of the hierarchy of our more general and abstract functional model of world learning (figure 8.2A).

14. For an analysis of brain regions associated with PE in rats, see Holland and Schiffino 2016.

15. If such reconstruction processes involve copying association circuits into the MEMU, Fernando, Szathmáry, and their colleagues might be right in their suggestion that in complex brains there are processes of "replication," where the architecture of reinforced neural circuits are copied with variation into higher levels of the neural hierarchy, and selection among them occurs through differential reinforcement. For expositions of this idea, see Fernando, Goldstein, and Szathmáry 2010; Fernando, Szathmáry, and Husbands 2012; de Vladar and Szathmáry 2015.

16. For example, Friston (2008), Friston and Kiebel (2009). Hohwy (2013), and Clark (2016) review the predictive-processing framework and its many implications; Seth and Friston (2016) present and review predictive processing in the context of the "emotional brain" and interoception (which is the sense of the body "from within," as they put it).

17. It is possible that the background, low-level fluctuations in dopamine levels, which are usually considered noise, play a role in the transient stabilization that occurs during updating.

18. Reported in Metzinger 2009 and attributed to something philosophers call the "ineffability" of conscious experience. This ineffability has been the driving force for the evolution of sensory discrimination.

19. The SIU, MIU, MEMU, and REIU in our model may require a neural network model similar to that constructed by Peng and Chittka (2017) (with fan-in, fan-out, and negative feedback interactions within each integrating unit, including the higher-level units, MEMU and AU). The basic architecture of the CNS of UAL animals may be based on repeated building blocks—repeated functional modules with similar internal architecture. To computationally implement our model, probably in a robotic mobile system, it would be necessary to specify how the different units are organized internally and how the arrows between units affect the organization of the top and bottom modules.

20. The same reasoning holds for second-order conditioning in LAL animals (Hawkins and Kandel 1984a, pp. 383–384).

21. Bronfman, Ginsburg, and Jablonka 2014.

22. For associative visual agnosia, see Carlesimo et al. 1998; for visual apperceptive agnosia, see Vecera and Gilds 1998.

23. For sustained goal-directed actions, in the sense of intending to reach a goal in the world by explicitly inferring the causal structure of the world, additional representation of the goal must be in place (de Wit and Dickinson 2009).

24. The explanatory role of the notion of representation and the ontological status of representations are subjects of heated debate; see, for example, discussions in Prinz 2012 and Shea 2018). Our position is closest to that of Shea.

25. For a discussion of a receiver-based notion of functional information, see Jablonka 2002.

26. The term "mental representation" is also commonly used to refer to nonconscious representations. In this interpretation, nonaccessed, once conscious but now suppressed memories are mental representations and so are other latent representations that have been subjectively experienced in the past. We usually use the term "mental" to imply "subjectively experienced," so in this book we employ the term "mental representation" to mean "neural representation that entails *actualized* subjective experiencing."

27. James 1890, vol. 2, chaps. 24 and 25.

28. This view is central to the notion of consciousness of Cabanac 1992; Morsella 2005; and Merker 2005, 2007.

29. We presented our ideas about this topic in Ginsburg and Jablonka 2007a, 2007b, 2010a. Our discussion here is based mainly on the latter paper.

30. See chapter 5 for an exposition of Denton's ideas.

31. See chapter 3 for a description of Tononi's IIT. For more recent developments of the theory, see Tononi et al. 2016.

32. The idea that there are distinct types of consciousness is emphasized and repeated throughout Feinberg and Mallatt's 2016 book.

33. For a discussion of the evolutionary primacy of sensory experience, favoring vision first, see Feinberg and Mallatt 2016, chap. 5 and the references therein.

34. Aristotle 1984b, A 1, p. 27.

35. Crisp and Burrell 2009. Some annelid species do, however, have mushroom bodies, which are also found in other invertebrate phyla; this suggests that the origin of these structures is ancient (Strausfeld and Hirth 2013; Wolff and Strausfeld 2015, 2016). Animals in these groups therefore have some of the brain structures required for UAL.

36. The comparative neuroanatomy of vertebrates is described, for example, in Mueller 2012; Barron and Klein 2016; and Feinberg and Mallatt 2016. Giurfa and Menzel (2013), Strausfeld and Hirth (2013), Barron and Klein (2016), and Wolff and Strausfeld (2016) describe the insect brain; the neuroanatomy of cephalopods is described in Hochner and Shomrat 2013.

37. Hindy, Ng, and Turk-Brownen 2016.

38. See Strausfeld and Hirth 2013. Even in Collembola, the smallest of arthropods with a brain of only seven hundred neurons, mushroom body–like and central complex–like structures have been found (Kollmann, Huetteroth, and Schachtner 2011).

39. Tomer et al. (2010) reported strong affinities between the cerebral cortex and arthropod mushroom bodies based on gene expression profiles. Wolff and Strausfeld (2016) describe mushroom bodies in some annelids and acoels, with neural architecture and an expression of proteins associated with memory similar to that found in insects.

40. Strausfeld and Hirth 2013.

41. Turner-Evans and Jayaraman 2016.

42. Strauss (2002), Sterling and Laughlin (2015, p. 99), and Shih et al. (2015) describe the *Drosophila* connectome (the connection map of the fly's brain) on the mesoscopic scale. The connectome is hierarchically organized, and there are many similarities between the functional units of the fly's brain and those of a vertebrate's brain, although, of course, there is no one-to-one mapping. Strausfeld and Hirth (2013, fig. 2) show the parallel connectivity patterns in the mammalian and the insect brain. These functional analogies are further discussed in Wolff

and Strausfeld (2016). The fossil evidence for the existence of these structures in the brains of vertebrates and arthropods is summarized in Feinberg and Mallatt (2016).

43. Shomrat et al. 2011; Hochner 2013; Hochner and Shomrat 2013; Hochner and Glanzman 2016.

44. This behavior is described in Finn, Tregenza, and Tregenza 2009.

45. Hague, Florini, and Andrews 2013.

46. Sumbre et al. 2001.

47. For the role of the brain in directing the arms, see Richter, Hochner, and Kuba 2015, 2016.

48. According to Benny Hochner (private communication, June 13, 2016), postdocs in his lab are very interested in these questions.

49. Godfrey Smith 2016c, p. 105.

50. Ginsburg and Jablonka 2007a, 2007b, 2010a, 2010b; Barron and Klein 2016; Feinberg and Mallatt 2016; Godfrey-Smith 2016a, 2016c.

51. Robert Hawkins, personal communication, January 12, 2016.

52. Gillette and Brown 2015.

53. Chase (2000), Benjamin (2013), and Hochner and Glanzman (2016, p. R968) note that "the protocerebrum has the glomerular organization that characterizes the neural structures dedicated to olfactory processing in nematodes, arthropods and vertebrates."

54. Watanabe, Kirino, and Gelperin (2008) describe neural circuits in *Limax* that employ, like mammals, cellular and network oscillations for computing various odor combinations; as in mammals, nitric oxide is used in these circuits to control the oscillations. These authors also showed that odor-memory formation occurs in a central circuit within the procerebral lobe and that selective gene activation is associated with odor learning. For additional information on associative memory mechanisms in terrestrial slugs, see Gelperin 2013.

55. Gomez-Marin et al. 2016. Margaret Yekutieli studied the same loss of rotation in Parkinson's patients and the reacquisition of axial rotation after dopamine therapy and/or physiotherapy (Margaret Yekutieli, private communication, June 14, 2017).

56. See Trestman 2013a, 2013b for a detailed analysis of complex embodied cognition and its evolutionary origins.

57. Animals belonging to other groups, such as the velvet worms (phylum Onychophora), which exhibit complex social behavior, may well be sentient too, but nothing is known about their learning abilities, and there is insufficient information

about the functional organization of their brains. Some annelid species that have mushroom bodies, distal senses, and segmented bodily organization may also be sentient.

58. This prediction is the same as that proposed by Dickinson (2008). We suggest that his reasoning and experiments, which showed goal-directed AL and causal thinking in rats, can be applied more generally, and that arthropods and cephalopods will also be shown capable of goal-directed AL.

59. For a detailed review, see Hartenstein and Stollewerk 2015. Interestingly, and in line with what we discussed in previous chapters about the reuse of ancient molecular factors and pathways, most of the modules of early neurogenesis are shared among all bilaterians and may have formed part of the neurogenetic "tool kit" of their common ancestor.

60. An example is the transport of the *ARC* gene products. The protein encoded by this gene, which is important for neural plasticity and cognition and has retroviral origins, forms virus-like capsids that encapsulate its mRNA. The proteins and RNAs are transported in extracellular vesicles either among neurons (in mice, Pastuzyn et al. 2018) or from neurons to muscle cells (in fruit flies, Ashley et al. 2018).

61. See Gallistel 2017 for a clear exposition of the neural coding question. Although we are not committed to Gallistel's view of the computer-like encoding of what he calls "numbers in the brain" and think that encoding in epigenetic engrams is network-specific at low levels of the neural hierarchy, it is plausible that during UAL and more complex forms of learning, high-level categories based on lower-level EEs may be constructed. The face cells in the inferior temporal (IT) cortex of monkeys, described by Chang and Tsao (2017), which encode fifty different axes of a face (such as the distance between the eyes or the width of the hairline), may store the causal correlates of these axes as EEs. We believe that exploring the role of EEs in memory is an exciting and important area of future research made possible by the current technologies of neuroepigenetics. We are currently investigating the theoretical aspects of this possibility.

62. Godfrey-Smith 2016a. The cephalopods have followed a unique evolutionary trajectory. This is reflected in their genomes, which have undergone massive rearrangements and regulatory changes (Albertin et al. 2015), and in their different mechanisms of long-term potentiation in the vertical lobe. Long-term memory in this lobe is stored in posttranslational modifications of proteins and does not depend on transcriptional activation, as in other mollusks.

Chapter 9

1. Ginsburg and Jablonka 2007a, 2007b, 2010a, 2010b.

2. For a comprehensive review and the suggestion that Ediacaran species are basal stem group animals, see Budd and Jensen 2017.

3. McMenamin 1986. For reviews of Ediacaran life, see Narbonne 2005; Xiao and Laflamme 2009; Droser and Gehling 2015; Dufour and McIlroy 2016.

4. Darwin 1872b, pp. 288–289. Darwin was partly right because subsequently, the pre-Cambrian Ediacaran fauna was discovered, and we know there were microbial life-forms 3.8 billion years ago. However, dramatic evolutionary changes, saltational in geological terms, did occur during the Cambrian.

5. See Fortey 2001 for the arguments on morphological complexity and Levinton 2008 for the comparative molecular analysis.

6. Janecka, Chowdhary, and Murphy 2012.

7. Pagel, Venditti, and Meade 2006.

8. The study of Lee, Soubrier, and Edgecombe (2013) supports this conclusion with regard to arthropod evolution, although they did not factor in speciation rates. They incorporated morphological data from both recent and fossil arthropods, as well as genetic data from living arthropod groups, and found that the average rate of morphological evolution in the Cambrian is about four times the rate after the Cambrian and that the average rate of molecular evolution in the Cambrian is about five times higher than the average rate after the Cambrian.

9. Valentine 2004; Marshall 2006; Brasier 2009.

10. For representative paleontologists' views, see Valentine, Jablonski, and Erwin 1999; Peterson, Dietrich, and McPeek 2009.

11. Brasier 2009; Cabej 2013, chap. 5.

12. Valentine 2004, p. 195.

13. Canfield, Poulton, and Narbonne 2006; Fike et al. 2006; Sperling et al. 2013, 2015; X. Chen et al. 2015; Doglioni, Pignatti, and Coleman 2016; Jin et al. 2016; Reinhard et al. 2016.

14. Butterfield 1997, 2011; Kirschvink and Raub 2003.

15. Brennan, Lowenstein, and Horita (2004); Porter (2011); and Peters and Gaines (2012) discuss the importance of biomineralization; Smith and Harper (2013) emphasize the feedback interactions between geochemical and biological factors.

16. Although it cannot be ruled out, we found no evidence for the comparable diversification of plants during the Cambrian.

17. For reviews, see Valentine 2004; Marshall 2006; Levinton 2008; Zhang et al. 2014; Fox 2016.

18. Butterfield 2011, p. 285.

19. Budd and Jensen 2017.

20. Cabej 2013, p. 269.

21. Cabej 2013, chap. 3.

22. West-Eberhard 2003; P. Bateson 2005; Newman, Forgacs, and Müller 2006; Newman and Bhat 2009; Pfennig et al. 2010. See also the introduction to part II of this book, where the advantage of exploration-stabilization strategies is discussed.

23. Baldwin 1896a.

24. Schmalhausen 1949; Waddington 1957.

25. Avital and Jablonka 2000.

26. West-Eberhard 2005, p. 610.

27. West-Eberhard 2003, 2005; see also Pfennig et al. 2010.

28. Hardy 1965.

29. Wyles, Kunkel, and Wilson 1983.

30. P. Bateson (2005) argued that selection driven by behavioral modifications can lead to increases in diversity.

31. Emes et al. (2008) compared the genomics and proteomics of postsynaptic density and membrane-associated guanyl kinase-signaling complexes that underlie memory and learning and found interesting differences between invertebrates and vertebrates. We extended this type of analysis to the synaptic complexity of different invertebrate taxa, including groups with a nerve net (cnidarians) and groups with a secondarily decentralized nervous system (echinoderms), and compared them with groups with fully centralized nervous systems and brains. We found, as did Emes and his colleagues, that vertebrates have a distinct genetic profile compared to all other taxa, with a large proportion of learning-related genes specific to them. We also found more similarity (more orthologues) between vertebrates and other deuterostomes (including echinoderms) than between vertebrates and less-related phyla such as arthropods. Predictably, we found extensive loss of learning-related genes in parasites, but surprisingly, there are many learning-related genes in placozoans and poriferans, indicating the multifunctionality of many of the synaptic and learning-specific genes (which are obviously employed for nonneural function in these neuronless groups). Hence, more detailed comparative analysis, which includes nonprotein regulators of protein-coding genes and comparisons of

gene networks, is required to unravel the molecular basis of associative learning (Rapoport 2016; Shlomo Shaul and Dvir Dahary contributed to the first stages of this work).

32. See Levenson and Sweatt 2005 and Wood, Hawk, and Abel 2006 for the role of epigenetic marks in memory formation.

33. Newman, Forgacs, and Müller 2006; Newman and Bhat 2009; Newman and Müller 2010.

34. See Friston et al. 2015 for an application of the HPP framework to the construction of simple morphology through self-assembly. It is interesting that a nondeterminate flexible morphology is seen in nonneural animals (sponges, placozoans) as well as in plants and fungi.

35. The proposal that arthropods are endowed with minimal consciousness and that it emerged in the Cambrian has also been discussed by Barron and Klein (2016) and Klein and Barron (2016).

36. For arguments about the role of social selection in speciation, see West-Eberhard (1983, p. 177), in which she argues: "Populations whose signals or appearance have diverged under social selection are preadapted for species recognition by the prior acquisition of species-specific markers, and need only be selected to distinguish them."

37. A. Parker 2003.

38. Darwin 1872b, pp. 35–36.

39. As Mary Jane West-Eberhard (personal communication, June 13, 2017) noted, "It would be fair to reason, that if an extant taxon representing a group that proliferated in the Cambrian explosion is sexually dimorphic and/or known to have socially/sexually competitive behaviors you could reasonably conclude that they did then as well." We know there is sexual selection among crustaceans, including horseshoe crabs, which are related to the Cambrian trilobites, so we assume that the selection of mating and other social behaviors was as common among Cambrian arthropods. The role of social selection in diversification is discussed by West-Eberhard 1983, 2003. See Kawase, Okata, and Ito 2013 for the original report of the artistic achievements of the male puffer fish *Torquigener* sp.

40. Hirth 2010.

41. Pennisi 2016.

42. Feinberg and Mallatt 2016, chap. 4.

43. We argued that associative learning was one of the factors that spurred the Cambrian explosion, leading to the radiation of the groups in which it emerged, as well as to learning-guided coevolutionary morphological radiation and a learning

arms race in some interacting taxa. Ginsburg and Jablonka 2007b, 2010a, 2010b, 2015; Bronfman, Ginsburg, and Jablonka 2016b, 2016c.

44. Nesse 2001.

45. Stefano et al. (2002, p. 85) define stress "as a type of stimulation that is stronger and lasts for a longer duration, upsetting a typical perturbation response given its dynamic parameters."

46. For an excellent discussion of innate and acquired responses in different taxa, see Rimer, Cohen, and Friedman 2014.

47. Ottaviani, Malagoli, and Franceschi 2007, p. 497.

48. For discussion, see Kioussis and Pachnis 2009; Ottaviani 2011. Marin and Kipnis (2013) review data showing that the disruption of the functioning of the immune system leads to impairments in cognition and in neurogenesis.

49. See Tort and Teles 2011.

50. Ottaviani and Franceschi 1996; Stefano et al. 2002; Ottaviani, Malagoli, and Franceschi 2007. For a detailed comparison between arthropod and vertebrate reward circuitries, see Scaplen and Kaun 2016.

51. Ottaviani and Franceschi 1996.

52. Luria 1968.

53. "Decay Happens" is part of a title of a review about active and passive forgetting (Hardt, Nader, and Nadel 2013). See Davis and Zhong 2017 for a review about active forgetting.

54. For studies showing retrieval-based forgetting, where retrieving a particular memory requires some degree of inhibitory control over interfering and competing memories, see Anderson et al. 1994; Wimber et al. 2015. Such inhibition is probably most important with complex percepts like those involved in event (or episodic) memory, which is described in the next section of this chapter.

55. Davis and Zhong (2017) review the studies on active forgetting and discuss neurogenesis-based forgetting in mammals. (We wonder whether neurogenesis in the mushroom body of arthropods also entails forgetting.) Other mechanisms, such as AMPA receptor endocytosis, seem to underlie active forgetting.

56. Sapolsky 2004.

57. For general reviews, see Lamm and Jablonka 2008; Jablonka and Lamb 2005/2014. Shapiro 2011 discusses the effects of stress on the generation of DNA variations. For the effects of stress on the epigenetic profiles in the nervous system, see Bronfman, Ginsburg, and Jablonka 2014; for the epigenetic effects of psychological stress in mammals, see Gapp et al. 2017.

58. For discussion and models of the evolution of transgenerational epigenetic inheritance in fluctuating conditions, see Lachman and Jablonka 1996 and Herman et al. 2013. The most general model exploring the conditions for the evolution of transgenerational inheritance was constructed by Rivoire and Leibler (2014). It shows that epigenetic inheritance of acquired variations is adaptive under many different environmental conditions.

59. One such mechanism, limiting between-generation epigenetic inheritance, has been discovered in *C. elegans* (Lev et al. 2017). When this mechanism, which is based on the histone methylating enzyme MET-2, is compromised, the worm can transmit epigenetic variations for more than thirty generations, but its germ line becomes mortal—it accumulates variations that eventually lead to sterility.

60. Philippidou and Dasen 2013.

61. Stetler et al. 2010.

62. For PTSD, see Zhou et al. 2014; for involvement of miRNAs in pain mechanisms, see Elramah, Landry, and Favereaux 2014; for earlier work, see Wang et al. 2003. Bronfman, Ginsburg, and Jablonka (2014, 2016a) provide a general review. Mehler (2008b) and Franklin, Saab, and Mansuy (2012) review the epigenetic mechanisms of neural resilience and vulnerability.

63. Mattick and Mehler (2008) and Mehler (2008a) review studies suggesting that aversion learning and fear conditioning involve DNA repair and recombination, especially during the consolidation of the memory. Mehler (2008a) further suggested that learned changes, involving RNA editing, are written back into DNA through RNA-mediated repair. See also Colón-Cesario et al. 2006; Saavedra-Rodríguez et al. 2009.

64. For the Nobel Prize lecture, see McClintock 1984.

65. McClintock 1984, p. 800.

66. Oliver and Greene 2009; Zeh, Zeh, and Ishida 2009.

67. For a general discussion of genome patterning following various types of stress, see Shapiro 2011. Haase (2016) reviews the relation between small RNA regulation and transposition. Feldman and Levy (2009) discuss genomic and epigenomic repatterning following hybridization.

68. For a discussion of the evolutionary implication of stress mediated through the heat shock response system, see Maresca and Schwartz 2006.

69. See Baker 2006. The experimental evidence showing that in stressful conditions the altered role of HSP90 leads to the expression of formerly cryptic genetic variation is summarized in this paper.

70. See, for example, Braun and David 2011. Stern et al. (2012) showed that this process also occurs in the fruit fly and leads to gametic inheritance.

71. See Cabej 2008, 2013.

72. For reviews, see Woldemichael et al. 2014; Bronfman, Ginsburg, and Jablonka 2016a; Razoux et al. 2016.

73. See, for example, Jablonka and Raz 2009; Yehuda et al. 2014; Sen et al. 2015.

74. Gapp et al. 2017.

75. Kaczer, Klappenbach, and Maldonado 2011.

76. Adamec, Holmes, and Blundell 2008.

77. For the involvement of JNKs in mice forgetting, see Sherrin et al. 2010; for studies on nematodes forgetting and the involvement of JNK proteins, see Inoue et al. 2013; for the involvement of translational regulators in forgetting, see Hadziselimovic et al. 2014. There are additional factors involved in active forgetting in mammals, such as NMDA receptors (Sachser et al. 2016) and the memory gene *KIRBA*, the overexpression of which in the hippocampus of mice increases long-term potentiation and decreases long-term depression (Heitz et al. 2016). Other factors and mechanisms with a role in active forgetting are described in Davis and Zhong 2017.

78. See Shuai et al. 2010 and 2015, which studied the role of the forgetting-enhancing protein (Rac) and the dedicated neuron types in the MB in active forgetting in fruit flies. See Davis and Zhong 2017 for a review of other aspects of active forgetting.

79. Kyle Patton put forward this suggestion in his final research project when he was an undergraduate student at Wheaton College. See Patton 2015.

80. Tatum et al. 2015.

81. For an argument about the canalizing effect of miRNA, see Hornstein and Shomron 2006; for the suppressing effects of piRNAs on transposition, see Halic and Moazed 2009.

82. Reviews about the contribution of epigenetic factors to learning and memory are published frequently (e.g., Woldemichael et al. 2014), and information about many additional factors that could have been targets of selection for regulating forgetting can be extracted from them; see, for example, Gräff et al. 2014 on the effect of one type of histone deacetylase (HDC2) on memory consolidation (the labile stage of recall at which engrams are reconstructed) in mice. It is of particular interest that knockdown of HDC2 improves associative memory and learning extinction (learning that a CS *is not* associated with a US) but not other types of memory (M. J. Morris et al. 2013).

83. Peyrache et al. 2009. Davis and Zhong (2017) suggest that sleep counters active forgetting.

84. We discussed the relation between sleep and learning in chapters 7 and 8.

85. For the involvement of the mushroom body in sleep regulation, see Campbell and Turner 2010; for the involvement of the hippocampus in sleep, see Jilg et al. 2010.

86. For evidence that the stress of domestication affects transgenerational epigenetic inheritance, see Belyaev and Borodin 1982; Trut, Oskina, and Kharlamova 2009.

87. Schacter, Addis, and Buckner 2007.

88. Tulving 1972.

89. For an excellent review of the work on scrub jays and a discussion of the criteria for episodic-like memory, see Clayton, Bussey, and Dickinson 2003. The first study of jays' ELM is Clayton and Dickinson 1998.

90. For general reviews of episodic-like memory (ELM), see Crystal 2009; Allen and Fortin 2013; Templer and Hampton 2013. For ELM in bees, see Pahl et al. 2007; in wrasses, Salwiczek and Bshary 2011; in zebra fish, Hamilton et al. 2016; in cuttlefish, Jozet-Alves, Bertin, and Clayton 2013.

91. Addis, Wong, and Schacter 2007.

92. The specific form that ELM takes in different species is likely to be tailored to their sensory-motor capacities and ecological conditions. In the special case of humans, episodic memory does seem to have unique features, as Tuvling originally suggested. The unique status of human episodic memory is also discussed by Suddendorf and Corballis (2007) and Suddendorf (2013), who focus on the subjective autonoetic mental time travel aspect of episodic memory in humans and argue that there is no evidence for the existence of this aspect in nonhuman animals. Although we are not in full agreement with their view, we do agree that human episodic memory, which is autobiographical, is very different from that of other animals (Jablonka, forthcoming). We come back to the implications of our approach to humans in the last chapter.

93. There are aspects of time measuring that are absolute and not relational, but here we are addressing the relation-dependent time sense as described in the ELM literature.

94. For these and other experiments describing rhesus monkeys' awareness of what they remember, see Templer and Hampton 2012; Basile et al. 2015. We realize that the behavior of the monkeys can be described in terms of complex meta-(semantic) memory, but an interpretation in terms of conscious recall is at least as plausible.

95. False memories have been shown in bumble bees and were implanted into mice. See Reinhard 2015; Liu, Ramirez, and Tonegawa 2014.

96. See Tomasello 2014. Tomasello defines thinking as involving these three elements. Great apes engage in nonverbal thinking that nevertheless requires imagination.

97. Allen and Fortin 2013; Templer and Hampton 2013.

98. Compare the analysis and conclusions of Allen and Fortin (2013), who suggest that EML evolved in the common ancestor of birds and mammals and has deep vertebrate roots, and Güntürkün and Bugnyar (2016), who suggest that the evidence points to parallel evolution in birds and mammals.

99. Allen and Fortin 2013.

100. Allen and Fortin 2013; Güntürkün and Bugnyar 2016.

101. Maybe highly intelligent predatory jumping spiders like *Portia* are similar in this respect to cephalopods.

102. See Avital and Jablonka 2000 for a discussion of the role of social learning in the cognitive and morphological evolution of birds and mammals.

103. Oudiette et al. (2012) describe non-REM dreaming.

104. For humans and other mammals, see Siegel 2005; Diekelmann and Born 2010; for birds, see Jackson et al. 2008; for bees, see Beyaert, Greggers, and Menzel 2012.

105. Bateson and Martin 2013.

106. See http://kellybulkeley.org/dreaming-is-play-a-new-theory-of-dream-psychology/.

107. Bateson and Martin 2013; Burghardt, Dinets, and Murphy 2015.

108. For a review of mirror neuron research, see Rizzolatti and Sinigaglia 2016.

109. Ibrahim (2009) described the selective impairment of one language in bilingual Hebrew and Arabic speakers. This dissociation suggests that the modules for Hebrew and Arabic usage are semiautonomous developmentally. Obviously, the genetic evolution of Arabic and Hebrew cannot be inferred since like all specific languages, they are products of cultural evolution.

110. Fedor et al. 2017. An important question is *which* neural patterns get replicated in the way suggested by this model. Since fan-in and fan-out interactions seem to characterize the mapping relations among parts of the nervous system and since such mapping profoundly modifies patterns of connectivity, the specification of the neural level or levels at which patterns of connectivity are replicated is crucial.

111. Darwin 1872b, p. 491.

Chapter 10

1. The Japanese monk Ryokan Taigu (1758–1831) believed that speaking about emotions and metaphysical issues (such as consciousness, we suspect) was senseless, yet he could not help it. See https://philosophadam.wordpress.com/2011/12/13/the -striking-honesty-of-ryokan/.

2. 1 Samuel, chap. 9 tells about Saul, son of Kish, who went to search for the lost she-asses of his father and found kingship: he was anointed as the first king of Israel.

3. Loeb 1900, p. 12. We cited this sentence in Ginsburg and Jablonka 2010a.

4. As the physiologist Denis Noble stressed, during both development and evolution, through the constraints higher levels of biological organization impose on lower ones, organisms harness stochasticity to generate new functional responses to environmental challenges (Noble 2017).

5. Einstein is supposed to have said that "everything should be made as simple as possible, but not simpler." Although the precise source of the quote is unknown, several suggestions have been put forward. See http://quoteinvestigator.com/2011/ 05/13/einstein-simple/.

6. Chalmers 1995, 1996. We referred to his view in chapter 1.

7. Maturana and Varela 1998, p. 234 (with reference to the role of language in the role of knowing).

8. Clark 2016, p. 122.

9. Maturana and Varela 1980, p. 13.

10. P. S. Churchland (1986) commented on changing intuitions (about consciousness and about concepts in general):

> Neurological studies are also vitally important for their capacity to unsettle our intuitions, our assumptions, and our commonplace verities about the nature of what the mind does. Unseated intuitions leave room for new theory. (p. 234)

> Nor of course do commonsense intuitions that two properties are substantially or even stunningly different entail anything about whether a future intertheoretic reduction might actually identify the two. Light may seem completely different from electromagnetic radiation, yet light turns out to *be* electromagnetic radiation. (p. 324)

11. Friston (2013a) shows that a dynamic system separated from its environment can be described in terms of predictive processing and lead to what looks like a simple form of autopoiesis; Van Duijn, Keijzer, and Franken (2006) identify sensory-motor coordination with cognition, and Lyon (2015) discusses bacterial capacities in terms encompassed by a general notion of cognition.

12. The description of this amazing "eye" is given in Gavelis et al. 2015. As the authors noted, a fascinating aspect of the study is that plastids and mitochondria can be structural elements.

13. See Maturana and Varela 1998. Moreno and Mossio (2015), who extend and qualify the Maturana and Varela project, ascribe minimal cognition to preneural organisms and full-blown cognition to neural animals. Following the arguments of Keijzer and Arnellos (2017), Godfrey-Smith (2016a, 2016b, 2016d) distinguishes between protocognitive (e.g., single cells) and cognitive (neural) organisms.

14. For a description of the effects of anesthetics on plants, see Yokawa et al. 2018, and for their effects on neurons in culture, see Platholi et al. 2014.

15. For the origins of Monod's aphorism, see Friedmann 2004.

16. For a discussion of epigenetic learning in nonneural organisms, see Ginsburg and Jablonka 2009. The evidence for ontogenetic learning by association in bacteria is, in our opinion, as yet unconvincing, since in all reported experiments, the bacteria did not ontogenetically learn to anticipate the US following the CS but evolved to anticipate it. The theoretical possibility that fairly simple chemical networks can support associative learning has been explored, however, and on the basis of this work, it is very likely that networks that allow ontogenetic learning by association in single cells (including bacteria) will be found (McGregor et al. 2012).

17. Gagliano et al. 2016. This fascinating study found no distinction between a Pavlovian type of limited AL and conditional sensitization because the effects of the US preceding the CS were not examined. Nevertheless, there is no doubt that learning by association did occur.

18. Schacter (2001) wrote a biography of Semon, emphasizing the neo-Lamarckian research tradition of which he was part. According to this tradition, memory and heredity form a continuum, and heredity can be conceived as a form of unconscious memory. It was believed that the repetition of activities that lead to memorization and to the formation of automatic habits during the lifetime of the individual can eventually, in some circumstances, become inherited through as yet unknown cell memory mechanisms. As we indicated in chapters 5 and 7, Semon developed the most sophisticated and systematic theory of heredity-as-memory within this tradition.

19. *Babylonian Talmud*, Tractate Sanhedrin, Folio 65b.

20. Holland et al. 2017.

21. Cangelosi and Schlesinger 2014. Machine learning, which explores the algorithms leading to intrinsic motivation and other cognitive abilities, is a very wide field employing increasingly sophisticated reinforcement machine-learning algorithms.

22. Metzinger 2013.

23. Godfrey Smith (2012) analyzes the notion of Darwinian individuals.

24. The notion of individuality is a central subject in the current philosophy of biology. For early ideas, see Buss 1987; Jablonka 1994; Maynard-Smith and Szathmáry 1995. For more recent philosophical discussions, see *Philosophy of Biology* 31 (December 2016), which is devoted to the analysis of the notion of biological individuality; see also Pradeu 2012 and Tauber 2017 for the role of the immune system in imparting individuality to a biological system. For a discussion of levels of autonomy (rather than levels of individuality), see Moreno and Mossio 2015. Interestingly, there is hardly anything in this literature, with the notable exception of Moreno and Mossio, about the individuating role of the nervous system, although it is self-evident that the individuality of neural organisms is far richer than that of nonneural ones.

25. For more writings of the Zen priest Ikkyū (1394–1481), see https://terebess.hu/zen/mesterek/IkkyuDoka.html.

26. Jablonka, Ginsburg, and Dor 2012.

27. Finlay 2015.

28. Jablonka, Ginsburg, and Dor (2012) and Ginsburg and Jablonka (2014) summarize some of the emotional, cognitive, and social preconditions for the evolution of language. For discussions on the learning of relation and rule-following behavior, see Hayes and Hayes 1989; Hayes, Blackledge, and Barnes-Holmes 2001; Zettle et al. 2016, chaps. 9 and 10.

29. For Dor's theory, see Dor 2015. The quotation is from Dor 2015, pp. 24–25.

30. Numbers 22:21–38 (King James Version).

31. Dor and Jablonka 2010; 2014.

32. Tomasello 2014, p. 95.

33. For a discussion of autobiographical memory as an evolutionary outcome of the problems generated by the instruction of imagination, see Jablonka (forthcoming). An important function of autobiographical memory may be to increase the reliability of communication because autobiographical memory enables a person to monitor her own and her interlocutors' linguistic information. Mahr and Csibra (2017) explored this idea from a different direction. Mercier and Sperber (2017) see human reason as reasoning about reasons, involving (mainly) post hoc rationalization of one's opinions and actions and a close scrutiny of the reasoning of others; this type of reasoning about reasons is tightly related to the evolution of linguistic communication.

34. Brembs 2011b.

35. James 1890, 1:141.

36. Hobson and Friston 2016. We presented our position on the issue of consciousness's "function" in Jablonka and Ginsburg 2013; Ginsburg and Jablonka 2015; Bronfman, Ginsburg, and Jablonka 2016c.

37. The quote is from R. W. Clark 1984, p. 141. Haldane expressed his view on animal experimentation in a lecture delivered in 1928, published in Haldane 1933.

References

Adamec, R., Holmes, A., and Blundell, J. (2008). Vulnerability to lasting anxiogenic effects of brief exposure to predator stimuli: Sex, serotonin and other factors—relevance to PTSD. *Neuroscience and Biobehavioral Reviews* 32, 1287–1292. doi:10.1016/j.neubiorev.2008.05.005.

Addis, D. R., Wong, A. T., and Schacter, D. L. (2007). Remembering the past and imagining the future: Common and distinct neural substrates during event construction and elaboration. *Neuropsychologia* 45, 1363–1377. doi:10.1016/j.neuropsychologia .2006.10.016.

Adrian, E. D. (1928). *The Basis of Sensation: The Action of the Sense Organs*. New York: W. W. Norton.

Agata, K. (2008). Planaria nervous system. *Scholarpedia* 3, 5558. doi:10.4249/scholar pedia.5558.

Agrillo, C., Piffer, L., and Bisazza, A. (2010). Large number discrimination by mosquito fish. *PLoS One* 5, e15232. doi:10.1371/journal.pone.0015232.

Ahissar, E., and Arieli, A. (2012). Seeing via miniature eye movements: A dynamic hypothesis for vision. *Frontiers in Computational Neuroscience* 6, 89. doi:10.3389/ fncom.2012.00089.

Ahissar, E., and Assa, E. (2016). Perception as a closed-loop convergence process. *eLife* 5, e12830. doi:10.7554/eLife.12830.

Ahissar, E., and Knutsen, P. M. (2008). Object localization with whiskers. *Biological Cybernetics* 98, 449–458. doi:10.1007/s00422-008-0214-4.

Akagi, M. (2016). Cognition in practice: Conceptual development and disagreement in cognitive science. PhD diss., University of Pittsburgh.

Akers, K. G., Martinez-Canabal, A., Restivo, L., Yiu, A. P., De Cristofaro, A., Hsiang, H.-L., Wheeler, A. L., Guskjolen, A., Niibori, Y., Shoji, H., Ohira, K., Richards, B. A., Miyakawa, T., Josselyn, S. A., and Frankland, P. W. (2014). Hippocampal neurogenesis regulates forgetting during adulthood and infancy. *Science* 344, 598–602. doi:10.1126/science.1248903.

Albert, D. J. (2011). What's on the mind of a jellyfish? A review of behavioural observations on *Aurelia sp.* jellyfish. *Neuroscience and Biobehavioral Reviews* 35, 474–482. doi:10.1016/j.neubiorev.2010.06.001.

Albertin, C. B., Simakov, O., Mitros, T., Wang, Z. Y., Pungor, J. R., Edsinger-Gonzales, E., Brenner, S., Ragsdale, C. W., and Rokhsar, D. S. (2015). The octopus genome and the evolution of cephalopod neural and morphological novelties. *Nature* 524, 220–237. doi:10.1038/nature14668.

Alcedo, J., and Zhang, Y. (2013). Molecular and cellular circuits underlying *Caenorhabditis elegans* olfactory plasticity. In R. Menzel and P. Benjamin (Eds.), *Invertebrate Learning and Memory*, pp. 112–123. London: Academic Press.

Alkire, M. T., Hudetz, A. G., and Tononi, G. (2008). Consciousness and anesthesia. *Science* 322, 876–880. doi:10.1126/science.1149213.

Allen, A. T., and Fortin, N. J. (2013). The evolution of episodic memory. *Proceedings of the National Academy of Sciences USA* 110, 10379–10386. doi:10.1073/pnas.1301199110.

Allen, C. (2010). Animal consciousness. In E. N. Zalta (Ed.), *The Stanford Encyclopedia of Philosophy*. Winter 2010 ed. https://plato.stanford.edu/archives/win2010/entries/consciousness-animal/.

Allen, C. (2012). Fish cognition and consciousness. *Journal of Agricultural and Environmental Ethics* 26, 25–39. doi:10.1007/s10806-011-9364-9.

Allen, C., Grau, J. W., and Meagher, M. W. (2009). The lower bounds of cognition: What do spinal cords reveal? In J. Bickle (Ed.), *The Oxford Handbook of Philosophy of Neuroscience*, pp. 129–142. Oxford: Oxford University Press.

Allen, C., and Trestman, M. (2015). Animal consciousness. In E. N. Zalta (Ed.), *The Stanford Encyclopedia of Philosophy*. Winter 2016 ed. https://plato.stanford.edu/archives/win2016/entries/consciousness-animal/.

Anctil, M. (2015). *Dawn of the Neuron: The Early Struggles to Trace the Origins of the Nervous System*. Montreal: McGill-Queen's University Press.

Anderson, M. C., Bjork, R. A., and Bjork, E. L. (1994). Remembering can cause forgetting: Retrieval dynamics in long-term memory. *Journal of Experimental Psychology* 20, 1063–1087.

Anderson, P. A. V. (2004). Cnidarian neurobiology: What does the future hold? *Hydrobiologia* 530/531, 107–116. doi:10.1007/978-1-4020-2762-8_13.

Antcliffe, J. B., Callow, R. H., and Brasier, M. D. (2014). Giving the early fossil record of sponges a squeeze. *Biological Reviews* 89, 972–1004. doi:10.1111/brv.12090.

Applewhite, P. B. (1968). Non-local nature of habituation in a rotifer and a protozoan. *Nature* 217, 287–288.

Ardiel, E. L., and Rankin, C. H. (2010). An elegant mind: Learning and memory in *Caenorhabditis elegans*. *Learning and Memory* 17, 191–201. doi:10.1101/lm.960510.

Arendt, D. (2008). The evolution of cell types in animals: Emerging principles from molecular studies. *Nature Reviews Genetics* 9, 868–882. doi:10.1038/nrg2416.

Arendt, D. (2018). Animal evolution: Convergent nerve cords? *Current Biology* 28, R225–R226. doi:10.1016/j.cub.2018.01.056.

Arendt, D., Denes, A. S., Jékely, G., and Tessmar-Raible, K. (2008). The evolution of nervous system centralization. *Philosophical Transactions of the Royal Society B* 363, 1523–1528. doi:10.1098/rstb.2007.2242.

Arendt, D., Tosches, M. A., and Marlow, H. (2016). From nerve net to nerve ring, nerve cord and brain—evolution of the nervous system. *Nature Reviews Neuroscience* 17, 61–72. doi:10.1038/nrn.2015.15.

Århem, P., and Liljenström, H. (2008). Beyond cognition—on consciousness transitions. In H. Liljenström and P. Århem (Eds.), *Consciousness Transitions: Phylogenetic, Ontogenetic, and Physiological Aspects*, pp. 1–25. Amsterdam, The Netherlands: Elsevier.

Aristotle. (1984a). *Generation of Animals* (J. Barnes, Ed., A. Platt, Trans.). In *The Complete Works of Aristotle: The Revised Oxford Translation*, vol. 1. Princeton, NJ: Princeton University Press.

Aristotle. (1984b). *Metaphysics* (J. Barnes, Ed., W. D. Ross, Trans.). In *The Complete Works of Aristotle: The Revised Oxford Translation*, vol. 2. Princeton, NJ: Princeton University Press.

Aristotle. (1984c). *On Memory* (J. Barnes, Ed., J. I. Beare, Trans.). *The Complete Works of Aristotle: The Revised Oxford Translation*, vol. 1. Princeton, NJ: Princeton University Press.

Aristotle. (1984d). *On the Soul* (J. Barnes, Ed., J. A. Smith, Trans.). *The Complete Works of Aristotle: The Revised Oxford Translation*, vol. 1. Princeton, NJ: Princeton University Press.

Arthur, W. (2000). *The Origin of Animal Body Plans*. Cambridge: Cambridge University Press.

Aru, J., Bachmann, T., Singer, W., and Mellon, L. (2012). Distilling the neural correlates of consciousness. *Neuroscience and Biobehavioral Reviews* 36, 737–746. doi:10.1016/j.neubiorev.2011.12.003.

Ashley, J., Cordy, B., Lucia, D., Fradkin, L. G., Budnik, V., and Thomson, T. (2018). Retrovirus-like Gag protein Arc1 binds RNA and traffics across synaptic boutons. *Cell* 172, 262–274.e11.

Aulsebrook, A. E., Jones, T. M., Niels, C., Rattenborg, N. C., Roth, T. C., and Lesku, J. A. (2016). Sleep ecophysiology: Integrating neuroscience and ecology. *Trends in Ecology and Evolution* 31, 590–599. doi:10.1016/j.tree.2016.05.004.

Avital, E., and Jablonka, E. (2000). *Animal Traditions: Behavioural Inheritance in Evolution.* Cambridge: Cambridge University Press.

Axel, R. (2004). Scents and sensibility: A molecular logic of olfactory perception. Nobelprize.org. http://www.nobelprize.org/nobel_prizes/medicine/laureates/2004/axel -lecture.html.

Baars, B. J. (1986). *The Cognitive Revolution in Psychology.* New York: Guilford Press.

Baars, B. J. (1997). *In the Theater of Consciousness: The Workspace of the Mind.* Oxford: Oxford University Press.

Baars, B. J. (2005a). Global workspace theory of consciousness: Toward a cognitive neuroscience of human experience. In S. Laureys (Ed.), *Progress in Brain Research* 150, 45–53. doi:10.1016/S0079-6123(05)50004-9.

Baars, B. J. (2005b). Subjective experience is probably not limited to humans: The evidence from neurobiology and behavior. *Consciousness and Cognition* 14, 7–21. doi: 10.1016/j.concog.2004.11.002.

Babenko, V. N., Rogozin, I. B., Mekhedov, S. L., and Koonin, E. V. (2004). Prevalence of intron gain over intron loss in the evolution of paralogous gene families. *Nucleic Acids Research* 32, 3724–3733. doi:10.1093/nar/gkh686.

Bain, T., and Strong, P. N. (1972). Classical conditioning in the leech *Macrobdella ditetra* as a function of CS and UCS intensity. *Conditional Reflex: A Pavlovian Journal of Research and Therapy* 7, 210–215. doi:10.1007/BF03000220.

Baker, M. E. (2006). The genetic response to Snowball Earth: Role of HSP90 in the Cambrian explosion. *Geobiology* 4, 11–14. doi:10.1111/j.1472-4669.2006.00067.x.

Balcombe, J. (2016). *What a Fish Knows: The Inner Lives of Our Underwater Cousins.* New York: Scientific American/Farrar, Straus and Giroux.

Baldwin, J. M. (1896a). A new factor in evolution. *American Naturalist* 30, 441–451. doi:10.1086/276408.

Baldwin, J. M. (1896b). Consciousness and evolution. *Psychological Review* 3, 300–309.

Barron, A. B., and Klein, C. (2016). What insects can tell us about the origins of consciousness. *Proceedings of the National Academy of Sciences USA* 113, 4900–4908. doi:10.1073/pnas.1520084113.

Barron, A. B., Søvik, E., and Cornish, J. L. (2010). The roles of dopamine and related compounds in reward-seeking behavior across animal phyla. *Frontiers in Behavioral Neuroscience* 4, 163. doi:10.3389/fnbeh.2010.00163.

Basile, B. M., Schroeder, G. R., Brown, E. K., Templer, V. L., and Hampton, R. R. (2015). Evaluation of seven hypotheses for metamemory performance in rhesus monkeys. *Journal of Experimental Psychology: General* 144, 85–102. doi:10.1037/xge 0000031.

Bateson, G. (1972). *Steps to an Ecology of Mind*. New York: Ballantine.

Bateson, G. (1979). *Mind and Nature: A Necessary Unity*. New York: Dutton.

Bateson, P. (2005). The return of the whole organism. *Journal of Biosciences* 30, 31–39. doi:10.1007/BF02705148.

Bateson, P., and Martin, P. (2013). *Play, Playfulness, Creativity and Innovation*. Cambridge: Cambridge University Press.

Baxter, D. A., and Byrne, J. H. (2006). Feeding behavior of *Aplysia*: A model system for comparing cellular mechanisms of classical and operant conditioning. *Learning and Memory* 13, 669–680. doi:10.1101/lm.339206.

Beaton, M. (2005). What RoboDennett still doesn't know. *Journal of Consciousness Studies* 12(12), 3–25.

Bedau, M. A. (2012). A functional account of degrees of minimal chemical life. *Synthese* 185, 73–88. doi:10.1007/s11229-011-9876-x.

Bédécarrats, A., Chen, S., Pearce, K., Cai, D., and Glanzman, D. L. (2018). RNA from trained *Aplysia* can induce an epigenetic engram for long term sensitization in untrained *Aplysia*. *eNeuro* 5(3). doi:10.1523/ENEURO.0038-18.2018.

Behbehani, M. B. (1995). Functional characteristics of the midbrain periaqueductal gray. *Progress in Neurobiology* 46, 575–605. doi:10.1016/0301-0082(95)00009-.

Bellingham, W. P., and Gillette, K. (1981). Spontaneous configuring to a tone-light compound using appetitive training. *Learning and Motivation* 12, 420–434. doi:10.1016/0023-9690(81)90003-5.

Belyaev, D. K., and Borodin, P. M. (1982). The influence of stress on variation and its role in evolution. *Biologisches Zentralblatt* 100, 705–714.

Bengtson, S. (2002). Origins and early evolution of predation. *Paleontological Society Papers* 8, 289–318. doi:10.1017/S1089332600001133.

Benjamin, P. R. (2013). A systems analysis of neural networks underlying gastropod learning and memory. In P. Benjamin and R. Menzel (Eds.), *Invertebrate Learning and Memory*, pp. 163–182. London: Academic Press.

Bennett, M. R., and Hacker, P. M. S. (2003). *Philosophical Foundations of Neuroscience*. Oxford: Blackwell.

Berg, S. (2000). *Ikkyu: Crow with No Mouth: 15th Century Zen Master*. Seattle: Copper Canyon Press.

Bergson, H. (1920). *Life and Consciousness* (W. Carr, Trans.). In *Mind-Energy: Lectures and Essays*. New York: Henry Holt.

Bergson, H. (1896/1991). *Matter and Memory* (N. M. Paul and W. S. Palmer, Trans.). New York: Zone Books.

Bergstrom, C. T., and Lachmann, M. (2004). Shannon information and biological fitness. In *Information Theory Workshop*, pp. 50–54. Piscataway, NJ: IEEE.

Bernal, J. D. (1949). The physical basis of life. *Proceedings of the Physical Society A* 62, 537–558. doi:10.1088/0370-1298/62/9/301.

Berridge, K. C. (2003). Pleasures of the brain. *Brain and Cognition* 52, 106–128. doi:10.1016/S0278-2626(03)0001.

Berridge, K. C., and Kringelbach, M. L. (2013). Neuroscience of affect: Brain mechanisms of pleasure and displeasure. *Current Opinion in Neurobiology* 23, 294–303. doi:10.1016/j.conb.2013.01.017.

Beyaert, L., Greggers, U., and Menzel, R. (2012). Honeybees consolidate navigation memory during sleep. *Journal of Experimental Biology* 215, 3981–3988. doi:10.1242/jeb.075499.

Bhagavan, S., and Smith, B. H. (1997). Olfactory conditioning in the honey bee, *Apis mellifera*: Effects of odor intensity. *Physiology and Behavior* 61, 107–117. doi:10.1016/S0031-9384(96)00357-5.

Bhatla, N. (2014). Tasting light through hydrogen peroxide: Molecular mechanisms and neural circuits. PhD diss., Massachusetts Institute of Technology.

Blanke, O., and Metzinger, T. K. (2009). Full-body illusions and minimal phenomenal selfhood. *Trends in Cognitive Sciences* 13, 7–13. doi:10.1016/j.tics.2008.10.003.

Blaze, J., and Roth, T. L. (2013). Epigenetic mechanisms in learning and memory. *WIREs Cognitive Science* 4, 105–115. doi:10.1002/wcs.1205.

Block, N. (2007). Consciousness, accessibility, and the mesh between psychology and neuroscience. *Behavioral and Brain Sciences* 30, 481–548. doi:10.1017/S0140525X07002786.

Block, N. (2008). Consciousness and cognitive access. *Proceedings of the Aristotelian Society* 108(part 3), 289–317. doi:10.1111/j.1467-9264.2008.00247.x.

Boakes, R. (1984). *From Darwin to Behaviourism: Psychology and the Minds of Animals*. Cambridge: Cambridge University Press.

Boal, J. G., Dunham, A. W., Williams, K. T., and Hanlon, R. T. (2000). Experimental evidence for spatial learning in octopuses (*Octopusbimaculoides*). *Journal of Comparative Psychology* 114, 246–252. doi:10.1037/0735-7036.114.3.246.

Bodelón, C., Fallah, M., and Reynolds, J. H. (2007). Temporal resolution for the perception of features and conjunctions. *Journal of Neuroscience* 27, 725–730. doi:10.1523/jneurosci.3860-06.2007.

Boden, M. A. (2006). *Mind as Machine: A History of Cognitive Science.* Oxford: Clarendon Press.

Boden, M. A. (2009). Life and mind. *Minds and Machines* 19, 453–463. doi:10.1007/s11023-009-9169-z.

Boisvert, M. J., and Sherry, D. F. (2006). Interval timing by an invertebrate, the bumble bee *Bombus impatiens. Current Biology* 16, 1636–1640. doi:10.1016/j.cub.2006.06.064.

Borges, J. L. (1942/1962). *Funes the memorious* (A. Kerrigan, Trans.). In *Ficciones*, pp. 107–115. New York: Grove Press.

Bourgine, P., and Stewart, J. (2004). Autopoiesis and cognition. *Artificial Life* 10, 327–345. doi:10.1162/1064546041255557.

Braithwaite, V. A. (2010). *Do Fish Feel Pain?* Oxford: Oxford University Press.

Brasier, M. (2009). *Darwin's Lost World.* Oxford: Oxford University Press.

Braun, E., and David, L. (2011).The role of cellular plasticity in the evolution of regulatory novelty. In S. B. Gissis and E. Jablonka (Eds.), *The Transformation of Lamarckism: From Subtle Fluids to Molecular Biology*, pp. 181–191. Cambridge, MA: MIT Press.

Brembs, B. (2011a). Spontaneous decisions and operant conditioning in fruit flies. *Behavioural Processes* 87, 157–164. doi:10.1016/j.beproc.2011.02.005.

Brembs, B. (2011b). Towards a scientific concept of free will as a biological trait: Spontaneous actions and decision-making in invertebrates. *Proceedings of the Royal Society B* 278, 930–939. doi:10.1098/rspb.2010.2325.

Brembs, B., and Heisenberg, M. (2000). The operant and the classical in conditioned orientation of *Drosophila melanogaster* at the flight simulator. *Learning and Memory* 7, 104–115. doi:10.1101/lm.7.2.104.

Brembs, B., and Heisenberg, M. (2001). Conditioning with compound stimuli in *Drosophila melanogaster* in the flight simulator. *Journal of Experimental Biology* 204, 2849–2859.

Brembs, B., Lorenzetti, F. D., Reyes, F. D., Baxter, D. A., and Byrne, J. H. (2002). Operant reward learning in *Aplysia*: Neuronal correlates and mechanisms. *Science* 296, 1706–1709. doi:10.1126/science.1069434.

Brennan, S. T., Lowenstein, T. K., and Horita, J. (2004). Seawater chemistry and the advent of biocalcification. *Geology* 32, 473–476. doi:10.1130/G20251.1.

Bronfman, Z. Z. (2017). The emergence of basic consciousness through the evolution of associative learning. PhD diss., Tel Aviv University.

Bronfman, Z. Z., Brezis, N., Jacobson, H., and Usher, M. (2014). We see more than we can report: "Cost free" color phenomenality outside focal attention. *Psychological Science* 25, 1394–1403. doi:10.1177/0956797614532656.

Bronfman, Z. Z., Ginsburg, S., and Jablonka, E. (2014). Shaping the learning curve: Epigenetic dynamics in neural plasticity. *Frontiers in Integrative Neuroscience* 8, 55. doi:10.3389/fnint.2014.00055.

Bronfman, Z. Z., Ginsburg, S., and Jablonka, E. (2016a). The epigenetics of neural learning. In R. Honey and R. Murphy (Eds.), *The Wiley-Blackwell Handbook on the Cognitive Neuroscience of Associative Learning*, pp. 136–176. Chichester, UK: Wiley.

Bronfman, Z. Z., Ginsburg, S., and Jablonka, E. (2016b). The evolutionary origins of consciousness: Suggesting a transition marker. *Journal of Consciousness Studies* 23(9/10), 7–34.

Bronfman, Z. Z., Ginsburg, S., and Jablonka, E. (2016c). The transition to minimal consciousness through the evolution of associative learning. *Frontiers in Psychology* 7, 1954. doi:10.3389/fpsyg.2016.01954.

Bronfman, Z. Z., Ginsburg, S., and Jablonka, E. (2018). Classical and operant conditioning: Evolutionarily distinct strategies? In D. S. Wilson and S. C. Hayes (Eds.), *Evolution and Contextual Behavioral Science: A Reunification*. Oakland, CA: Context Press/New Harbinger.

Brunet, T., and Arendt, D. (2016). From damage response to action potentials: Early evolution of neural and contractile modules in stem eukaryotes. *Philosophical Transactions of the Royal Society B* 371, 20150043. doi:10.1098/rstb.2015.0043.

Bryan, J. S., and Krasne, F. B. (1977). Protection from habituation of the crayfish lateral giant fibre escape response. *Journal of Physiology* 271, 351–368.

Budd, G. E., and Jensen, S. (2017). The origin of the animals and a "Savannah" hypothesis for early bilaterian evolution. *Biological Reviews* 92, 446–473. doi:10.1111/brv.12239.

Bullmore, E., and Sporns, O. (2012). The economy of brain network organization. *Nature Reviews Neuroscience* 13, 336–349. doi:10.1038/nrn3214.

Burghardt, G. M. (1985). Animal awareness: Current perceptions and historical perspective. *American Psychologist* 40, 905–919. doi:10.1037/0003-066X.40.8.905.

Burghardt, G. M. (2005). *The Genesis of Play*. Cambridge, MA: MIT Press.

Burghardt, G. M., Dinets, V., and Murphy, J. B. (2015). Highly repetitive object play in a cichlid fish (*Tropheus duboisi*). *Ethology* 121, 38–44. doi:10.1111/eth.12312.

Burkhardt, R. W., Jr. (1977). *The Spirit of System: Lamarck and Evolutionary Biology.* Cambridge, MA: Harvard University Press.

Buss, L. W. (1987). *The Evolution of Individuality.* Princeton, NJ: Princeton University Press.

Buss, R. R., Sun, W., and Oppenheim, R. W. (2006). Adaptive roles of programmed cell death during nervous system development. *Annual Review of Neuroscience* 29, 1–35. doi:10.1146/annurev.neuro.29.051605.112800.

Butler, A. B., Manger, P. R., Lindahl, B. I., and Århem, P. (2005). Evolution of the neural basis of consciousness: A bird-mammal comparison. *BioEssays* 27, 923–936. doi:10.1002/bies.20280.

Butterfield, N. J. (1997). Plankton ecology and the Proterozoic-Phanerozoic transition. *Paleobiology* 23, 247–262. doi:10.1017/S009483730001681X.

Butterfield, N. J. (2011). Animals and the invention of the Phanerozoic Earth system. *Trends in Ecology and Evolution* 26, 281–287. doi:10.1016/j.tree.2010.11.012.

Byrne, J. H., and Hawkins, R. D. (2015). Nonassociative learning in invertebrates. *Cold Spring Harbor Perspectives in Biology* 7, a021675. doi:10.1101/cshperspect.a021675.

Cabanac, M. (1992). Pleasure: The common currency. *Journal of Theoretical Biology* 155, 173–200.

Cabanac, M., Cabanac, A. J., and Parent, A. (2009). The emergence of consciousness in phylogeny. *Behavioural Brain Research* 198, 267–272. doi:10.1016/j.bbr.2008.11.028.

Cabanis, P-J. G. (1789–1824/1956). *Oeuvres philosophique de Cabanis* (C. Lehec and J. Cazeneuve, Eds.). 2 vols. Paris: Paris Presses Universitaires de France.

Cabanis, P-J. G. (1802/1981). *On the Relations between the Physical and Moral Aspects of Man* (G. Mora, Ed., M. Duggan Saidi, Trans.). Baltimore: Johns Hopkins University Press.

Cabej, R. N. (2008). *Epigenetic Principles of Evolution.* Dumont, NJ: Albanet.

Cabej, R. N. (2013). *Building the Most Complex Structure on Earth: An Epigenetic Narrative of Development and Evolution of Animals.* Amsterdam, The Netherlands: Elsevier.

Cai, X. (2012). Evolutionary genomics reveals the premetazoan origin of opposite gating polarity in animal-type voltage-gated ion channels. *Genomics* 99, 241–245. doi:10.1016/j.ygeno.2012.01.007.

Cairns-Smith, A. G. (1996). *Evolving the Mind: On the Nature of Matter and the Origin of Consciousness.* New York: Cambridge University Press.

Calvin, W. H. (1996). *The Cerebral Code: Thinking a Thought in the Mosaics of the Mind.* Cambridge, MA: MIT Press.

Campbell, D. T. (1965). Variation and selective retention in socio-cultural evolution. In H. R. Barringer, G. I. Blanksten, and R. W. Mack (Eds.), *Social Change in Developing Areas: A Reinterpretation of Evolutionary Theory*, pp. 19–49. Cambridge, MA: Schenkman.

Campbell, D. T. (1990). Epistemological roles for selection theory. In N. Rescher (Ed.), *Evolution, Cognition, and Realism: Studies in Evolutionary Epistemology*, pp. 1–19. Lanham, MD: University Press of America.

Campbell, R. A. A., and Turner, G. C. (2010). The mushroom body. *Current Biology* 20, R11–R12. doi:10.1016/j.cub.2009.10.031.

Canfield, D. E., Poulton, S. W., and Narbonne, G. M. (2006). Late Neoproterozoic deep-ocean oxidation and the rise of animal life. *Science* 315, 92–95. doi:10.1126/science.1135013.

Cangelosi, A., and Schlesinger, M. (2014). *Developmental Robotics: From Babies to Robots.* Cambridge, MA: MIT Press.

Cannon, W. B. (1927). The James-Lange theory of emotions: A critical examination and an alternative theory. *American Journal of Psychology* 39, 106–124. doi:10.2307/1415404.

Carbone, C., and Narbonne, G. M. (2014). When life got smart: The evolution of behavioral complexity through the Ediacaran and early Cambrian of NW Canada. *Journal of Paleontology* 88, 309–330. doi:10.1666/13-066.

Carlesimo, G. A., Casadio, P., Sabbadini, M., and Caltagirone, C. (1998). Associative visual agnosia resulting from a disconnection between intact visual memory and semantic systems. *Cortex* 34, 563–576. doi:10.1016/S0010-9452(08)70514-8.

Carpenter, W. J. (1852). On the influence of suggestion in modifying and directing muscular movement, independently of volition. *Proceedings of the Royal Institution of Great Britain* 1, 147–153.

Carruthers, P. (2000). *Phenomenal Consciousness: A Naturalistic Theory.* Cambridge: Cambridge University Press.

Cavalier-Smith, T. (2006). Cell evolution and earth history: Stasis and revolution. *Philosophical Transactions of the Royal Society B* 361, 969–1006. doi:10.1098/rstb.2006.184.

Cavalier-Smith, T. (2017). Origin of animal multicellularity: Precursors, causes, consequences—the choanoflagellate/sponge transition, neurogenesis and the Cambrian explosion. *Philosophical Transactions of the Royal Society B* 372, 20150476. doi:10.1098/rstb.2015.0476.

Chalmers, D. J. (1995). Facing up to the problem of consciousness. *Journal of Consciousness Studies* 2(3), 200–219.

Chalmers, D. J. (1996). *The Conscious Mind: In Search of a Fundamental Theory.* Oxford: Oxford University Press.

Chalmers, D. J. (1997). Moving forward on the problem of consciousness. *Journal of Consciousness Studies* 4(1), 3–46.

Chang, L., and Tsao, D. Y. (2017). The code for facial identity in the primate brain. *Cell* 169, 1013–1028. e14. doi:10.1016/j.cell.2017.05.0.

Changeux, J.-P. (1983). *L'Homme neuronal.* Paris: Fayard.

Changeux, J.-P. (1997). *Neuronal Man: The Biology of the Mind.* Princeton, NJ: Princeton University Press.

Changeux, J.-P. (2010). Allosteric receptors: From electric organ to cognition. *Annual Review of Pharmacology and Toxicology* 50, 1–38. doi:10.1146/annurev.pharmtox.010909.105741.

Changeux, J.-P., Courrége, P., and Danchin, A. (1973). A theory of the epigenesis of neuronal networks by selective stabilization of synapses. *Proceedings of the National Academy of Sciences USA* 70, 2974–2978. doi:10.1073/pnas.70.10.2974.

Changeux, J.-P., and Danchin, A. (1976). Selective stabilization of developing synapses as a mechanism for the specification of neuronal networks. *Nature* 264, 705–712. doi:10.1038/264705a0.

Chase, R. (2000). Structure and function in the cerebral ganglion. *Microscopy Research and Technique* 49, 511–520. doi:10.1002/1097-0029(20000615)49:6<511::AID-JEMT2>3.0.CO;2-L.

Chau, A., Salazar, A. M., Krueger, F., Cristofori, I., and Grafman, J. (2015). The effect of claustrum lesions on human consciousness and recovery of function. *Consciousness and Cognition* 36, 256–264. doi:10.1016/j.concog.2015.06.017.

Chekhov, A. (1891/1916). *The Duel and Other Stories* (C. Garnett, Trans.). New York: Macmillan.

Chen, S., Cai, D., Pearce, K., Sun, P. Y. W., Roberts, A. C., and Glanzman, D. L. (2014). Reinstatement of long-term memory following erasure of its behavioral and synaptic expression in *Aplysia*. *eLife* 3, e03896. doi:10.7554/eLife.03896.

Chen, X., Ling, H.-F., Vance, D., Shields-Zhou, G. A., Zhu, M., Poulton, S. W., Och, L. M., Jiang, S.-Y., Li, D., Cremonese, L., and Archer, C. (2015). Rise to modern levels of ocean oxygenation coincided with the Cambrian radiation of animals. *Nature Communications* 6, 7142. doi:10.1038/ncomms8142.

Chirimuuta, M. (2015). *Outside Color: Perceptual Science and the Puzzle of Color in Philosophy*. Cambridge, MA: MIT Press.

Chittka, L., and Niven, J. (2009). Are bigger brains better? *Current Biology* 19, R995–R1008. doi:10.1016/j.cub.2009.08.023.

Churchland, P. M. (1985). Reduction, qualia, and the direct introspection of brain states. *Journal of Philosophy* 82, 8–28. doi:10.2307/2026509.

Churchland, P. M. (1988). *Matter and Consciousness: A Contemporary Introduction to the Philosophy of Mind*. Cambridge, MA: MIT Press.

Churchland, P. M. (1989). Knowing qualia: A reply to Jackson. In *A Neurocomputational Perspective: The Nature of Mind and the Structure of Science*, pp. 67–76. Cambridge, MA: MIT Press.

Churchland, P. S. (1986). *Neurophilosophy: Toward a Unified Science of the Mind-Brain*. Cambridge, MA: MIT Press.

Churchland, P. S. (2002). *Brain-Wise: Studies in Neurophilosophy*. Cambridge, MA: MIT Press.

Claparède, E. (1911). Récognition et moite. *Archives de psychologic Geneve* 11, 79–90.

Clark, A. (2008). *Supersizing the Mind: Embodiment, Action and Cognitive Extension*. Oxford: Oxford University Press.

Clark, A. (2013). Whatever next? Predictive brains, situated agents, and the future of cognitive science. *Behavioral and Brain Sciences* 36, 181–204. doi:10.1017/S0140525X12000477.

Clark, A. (2016). *Surfing Uncertainty: Prediction, Action, and the Embodied Mind*. New York: Oxford University Press.

Clark, R. W. (1984). *J.B.S.: The Life and Work of J.B.S. Haldane*. Oxford: Oxford University Press.

Clayton, N. S., Bussey, T. J., and Dickinson, A. (2003). Can animals recall the past and plan for the future? *Nature Reviews Neuroscience* 4, 685–691. doi:10.1038/nrn1180.

Clayton, N. S., and Dickinson, A. (1998). Episodic-like memory during cache recovery by scrub jays. *Nature* 395, 272–278. doi:10.1038/26216.

Cleeremans, A. (2008). Consciousness: The radical plasticity thesis. In R. Banerjee and B. K. Chakrabarti (Eds.), *Models of Brain and Mind: Physical, Computational and Psychological Approaches*, pp. 19–33. Amsterdam, The Netherlands: Elsevier.

Cleeremans, A. (2011). The radical plasticity thesis: How the brain learns to be conscious. *Frontiers in Psychology* 2, 86. doi:10.3389/fpsyg.2011.00086.

Coates, M. M. (2003). Visual ecology and functional morphology of Cubozoa (Cnidaria). *Integrative and Comparative Biology* 43, 542–548. doi:10.1093/icb/43.4.542.

Colbran, R. J., and Brown, A. M. (2004). Calcium/calmodulin-dependent protein kinase II and synaptic plasticity. *Current Opinion in Neurobiology* 14, 318–327.

Collett, M., Chittka, L., and Collett, T. S. (2013). Spatial memory in insect navigation. *Current Biology* 23, R789–R800. doi:10.1016/j.cub.2013.07.020.

Collett, M., and Collett T. S. (2009). The learning and maintenance of local vectors in desert ant navigation. *Journal of Experimental Biology* 212, 895–900. doi:10.1242/jeb.024521.

Colomb, J., and Brembs, B. (2010). The biology of psychology: "Simple" conditioning? *Communicative and Integrative Biology* 3, 142–145. doi:10.4161/cib.3.2.10334.

Colón-Cesario, M., Wang, J., Ramos, X., García, H. G., Dávila, J. J., Laguna, J., Rosado, C., and Peña de Ortiz, S. (2006). An inhibitor of DNA recombination blocks memory consolidation, but not reconsolidation, in context fear conditioning. *Journal of Neuroscience* 26, 5524–5533. doi:10.1523/JNEUROSCI.3050-05.2006.

Colón-Ramos, D. A. (2009). Synapse formation in developing neural circuits. *Current Topics in Developmental Biology* 87, 53–79. doi:10.1016/S0070-2153(09)01202-2.

Colwill, R. M., Absher, R. A., and Robert, M. L. (1988). Context-US learning in *Aplysia californica*. *Journal of Neuroscience* 8, 4434–4439.

Corning, W. C., Dyal, J. A., and Willows, A. O. D. (Eds.). (1973). *Invertebrate Learning.* 3 vols. New York: Plenum.

Corsi, P. (1988). *The Age of Lamarck: Evolutionary Theories in France, 1790–1830.* Berkeley: University of California Press.

Corsi, P. (2011). Jean-Baptiste Lamarck. From myth to history. In S. Gissis and E. Jablonka (Eds.), *Transformations of Lamarckism: From Subtle Fluids to Molecular Biology*, pp. 12–28. Cambridge, MA: MIT Press.

Craig, A. D. (2002). How do you feel? Interoception: The sense of the physiological condition of the body. *Nature Reviews Neuroscience* 3, 655–666. doi:10.1038/nrn894.

Craig, A. D. (2009). How do you feel—now? The anterior insula and human awareness. *Nature Reviews Neuroscience* 10, 59–70. doi:10.1038/nrn2555.

Crapse, T. B., and Sommer, M. A. (2008). Corollary discharge across the animal kingdom. *Nature Reviews Neuroscience* 9, 587–600. doi:10.1038/nrn2457.

Crick, F. (1988). *What Mad Pursuit: A Personal View of Scientific Discovery.* New York: Basic Books.

Crick, F. (1989). Neural Edelmanism. *Trends in Neuroscience* 12, 240–248. doi:10.1016/0166-2236(89)90019-2.

Crick, F. (1994). *The Astonishing Hypothesis*. New York: Charles Scribner's Sons.

Crick, F., and Koch, C. (1990). Towards a neurobiological theory of consciousness. *Seminars in the Neurosciences* 2, 263–275.

Crick, F., and Koch, C. (2003). A framework for consciousness. *Nature Neuroscience* 6, 119–126. doi:10.1038/nn0203-119.

Crisp, K. M., and Burrell, B. D. (2009). Cellular and behavioral properties of learning in leech and other annelids. In D. H. Shain (Ed.), *Annelids in Modern Biology*, pp. 135–155. Hoboken, NJ: Wiley-Blackwell.

Crook, R. J., and Basil, J. A. (2008). A biphasic memory curve in chambered nautilus, *Nautilus pompilius*. *Journal of Experimental Biology* 211, 1992–1998. doi:10.1242/jeb.018531.

Crystal, J. D. (2009). Elements of episodic-like memory in animal models. *Behavioural Processes* 80, 269–277. doi:10.1016/j.beproc.2008.09.009.

Cummins, R. (1975). Functional analysis. *Journal of Philosophy* 72, 741–765. doi:10.2307/2024640.

Damasio, A. R. (1994). *Descartes' Error: Emotion, Reason, and the Human Brain*. New York: Grosset/Putnam.

Damasio, A. (1999). *The Feeling of What Happens—Body, Emotion and the Making of Consciousness*. London: Vintage.

Damasio, A. (2010). *Self Comes to Mind—Constructing the Conscious Brain*. London: Heinemann.

Damasio, A., and Carvalho, G. B. (2013). The nature of feelings: Evolutionary and neurobiological origins. *Nature Reviews Neuroscience* 14, 143–152. doi:10.1038/nrn3403.

Damasio, A., Damasio, H., and Tranel, D. (2013). Persistence of feelings and sentience after bilateral damage of the insula. *Cerebral Cortex* 23, 833–846. doi:10.1093/cercor/bhs077.

Danziger, K. (2008). *Marking the Mind: A History of Memory*. Cambridge: Cambridge University Press.

Darwin, C. (1859). *On the Origin of Species by Means of Natural Selection, or the Preservation of Favoured Races in the Struggle for Life*. 1st ed. London: John Murray.

Darwin, C. (1871). *The Descent of Man, and Selection in Relation to Sex*. 1st ed. 2 vols. London: John Murray.

Darwin, C. (1872a). *The Expression of the Emotions in Man and Animals*. 1st ed. London: John Murray.

Darwin, C. (1872b). *The Origin of Species by Means of Natural Selection, or the Preservation of Favoured Races in the Struggle for Life*. 6th ed. London: John Murray.

Darwin, C. (1890/1998). *The Expression of the Emotion in Man and Animals*. 3rd ed. (P. Ekman, Ed.). London: Harper Collins.

Darwin, F. (Ed.). (1887). *The Life and Letters of Charles Darwin, Including an Autobiographical Chapter*. London: John Murray.

Davidson, E. H., and Erwin, D. H. (2006). Gene regulatory networks and the evolution of animal body plans. *Science* 311, 796–800. doi:10.1126/science.1113832.

Davidson, E. H., Peterson, K. J., and Cameron, R. A. (1995). Origin of bilaterian body plans: Evolution of developmental regulatory mechanisms. *Science* 270, 1319–1325. doi:10.1126/science.270.5240.1319.

Davis, R. L., and Zhong, D. Y. (2017). The biology of forgetting—a perspective. *Neuron* 95, 490–503. doi:10.1016/j.neuron.2017.05.039.

Dayan, P., and Huys, Q. J. (2009). Serotonin in affective control. *Annual Review of Neuroscience* 32, 95–126. doi:10.1146/annurev.neuro.051508.135607.

Deacon, T. (2011). *Incomplete Nature: How Mind Emerged from Matter*. New York: W. W. Norton.

de Duve, C. (1991). *Blueprint for a Cell: The Nature and Origin of Life*. Burlington, NC: Neil Patterson.

de Duve, C. (2002). *Life Evolving: Molecules, Mind, and Meaning*. New York: Oxford University Press.

de Graaf, T. A., Hsieh, P. J., and Sack, A. T. (2012). The "correlates" in neural correlates of consciousness. *Neuroscience and Biobehavioral Reviews* 36, 191–197. doi:10.1016/j.neubiorev.2011.05.012.

Dehaene, S. (2014). *Consciousness and the Brain: Deciphering How the Brain Codes Our Thoughts*. New York: Viking.

Dehaene, S., and Changeux, J.-P. (2011). Experimental and theoretical approaches to conscious processing. *Neuron* 70, 201–227. doi:10.1016/j.neuron.2011.03.018.

de Laguna, G. A. (1918). The empirical correlation of mental and bodily phenomena. *Journal of Philosophy, Psychology and Scientific Methods* 15, 533–541.

Denes, A., Jékely, G., Steinmetz, P. R. H., Raible, F., Snyman, H., Prud'homme, B., Ferrier, D. E. K., Balavoine, G., and Arendt, D. (2007). Molecular architecture of annelid nerve cord supports common origin of nervous system centralization in Bilateria. *Cell* 129, 277–288. doi:10.1016/j.cell.2007.02.040.

Deneubourg, J. L., Goss, S., Franks, N., and Pasteels, J. M. (1989). The blind leading the blind: Modeling chemically mediated morphogenesis and army ant raid patterns. *Journal of Insect Behavior* 2, 719–725. doi:10.1007/BF01065789.

Dennett, D. C. (1978). *Brainstorms: Philosophical Essays on Mind and Psychology*. Cambridge, MA: MIT Press.

Dennett, D. C. (1991). *Consciousness Explained*. London: Penguin Books.

Dennett, D. C. (1995). *Darwin's Dangerous Idea*. New York: Simon and Schuster.

Dennett, D. C. (1996). Facing backwards on the problem of consciousness. *Journal of Consciousness Studies* 3(1), 4–6.

Dennett, D. C. (1997). *Kinds of Minds: Towards an Understanding of Consciousness*. New York: Basic Books.

Dennett, D. C. (2005). *Sweet Dreams: Philosophical Obstacles to a Theory of Consciousness*. Cambridge, MA: MIT Press.

Denton, D. (2006). *The Primordial Emotions: The Dawning of Consciousness*. Oxford: Oxford University Press.

Descartes, R. (1637/1976). *Discourse on Method*. London: Penguin Books.

de Vladar, H. P., and Szathmary, E. (2015). Neuronal boost to evolutionary dynamics. *Interface Focus* 5, 20150074. doi:10.1098/rsfs.2015.0074.

de Vladar, H. P., Santos, M., and Szathmáry, E. (2017). Grand views of evolution. *Trends in Ecology and Evolution* 32, 324–334. doi:10.1016/j.tree.2017.01.008.

de Wit, S., and Dickinson, A. (2009). Associative theories of goal-directed behavior: A case for animal-human translational models. *Psychological Research* 73, 463–476. doi:10.1007/s00426-009-0230-6.

Dickinson, A. (2008). Why a rat is not a beast machine. In L. Weiskrantz and M. Davies (Eds.), *Frontiers of Consciousness*, pp. 275–288. Oxford: Oxford University Press.

Dickinson, A., and Balleine, B. (1994). Motivational control of goal-directed action. *Animal Learning and Behavior* 22, 1–18. doi:10.3758/BF03199951.

Diekelmann, S., and Born, J. (2010). The memory function of sleep. *Nature Reviews Neuroscience* 11, 114–126. doi:10.1038/nrn2762-c2.

Di Paolo, E. (2009). Extended life. *Topoi* 28, 9–21. doi:10.1007/s11245-008-9042-3.

Doglioni, C., Pignatti, J., and Coleman, M. (2016). Why did life develop on the surface of the earth in the Cambrian? *Geoscience Frontiers* 7, 865–873. doi:10.1016/j.gsf.2016.02.001.

Donald, M. (1991). *The Origins of the Modern Human Mind*. Cambridge: Cambridge University Press.

Donaldson-Matasci, M. C., Bergstrom, C. T., and Lachmann, M. (2010). The fitness value of information. *Oikos* 119, 219–230. doi:10.1111/j.1600-0706.2009.17781.x.

Dor, D. (2015). *The Instruction of Imagination: Language as a Social Communication Technology*. Oxford: Oxford University Press.

Dor, D., and Jablonka, E. (2010). Canalization and plasticity in the evolution of linguistic communication. In R. K. Larson, V. Deprez, and H. Yamakido (Eds.), *The Evolution of Human Language*, pp. 135–147. Cambridge: Cambridge University Press.

Dor, D., and Jablonka, E. (2014). Why we need to move from gene-culture co-evolution to culturally-driven co-evolution. In D. Dor, C. Knight, and J. Lewis (Eds.), *Social Origins of Language*, pp. 15–30. Oxford: Oxford University Press.

Dr. Seuss. *See* Seuss, Dr.

Dretske, F. I. (1988). *Explaining Behavior: Reasons in a World of Causes*. Cambridge, MA: MIT Press.

Dretske, F. (1997). What good is consciousness? *Canadian Journal of Philosophy* 27, 1–17. doi:10.1080/00455091.1997.10717471.

Dretske, F. I. (1999). Machines, plants and animals: The origins of agency. *Erkenntnis* 51, 523–535. doi:10.1023/A:1005541307925.

Driesch, H. (1914). *The History and Theory of Vitalism*. London: Macmillan.

Dror, O. E. (2001). Techniques of the brain and the paradox of emotions, 1880–1930. *Science in Context* 14, 643–660. doi:10.1017/S026988970100028X.

Droser, M. L., and Gehling, J. G. (2015). The advent of animals: The view from the Ediacaran. *Proceedings of the National Academy of Sciences USA* 112, 4865–4870. doi:10.1073/pnas.1403669112.

Du Bois-Reymond, E. (1874). On the limits of our knowledge of nature. *Popular Science Monthly* 5(May), pp. 17–32.

Dudai, Y., Roediger III, H. L., and Tulving, E. (2007). Memory concepts. In Y. Dudai, H. L. Roediger III, and S. M. Fitzpatrick (Eds.), *Science of Memory: Concepts*, pp. 1–9. Oxford: Oxford University Press.

Dufour, S. C., and McIlroy, D. (2016). Ediacaran pre-placozoan diploblasts in the Avalonian biota: The role of chemosynthesis in the evolution of early animal life. In A. T. Brasier, D. McIlroy, and N. McLoughlin (Eds.), *Earth System Evolution and Early Life: A Celebration of the Work of Martin Brasier*, pp. 211–219. London: Geological Society.

Dunn, C. W., Hejnol, A., Matus, D. Q., Pang, K., Browne, W. E., Smith, S. A., Seaver, E., Rouse, G. W., Obst, M., Edgecombe, G. D., Sørensen, M. V., Haddock, H. D., Schmidt-Rhaesa, A., Okusu, A., Kristensen, R. M., Wheeler, W. C., Martindale, M. Q., and Giribet, G. (2008). Broad phylogenomic sampling improves resolution of the animal tree of life. *Nature* 452, 745–749. doi:10.1038/nature06614.

Dyal, J. A., and Corning, W. C. (1973). Invertebrate learning and behavior taxonomies. In W. C. Corning, J. A. Dyal, and A. O. D. Willows (Eds.), *Invertebrate Learning*, vol. 1, *Protozoans through Annelids*, pp. 1–48. New York: Plenum.

Dylla, K. V., Galili, D. S., Szyszka, P., and Lüdke, A. (2013). Trace conditioning in insects—keep the trace! *Frontiers in Physiology* 4, 67. doi:10.3389/fphys.2013.00067.

Dyson, F. (1984). *Origins of Life*. Cambridge: Cambridge University Press.

Dzik, J. (2005). Behavioral and anatomical unity of the earliest burrowing animals and the cause of the "Cambrian explosion." *Paleobiology* 31, 503–521. doi:10.1666/0094-8373(2005)031[0503:BAAUOT]2.0.CO;2.

Eban-Rothschild, A., and Bloch, G. (2012). Social influences on circadian rhythms and sleep in insects. *Advances in Genetics* 77, 1–32. doi:10.1016/b978-0-12-387687-4.

Edelman, D. B., Baars, B. J., and Seth, A. K. (2005). Identifying hallmarks of consciousness in non-mammalian species. *Consciousness and Cognition* 14, 169–187. doi:10.1016/j.concog.2004.09.001.

Edelman, D. B., and Seth, A. K. (2009). Animal consciousness: A synthetic approach. *Trends in Neuroscience* 32, 476–484. doi:10.1016/j.tins.2009.05.008.

Edelman, G. M. (1987). *Neural Darwinism: The Theory of Neuronal Group Selection*. New York: Basic Books.

Edelman, G. M. (2003). Naturalizing consciousness: A theoretical framework. *Proceedings of the National Academy of Sciences USA* 100, 5520–5524. doi:10.1073/pnas.0931349100.

Edelman, G. M. (2005). *Wider than the Sky: The Phenomenal Gift of Consciousness*. New Haven, CT: Yale University Press.

Edelman, G. M., Gally, J. A., and Baars, B. J. (2011). Biology of consciousness. *Frontiers in Psychology* 2, 4. doi:10.3389/fpsyg.2011.00004.

Edelman, G. M., and Tononi, G. A. (2000). *Universe of Consciousness: How Matter Becomes Imagination*. New York: Basic Books.

Edgecombe, G. D., Giribet, G., Dunn, C. W., Hejnol, A., Kristensen, R. M., Neves, R. C., Rouse, G. W., Worsaae, K., and Sørensen, M. V. (2011). Higher-level metazoan relationships: Recent progress and remaining questions. *Organisms, Diversity and Evolution* 11, 151–172. doi:10.1007/s13127-011-0044-4.

Edwards, D. H., Heitler, W. J., and Krasne, F. B. (1999). Fifty years of a command neuron: The neurobiology of escape behavior in the crayfish. *Trends in Neuroscience* 22, 153–161. doi:10.1016/S0166-2236(98)01340-X.

Eigen, M., and Schuster, P. (1979). *The Hypercycle: A Principle of Natural Self-Organization*. Berlin: Springer.

Elramah, S., Landry, M., and Favereaux, A. (2014). MicroRNAs regulate neuronal plasticity and are involved in pain mechanisms. *Frontiers in Cellular Neuroscience* 8, 31. doi:10.3389/fncel.2014.00031.

Elwood, R. W., and Adams, L. (2015). Electric shock causes physiological stress responses in shore crabs, consistent with prediction of pain. *Biology Letters* 11, 20150800. doi:10.1098/rsbl.2015.0800.

Emes, R. D., and Grant, S. G. (2012). Evolution of synapse complexity and diversity. *Annual Review of Neuroscience* 35, 111–131. doi:10.1146/annurev-neuro-062111-150433.

Emes, R. D., Pocklington, A. J., Anderson, C. N. G., Bayes, A., Collins, M. O., Vickers, C. A., Croning, M. D. R., Malik, B. R., Choudhary, J. S., Armstrong, J. D., and Grant, S. G. N. (2008). Evolutionary expansion and anatomical specialization of synapse-proteome complexity. *Nature Neuroscience* 11, 799–806. doi:10.1038/nn.2135.

Engel, A. K., Fries, P., and Singer, W. (2001). Dynamic predictions: Oscillations and synchrony in top-down processing. *Nature Reviews Neuroscience* 2, 704–71610. doi:10.1038/35094565.

Enquist, M., Lind, J., and Ghirlanda, S. (2016). The power of associative learning and the ontogeny of optimal behaviour. *Royal Society Open Science* 3, 160734. doi:10.1098/rsos.160734.

Erwin, D. H. (2015). Early metazoan life: Divergence, environment and ecology. *Philosophical Transactions of the Royal Society B* 370, 20150036. doi:10.1098/rstb.2015.0036.

Erwin, D. H., and Davidson, E. H. (2002). The last common bilaterian ancestor. *Development* 129, 3021–3032.

Erwin, D. H., and Davidson, E. H. (2009). The evolution of hierarchical gene regulatory networks. *Nature Reviews Genetics* 10, 141–148. doi:10.1038/nrg2499.

Erwin, D. H., Laflamme, M., Tweedt, S. M., Sperling, E. A., Pisani, D., and Peterson, K. J. (2011). The Cambrian conundrum: Early divergence and later ecological success in the early history of animals. *Science* 334, 1091–1097. doi:10.1126/science.1206375.

Ettensohn, C. A. (2014). Horizontal transfer of the msp130 gene supported the evolution of metazoan biomineralization. *Evolution and Development* 16, 139–148. doi:10.1111/ede.12074.

Etxeberria, A. (2004). Autopoiesis and natural drift: Genetic information, reproduction, and evolution revisited. *Artificial Life* 10, 347–360. doi:10.1162/1064546041255575.

Fabbro, F., Aglioti, S. M., Bergamasco, M., Clarici, A., and Panksepp, J. (2015). Evolutionary aspects of self- and world consciousness in vertebrates. *Frontiers in Human Neuroscience* 9, 157. doi:10.3389/fnhum.2015.00157.

Farris, S. M. (2015). Evolution of brain elaboration. *Philosophical Transactions of the Royal Society B* 370, 20150054. doi:10.1098/rstb.2015.0054.

Fazelpour, S., and Thompson, E. (2015). The Kantian brain: Brain dynamics from a neurophenomenological perspective. *Current Opinions in Neurobiology* 31, 223–229. doi:10.1016/j.conb.2014.12.006.

Fedor, A., Zachar, I., Szilágyi, A., Öllinger, M., de Vladar, H. P., and Szathmáry, E. (2017). Cognitive architecture with evolutionary dynamics solves insight problem. *Frontiers in Psychology* 8, 427. doi:10.3389/fpsyg.2017.00427.

Feinberg, T. E. (2012). Neuroontology, neurobiological naturalism, and consciousness: A challenge to scientific reduction and a solution. *Physics of Life Reviews* 9, 13–34. doi:10.1016/j.plrev.2011.10.019.

Feinberg, T. E., and Mallatt, J. (2013). The evolutionary and genetic origins of consciousness in the Cambrian Period over 500 million years ago. *Frontiers in Psychology* 4, 667. doi:10.3389/fpsyg.2013.00667.

Feinberg, T. E., and Mallatt, J. (2016). *The Ancient Origins of Consciousness: How the Brain Created Experience*. Cambridge, MA: MIT Press.

Feldman, M., and Levy, A. A. (2009). Genome evolution in allopolyploid wheat—a revolutionary reprogramming followed by gradual change. *Journal of Genetics and Genomics* 36, 511–518. doi:10.1016/S1673-8527(08)60142-3.

Fernando, C., Goldstein, R., and Szathmáry, E. (2010). The neuronal replicator hypothesis. *Neural Computation* 22, 2809–2857. doi:10.1162/NECO_a_00031.

Fernando, C., Szathmary, E., and Husbands, P. (2012). Selectionist and evolutionary approaches to brain function: A critical appraisal. *Frontiers in Computational Neuroscience* 6, 24. doi:10.3389/fncom. 2012.00024.

Fike, D. A., Grotzinger, J. P., Pratt, L. M., and Summons, R. E. (2006). Oxidation of the Ediacaran ocean. *Nature* 444, 744–747. doi:10.1038/nature05345.

Finlay, B. (2015). The unique pain of being human. *New Scientist* 226, 28–29. doi:10.1016/S0262-4079(15)30311-0.

Finn, J. K., Tregenza, T., and Tregenza, N. (2009). Defensive tool use in a coconut-carrying octopus. *Current Biology* 19, R1069–R1070. doi:10.1016/j.cub.2009.10.052.

Fiorito, G., and Scotto, P. (1992). Observational learning in *Octopus vulgaris*. *Science* 256, 545–546. doi:10.1126/science.256.5056.545.

Fisch, L., Privman, E., Ramot, M., Harel, M., Nir, Y., Kipervasser, S., Andelman, F., Neufeld, M. Y., Kramer, U., Fried, I., and Malach, R. (2009). Neural "ignition": Enhanced activation linked to perceptual awareness in human ventral stream visual cortex. *Neuron* 64, 562–574. doi:10.1016/j.neuron.2009.11.001.

Fodor, J. (1981). The mind-body problem. *Scientific American* 244, 114–123.

Fortey, R. (2001). The Cambrian explosion exploded? *Science* 293, 438–439. doi:10 .1126/science.1062987.

Fox, D. (2016). What sparked the Cambrian explosion? *Nature* 530, 268–270. doi:10 .1038/530268a.

Fox, S. W., and Dose, K. (1977). *Molecular Evolution and the Origin of Life*. Rev. ed. New York: Marcel Dekker.

Francis, M. (2007). *Herbert Spencer and the Invention of Modern Life*. Stockfield, UK: Acumen.

Frank, S. A. (2012). Natural selection. IV. The Price equation. *Journal of Evolutionary Biology* 25, 1002–1019. doi:10.1111/j.1420-9101.2012.02498.x.

Franklin, T. B., Saab, B. J., and Mansuy I. M. (2012). Neural mechanisms of stress resilience and vulnerability. *Neuron* 75, 747–761. doi:10.1016/j.neuron.2012.08.016.

Freeman, W. J. (2000). *How Brains Make Up Their Minds*. London: Phoenix.

Friedman, B. H. (2010). Feelings and the body: The Jamesian perspective on autonomic specificity of emotion. *Biological Psychology* 84, 383–393. doi:10.1016/j.bio psycho.2009.10.006.

Friedmann, H. C. (2004). From "Butyribacterium" to "*E. coli*": An essay on unity. *Biochemistry Perspectives in Biology and Medicine* 47, 47–66. doi:10.1353/pbm.2004 .0007.

Friston, K. (2005). A theory of cortical responses. *Philosophical Transactions of the Royal Society B* 360, 815–836. doi:10.1098/rstb.2005.1622.

Friston, K. (2008). Hierarchical models in the brain. *PLoS Computational Biology* 4, e1000211. doi:10.1371/journal.pcbi.1000211.

Friston, K. (2010). The free-energy principle: A unified brain theory? *Nature Reviews Neuroscience* 11, 127–138. doi:10.1038/nrn2787.

Friston, K. (2012). A free energy principle for biological systems. *Entropy* 14, 2100–2121. doi:10.3390/e14112100.

Friston, K. (2013a). Life as we know it. *Journal of the Royal Society Interface* 10, 20130475. doi:10.1098/rsif.2013.0475.

Friston, K. (2013b). The fantastic organ. *Brain* 136, 1328–1332. doi:10.1093/brain/awt038.

Friston, K., and Kiebel, S. (2009). Predictive coding under the free-energy principle. *Philosophical Transactions of the Royal Society B* 364, 1211–1221. doi:10.1098/rstb.2008.0300.

Friston, K., Levin, M., Sengupta, B., and Pezzulo, G. (2015). Knowing one's place: A free-energy approach to pattern regulation. *Journal of the Royal Society Interface* 12, 20141383. doi:10.1098/rsif.2014.1383.

Frost, W. N., and Kandel, E. R. (1995). Structure of the network mediating siphon-elicited siphon withdrawal in *Aplysia*. *Journal of Neurophysiology* 73, 2413–2427.

Fry, I. (2000). *The Emergence of Life on Earth—a Historical and Scientific Overview*. New Brunswick, NJ: Rutgers University Press.

Gabriel, M. (2015). *Why the World Does Not Exist*. Cambridge: Polity Press.

Gagliano, M., Vyazovskiy, V. V., Borbély, A. A., Grimonprez, M., and Depczynski, M. (2016). Learning by association in plants. *Scientific Reports* 6, 38427. doi:10.1038/srep38427.

Galili, D. S., Lüdke, A., Galizia, C. G., Szyszka, P., and Tanimoto, H. (2011). Olfactory trace conditioning in *Drosophila*. *Journal of Neuroscience* 31, 7240–7248. doi:10.1523/JNEUROSCI.6667-10.2011.

Galizia, C. G., Eisenhardt, D., and Giurfa, M. (Eds.). (2012). *Honeybee Neurobiology and Behavior: A Tribute to Randolf Menzel*. Dordrecht, The Netherlands: Springer.

Gallagher, S., and Zahavi, D. (2012). *The Phenomenological Mind*. 2nd ed. London: Routledge.

Gallistel, C. R. (2017). The coding question. *Trends in Cognitive Sciences* 21, 498–508. doi:10.1016/j.tics.2017.04.012.

Gallotti, M., and Frith, C. D. (2013). Social cognition in the we-mode. *Trends in Cognitive Sciences* 17, 160–165. doi:10.1016/j.tics.2013.02.002.

Gánti, T. (1971/1987). *The Principle of Life* (L. Vekerd, Trans.). Budapest: Omikk.

Gánti, T. (2003). *The Principles of Life, with a Commentary by James Griesemer and Eörs Szathmáry*. New York: Oxford University Press.

Gapp, K., Corcoba, A., van Steenwyk, G., Mansuy, I. M., and Duarte, J. M. N. (2017). Brain metabolic alterations in mice subjected to postnatal traumatic stress and in their offspring. *Journal of Cerebral Blood Flow and Metabolism* 37, 2423–2432. doi:10.1177/0271678X16667525.

Garcia, J., Hankins, W. G., and Rusiniak, K.W. (1974). Behavioral regulation of the milieu interne in man and rat. *Science* 185, 824–831. doi:10.1126/science .185.4154.824.

Gardner, A. (2008). The Price equation. *Current Biology* 18, R198–R202. doi:10.1016 /j.cub.2008.01.005.

Garm, A., Oskarsson, M., and Nilsson, D. E. (2011). Box jellyfish use terrestrial visual cues for navigation. *Current Biology* 21, 798–803. doi:10.1016/j.cub.2011.03.054.

Garrett, B. J. (2006). What the history of vitalism teaches us about consciousness and the "Hard Problem." *Philosophy and Phenomenological Research* 72, 616–628. doi:10.1111/j.1933-1592.2006.tb00584.x.

Garson, J. (2003). The introduction of information in neurobiology. *Philosophy of Science* 70, 926–936. doi:10.1086/377378.

Gavelis, G. S., Hayakawa, S., White, R. A., III, Gojobori, T., Suttle, C. A., Keeling, P. J., and Leander, B. S. (2015). Eye-like ocelloids are built from different endosymbiotically acquired components. *Nature* 523, 204–207. doi:10.1038/nature14593.

Gehrmann, J., Matsumoto, Y., and Kreutzberg, G. W. (1995). Microglia: Intrinsic immune effector of the brain. *Brain Research Reviews* 20, 269–287. doi: 10.1016/0165-0173(94)00015-H. PMID 7550361.

Gelperin, A. (2013). Associative memory mechanisms in terrestrial slugs and snails. In P. Benjamin and R. Menzel (Eds.), *Invertebrate Learning and Memory*, pp. 280–290. London: Academic Press.

Gerhart, J. C., and Kirschner, M. W. (1997). *Cells, Embryos, and Evolution: Toward a Cellular and Developmental Understanding of Phenotypic Variation and Evolutionary Adaptability*. Cambridge, MA: Blackwell.

Gerhart, J. C., and Kirschner, M. W. (2007). The theory of facilitated variation. *Proceedings of the National Academy of Sciences USA* 104(suppl. 1), 8582–8589. doi:10 .1073/pnas.0701035104.

Gilbert, F. S. (2003). The morphogenesis of evolutionary developmental biology. *International Journal of Developmental Biology* 47, 467–477.

Gili, T., Saxena, N., Diukova, A., Murphy, K., Hall, J. E., and Wise, R. G. (2013). The thalamus and brainstem act as key hubs in alterations of human brain network connectivity induced by mild propofol sedation. *Journal of Neuroscience* 33, 4024–4031. doi:10.1523/JNEUROSCI.3480-12.2013.

Gilissen, E. (2007). Cognitive achievements with a miniature brain: The lesson of jumping spiders. *Behavioral and Brain Sciences* 30, 93–95. doi:10.1017/S0140525 X07001021.

Gillam, B. (2000). Perceptual constancy. In A. E. Kazdin (Ed.), *Encyclopedia of Psychology* 6, pp. 89–93. New York: American Psychological Association/Oxford University Press.

Gillette R., and Brown, J. W. (2015). The sea slug, *Pleurobranchaea californica*: A signpost species in the evolution of complex nervous systems and behavior. *Integrative and Comparative Biology* 55, 1058–1069. doi:10.1093/icb/icv081.

Ginsburg, S., and Jablonka, E. (2007a). The transition to experiencing: I. Limited learning and limited experiencing. *Biological Theory* 2, 218–230. doi:10.1162/biot .2007.2.3.218.

Ginsburg, S., and Jablonka, E. (2007b). The transition to experiencing: II. The evolution of associative learning based on feelings. *Biological Theory* 2, 231–243. doi:10.1162/biot.2007.2.3.231.

Ginsburg, S., and Jablonka, E. (2009). Epigenetic learning in non-neural organisms. *Journal of Bioscience* 34, 633–646. doi:10.1007/s12038-009-0081-8.

Ginsburg, S., and Jablonka, E. (2010a). Experiencing: A Jamesian approach. *Journal of Consciousness Studies* 17(5/6), 102–124.

Ginsburg, S., and Jablonka, E. (2010b). The evolution of associative learning: A factor in the Cambrian explosion. *Journal of Theoretical Biology* 266, 11–20. doi:10.1016/ j.jtbi.2010.06.017.

Ginsburg, S., and Jablonka, E. (2014). Memory, imagination, and the evolution of modern language. In J. Lewis, C. Knight, and D. Dor (Eds.), *The Social Origins of Language: Early Society, Communication and Polymodality*, pp. 317–324. New York: Oxford University Press.

Ginsburg, S., and Jablonka, E. (2015). The teleological transitions in evolution: A Gántian view. *Journal of Theoretical Biology* 381, 55–60. doi:10.1016/j.jtbi.2015.04.007.

Gissis, S. B. (2010). Lamarck on feelings: From worms to humans. In C. T. Wolfe and O. Gal (Eds.), *The Body as an Object and Instrument of Knowledge: Embodied Empiricism in Early Modern Science*, pp. 211–239. Berlin: Springer.

Gissis, S. B., and Jablonka, E. (Eds.). (2011). *Transformations of Lamarckism: From Subtle Fluids to Molecular Biology*. Cambridge, MA: MIT Press.

Giurfa, M., Devaud, J. M., and Sandoz, J. C. (Eds.). (2011). *Invertebrate Learning and Memory*. Lausanne, Switzerland: Frontiers in Behavioral Neuroscience.

Giurfa, M., and Menzel, R. (2013). Cognitive components of insect behavior. In R. Menzel and P. R. Benjamin (Eds.), *Invertebrate Learning and Memory*, pp. 14–25. London: Academic Press.

Glanzman, D. L. (1995).The cellular basis of classical conditioning in *Aplysia californica*—it's less simple than you think. *Trends in Neurosciences* 18, 30–36.

Glanzman, D. L. (2010). Common mechanisms of synaptic plasticity in vertebrates and invertebrates. *Current Biology* 20, R31–R36. doi:10.1016/j.cub.2009.10.023.

Godfrey-Smith, P. (2012). Darwinian individuals. In F. Bouchard and P. Huneman (Eds.), *From Groups to Individuals: Perspectives on Biological Associations and Emerging Individuality*, pp. 17–36. Cambridge, MA: MIT Press.

Godfrey-Smith, P. (2016a). Animal evolution and the origins of experience. In D. Livingstone Smith (Ed.), *How Biology Shapes Philosophy: New Foundations for Naturalism*, pp. 51–71. Cambridge: Cambridge University Press.

Godfrey-Smith, P. (2016b). Individuality, subjectivity, and minimal cognition. *Biology and Philosophy* 31, 775–796. doi:10.1007/s10539-016-95.

Godfrey-Smith, P. (2016c). *Other Minds: The Octopus, the Sea, and the Deep Origins of Consciousness*. New York: Farrar, Straus and Giroux.

Godfrey-Smith, P. (2016d). Mind, matter, and metabolism. *Journal of Philosophy* 113, 481–506. doi:10.5840/jphil20161131034.

Gomez-Marin, A., Oron, E., Gakamsky, A., Valente, D., Benjamini, Y., and Golani, I. (2016). Generative rules of *Drosophila* locomotor behavior as a candidate homology across phyla. *Nature Scientific Reports* 6, 27555. doi:10.1038/srep27555.

Gopnik, A. (2010). How weird is consciousness? Scientists may not even be asking the right questions. *Slate*, November 29. http://www.slate.com/articles/arts/books/2010/11/how_weird_is_consciousness.html.

Gräff, J., Joseph. N. F., Horn, M. E., Samiei, A., Meng, J., Seo, J., Rei, D., Bero, A. W., Phan, T. X., Wagner, F., Holson, E., Xu, J., Sun, J., Neve, R. L., Mach, R. H., Haggarty, S. J., and Tsai, L. H. (2014). Epigenetic priming of memory updating during reconsolidation to attenuate remote fear memories. *Cell* 156, 261–276. doi:10.1016/j.cell.2013.12.020.

Grant, S. G. N. (2016). The molecular evolution of the vertebrate behavioural repertoire. *Philosophical Transactions of the Royal Society B* 371, 20150051. doi:10.1098/rstb.2015.0051.

Gray, C. M., König, P., Engel, A. K., and Singer, W. (1989). Oscillatory responses in cat visual cortex exhibit inter-columnar synchronization which reflects global stimulus properties. *Nature* 338, 334–337. doi:10.1038/338334a0.

Griffin, D. R. (1976). *The Question of Animal Awareness: Evolutionary Continuity of Mental Experience*. New York: Rockefeller University Press.

Griffin, D. R. (2001). *Animal Minds: Beyond Cognition to Consciousness*. Chicago: University of Chicago Press.

Griffin, D. R., and Speck, G. B. (2004). New evidence of animal consciousness. *Animal Cognition* 7, 5–18. doi:10.1007/s10071-003-0203-x.

Griffiths, P. E. (2004). Is emotion a natural kind? In R. C. Solomon (Ed.), *Thinking about Emotion: Contemporary Philosophers on Emotion*, pp. 233–249. Oxford: Oxford University Press.

Groening, J., Venini, D., and Srinivasan, M. V. (2017). In search of evidence for the experience of pain in honeybees: A self-administration study. *Scientific Reports 7*, 45825. doi:10.1038/srep45825.

Grosvenor W., Rhoads, D. E., and Kass-Simon, G. (1996). Chemoreceptive control of feeding processes in hydra. *Chemical Senses 21*, 313–321.

Güntürkün, O., and Bugnyar, T. (2016). Cognition without cortex. *Trends in Cognitive Sciences 20*, 291–303. doi:10.1016/j.tics.2016.02.001.

Haase, A. D. (2016). A small RNA-based immune system defends germ cells against mobile genetic elements. *Stem Cells International 2016*, 7595791. doi:10.1155/2016/7595791.

Haddock, S. H. D., Moline, M. A., and Case, J. F. (2010). Bioluminescence in the sea. *Annual Review of Marine Science 2*, 443–493. doi:10.1146/annurev-marine-120308-081028.

Hadziselimovic, N., Vukojevic, V., Peter, F., Milnik, A., Fastenrath, M., Fenyves, B. G., Hieber, P., Demougin P., Vogler, C., de Quervain, D. J., Papassotiropoulos, A., and Stetak, A. (2014). Forgetting is regulated via Musashi-mediated translational control of the Arp2/3 complex. *Cell 156*, 1153–66. doi:10.1016/j.cell.2014.01.054.

Hague, T., Florini, M., and Andrews, P. L. R. (2013). Preliminary in vitro functional evidence for reflex responses to noxious stimuli in the arms of *Octopus vulgaris*. *Journal of Experimental Marine Biology and Ecology 447*, 100–105. doi:10.1016/j.jembe.2013.02.016.

Halas, E. S., James, R. L., and Knutson, C. S. (1962). An attempt at classical conditioning in the planarian. *Journal of Comparative and Physiological Psychology 55*, 969–971.

Haldane, J. B. S. (1933). *Science and Human Life*. New York: Harper.

Halic, M., and Moazed, D. (2009). Transposon silencing by piRNAs. *Cell 138*, 1059–1060. doi:10.1016/j.cell.2009.08.030.

Hameroff, S. R. (1994). Quantum coherence in microtubules: A neural basis for emergent consciousness? *Journal of Consciousness Studies 1*(1), 91–118.

Hameroff, S. R. (1998). Did consciousness cause the Cambrian explosion? In S. Hameroff and A. Kaszniak (Eds.), *Towards a Science of Consciousness II: The 1996 Tuscon Discussions and Debates*, pp. 421–37. Cambridge, MA: MIT Press.

Hameroff, S. R. (2007). The brain is both neurocomputer and quantum computer. *Cognitive Science 31*, 1035–1045. doi:10.1080/03640210701704004.

Hamilton, E. (1930). *The Greek Way*. New York: W. W. Norton.

Hamilton, T. J., Myggland, A., Duperreault, E., May, Z., Gallup, J., Powell, R. A., Schalomon, M., and Digweed, S. M. (2016). Episodic-like memory in zebrafish. *Animal Cognition* 19, 1071–1079. doi:10.1007/s10071-016-1014-1.

Handford, C. E., Tan, S., Lawrence, A. J., and Kim, J. H. (2014). The effect of the mGlu5 negative allosteric modulator MTEP and NMDA receptor partial agonist D-cycloserine on Pavlovian conditioned fear. *International Journal of Neuropsychopharmacology* 17, 1521–1532. doi:10.1017/S1461145714000303.

Haralson, J. V., Groff, C. I., and Haralson, S. J. (1975). Classical conditioning in the sea anemone, *Cribrina xanthogrammica*. *Physiology and Behavior* 15, 455–460. doi:10.1016/0031-9384(75)90259-0.

Hardt, O., Nader, K., and Nadel, L. (2013). Decay happens: The role of active forgetting in memory. *Trends in Cognitive Sciences* 17, 111–120. doi:10.1016/j.tics.2013.01.001.

Hardy, A. (1965). *The Living Stream: A Restatement of Evolution Theory and Its Relation to the Spirit of Man*. London: Collins.

Harnad, S. (2011). Doing, feeling, meaning and explaining. In G. Comstock, P. Barron, P. Shipton, and S. Haslanger (Eds.), *On the Human Forum*. http://eprints.ecs.soton.ac.uk/22243/.

Hartenstein, V., and Stollewerk, A. (2015). The evolution of early neurogenesis. *Developmental Cell Review* 32, 390–407. doi:10.1016/j.devcel.2015.02.004.

Hartley, D. (1749). *Observations on Man, his Frame, his Duty, and his Expectations*. London: Samuel Richardson.

Hartline, D. K. (2011). The evolutionary origins of glia. *Glia* 59, 1215–1236. doi:10.1002/glia.21149.

Hatfield, E., Cacioppo, J. T., and Rapson, R. L. (1994). *Emotional Contagion*. New York: Cambridge University Press.

Hawkins, R. D., Abrams, T. W., Carew, T. J., and Kandel, E. R. (1983). A cellular mechanism of classical conditioning in *Aplysia*: Activity-dependent amplification of presynaptic facilitation. *Science* 219, 400–405. doi:10.1126/science.6294833.

Hawkins, R. D., and Byrne, J. H. (2015). Associative learning in invertebrates. *Cold Spring Harbor Perspectives in Biology* 7, a021709. doi:10.1101/cshperspect.a021709.

Hawkins, R. D., Greene, W., and Kandel, E. R. (1998). Classical conditioning, differential conditioning, and second-order conditioning of the *Aplysia* gill-withdrawal reflex in a simplified mantle organ preparation. *Behavioral Neuroscience* 112, 636–645. doi:10.1037/0735-7044.112.3.636.

Hawkins, R. D., and Kandel, E. R. (1984a). Is there a cell-biological alphabet for simple forms of learning? *Psychological Review* 91, 375–391. doi:10.1037/0033-295X.91.3.375.

Hawkins, R. D., and Kandel, E. R. (1984b). Steps toward a cell-biological alphabet for elementary forms of learning. In G. Lynch, J. L. McGaugh, and N. M. Weinberger (Eds.), *Neurobiology of Learning and Memory*, pp. 385–404. New York: Guilford Press.

Hawkins, R. D., Kandel, E. R., and Bailey, C. H. (2006). Molecular mechanisms of memory storage in *Aplysia*. *Biological Bulletin* 210, 174–191. doi:10.2307/4134556.

Hayes, S. C., Blackledge, J. T., and Barnes-Holmes, D. (2001). Language and cognition: Constructing an alternative approach within the behavioral tradition. In S. C. Hayes, D. Barnes-Holmes, and B. Roche (Eds.), *Relational Frame Theory: A Post-Skinnerian Account of Human Language and Cognition*, pp. 3–20. New York: Plenum.

Hayes, S. C., and Hayes, L. J. (1989). The verbal action of the listener as a basis for rule-governance. In S. C. Hayes (Ed.), *Rule-Governed Behavior: Cognition, Contingencies, and Instructional Control*, pp. 153–190. New York: Plenum.

Head, B. W. (1980). The origin of idéologue and idéologie. *Studies on Voltaire and the Eighteenth Century* 183, 257–264.

Hebb, D. O. (1949). *The Organization of Behavior*. New York: Wiley.

Heger, P., Marin, B., Bartkuhn, M., Schierenberg, E., and Wiehe, T. (2012). The chromatin insulator CTCF and the emergence of metazoan diversity. *Proceedings of the National Academy of Sciences USA* 109, 17507–17512. doi:10.1073/pnas.1111941109.

Heimberg, A. M., Sempere, L. F., Moy, V. N., Donoghue, P. C. J., and Peterson, K. J. (2008). MicroRNAs and the advent of vertebrate morphological complexity. *Proceedings of the National Academy of Sciences USA* 105, 2496–2950. doi:10.1073/pnas.0712259105.

Heisenberg, M. (2013). Action selection: The brain as a behavioral organizer. In R. Menzel and P. R. Benjamin (Eds.), *Invertebrate Learning and Memory*, pp. 9–13. London: Academic Press.

Heitz, F. D., Farinelli, M., Mohanna, S., Kahn, M., Duning, K., Frey, M. C., Pavenstädt, H., and Mansuy, I. M. (2016). The memory gene KIBRA is a bidirectional regulator of synaptic and structural plasticity in the adult brain. *Neurobiology of Learning and Memory* 135, 100–114. doi:10.1016/j.nlm.2016.07.028.

Helmholtz, H. V. (1867). *Handbuch der Physiologischen Optik*. Leipzig: Leopold Voss.

Herman, J. J., Spencer, H. G., Donohue, K., and Sultan, S. E. (2013). How stable "should" epigenetic modifications be? Insights from adaptive plasticity and bet hedging. *Evolution* 68, 632–643. doi:10.1111/evo.12324.

Heyes, C. M. (2010). Where do mirror neurons come from? *Neuroscience and Biobehavioral Reviews* 34, 575–583. doi:10.1016/j.neubiorev.2009.11.007.

Heyes, C. M. (2012). Simple minds: A qualified defence of associative learning. *Philosophical Transactions of the Royal Society of London B* 367, 2695–2703. doi:10.1098/rstb.2012.0217.

Hige, T., Aso, Y., Modi, M. N., Rubin, G. M., and Turner, G. C. (2015). Heterosynaptic plasticity underlies aversive olfactory learning in *Drosophila*. *Neuron* 88, 985–998. doi:10.1016/j.neuron.2015.11.003.

Hinde, R. A. (1966). *Animal Behavior: A Synthesis of Ethology and Comparative Psychology*. New York: McGraw-Hill.

Hindy, N. C., Ng, F. Y., and Turk-Brownen, N. B. (2016). Linking pattern completion in the hippocampus to predictive coding in visual cortex. *Nature Neuroscience* 19, 665–667. doi:10.1038/nn.4284.

Hintzman, D. L. (2003). Robert Hooke's model of memory. *Psychonomic Bulletin and Review* 10, 3–14. doi:10.3758/BF03196465.

Hirth, F. (2010). On the origin and evolution of the tripartite brain. *Brain Behavior and Evolution* 76, 3–10. doi:10.1159/000320218.

Hobson, A. J. (2009). REM sleep and dreaming: Towards a theory of protoconsciousness. *Nature Reviews Neuroscience* 10, 803–813. doi:10.1038/nrn2716.

Hobson, J. A., and Friston, K. J. (2016). A response to our theatre critics. *Journal of Consciousness Studies* 23(3/4), 245–254.

Hobson, J. A., Goin, O. B., and Goin, C. J. (1968). Electrographic correlates of behaviour in tree frogs. *Nature* 220, 386–387. doi:10.1038/220386a0.

Hochner, B. (2013). How nervous systems evolve in relation to their embodiment: What we can learn from octopuses and other molluscs. *Brain Behavior and Evolution* 82, 19–30. doi:10.1159/000353419.

Hochner, B., and Glanzman, D. L. (2016). Evolution of highly diverse forms of behavior in molluscs. *Current Biology* 26, R965–R971. doi:10.1016/j.cub.2016.08.047.

Hochner, B., and Shomrat, T. (2013). The neurophysiological basis of learning and memory in advanced invertebrates: The octopus and the cuttlefish. In R. Menzel and P. R. Benjamin (Eds.), *Invertebrate Learning and Memory*, pp. 303–317. London: Academic Press.

Hochner, B., Shomrat, T., and Fiorito, G. (2006). The octopus: A model for a comparative analysis of the evolution of learning and memory mechanisms. *Biological Bulletin* 210, 308–317. doi:10.2307/4134567.

Hoffman, P. (2012). *Life's Ratchet: How Molecular Machines Extract Order from Chaos.* New York: Basic Books.

Hofstadter, D. (1979). *Gödel, Escher, Bach: An Eternal Golden Braid.* New York: Basic Books.

Hofstadter, D. R. (2007). *I Am a Strange Loop.* New York: Basic Books.

Hofstadter, D., and Dennett, D. (1981). *The Mind's I: Fantasies and Reflections on Self and Soul.* New York: Basic Books.

Hohwy, J. (2013). *The Predictive Mind.* Oxford: Oxford University Press.

Holland, D., Abah, C., Enriquez, M. V., Herman M., Bennett, G. J., Vela, E., and Walsh, C. J. (2017). The soft robotics toolkit: Strategies for overcoming obstacles to the wide dissemination of soft-robotic hardware. *IEEE Robotics and Automation Magazine, Special Issue on Open Source and Widely Disseminated Robot Hardware* 24, 57–64. doi:10.1109/MRA.2016.2639067.

Holland, P. C., and Schiffino, F. L. (2016). Mini-review: Prediction errors, attention and associative learning. *Neurobiology of Learning and Memory* 131, 207–215. doi:10.1016/j.nlm.2016.02.014.

Holland, P. W. H. (2015). Did homeobox gene duplications contribute to the Cambrian explosion? *Zoological Letters* 1, 1. doi:10.1186/s40851-014-0004-x.

Holló, G. (2015). A new paradigm for animal symmetry. *Interface Focus* 5, 20150032. doi:10.1098/rsfs.2015.0032.

Holló, G., and Novák, M. (2012). The manoeuvrability hypothesis to explain the maintenance of bilateral symmetry in animal evolution. *Biology Direct* 7, 22. doi:10.1186/1745-6150-7-22.

Homberg, U. (2008). Evolution of the central complex in the arthropod brain with respect to the visual system. *Arthropod Structure and Development* 37, 347–362. doi:10.1016/j.asd.2008.01.008.

Honey, R. C., Iordanova, M. D., and Good, M. (2014). Associative structures in animal learning: Dissociating elemental and configural processes. *Neurobiology of Learning and Memory* 108, 96–103. doi:10.1016/j.nlm.2013.06.002.

Hooper, S. L. (2001). Central pattern generators. *Encyclopedia of Life Sciences.* New York: John Wiley and Sons.

Hoover, B. A., Nguyen, H., Thompson, L., and. Wright, W. G. (2006). Associative memory in three aplysiids: Correlation with heterosynaptic modulation. *Learning and Memory* 13, 820–826. doi:10.1101/lm.284006.

Hornstein E., and Shomron, N. (2006). Canalization of development by microRNAs. *Nature Genetics* 38, S20–S24. doi:10.1038/ng1803.

Hoy, A. M., and Buck, A. H. (2012). Extracellular small RNAs: What, where, why? *Biochemical Society Transactions* 40, 886–890. doi:10.1042/BST20120019.

Hu, H. (2016). Reward and aversion. *Annual Review of Neuroscience* 39, 297–324. doi:10.1146/annurev-neuro-070815-014106.

Hull, C. L. (1943). *Principles of Behavior.* New York: Appleton-Century-Crofts.

Hume, D. (1748/1993). *An Enquiry Concerning Human Understanding* (E. Steinberg, Ed.). Indianapolis: Hackett.

Humphrey, N. (1992). *A History of the Mind.* New York: Simon and Schuster.

Humphrey, N. (2011). *Soul Dust: The Magic of Consciousness.* Princeton, NJ: Princeton University Press.

Humphrey, N. (2015). Consciousness as art. *Scientific American Mind* 26, 64–69.

Huxley, T. H., and Youmans, W. J. (1868). *The Elements of Physiology and Hygiene: A Text-book for Educational Institutions.* New York: Appleton.

Ibrahim, R. (2009). Selective deficit of second language: A case study of a brain-damaged Arabic-Hebrew bilingual patient. *Behavioral and Brain Functions* 5, 17. doi:10.1186/1744-9081-5-17.

Inoue, A., Sawatari, E., Hisamoto, N., Kitazono, T., Teramoto, T., Fujiwara, M., Matsumoto, K., and Ishihara, T. (2013). Forgetting in *C. elegans* is accelerated by neuronal communication via the TIR-1/JNK-1 pathway. *Cell Reports* 3, 808–819. doi:10.1016/j.celrep.2013.02.019.

Jabbi, M., Swart, M., and Keysers, C. (2007). Empathy for positive and negative emotions in the gustatory cortex. *NeuroImage* 34, 1744–53. doi:10.1016/j.neuroimage.2006.10.032.

Jablonka, E. (1994). Inheritance systems and the evolution of new levels of individuality. *Journal of Theoretical Biology* 170, 301–309. doi:10.1006/jtbi.1994.1191.

Jablonka, E. (2002). Information: Its interpretation, its inheritance and its sharing. *Philosophy of Science* 69, 578–605. doi:10.1086/344621.

Jablonka, E. (2017). Collective narratives, false memories, and the origins of autobiographical memory. *Biology and Philosophy* 32, 839–853. doi:10.1007/s10539-017-9593-z.

Jablonka, E., and Ginsburg, S. (2013). The major teleological transitions in evolution: Why the materialistic evolutionary conception of nature is almost certainly right. *Journal of Consciousness Studies* 20(9/10), 177–189.

Jablonka, E., Ginsburg, S., and Dor, D. (2012). The co-evolution of language and emotions. *Philosophical Transactions of the Royal Society B* 367, 2152–2159. doi:10.1098/rstb.2012.0117.

Jablonka, E., and Lamb, M. J. (2005/2014). *Evolution in Four Dimensions: Genetic, Epigenetic, Behavioral, and Symbolic Variation in the History of Life*. 2nd ed. Cambridge, MA: MIT Press.

Jablonka, E., and Lamb, M. J. (2006). The evolution of information in the major transitions. *Journal of Theoretical Biology* 239, 236–246. doi:10.1016/j.jtbi.2005.08.038.

Jablonka, E., and Raz, G. (2009). Transgenerational epigenetic inheritance: Prevalence, mechanisms, and implications for the study of heredity and evolution. *Quarterly Review of Biology* 84, 131–176. doi:10.1086/598822.

Jacobson, A. L., Horwitz, S. D., and Fried, C. (1967). Classical conditioning, pseudo-conditioning or sensitization in the planarian. *Journal of Comparative and Physiological Psychology* 64, 73–79. doi:10.1037/h0024808.

Jackson, C., McCabe, B. J., Nicol, A. U., Grout, A. S., Brown, M. W., and Horn, G. (2008). Dynamics of a memory trace: Effects of sleep on consolidation. *Current Biology* 18, 393–400. doi:10.1016/j.cub.2008.01.062.

Jackson, D. J., Macis, L., Reitner, J., and Wörheide, G. (2011). A horizontal gene transfer supported the evolution of an early metazoan biomineralization strategy. *BMC Evolutionary Biology* 11, 238. doi:10.1186/1471-2148-11-238.

Jackson, F. (1982). Epiphenomenal qualia. *Philosophical Quarterly* 32, 127–136.

James, J., Gross, J. J., and Barrett, L. F. (2011). Emotion generation and emotion regulation: One or two depends on your point of view. *Emotion Review* 3, 8–16. doi:10.1177/1754073910380974.

James, W. (1878). Remarks on Spencer's definition of the mind as correspondence. *Journal of Speculative Philosophy* 12, 1–18.

James, W. (1890). *The Principles of Psychology*. 2 vols. New York: Dover.

Jameson, K., and Highnote, S. (2001). Richer color experience in observers with multiple photopigment opsin genes. *Psychonomic Bulletin and Review* 8, 244–261. doi:10.3758/BF03196159.

Janecka, J., Chowdhary, B., and Murphy, W. (2012). Exploring the correlations between sequence evolution rate and phenotypic divergence across the mammalian tree provides insights into adaptive evolution. *Journal of Biosciences* 37, 897–909. doi:10.1007/s12038-012-9254-y.

Jékely, G. (2011). Origin and early evolution of neural circuits for the control of ciliary locomotion. *Proceedings of the Royal Society B* 278, 914–922. doi:10.1098/rspb.2010.2027.

Jékely, G., Keijzer, F., and Godfrey-Smith, P. (2015). An option space for early neural evolution. *Philosophical Transactions of the Royal Society B* 370, 20150181. doi:10.1098/rstb.2015.0181.

Jeltsch, A. (2013). Oxygen, epigenetic signaling, and the evolution of early life. *Trends in Biochemical Sciences* 38, 172–176. doi:10.1016/j.tibs.2013.02.001.

Jilg, A., Lesny, S., Peruzki, N., Schwegler, H., Selbach, O., Dehghani, F., and Stehle, J. H. (2010). Temporal dynamics of mouse hippocampal clock gene expression support memory processing. *Hippocampus* 20, 377–388. doi:10.1002/hipo.20637.

Jin, C., Li, C., Algeo, T. J., Planavsky, N. J., Cui, H., Yang, X., Zhao, Y., Zhang, X., and Xie, S. (2016). A highly redox-heterogeneous ocean in South China during the early Cambrian (~529–514Ma): Implications for biota-environment co-evolution. *Earth and Planetary Science Letters* 441, 38–51. doi:10.1016/j.epsl.2016.02.019.

Jirenhed, D.-A., Rasmussen, A., Johansson, F., and Hesslow, G. (2017). Learned response sequences in cerebellar Purkinje cells. *Proceedings of the National Academy of Sciences USA* 114, 6127–6132. doi:10.1073/pnas.1621132114.

Johansson, F., Carlsson, H. A. E., Rasmussen, A., Yeo, C. H., and Hesslow, G. (2015). Activation of a temporal memory in Purkinje cells by the mGluR7 receptor. *Cell Reports* 13, 1741–1746. doi:10.1016/j.celrep.2015.10.047.

Johnson, M. C., and Wuensch, K. L. (1994). An investigation of habituation in the jellyfish *Aurelia aurita*. *Behavioral and Neural Biology* 61, 54–59. doi:10.1016/S0163-1047(05)80044-5.

Jolley, N. (2000). Malebranche on the soul. In S. Nadler (Ed.), *The Cambridge Companion to Malebranche*, pp. 31–58. Cambridge: Cambridge University Press.

Jones, B. W., and Nishiguchi, M. K. (2004). Counterillumination in the Hawaiian bobtail squid, *Euprymna scolopes* Berry (Mollusca: Cephalopoda). *Marine Biology* 144, 1151–1155. doi:10.1007/s00227-003-1285-3.

Jordan, G., Deeb, S. S., Bosten, J. M., and Mollon, J. D. (2010). The dimensionality of color vision in carriers of anomalous trichromacy. *Journal of Vision* 10, 12. doi:10.1167/10.8.12.

Josselyn, S. A., Stefan Köhler, S., and Frankland, P. W (2017). Heroes of the engram. *Journal of Neuroscience* 37, 4647–4657. doi:10.1523/JNEUROSCI.0056-17.2017.

Jozet-Alves, C., Bertin, M., and Clayton, N. S. (2013). Evidence of episodic-like memory in cuttlefish. *Current Biology* 23, 1033–1035. doi:10.1016/j.cub.2013.10.021.

Juarrero, A. (1999). *Dynamics in Action: Intentional Behavior as a Complex System*. Cambridge, MA: MIT Press.

Kaczer, L., Klappenbach, M., and Maldonado, H. (2011). Dissecting mechanisms of reconsolidation: Octopamine reveals differences between appetitive and aversive memories in the crab *Chasmagnathus*. *European Journal of Neuroscience* 34, 1170–1178. doi:10.1111/j.1460-9568.2011.07830.x.

Kaczer, L., and Maldonado, H. (2009). Contrasting role of octopamine in appetitive and aversive learning in the crab *Chasmagnathus*. *PLoS One* 4, e6223. doi:10.1371/journal.pone.0006223.

Kamin, L. J. (1969). Predictability, surprise, attention and conditioning. In B. A. Campbell and R. M. Church (Eds.), *Punishment and Aversive Behavior*, pp. 279–296. New York: Appleton-Century-Crofts.

Kandel, E. R. (2000). A biographical note prepared for the Nobel prize. Nobelprize.org, http://www.nobelprize.org/nobel_prizes/medicine/laureates/2000/kandel-bio.html.

Kandel, E. R. (2001). The molecular biology of memory storage: A dialogue between genes and synapses. *Science* 294, 1030–1038.

Kandel, E. R. (2006). *In Search of Memory: The Emergence of a New Science of Mind*. New York: W. W. Norton.

Kandel, E. R. (2012). The molecular biology of memory: cAMP, PKA, CRE, CREB-1, CREB-2, and CPEB. *Molecular Brain* 5, 14. doi:10.1186/1756-6606-5-14.

Kandel, E. R., Dudai, Y., and Mayford, M. R. (2014). The molecular and systems biology of memory. *Cell* 157, 163–186. doi:10.1016/j.cell.2014.03.001.

Kandel, E. R., and Tauc, L. (1964). Mechanism of prolonged heterosynaptic facilitation. *Nature* 202, 145–147. doi:10.1038/202145a0.

Kant, I. (1790/1987). *Critique of Judgment* (W. S. Pulhar, Trans.). Indianapolis: Hackett.

Kass-Simon, G. (1988). Towards a neurobiology of nematocyst discharge in the tentacles of hydra. In D. A. Hessinger and H. M. Lenhoff (Eds.), *The Biology of Nematocysts*, pp. 531–541. San Diego: Academic Press.

Kass-Simon G., and Scappaticci, A. A., Jr. (2002). The behavioral and developmental physiology of nematocysts. *Canadian Journal of Zoology* 80, 1772–1794. doi:10.1139/Z02-135.

Kauffman, S. A. (1969). Metabolic stability and epigenesis in randomly constructed genetic nets. *Journal of Theoretical Biology* 22, 437–467. doi:10.1016/0022-5193(69)90015-0.

Kavanau, J. L. (1998). Vertebrates that never sleep: Implications for sleep's basic function. *Brain Research Bulletin* 46(4), 269–279. doi:10.1016/S0361-9230(98)00018-5.

Kavanau, J. L. (2006). Is sleep's "supreme mystery" unraveling? An evolutionary analysis of sleep encounters no mystery; nor does life's earliest sleep, recently discovered in jellyfish. *Medical Hypotheses* 66, 3–9. doi:10.1016/j.mehy.2005.08.036.

Kawase, H., Okata, Y., and Ito, K. (2013). Role of huge geometric circular structures in the reproduction of a marine pufferfish. *Scientific Reports* 3, 2106. doi:10.1038/srep02106.

Keijzer, F. A., and Arnellos, A. (2017). The animal sensorimotor organization: A challenge for the environmental complexity thesis. *Biology and Philosophy* 32, 421–441. doi:10.1007/s10539-017-9565-3.

Keijzer, F. A., van Duijn, M., and Lyon, P. (2013). What nervous systems do: Early evolution, input-output, and the skin brain thesis. *Adaptive Behavior* 21, 67–85. doi:10.1177/105971231246533.

Keller, E. F. (2011). Self-organization, self-assembly, and the inherent activity of matter. In S. B. Gissis and E. Jablonka (Eds.), *Transformations of Lamarckism: From Subtle Fluids to Molecular Biology*, pp. 357–364. Cambridge, MA: MIT Press.

Key, B. (2016). Why fish do not feel pain. *Animal Sentience* 1(3), 1–33.

Kioussis, D., and Pachnis, V. (2009). Immune and nervous systems: More than just a superficial similarity? *Cell* 31, 705–710. doi:10.1016/j.immuni.2009.09.009.

Kirschner, M. W., and Gerhart, J. C. (2005). *The Plausibility of Life: Resolving Darwin's Dilemma*. New Haven, CT: Yale University Press.

Kirschvink, J. L., and Raub, T. D. (2003). A methane fuse for the Cambrian explosion: True polar wander. *Comptes Rendus Geosciences* 335, 71–83. doi:10.1016/S1631-0713(03)00011-7.

Kitcher, P. S. (1993). Function and design. *Midwest Studies in Philosophy* 18, 379–397. doi:10.1111/j.1475-4975.1993.tb00274.x.

Klein, C., and Barron, A. B. (2016). Insects have the capacity for subjective experience. *Animal Sentience* 9(1), 1–19.

Kleinenberg, N. (1872). *Hydra—eine anatomisch-entwicklungsgeschichtliche Untersuchung* [An anatomical-evolutionary investigation of Hydra]. Leipzig, Germany: Wilhelm Engelmann.

Knoll, A. H., and Carroll, S. B. (1999). Early animal evolution: Emerging views from comparative biology and geology. *Science* 284, 2129–2137. doi:10.1126/science.284.5423.2129.

Koch, C. (2004). *The Quest for Consciousness: A Neurobiological Approach*. Englewood, CO: Roberts.

Koch, C. (2012). *Consciousness—Confessions of a Romantic Reductionist*. Cambridge, MA: MIT Press.

Koch, C., Massimini, M., Boly, M., and Tononi, G. (2016). Neural correlates of consciousness: Progress and problems. *Nature Reviews Neuroscience* 17, 307–321. doi:10.1038/nrn.2016.22.

Koch, C., and Tsuchiya, N. (2007). Attention and consciousness: Two distinct brain processes. *Trends in Cognitive Sciences* 11, 16–22. doi:10.1016/j.tics.2006.10.012.

Kocot, K. M., Cannon, J. T., Todt, C., Citarella, M. R., Kohn, A. B., Meyer, A., Santos, S. R., Schander, C., Moroz, L. L., Lieb, B., and Halanych, K. M. (2011). Phylogenomics reveals deep molluscan relationships. *Nature* 477, 452–456. doi:10.1038/nature10382.

Koizumi, O., Hamada, S., Minobe, S., Hamaguchi-Hamada, K., Kurumata-Shigeto, M., Nakamura, M., and Namikawa, H. (2015). The nerve ring in cnidarians: Its presence and structure in hydrozoan medusae. *Zoology* 118, 79–88. doi:10.1016/j. zool.2014.10.001.

Kollmann, M., Huetteroth, W., and Schachtner, J. (2011). Brain organization in Collembola (springtails). *Arthropod Structure and Development* 40, 304–316. doi:10.1016/j. asd.2011.02.003).

Konorski, J. (1967). *Integrative Activity of the Brain: An Interdisciplinary Approach.* Chicago: University of Chicago.

Kosik, K. S. (2009). Exploring the early origins of the synapse by comparative genomics. *Biology Letters* 5, 108–111. doi:10.1098/rsbl.2008.0594.

Kravitz, E. A., and Huber, R. (2003). Aggression in invertebrates. *Current Opinion in Neurobiology* 13, 736–743. doi:10.1016/j.conb.2003.10.003.

Kristan, W. B. (2016). Early evolution of neurons. *Current Biology* 26, R949–R954. doi:10.1016/j.cub.2016.05.030.

Lachmann, M. (2013). The value of information in signals and cues. In U. E. Stegmann (Ed.), *Animal Communication Theory: Information and Influence*, pp. 357–376. Cambridge: Cambridge University Press.

Lachmann, M., and Jablonka, E. (1996). The inheritance of phenotypes: An adaptation to fluctuating environments. *Journal of Theoretical Biology* 181, 1–9. doi:10.1006/jtbi.1996.0109.

Lamarck, J. B. (1809/1914). *Zoological Philosophy* (H. Elliot, Trans.). London: Macmillan.

Lamm, E., and Jablonka, E. (2008). The nurture of nature: Hereditary plasticity in evolution. *Philosophical Psychology* 21, 305–319. doi:10.1080/09515080802170093.

Lamme, V. A. F. (2006). Towards a true neural stance on consciousness. *Trends in Cognitive Sciences* 10, 494–501. doi:10.1016/j.tics.2006.09.001.

Lamme, V. A. F. (2010). How neuroscience will change our view on consciousness: Discussion paper. *Cognitive Neuroscience* 1, 204–220, 235–240. doi:10.1080/17588921003731586.

Lang, P. J. (1994). The varieties of emotional experience: A meditation on James-Lange theory. *Psychological Review* 101, 211–221. doi:10.1037/0033-295X.101.2.211.

Långsjö, J. W., Alkire, M. T., Kaskinoro, K., Hayama, H., Maksimow, A., Kaisti, K. K., Aalto, S., Aantaa, R., Jääskeläinen, S. K., Revonsuo, A., and Scheinin, H. (2012). Returning from oblivion: Imaging the neural core of consciousness. *Journal of Neuroscience* 32, 4935–4943. doi:10.1523/JNEUROSCI.4962-11.2012.

LeDoux, J. (1996). *The Emotional Brain: The Mysterious Underpinnings of Emotional Life*. New York: Simon and Schuster.

LeDoux, J. (2012). Rethinking the emotional brain. *Neuron* 73, 653–676. doi:10.1016/j.neuron.2012.02.004.

Lee, M. S. Y., Soubrier, J., and Edgecombe, G. D. (2013). Rates of phenotypic and genomic evolution during the Cambrian explosion. *Current Biology* 23, 1889–1895. doi:10.1016/j.cub.2013.07.055.

Leibowitz, Y. (1985). *On Science and Values*. Jerusalem: Ministry of Defense Press (in Hebrew).

Lenggenhager, B., Tadi, T., Metzinger, T., and Blanke, O. (2007). Video ergo sum: Manipulating bodily self-consciousness. *Science* 317, 1096–1099. doi:10.1126/science.1143439.

Lennox, J. G. (2000). *Aristotle's Philosophy of Biology: Studies in the Origins of Life Science*. Cambridge: Cambridge University Press.

Lennox, J. G. (2017). An Aristotelian philosophy of biology: Form, function and development. *Acta Philosophica* 26(1), 33–52. doi:10.19272/201700701003.

Lesku, J. A., Martinez-Gonzalez, D., and Rattenborg, N. C. (2009). Sleep and sleep states: Phylogeny and ontogeny. In L. R. Squire (Ed.), *The New Encyclopedia of Neuroscience*, pp. 963–971. Oxford: Academic Press.

Lev, I., Seroussi, U., Gingold, H., Bril, R., Anava, S., and Rechavi, O. (2017). MET-2-dependent H3K9 methylation suppresses transgenerational small RNA inheritance. *Current Biology* 27, 1138–1147. doi:10.1016/j.cub.2017.03.008.

Levenson, J. M., and Sweatt, J. D. (2005). Epigenetic mechanisms in memory formation. *Nature Reviews Neuroscience* 6, 108–118. doi:10.1038/nrn1604.

Levi-Montalcini, R., Skaper, S. D., Dal Toso, R., Petrelli, L., and Leon, A. (1996). Nerve growth factor: From neurotrophin to neurokine. *Trends in Neurosciences* 19, 514–520. doi:10.1016/S0166-2236(96)10058-8.

Levin, M. (2012). Morphogenetic fields in embryogenesis, regeneration, and cancer: Non-local control of complex patterning. *BioSystems* 109, 243–261. doi:10.1016/j.biosystems.

Levin, M. (2013). Reprogramming cells and tissue patterning via bioelectrical pathways: Molecular mechanisms and biomedical opportunities. *WIREs Systems Biology and Medicine* 5, 657–676. doi:10.1002/wsbm.1236.

Levine, J. (1983). Materialism and qualia: The explanatory gap. *Pacific Philosophical Quarterly* 64, 354–361. doi:10.1111/j.1468-0114.1983.tb00207.x.

Levinton, J. S. (2008). The Cambrian explosion: How do we use the evidence? *BioScience* 58, 855–864. doi:10.1641/B580912.

Leys, S. P. (2007). Sponge coordination, tissues, and the evolution of gastrulation. In *Porifera Research: Biodiversity, Innovation and Sustainability.* Buzios, Brazil: International Sponge Symposium, pp. 53–59.

Leys, S. P. (2015). Elements of a "nervous system" in sponges. *Journal of Experimental Biology* 218, 581–591. doi:10.1242/jeb.110817.

Leys, S. P., Mackie, G. O., and Meech, R. W. (1999). Impulse conduction in a sponge. *Journal of Experimental Biology* 202, 1139–1150.

Liebeskind, B. J., Hillis, D. M., and Zakon, H. H. (2011). Evolution of sodium channels predates the origin of nervous systems in animals. *Proceedings of the National Academy of Sciences USA* 108, 9154–9159. doi:10.1073/pnas.1106363108.

Liljenström, H., and Århem, P. (Eds.). (2008). *Consciousness Transitions: Phylogenetic, Ontogenetic, and Physiological Aspects.* Amsterdam, The Netherlands: Elsevier.

Liu, X., Ramirez, S., and Tonegawa, S. (2014). Inception of a false memory by optogenetic manipulation of a hippocampal memory engram. *Philosophical Transactions of the Royal Society B* 369, 20130142. doi:10.1098/rstb.2013.0142.

Livingstone, M. S., Sziber, P. P., and Quinn, W. G. (1984). Loss of calcium/calmodulin responsiveness in adenylate cyclase of *rutabaga*, a Drosophila learning mutant. *Cell* 37, 205–215. doi:10.1016/0092-8674(84)90316-7.

Llinás, R. R. (2002). *I of the Vortex: From Neurons to Self.* Cambridge, MA: MIT Press.

Llinás, R. R., and Ribary, U. (1993). Coherent 40-Hz oscillation characterizes dream state in humans. *Proceedings of the National Academy of Sciences USA* 90, 2078–2081.

Locke, J. (1690). *An Essay Concerning Human Understanding.* 2 vols. London. Rev. ed., London: Dent, 1961 (J. W. Yolton, Ed.).

Loeb, J. (1900). *Comparative Physiology of the Brain and Comparative Psychology.* New York: G. P. Putnam's Sons.

Loesel, R., Nässel, D. R., and Strausfeld, N. J. (2002). Common design in a unique midline neuropil in the brains of arthropods. *Arthropod Structure and Development* 31, 77–91. doi:10.1016/S1467-8039(02)00017-8.

Logan, C. A., and Beck, H. P. (1977). Persistent sensitization following habituation in the sea anemone (*Anthopleura elegantissima*). *Journal of Biological Psychology* 19, 22–24.

Logan, C. A., and Beck, H. P. (1978). Long-term retention of habituation in the sea anemone (*Anthopleura elegantissima*). *Journal of Comparative and Physiological Psychology* 92, 928–936. doi:10.1037/h0077532.

Logothetis, N. (1998). Single units and conscious vision. *Philosophical Transactions of the Royal Society of London B* 353, 1801–1818. doi:10.1098/rstb.1998.0333.

Lorenzetti, F. D., Baxter, D. A., and Byrne, J. H. (2011). Classical conditioning analog enhanced acetylcholine responses but reduced excitability of an identified neuron. *Journal of Neuroscience* 31, 14789–14793. doi:10.1523/JNEUROSCI.1256-11.2011.

Lorenzetti, F. D., Mozzachiodi, R., Baxter, D. A., and Byrne, J. H. (2006). Classical and operant conditioning differentially modify the intrinsic properties of an identified neuron. *Nature Neuroscience* 9, 17–29. doi:10.1038/nn1593.

Love, G. D., Grosjean, E., Stalvies, C., Fike, D. A., Grotzinger, J. P., Bradley, A. S., Kelly, A. E., Bhatia, M., Meredith, W., Snape, C. E., Bowring, S. A., Condon, D. J., and Summons, R. E. (2009). Fossil steroids record the appearance of Demospongiae during the Cryogenian period. *Nature* 457, 718–721. doi:10.1038/nature07673.

Lovibond, P. F., and Shanks, D. R. (2002). The role of awareness in Pavlovian conditioning: Empirical evidence and theoretical implications. *Journal of Experimental Psychology: Animal Behavior Processes* 28, 3–26. doi:10.1037//0097-7403.28.1.3.

Loy, I., Fernández, V., and Acebes, F. (2006). Conditioning of tentacle lowering in the snail (*Helix aspersa*): Acquisition, latent inhibition, overshadowing, second-order conditioning, and sensory preconditioning. *Learning and Behavior* 34, 305–314. doi:10.3758/BF03192885.

Lüersen, K., Faust, U., Gottschling, D.-C., and Döring, F. (2014). Gait-specific adaptation of locomotor activity in response to dietary restriction in *Caenorhabditis elegans*. *Journal of Experimental Biology* 217, 2480–2488. doi:10.1242/jeb.099382.

Lundin, L. G. (1999). Gene duplications in early metazoan evolution. *Seminars in Cell and Developmental Biology* 10, 523–530. doi:10.1006/scdb.1999.0333.

Luria, A. R. (1968). *The Mind of a Mnemonist: A Little Book about a Vast Memory*. Cambridge, MA: Harvard University Press.

Lyon, P. (2015). The cognitive cell: Bacterial behavior reconsidered. *Frontiers in Microbiology* 6, 264. doi:10.3389/fmicb.2015.00264.

Mackie, G. O. (1970). Neuroid conduction and the evolution of conducting tissues. *Quarterly Review of Biology* 45, 319–332.

Mackie, G. O., and Meech, R. W. (1995a). Central circuitry in the jellyfish *Aglantha digitale*: I. The relay system. *Journal of Experimental Biology* 198, 2261–2270.

Mackie, G. O., and Meech, R. W. (1995b). Central circuitry in the jellyfish *Aglantha digitale*: II. The ring giant and carrier systems. *Journal of Experimental Biology* 198, 2271–2278.

Mackie, G. O., and Meech, R. W. (2000). Central circuitry in the jellyfish *Aglantha digitale*: III. The rootlet and pacemaker systems. *Journal of Experimental Biology* 203, 1797–1807.

Macphail, E. M. (1982). *Brain and Intelligence in Vertebrates*. New York: Oxford University Press.

Macphail, E. M. (1987). The comparative psychology of intelligence. *Behavioral and Brain Sciences* 10, 645–696.

Macphail, E. M. (1998). *The Evolution of Consciousness*. Oxford: Oxford University Press.

Macphail, E. M. (2008). A bird's eye view of consciousness. In H. Liljenström and P. Århem (Eds.), *Consciousness Transitions: Phylogenetic, Ontogenetic, and Physiological Aspects*, pp. 97–121. Amsterdam, The Netherlands: Elsevier.

Macphail, E. M., and Bolhuis, J. J. (2001). The evolution of intelligence: Adaptive specializations versus general process. *Biological Reviews* 76, 341–364. doi:10.1017/S146479310100570X.

Magee, B., and Elwood, R. W. (2013). Shock avoidance by discrimination learning in the shore crab (*Carcinus maenas*) is consistent with a key criterion for pain. *Journal of Experimental Biology* 216, 353–358. doi:10.1242/jeb.072041.

Mahr, J., and Csibra G. (2017). Why do we remember? The communicative function of episodic memory. *Behavioral and Brain Sciences* 19, 1–93. doi:10.1017/S0140525X17000012.

Malach, R. (2012). Neuronal reflections and subjective awareness. In S. Edelman, T. Fekete, and N. Zach (Eds.), *Being in Time: Dynamical Models of Phenomenal Experience*, pp. 22–36. Amsterdam, The Netherlands: John Benjamins.

Malec, M. (2015). Yet another look at the conceivability and possibility of zombies. *Balkan Journal of Philosophy* 2, 115–124. doi:10.5840/bjp20157215.

Maresca, B., and Schwartz, J. H. (2006). Sudden origins: A general mechanism of evolution based on stress protein concentration and rapid environmental change. *Anatomical Record* 289B, 38–46. doi:10.1002/ar.b.20089.

Margulis, L. (2001). The conscious cell. *Annals of the New York Academy of Sciences* 929, 55–70. doi:10.1111/j.1749-6632.2001.tb05707.x.

Marin, I., and Kipnis, J. (2013). Learning and memory ... and the immune system. *Learning and Memory* 20, 601–606. doi:10.1101/lm.028357.112.

Marinesco, S., Duran, K. L., and Wright, W. G. (2003). Evolution of learning in three aplysiid species: Differences in heterosynaptic plasticity contrast with conservation in serotonergic pathways. *Journal of Physiology* 550, 241–253. doi:10.1113/jphysiol.2003.038356.

Marlow, H., and Arendt, D. (2014). Evolution: Ctenophore genomes and the origin of neurons. *Current Biology* 24, R757–R761. doi:10.1016/j.cub.2014.06.057.

Marshall, C. R. (2006). Explaining the Cambrian "explosion" of animals. *Annual Review of Earth and Planetary Sciences* 34, 355–384. doi:10.1146/annurev.earth.33.031504.103001.

Martindale, M. Q., Finnerty, J. R., and Henry, J. Q. (2002). The Radiata and the evolutionary origins of the bilaterian body plan. *Molecular Phylogenetics and Evolution* 24, 358–365. doi:10.1016/S1055-7903(02)00208-7.

Martín-Durán, J. M., Pang, K., Børve, A., Lê, H. S., Furu, A., Cannon, J. T., Jondelius, U., and Hejnol, A. (2018). Convergent evolution of bilaterian nerve cords. *Nature* 553, 45–50. doi:10.1038/nature25030.

Mashour, G. A., and Alkireb, M. T. (2013). Evolution of consciousness: Phylogeny, ontogeny, and emergence from general anesthesia. *Proceedings of the National Academy of Sciences USA* 110(suppl. 2), 10357–10364. doi:10.1073/pnas.1301188110.

Mather, J. A. (2008). Cephalopod consciousness: Behavioural evidence. *Consciousness and Cognition* 17, 37–48. doi:10.1016/j.concog.2006.11.006.

Mather, J. A., Anderson, R. C., and Wood, J. B. (2010). *Octopus: The Ocean's Intelligent Invertebrate*. Portland, OR: Timber Press.

Matheson, T. (2002). Invertebrate nervous systems. In *Encyclopedia of Life Sciences*. London: Macmillan.

Mattick, J. S., and Mehler M. F. (2008). RNA editing, DNA recoding and the evolution of human cognition. *Trends in Neuroscience* 31, 277–233. doi:10.1016/j.tins.2008.02.003.

Maturana, H., and Varela, F. J. (1980). *Autopoiesis and Cognition: The Realization of the Living*. Dordrecht, The Netherlands: Reidel.

Maturana, H., and Varela, F. J. (1998). *The Tree of Knowledge: The Biological Roots of Human Understanding*. London: Shambala.

Maxmen, A. (2016). Comb jelly "anus" guts ideas on origin of through-gut. *Science* 351, 1378–1380. doi:10.1126/science.351.6280.1378.

Maxwell, E. K., Ryan, J. F., Schnitzler, C. E., Browne, W. E., and Baxevanis, A. D. (2012). MicroRNAs and essential components of the microRNA processing machinery are not encoded in the genome of the ctenophore *Mnemiopsis leidyi*. *BMC Genomics* 13, 714. doi:10.1186/1471-2164-13-714.

Maxwell, N. (1968). Understanding sensations. *Australasian Journal of Philosophy* 46, 127–145. doi:10.1080/00048406812341111.

Mayford, M., Siegelbaum, S. A., and Kandel, E. R. (2012). Synapses and memory storage. *Cold Spring Harbor Perspectives in Biology* 4, a005751. doi:10.1101/cshperspect.a005751.

Maynard Smith, J. (1986). *The Problems of Biology*. Oxford: Oxford University Press.

Maynard Smith, J., and Szathmáry, E. (1995). *The Major Transitions in Evolution*. Oxford: Oxford University Press.

Mayr, E. (1982). *The Growth of Biological Thought*. Cambridge, MA: Harvard University Press.

McClintock, B. (1984). The significance of the responses of the genome to challenge. *Science* 226, 792–801. doi:10.1126/science.15739260.

McClintock, J. B., and Lawrence, J. M. (1982). Photoresponse and associative learning in *Luidia clathrata Say* (Echinodermata: Asteroidea). *Marine and Freshwater Behaviour and Physiology* 9, 13–21. doi:10.1080/10236248209378580.

McEwan, A. H., and Rankin, C. H. (2013). Mechanosensory learning and memory in *Caenorhabditis elegans*. In R. Menzel and P. R. Benjamin (Eds.), *Invertebrate Learning and Memory*, pp. 91–111. London: Academic Press.

McFarlane, I. D. (1982). *Calliactis parasitica*. In G. A. B. Shelton (Ed.), *Electrical Conduction and Behaviour in "Simple" Invertebrates*, pp. 243–265. Oxford: Clarendon Press.

McFarlane, I. D., Graff, D., and Grimmelikhuijzen, C. J. P. (1989). Evolution of conducting systems and neurotransmitters in the Anthozoa. In P. A. V. Anderson (Ed.), *Evolution of the First Nervous Systems*, pp.111–127. New York: Plenum.

McGinn, C. (1989). Can we solve the mind-body problem? *Mind* 98, 349–366. doi:10.1093/mind/XCVIII.391.349.

McGinn, C. (1999). *The Mysterious Flame*. New York: Basic Books.

McGregor, S., Vasas, V., Husbands, P., and Fernando, C. (2012). Evolution of associative learning in chemical networks. *PLoS Computational Biology* 8, e1002739. doi:10.1371/journal.pcbi.1002739.

McHugh, S. B., Barkus, C., Huber, A., Capitão, L., Lima, J., Lowry, J. P., and Bannerman, D. M. (2014). Aversive prediction error signals in the amygdala. *Journal of Neuroscience* 34, 9024–9033. doi:10.1523/JNEUROSCI.4465-13.2014.

McLaughlin, P. (1990). *Kant's Critique of Teleology in Biological Explanation: Antimony and Teleology*. Lewiston, NY: Edwin Mellen Press.

McMenamin, M. A. S. (1986). The garden of Ediacara. *Palaios* 1, 178–182. doi:10.2307/3514512.

Meech, R. W., and Mackie, G. O. (2007). Evolution of excitability in lower metazoans. In G. North and R. J. Greenspan (Eds.), *Invertebrate Neurobiology*, pp. 581–615. Cold Spring Harbor, NY: Cold Spring Harbor Laboratory Press.

Mehler, M. F. (2008a). Epigenetics and the nervous system. *Annals of Neurology* 64, 602–617. doi:10.1002/ana.21595.

Mehler, M. F (2008b). Epigenetic principles and mechanisms underlying nervous system functions in health and disease. *Progress in Neurobiology* 86, 305–341. doi:10.1016/j.pneurobio.2008.10.001.

Mendl, M., Burman, O. H. P., and Paul, E. S. (2010). An integrative and functional framework for the study of animal emotions and mood. *Proceedings of the Royal Society B* 277, 2895–2904. doi:10.1098/rspb.2010.0303.

Menzel, R., and Benjamin, P. R. (Eds.). (2013). *Invertebrate Learning and Memory*. London: Academic Press.

Menzel, R., and Giurfa, M. (2001). Cognitive architecture of a mini-brain: The honeybee. *Trends in Cognitive Sciences* 5, 62–71.

Mercier, H., and Sperber, D. (2017). *The Enigma of Reason: A New Theory of Human Reasoning*. London: Allen Lane.

Merker, B. (2005). The libilities of mobility: A selection pressure for the transition to consciousness in animal evolution. *Consciousness and Cognition* 14, 89–114. doi:10.1016/S1053-8100(03)00002-3.

Merker, B. (2007). Consciousness without a cerebral cortex: A challenge for neuroscience and medicine. *Behavioral and Brain Sciences* 30, 63–134. doi:10.1017/S0140525X07000891.

Merker, B. (2012). From probabilities to percepts: A subcortical "global best estimate buffer" as locus of phenomenal experience. In S. Edelman, T. Fekete, and N. Zach (Eds.), *Being in Time: Dynamical Models of Phenomenal Experience*, pp. 37–79. Amsterdam, The Netherlands: John Benjamins.

Merker, B. (2013). The efference cascade, consciousness, and its self: Naturalizing the first person pivot of action control. *Frontiers in Psychology* 4, 501. doi:10.3389/fpsyg.2013.00501.

Merker, B. (2016). Drawing the line on pain. *Animal Sentience* 3(23), 1–10.

Metzinger, T. (2003). *Being No One: The Self-Model Theory of Subjectivity*. Cambridge, MA: MIT Press.

Metzinger, T. (2007). Self models. *Scholarpedia* 2, 4174. doi:10.4249/scholarpedia.4174.

Metzinger, T. (2009). *The Ego Tunnel: The Science of the Mind and the Myth of the Self.* New York: Basic Books.

Metzinger, T. (2013). Two principles for robot ethics. In E. Hilgendorf and J.-P. Günther (Eds.), *Robotik und Gesetzgebung*, pp. 247–286. Baden-Baden, Germany: Nomos.

Michel, M., and Lyons, L. C. (2014). Unraveling the complexities of circadian and sleep interactions with memory formation through invertebrate research. *Frontiers in Systems Neuroscience* 8, 133. doi:10.3389/fnsys.2014.00133.

Migita, M., Mizukami, E., and Gunji, Y.-P. (2005). Flexibility in starfish behavior by multi-layered mechanism of self-organization. *BioSystems* 82, 107–115. doi:10.1016/j.biosystems.2005.05.012.

Mikulka, P., Vaughan, P., and Hughes, J. (1981). Lithium chloride-produced prey aversion in the toad (Bufo americanus). *Behavioral and Neural Biology* 33, 220–229. doi:10.1016/S0163-1047(81)91664-2.

Mill, J. S. (1843). *A System of Logic, Ratiocinative and Inductive.* 2 vols. London: John W. Parker.

Miller, C. A., Gavin, C. F., White, J. A., Parrish, R. R., Honasoge, A., Yancey, C. R., Rivera, I. M., Rubio, M. D., Rumbaugh, G., and Sweatt, J. D. (2010). Cortical DNA methylation maintains remote memory. *Nature Neuroscience* 13, 664–666. doi:10.1038/nn.2560.

Miller, C. A., and Sweatt, J. D. (2007). Covalent modification of DNA regulates memory formation. *Neuron* 53, 857–869. doi:10.1016/j.neuron.2007.02.022.

Miller, G. (2012). Sex, mutations and marketing. *EMBO Reports* 13, 880–884. doi:10.1038/embor.2012.131.

Miller, R. R., and Grace, R. C. (2003). Conditioning and learning. In A. F. Healy and R. W. Proctor (Eds.), *Handbook of Psychology*, vol. 4, pp. 357–198. Hoboken, NJ: John Wiley and Sons.

Milner, P. M. (1974). A model for visual shape recognition. *Psychological Review* 81, 521–535. doi:10.1037/h0037149.

Miyata, T., and Suga, H. (2001). Divergence pattern of animal gene families and relationship with the Cambrian explosion. *BioEssays* 23, 1018–1027. doi:10.1002/bies.1147.

Mizunami, M., Matsumoto, Y., Watanabe, H., and Nishino, H. (2013). Olfactory and visual learning in cockroaches and crickets. In R. Menzel and P. R. Benjamin (Eds.), *Invertebrate Learning and Memory*, pp. 549–560. London: Academic Press.

Monk, T., and Paulin, M. G. (2014). Predation and the origin of neurones. *Brain, Behavior and Evolution* 84, 246–261. doi:10.1159/000368177.

Monod, J. (1971). *Chance and Necessity: Essay on the Natural Philosophy of Modern Biology*. New York: Alfred A. Knopf.

Montgomery, S. (2015). *The Soul of an Octopus: A Surprising Exploration into the Wonder of Consciousness*. New York: Atria Books.

Moore, B. R. (2004). The evolution of learning. *Biological Reviews* 79, 301–335. doi:10.1017/S1464793103006225.

Moore, G. E. (1903). The refutation of idealism. *Mind* 12, 433–453. doi:10.1093/mind/XII.4.433.

Moreno, A., and Mossio, M. (2015). *Biological Autonomy: A Philosophical and Theoretical Enquiry*. Heidelberg, Germany: Springer.

Morgan, C. L. (1896). *Habit and Instinct*. London: Arnold.

Morgan, C. L. (1903). *Introduction to Comparative Psychology*. 2nd ed. London: Walter Scott.

Moroz, L. L. (2009). On the independent origins of complex brains and neurons. *Brain, Behavior and Evolution* 74, 177–190. doi:10.1159/000258665.

Moroz, L. L. (2012). Phylogenomics meets neuroscience: How many times might complex brains have evolved? *Acta Biologica Hungarica* 63, 3–19. doi:10.1556/ABiol .63.2012.Suppl.2.1.

Moroz, L. L., Kocot, K. M., Citarella, M. R., Dosung, S., Norekian, T. P., Povolotskaya, I. S., Grigorenko, A. P., Dailey, C., Berezikov, E., Buckley, K. M., Ptitsyn, A., Reshetov, D., Mukherjee, K., Moroz, T. P., Bobkova, Y., Yu, F., Kapitonov, V. V., Jurka, J., Bobkov, Y. V., Swore, J. J., Girardo, D. O., Fodor, A., Gusev, F., Sanford, R., Bruders, R., Kittler, E., Mills, C. E., Rast, J. P., Derelle, R., Solovyev, V. V., Kondrashov, F. A., Swalla, B. J., Sweedler, J. V., Rogaev, E. I., Halanych, K. M., and Kohn, A. B. (2014). The ctenophore genome and the evolutionary origins of neural systems. *Nature* 510, 109–114. doi:10.1038/nature13400.

Morris, G. P., Clark, I. A., Zinn, R., and Vissel, B. (2013). Microglia: A new frontier for synaptic plasticity, learning and memory, and neurodegenerative disease research. *Neurobiology of Learning and Memory* 105, 40–53. doi:10.1016/j.nlm.2013.07.002.

Morris, M. J., Mahgoub, M., Na, E. S., Pranav, H., and Monteggia, L. M. (2013). Loss of histone deacetylase 2 improves working memory and accelerates extinction learning. *Journal of Neuroscience* 33, 6401–6411. doi:10.1523/JNEUROSCI.1001-12.2013.

Morsella, E. (2005). The function of phenomenal states: Supramodular interaction theory. *Psychological Review* 112, 1000–1021. doi:10.1037/0033-295X.112.4.1000.

Mourant, A. E. (1971). Transduction and skeletal evolution. *Nature* 231, 486–487. doi:10.1038/231466a0.

Mudrik, L., Breska, A., Lamy, D., and Deouell, L. Y. (2011). Integration without awareness: Expanding the limits of unconscious processing. *Psychological Science* 22, 764–770. doi:10.1177/0956797611408736.

Mueller, T. (2012). What is the thalamus in zebrafish? *Frontiers in Neuroscience* 6, 64. doi:10.3389/fnins.2012.00064.

Munno, D. W., Naweed, I., and Syed, N. I. (2003). Synaptogenesis in the CNS: An odyssey from wiring together to firing together. *Journal of Physiology* 552, 1–11. doi:10.1113/jphysiol.2003.045062.

Murphy, G. G., and Glanzman, D. L. (1997). Mediation of classical conditioning in *Aplysia californica* by long-term potentiation of sensorimotor synapses. *Science* 278, 467–471. doi:10.1126/science.278.5337.467.

Myhrer, T. (2003). Neurotransmitter systems involved in learning and memory in the rat: A meta-analysis based on studies of four behavioral tasks. *Brain Research Reviews* 41, 268–287. doi:10.1016/S0165-0173(02)00268-0.

Nagel, T. (1974). What is it like to be a bat? *Philosophical Review* 83, 435–450.

Nagel, T. (2012). *Mind and Cosmos: Why the Materialist Neo-Darwinian Conception of Nature Is Almost Certainly False.* New York: Oxford University Press.

Narbonne, G. M. (2005). The Ediacara biota: Neoproterozoic origin of animals and their ecosystems. *Annual Review of Earth and Planetary Sciences* 33, 421–442. doi:10.1146/annurev.earth.33.092203.122519.

Nargeot, R., and Simmers, J. (2012). Functional organization and adaptability of a decision-making network in *Aplysia*. *Frontiers in Neuroscience* 6, 113. doi:10.3389/fnins.2012.00113.

Nath, R. D., Bedbrook, C. N., Abrams, M. J., Basinger, T., Bois, J. S., Prober, D. A., Sternberg, P. W., Gradinaru, V., and Goentoro, L. (2017). The jellyfish *Cassiopea* exhibits a sleep-like state. *Current Biology* 27, 2984–2990. doi:10.1016/j.cub.2017.08.014.

Nelson, B., Fornito, A., Harrison, B. J., Yücel, M., Sass, L. A., Yung, A. R., Thompson, A., Wood, S. J., Pantelis, C., and McGorry, P. D. (2009). A disturbed sense of self in the psychosis prodrome: Linking phenomenology and neurobiology. *Neuroscience and Biobehavioral Reviews* 33, 807–817. doi:10.1016/j.neubiorev.2009.01.002.

Nelson, K., and Fivush, R. (2004). The emergence of autobiographical memory: A social cultural developmental theory. *Psychological Review* 111, 486–511. doi:10.1037/0033-295X.111.2.486.

Nesse, R. M. (2001). The smoke detector principle. *Annals of the New York Academy of Science* 935, 75–85. doi:10.1111/j.1749-6632.2001.tb03472.x.

Newman, S. A., and Bhat, R. (2009). Dynamical patterning modules: A "pattern language" for development and evolution of multicellular form. *International Journal of Developmental Biology* 53, 693–705. doi:10.1088/1478-3975/5/1/015008.

Newman, S. A., Forgacs, G., and Müller, G. B. (2006). Before programs: The physical origination of multicellular forms. *International Journal of Developmental Biology* 50, 289–299. doi:10.1387/ijdb.052049sn.

Newman, S. A., and Müller, G. B. (2010). Morphological evolution: Epigenetic mechanisms. In *Encyclopedia of Life Sciences*. Chichester, UK: John Wiley and Sons.

Newport, C., Wallis, G., and Siebeck, U. E. (2014). Concept learning and the use of three common psychophysical paradigms in the archerfish (*Toxotes chatareus*). *Frontiers in Neural Circuits* 8, 39. doi:10.3389/fncir.2014.00039.

Nickel, M. (2010). Evolutionary emergence of synaptic nervous systems: What can we learn from the non-synaptic, nerveless Porifera? *Invertebrate Biology* 129, 1–16. doi:10.1111/j.1744-7410.2010.00193.x.

Nicolas, C. L., Abramson, C. I., and Levin, M. (2008). Analysis of behavior in the planarian model. In R. B. Raffa and S. M. Rawls (Eds.), *Planaria: A Model for Drug Action and Abuse*, pp. 83–94. Austin, TX: RG Landes.

Nilsson, D. E., Gislén, L., Coates, M. M., Skogh, C., and Garm, A. (2005). Advanced optics in a jellyfish eye. *Nature* 435, 201–205. doi:10.1038/nature03484.

Niv, Y. (2009). Reinforcement learning in the brain. *Journal of Mathematical Psychology* 53, 139–154. doi:10.1016/j.jmp.2008.12.005.

Noble, D. (2017). *Dance to the Tune of Life: Biological Relativity*. Cambridge: Cambridge University Press.

Noë, A. (2009). *Out of Our Heads: Why You Are Not Your Brain and Other Lessons from the Biology of Consciousness*. New York: Hill and Wang.

Northcutt, R. G. (2002). Understanding vertebrate brain evolution. *Integrative and Comparative Biology* 42, 743–756. doi:10.1093/icb/42.4.743.

Northoff, G. (2012). From emotions to consciousness—a neuro-phenomenal and neuro-relational approach. *Frontiers in Psychology* 3, 303. doi:10.3389/fpsyg.2012.00303.

Nussbaum, M. (1986). *Aristotle's De Motu Animalium*. Princeton, NJ: Princeton University Press.

Nussbaum, M., and Rorty, A. O. (1992). *Essays on Aristotle's De Anima*. Oxford: Clarendon Press.

Öhman, A., and Soares, J. J. (1993). On the automatic nature of phobic fear: Conditioned electrodermal responses to masked fear-relevant stimuli. *Journal of Abnormal Psychology* 102, 121–132. doi:10.1037/0021-843X.102.1.121.

Olds, J., and Milner, P. (1954). Positive reinforcement produced by electrical stimulation of septal area and other regions of rat brain. *Journal of Comparative and Physiological Psychology* 47, 419–427. doi:10.1037/h0058775.

Oliver, K. R., and Greene, W. K. (2009). Transposable elements: Powerful facilitators of evolution. *BioEssays* 31, 703–714. doi:10.1002/bies.200800219.

Orgel, L. E. (1973). *The Origins of Life: Molecules and Natural Selection.* New York: John Wiley & Sons.

Ottaviani, E. (2011). Evolution of immune-neuroendocrine integration from an ecological immunology perspective. *Cell and Tissue Research* 344, 213–215. doi:10.1007/s00441-011-1147-0.

Ottaviani, E., and Franceschi, C. (1996). The neuroimmunology of stress from invertebrates to man. *Progress in Neurobiology* 48, 421–440. doi:10.1016/0301-0082(95)00049-6.

Ottaviani, E., Malagoli, D., and Franceschi, C. (2007). Common evolutionary origin of the immune and neuroendocrine systems: From morphological and functional evidence to in silico approaches. *Trends in Immunology* 28, 497–502. doi:10.1016/j.it.2007.08.007.

Oudiette, D., Dealberto, M.-J., Uguccioni, G., Golmard, J.-L., Merino-Andreu, M., Tafti, M., Garma, L., Schwartz, S., and Arnu, I. (2012). Dreaming without REM sleep. *Consciousness and Cognition* 21, 1129–1140. doi:10.1016/j.concog.2012.04.010.

Pagel, M., Venditti, C., and Meade, A. (2006). Large punctuational contribution of speciation to evolutionary divergence at the molecular level. *Science* 314, 119–121. doi:10.1126/science.1129647.

Pahl, M., Zhu, H., Pix, W., Tautz, J., and Zhang, S. (2007). Circadian timed episodic-like memory—a bee knows what to do when, and also where. *Journal of Experimental Biology* 210, 3559–3567. doi:10.1242/jeb.005488.

Paley, W. (1802). *Natural Theology: Or, Evidences of the Existence and Attributes of the Deity.* London: J. Faulder.

Pang, K., Ryan, J. F., Mullikin. J. C., Baxevanis, A. D., and Martindale, M. Q. (2010). Genomic insights into Wnt signaling in an early diverging metazoan, the ctenophore *Mnemiopsis leidyi. EvoDevo* 1, 10. doi:10.1186/2041-9139-1-10.

Panksepp, J. (1998). *Affective Neuroscience: The Foundations of Human and Animal Emotions.* Oxford: Oxford University Press.

Panksepp, J. (2005). Affective consciousness: Core emotional feelings in animals and humans. *Consciousness and Cognition* 14, 30–80. doi:10.1016/j.concog.2004.10.004.

Panksepp, J. (2011a). Toward a cross-species neuroscientific understanding of the affective mind: Do animals have emotional feelings? *American Journal of Primatology* 73, 1–17. doi:10.1002/ajp.20929.

Panksepp, J. (2011b). Cross-species affective neuroscience: Decoding of the primal affective experiences of humans and related animals. *PLoS One* 6, e21236. doi:10.1371/journal.pone.0021236.

Panksepp, J., and Biven, L. (2011). *The Archeology of Mind: Neuroevolutionary Origins of Human Consciousness*. New York: W. W. Norton.

Pantin, C. F. A. (1956). The origin of the nervous system. *Pubblicazioni della Stazione Zoologica di Napoli* 28, 171–181.

Pantin, C. F. A. (1965). Capabilities of the coelenterate behavior machine. *American Zoologist* 5, 581–589. doi:10.1093/icb/5.3.581.

Papineau, D. (2013). Causation is macroscopic but not irreducible. In S. C. Gibb, E. J. Lowe, and R. Ingthorsson (Eds.), *Mental Causation and Ontology*, pp. 127–151. Oxford: Oxford University Press.

Papini, M. R. (2002). Pattern and process in the evolution of learning. *Psychological Review* 109, 186–201. doi:10.1037//0033-295X.109.1.186.

Parker, A. (2003). *In the Blink of an Eye: How Vision Kick-Started the Big Bang of Evolution*. Sydney, Australia: Free Press.

Parker, G. H. (1919). *The Elementary Nervous System*. Philadelphia: J. B. Lippincott.

Passano, L. (1963). Primitive nervous systems. *Proceedings of the National Academy of Sciences USA* 50, 306–313.

Pastuzyn, E. D., Day, C. E., Kearns, R. B., Kyrke-Smith, M., Taibi, A. V., McCormick, J., Yoder, N., Belnap, D. M., Erlendsson, S., and Morado, D. R. (2018). The neuronal gene Arc encodes a repurposed retrotransposon Gag protein that mediates intercellular RNA transfer. *Cell* 172, 275–288.e18.

Patton, K. A. (2015). *Exaptation: The Rooting of Learning and Memory in the Stress Response*. Unpublished undergraduate diss., Wheaton College, Illinois.

Pavlov, I. P. (1927). *Conditioned Reflexes: An Investigation of the Physiological Activity of the Cerebral Cortex* (G. V. Anrep, Trans.). Oxford: Oxford University Press.

Peng, F., and Chittka, L. (2017). A simple computational model of the bee mushroom body can explain seemingly complex forms of olfactory learning and memory. *Current Biology* 27, 224–230. doi:10.1016/j.cub.2016.10.054.

Pennisi, E. (2016). Shaking up the tree of life. *Science* 354, 817–821. doi:10.1126/science.354.6314.817.

Pepperberg, I. (2002). *The Alex Studies: Cognitive and Communicative Abilities of Grey Parrots*. Cambridge, MA: Harvard University Press.

Perry, C. J., Barron, A. B., and Cheng, K. (2013). Invertebrate learning and cognition: Relating phenomena to neural substrate. *WIREs Cognitive Science* 4, 561–582. doi:10.1002/wcs.1248.

Perry, C. J., Barron, A. B., and Chittka, L. (2017). The frontiers of insect cognition. *Current Opinion in Behavioral Sciences* 16, 111–118. doi:10.1016/j.cobeha.2017.05 .011.

Perry, R. B. (1935). *The Thought and Character of William James*. Boston: Little, Brown.

Pessiglione, M., Petrovic, P., Daunizeau, J., Palminteri, S., Dolan, R. J., and Frith, C. D. (2008). Subliminal instrumental conditioning demonstrated in the human brain. *Neuron* 59, 561–567. doi:10.1016/j.neuron.2008.07.005.

Peters, S. E., and Gaines, R. R. (2012). Formation of the "Great Unconformity" as a trigger for the Cambrian explosion. *Nature* 484, 363–366. doi:10.1038/nature10969.

Peterson, K. J., Cameron, R. A., and Davidson, E. H. (1997). Set aside cells in maximal indirect development: Evolutionary and developmental significance. *BioEssays* 19, 623–631. doi:10.1002/bies.950190713.

Peterson, K. J., Dietrich, M. R., and McPeek, M. A. (2009). MicroRNAs and metazoan macroevolution: Insights into canalization, complexity, and the Cambrian explosion. *BioEssays* 31, 736–747. doi:10.1002/bies.200900033.

Petitot, J., Varela, F. J., Pachoud, B., and Roy, J.-M. (Eds.). (1999). *Naturalizing Phenomenology: Contemporary Issues in Phenomenology and Cognitive Science*. Stanford, CA: Stanford University Press.

Peyrache, A., Khamassi, M., Benchenane, K., Wiener, S. I., and Battaglia, F. P. (2009). Replay of rule-learning related neural patterns in the prefrontal cortex during sleep. *Nature Neuroscience* 12, 919–926. doi:10.1038/nn.2337.

Pfeiffer, K., and Homberg, U. (2014). Organization and functional roles of the central complex in the insect brain. *Annual Review of Entomology* 59, 165–184. doi:10.1146/ annurev-ento-011613-162031.

Pfennig, D. W., Wund, M. A., Snell-Rood, E. C., Cruickshank, T., Schlichting, C. D., and Moczek, A. P. (2010). Phenotypic plasticity's impacts on diversification and speciation. *Trends in Ecology and Evolution* 25, 459–467. doi:10.1016/j.tree.2010.05.006.

Philippe, H., Brinkmann, H., Copley, R. R., Moroz, L. L., Nakano, H., Poustka, A. J., Wallberg, A., Peterson, K. J., and Telford, M. J. (2011). Acoelomorph flatworms are deuterostomes related to *Xenoturbella*. *Nature* 470, 255–258. doi:10.1038/nature 09676.

Philippidou, P., and Dasen, J. S. (2013). Hox genes: Choreographers in neural development, architects of circuit organization. *Neuron* 80, 12–34. doi:10.1016/j.neuron.2013.09.020.

Pisani, D., Pett, W., Dohrmann, M., Feuda, R., Rota-Stabelli, O., Philippe, H., Lartillot, N., and Lartillot, G. (2015). Genomic data do not support comb jellies as the sister group to all other animals. *Proceedings of the National Academy of Sciences USA* 112, 15402–15407. doi:10.1073/pnas.1518127112.

Platholi, J., Herold, K. F., Hemmings, H. C., Jr., and Halpain, S. (2014). Isoflurane reversibly destabilizes hippocampal dendritic spines by an actin-dependent mechanism. *PLoS One* 9(7), e102978. doi:10.1371/journal.pone.0102978.

Plato. (1961). *Philebus* (E. Hamilton and H. Cairns, Eds.; R. Hoforth, Trans.). In *The Collected Dialogues of Plato*, pp. 1086–1150. Princeton, NJ: Princeton University Press.

Plotkin, H. (1997). *Darwin Machines and the Nature of Knowledge*. Cambridge, MA: Harvard University Press.

Poo, M.-M., Pignatelli, M., Ryan, T. J., Tonegawa, S., Bonhoeffer, T., Martin, K. C., Rudenko, A., Tsai, L.-H., Tsien, R. W., Fishell, G., Mullins, C., Gonçalves, J. T., Shtrahman, M., Johnston, S. T., Gage, F. H., Dan, Y., Long, J., Buzsáki, G., and Stevens, C. (2016). What is memory? The present state of the engram. *BMC Biology* 14, 40. doi:10.1186/s12915-016-0261-6.

Popper, K. R. (1972). *Objective Knowledge: An Evolutionary Approach*. Oxford: Oxford University Press.

Popper, K. R. (1978). Natural selection and the emergence of mind. *Dialectica* 32, 339–355. doi:10.1111/j.1746-8361.1978.tb01321.x.

Porter, S. (2011). The rise of predators. *Geology* 39, 607–608. doi:10.1130/focus 062011.1.

Potdar, S., and Sheeba, V. (2013). Lessons from sleeping flies: Insights from *Drosophila melanogaster* on the neuronal circuitry and importance of sleep. *Journal of Neurogenetics* 27, 23–42. doi:10.3109/01677063.2013.791692.

Poulet, J. F. A., and Hedwig, B. (2007). New insights into corollary discharges mediated by identified neural pathways. *Trends in Neurosciences* 30, 14–21. doi:10.1016/j.tins.2006.11.005.

Pradeu, T. (2012). *The Limits of the Self: Immunology and Biological Identity*. Oxford: Oxford University Press.

Prados, J., Alvarez, B., Howarth, J., Stewart, K., Gibson, C. L., Hutchinson, C. V., Young, A. M., and Davidson, C. (2013). Cue competition effects in the planarian. *Animal Cognition* 16, 177–186. doi:10.1007/s10071-012-0561-3.

Price, G. R. (1971/1995). The nature of selection. *Journal of Theoretical Biology* 175, 389–396. doi:10.1006/jtbi.1995.0149.

Prinz, J. (2005). A neurofunctional theory of consciousness. In A. Brook and K. Akins (Eds.), *Cognition and the Brain: The Philosophy and Neuroscience Movement*, pp. 381–396. Cambridge: Cambridge University Press.

Prinz, J. J. (2012). *The Conscious Brain: How Attention Engenders Experience*. New York: Oxford University Press.

Putnam, H. (1975). Philosophy and our mental life. In *Mind, Language and Reality: Philosophical Papers*, vol. 2, pp. 291–303. Cambridge: Cambridge University Press.

Rahn, E. J., Guzman-Karlsson, M. C., and Sweatt, J. D. (2013). Cellular, molecular, and epigenetic mechanisms in non-associative conditioning: Implications for pain and memory. *Neurobiology of Learning and Memory* 105, 133–150. doi:10.1016/j.nlm.2013.06.008.

Raizen, D. M., Zimmerman, J. E., Maycock, M. H., Ta, U. D., You, Y. J., Sundaram, M. V., and Pack, A. I. (2008). Lethargus is a *Caenorhabditis elegans* sleep-like state. *Nature* 451, 569–572. doi:10.1038/nature06535.

Ramachandran, V. S. (2005). *A Brief Tour of Human Consciousness: From Impostor Poodles to Purple Numbers*. New York: Pi Press.

Ramachandran, V. S., and Rogers-Ramachandran, D. (1996). Synaesthesia in phantom limbs induced with mirrors. *Proceedings of the Royal Society of London B* 263, 377–386. doi:10.1098/rspb.1996.0058.

Ramón y Cajal, S. (1906). The structure and connexions of neurons. Nobelprize.org. http://www.nobelprize.org/nobel_prizes/medicine/laureates/1906/cajal-lecture.html.

Rankin, C. H., and Carew, T. J. (1988). Dishabituation and sensitization emerge as separate processes during development in *Aplysia*. *Journal of Neuroscience* 8, 197–211.

Rao, R. P., and Ballard, D. H. (1999). Predictive coding in the visual cortex: A functional interpretation of some extra-classical receptive-field effects. *Nature Neuroscience* 2, 79–87.

Rapoport, E. (2016). RecBlast: A computational cloud-based method for large scale orthology detection. Master's thesis, Tel Aviv University.

Rapoport, E., and Neuhof, M. (2017). RecBlast: Cloud-based large scale orthology detection. *bioRxiv*. doi:doi:10.1101/112946.

Rauss, K., and Pourtois, G. (2013). What is bottom-up and what is top-down in predictive coding? *Frontiers in Psychology* 4, 276. doi:10.3389/fpsyg.2013.00276.

Razoux, F., Russig, R., Baltes, C., Mueggler, T., Dikaiou, K., Rudin, M., and Mansuy I. M. (2016). Transgenerational disruption of functional 5-HT$_{1A}$R-induced connectivity in the adult mouse brain by traumatic stress in early life. *Molecular Psychiatry* 22, 519–526. doi:10.1038/mp.2016.146.

Razran, G. (1965). Empirical codifications and specific theoretical implications of compound-stimulus conditioning: Perception. In W. F. Prokasy (Ed.), *Classical Conditioning*, pp. 226–248. New York: Appleton-Century-Crofts.

Razran, G. (1971). *Mind in Evolution: An East-West Synthesis of Learnt Behavior and Cognition*. Boston: Houghton Mifflin.

Reddy, L., Reddy, L., and Koch, C. (2006). Face identification in the near-absence of focal attention. *Vision Research* 46, 2336–2343. doi:10.1016/j.visres.2006.01.020.

Red'ko, V. G., Prokhorov, D. V., and Burtsev, M. S. (2004). Theory of functional systems, adaptive critics and neural networks. In *Proceedings of the International Joint Conference on Neural Networks*, pp. 1787–1792. Budapest: IJCNN.

Rees, G., and Frith, C. (2007). Methodologies for identifying the neural correlates of consciousness. In M. Velmans and S. Schneider (Eds.), *The Blackwell Companion to Consciousness*, pp. 553–566. Oxford: Blackwell.

Reinhard, C. T., Planavsky, N. J., Olson, S. L., Lyons, T. W., and Erwin, D. H. (2016). Earth's oxygen cycle and the evolution of animal life. *Proceedings of the National Academy of Sciences USA* 113, 8933–8938. doi:10.1073/pnas.1521544113.

Reinhard, J. (2015). Animal cognition: bumble bees suffer "false memories." *Current Biology* 25, R236–R238. doi:10.1016/j.cub.2015.01.072.

Reinhard, J., and Rowell, D. M. (2005). Social behaviour in an Australian velvet worm, *Euperipatoides rowelli* (Onychophora: Peripatopsidae). *Journal of Zoology* 267, 1–7. doi:10.1017/S095283690500709.

Renard, E., Vacelet, J., Gazave, E., Lapebie, P., Borchiellini, C., and Ereskovsky, A. V. (2009). Origin of the neuro-sensory system: New and expected insights from sponges. *Integrative Zoology* 4, 294–308. doi:10.1111/j.1749-4877.2009.00167.x.

Rescorla, M. (2016). Bayesian sensorimotor psychology. *Mind and Language* 31, 3–36. doi:10.1111/mila.12093.

Rescorla, R. A. (1976). Second-order conditioning of Pavlovian conditioned inhibition. *Learning and Motivation* 7, 161–172. doi:10.1016/0023-9690(76)90025-4.

Rescorla, R. A., and Solomon, R. L. (1967). Two-process learning theory: Relationships between Pavlovian conditioning and instrumental learning. *Psychology Review* 74, 151–182.

Rescorla, R. A., and Wagner, A. R. (1972). A theory of Pavlovian conditioning: Variations in the effectiveness of reinforcement and non reinforcement. In A. H. Black and W. F. Prokasy (Eds.), *Classical Conditioning II: Current Research and Theory*, pp. 64–99. New York: Appleton-Century-Crofts.

Revonsuo, A. (2005). *Inner Presence: Consciousness as a Biological Phenomenon.* Cambridge, MA: MIT Press.

Revonsuo, A. (2010). *Consciousness: The Science of Subjectivity.* Hove, UK: Taylor and Francis.

Rial, R. V., Nicolau, M. C., Gamundí, A., Akaârir, M., Garau, C., and Esteban, S. (2008). The evolution of consciousness in animals. In H. Liljenström and P. Århem (Eds.), *Consciousness Transitions: Phylogenetic, Ontogenetic, and Physiological Aspects,* pp 45–76. Amsterdam, The Netherlands: Elsevier.

Ribary, U. (2005). Dynamics of thalamo-cortical network oscillations and human perception. In S. Laureys (Ed.), *Progress in Brain Research* 150, 127–142. doi:10.1016/S0079-6123(05)50010-4.

Richards, R. J. (1979). Influence of sensationalist tradition on early theories of the evolution of behavior. *Journal of the History of Ideas* 40, 85–105.

Richards, R. J. (1987). *Darwin and the Emergence of Evolutionary Theories of Mind and Behavior.* Chicago: University of Chicago Press.

Richards, R. J. (1993). Ideology and the history of science. *Biology and Philosophy* 8, 103–108.

Richardson, R. D. (2006). *William James: In the Maelstrom of American Modernism.* Boston: Houghton Mifflin.

Richter, J. N., Hochner, B., and Kuba, M. J. (2015). Octopus arm movements under constrained conditions: Adaptation, modification and plasticity of motor primitives. *Journal of Experimental Biology* 218, 1069–1076.

Richter, J. N., Hochner, B., and Kuba, M. J. (2016). Pull or push? Octopuses solve a puzzle problem. *PloS One* 11, e0152048.

Rimer, J., Cohen, I. R., and Friedman, N. (2014). Do all creatures possess an acquired immune system of some sort? *Bioessays* 36, 273–281. doi:10.1002/bies.201300124.

Riskin, J. (2002). *Science in the Age of Sensibility.* Chicago: University of Chicago Press.

Riskin, J. (2016). *The Restless Clock.* Chicago: University of Chicago Press.

Rivoire, O., and Leibler, S. (2014). A model for the generation and transmission of variations in evolution. *Proceedings of the National Academy of Sciences USA* 111, E1940–E1949. doi:10.1073/pnas.1323901111.

Rizzolatti, G., and Craighero, L. (2004). The mirror-neuron system. *Annual Review of Neuroscience* 27, 169–192. doi:10.1146/annurev.neuro.27.070203.144230. PMID 15217330.

Rizzolatti, G., and Sinigaglia, C. (2016). The mirror mechanism: A basic principle of brain function. *Nature Reviews Neuroscience* 17, 757–765. doi:10.1038/nrn.2016.135.

Roediger H. L., III, Dudai, Y., and Fitzpatrick, S. M. (2007). *Science of Memory: Concepts*. New York: Oxford University Press.

Rolls, E. T. (1999). *The Brain and Emotion*. Oxford: Oxford University Press.

Rolls, E. T. (2005). *Emotion Explained*. Oxford: Oxford University Press.

Rolls, E. T. (2013). What are emotional states, and why do we have them? *Emotion Review* 5, 241–247.

Romanes, E. (1896). *Life and Letters of George John Romanes*. London: Longmans, Green.

Romanes, G. J. (1882). *Animal Intelligence*. New York: D. Appleton.

Romanes, G. J. (1883). *Mental Evolution in Animals, with a Posthumous Essay on Instinct by Charles Darwin*. London: Kegan Paul, Trench.

Romanes, G. J. (1888). *Mental Evolution in Man*. London: Kegan Paul, Trench.

Rose, J. D. (2002). The neurobehavioral nature of fishes and the question of awareness and pain. *Reviews in Fisheries Science* 10, 1–38.

Rose, J. D., Arlinghaus, R., Cooke, S. J., Diggles B. K., Sawynok, W., Stevens E. D., and Wynne C. D. L. (2014). Can fish really feel pain? *Fish and Fisheries* 15, 97–133.

Rosen, R. (1970). *Dynamical Systems Theory in Biology*. New York: Wiley Interscience.

Rosen, R. (1991). *Life Itself: A Comprehensive Inquiry into the Nature, Origin, and Fabrication of Life*. New York: Columbia University Press.

Rosenthal, D. M. (2005). *Consciousness and Mind*. Oxford: Oxford University Press.

Rota, G-C. (1986). In memoriam of Stan Ulam: The barrier of meaning. *Physica* 22D, 1–3.

Röttinger, E., Dahlin, P., and Martindale, M. Q. (2012). A framework for the establishment of a cnidarian gene regulatory network for "endomesoderm" specification: The inputs of β-Catenin/TCF signaling. *PLoS Genetics* 8, e1003164. doi:10.1371/journal.pgen.1003164.

Rouse, G. W., Wilson, N. G., Carvajal, J. I., and Vrijenhoek, R. C. (2016). New deep-sea species of *Xenoturbella* and the position of Xenoacoelomorpha. *Nature* 530, 94–97.

Ruby, E. G., and McFall-Ngai, M. J. (1992). A squid that glows in the night: Development of an animal-bacterial mutualism. *Journal of Bacteriology* 174, 4865–4870.

Ruiz-Mirazo, K., and Luisi, P. L. (Eds.). (2010). *Origins of Life and Evolution of Biospheres*. Special issue, *Workshop OQOL'09: Open Questions on the Origins of Life* 40, 347–497.

Rushforth, N. D. (1973). Behavioral modifications in coelenterates. In W. C. Corning, J. A. Dyal, and A. O. D. Willows (Eds.), *Invertebrate Learning*, vol. 1, *Protozoans through Annelids*, pp. 123–169. New York: Plenum.

Russell, B. (1905). On denoting. *Mind* 14, 479–493.

Russell, J. A. (2003). Core affect and the psychological construction of emotion. *Psychological Review* 110, 145–172.

Rusten, T. E., Cantera, R., Kafatos, F. C., and Barrio, R. (2002). The role of TGFβ signaling in the formation of the dorsal nervous system is conserved between *Drosophila* and chordates. *Development* 129, 3575–3584.

Ryan, J. F., and Chiodin, M. (2015). Where is my mind? How sponges and placozoans may have lost neural cell types. *Philosophical Transactions of the Royal Society B* 370, 20150059. doi:10.1098/rstb.2015.0059.

Ryan, J. F., Pang, K., Mullikin, J. C., Martindale, M. Q., and Baxevanis, A. D. (2010). The homeodomain complement of the ctenophore *Mnemiopsis leidyi* suggests that Ctenophora and Porifera diverged prior to the ParaHoxozoa. *EvoDevo* 1, 9. doi:10.1186/2041-9139-1-9.

Ryan, J. F., Pang, K., Schnitzler, C. E., Nguyen, A.-D., Moreland, T., Simmons, D. K., Koch, B. J., Francis, W. R., Havlak, P., Smith, S. A., Putnam, N. H., Haddock, S. H. D., Dunn, C. W., Wolfsberg, T. G., Mullikin, J. C., Martindale, M. Q., and Baxevanis, A. D. (2013). The genome of the ctenophore *Mnemiopsis leidyi* and its implications for cell type evolution. *Science* 342, 1336–1340. doi:10.1126/science.1242592.

Saavedra-Rodríguez L., Vázquez, A., Ortiz-Zuazaga, H. G., Chorna, N. E., González, F. A., Andrés, L., Rodríguez, K., Ramírez, F., Rodríguez, A., and Peña de Ortiz, S. (2009). Identification of flap structure-specific endonuclease 1 as a factor involved in long-term memory formation of aversive learning. *Journal of Neuroscience* 29, 5726–5737. doi:10.1523/JNEUROSCI.4033-08.2009.

Sachser, R. M., Santana, F., Crestani, A. P., Lunardi, P., Pedraza, L. K., Quillfeldt, J. A., Hardt, O., and de Oliveira Alvares, L. (2016). Forgetting of long-term memory requires activation of NMDA receptors, L-type voltage-dependent Ca^{2+} channels, and calcineurin. *Scientific Reports* 6, 22771.

Saeki, S., Yamamoto, M., and Iino, Y. (2001). Plasticity of chemotaxis revealed by paired presentation of a chemoattractant and starvation in the nematode *Caenorhabditis elegans*. *Journal of Experimental Biology* 204, 1757–1764.

Sahley, C. L., and Ready, D. F. (1988). Associative learning modifies two behaviors in the leech, *Hirudo medicinalis*. *Journal of Neuroscience* 8, 4812–4820.

Sainte-Beuve, C. A. (1834/1986). *Volupté* (André Guyaux, Ed.). Paris: Gallimard, pp. 136–137.

Sakarya, O., Armstrong, K. A., Adamska, M., Adamski, M., Wang, I. F., Tidor, B., Degnan, B. M., Oakley, T. H., and Kosik, K. S. (2007). A post synaptic-scaffold at the origin of the animal kingdom. *PLoS One* 2, e506. doi:10.1371/journal.pone.0000506.

Salwiczek, L. H., and Bshary, R. (2011). Cleaner wrasses keep track of the "when" and "what" in a foraging task. *Ethology* 117, 939–948.

Santello, M., and Volterra, A. (2009). Synaptic modulation by astrocytes via Ca2+-dependent glutamate release. *Journal of Neuroscience* 158, 253–259. doi:10.1016/j.neuroscience.

Sapolsky, R. M. (2004). *Why Zebras Don't Get Ulcers*. 3rd ed. New York: Henry Holt.

Satterlie, R. A. (2011). Do jellyfish have central nervous systems? *Journal of Experimental Biology* 214, 1215–1223. doi:10.1242/jeb.043687.

Scaplen, K. M., and Kaun, K. R. (2016). Reward from bugs to bipeds: A comparative approach to understanding how reward circuits function. *Journal of Neurogenetics* 30, 133–148. doi:10.1080/01677063.2016.1180385.

Schacter, D. L. (2001). *Forgotten Ideas, Neglected Pioneers: Richard Semon and the Story of Memory*. Philadelphia: Psychology Press.

Schacter, D. L., Addis, D. R., and Buckner, R. L. (2007). Remembering the past to imagine the future: The prospective brain. *Nature Reviews Neuroscience* 8, 657–661. doi:10.1038/nrn2213.

Schapiro, A. C., Kustner, L. V., and Turk-Browne, N. B. (2012). Shaping of object representations in the human medial temporal lobe based on temporal regularities. *Current Biology* 22, 1622–1627.

Scherer, K. R. (2009). Emotions are emergent processes: They require a dynamic computational architecture. *Philosophical Transactions of the Royal Society B* 364, 3459–3474. doi:10.1098/rstb.2009.0141.

Schleyer, M., Miura, D., Tanimura, T., and Gerber, B. (2015). Learning the specific quality of taste reinforcement in larval *Drosophila*. *eLife* 4, e04711. doi:10.7554/eLife.04711.

Schmalhausen, I. I. (1949). *Factors of Evolution: The Theory of Stabilizing Selection* (I. Dordick, Trans.). Philadelphia: Blakiston.

Schneider, S. M. (2012). *The Science of Consequences: How They Affect Genes, Change the Brain and Impact Our World*. Amherst, NY: Prometheus Books.

Schultz, W. (1998). Predictive reward signal of dopamine neurons. *Journal of Neurophysiology* 80, 1–27.

Schultz, W. (2015). Neuronal reward and decision signals: From theories to data. *Physiological Reviews* 95, 853–951.

Schultz, W., Dayan, P., and Montague, P. R. (1997). A neural substrate of prediction and reward. *Science* 275, 1593–1599.

Schultz, W., and Dickinson, A. (2000). Neuronal coding of prediction errors. *Annual Review of Neuroscience* 23, 473–500.

Schumacher, S., de Perera, T. B., Thenert, J., and von der Emde, G. (2016). Cross-modal object recognition and dynamic weighting of sensory inputs in a fish. *Proceedings of the National Academy of Sciences USA* 113, 7638–7643. doi:10.1073/pnas.1603120113.

Scott, R. B., Samaha, J., Chrisley, R., and Dienes, Z. (2018). Prevailing theories of consciousness are challenged by novel cross-modal associations acquired between subliminal stimuli. *Cognition* 175, 169–185. doi: 10.1016/j.cognition.2018.02.008.

Searle, J. (1980). Minds, brains and programs. *Behavioral and Brain Sciences* 3, 417–424.

Searle, J. (1999). *Mind, Language, and Society: Philosophy in the Real World.* New York: Basic Books.

Searle, J. (2004). *Mind: A Brief Introduction.* New York: Oxford University Press.

Searle, J. (2013). Theory of mind and Darwin's legacy. *Proceedings of the National Academy of Sciences USA* 110(suppl. 2), 10343–10348. doi:10.1073/pnas.

Semon, R. (1904/1921). *The Mneme* (L. Simon, Trans.). London: George Allen and Unwin.

Sen, A., Heredia, N., Senut, M-C., Land, S., Hollocher, K., Lu, X., Dereski, M. O., and Ruden, D. M. (2015). Multigenerational epigenetic inheritance in humans: DNA methylation changes associated with maternal exposure to lead can be transmitted to the grandchildren. *Scientific Reports* 5, 14466. doi:10.1038/srep14466.

Sergent, C., and Naccache, L. (2012). Imaging neural signatures of consciousness: "What'," "Where" and "How" does it work? *Archives Italiennes de biologie* 150, 91–106. doi:10.4449/aib.v150i2.1270.

Seth, A. (2009a). Explanatory correlates of consciousness: Theoretical and computational challenges. *Cognitive Computation* 1, 50–63. doi:10.1007/s12559-009-9007-x.

Seth, A. K. (2009b). Functions of consciousness. In W. P. Banks (Ed.), *Encyclopedia of Consciousness*, pp. 279–293. Oxford: Elsevier.

Seth, A. K. (2011). On the Human, a forum of the National Humanities Center. Comment following Stevan Harnad's Doing, feeling, meaning and explaining. http://nationalhumanitiescenter.org/on-the-human/2011/04/doing-feeling-meaning-explaining/comment-page-1/#comment-6690.

Seth, A. K. (2013). Interoceptive inference, emotion, and the embodied self. *Trends in Cognitive Sciences* 17, 565–573.

Seth, A. K. (2015). The cybernetic Bayesian brain—from interoceptive inference to sensorimotor contingencies. In T. Metzinger and J. M. Windt (Eds.), *Open MIND*, p. e35(T). Frankfurt am Main, Germany: MIND Group. doi:10.15502/9783958570108.

Seth, A. K., Baars, B. J., and Edelman, D. B. (2005). Criteria for consciousness in humans and other mammals. *Consciousness and Cognition* 14, 119–139.

Seth, A., Barrett, A. B., and Barnet, L. (2011). Causal density and integrated information as measures of conscious level. *Philosophical Transactions of the Royal Society A* 369, 3748–3767.

Seth, A. K., and Friston, K. J. (2016). Active interoceptive inference and the emotional brain. *Philosophical Transactions of the Royal Society B* 371, 20160007. doi:10.1098/rstb.2016.0007.

Seuss, Dr. [T. S. Geisel]. (2003). *Oh, the Thinks You Can Think!* New York: Random House.

Shadlen, M. N., and Kiani, R. (2013). Decision making as a window on cognition. *Neuron* 80, 791–806. doi:10.1016/j.neuron.2013.10.047.

Shadlen, M. N., and Movshon, J. A. (1999). Synchrony unbound: A critical evaluation of the temporal binding hypothesis. *Neuron* 24, 67–77. doi:10.1016/S0896-6273(00)80822-3.

Shahaf, G., and Marom, S. (2001). Learning in networks of cortical neurons. *Journal of Neuroscience* 21, 8782–8788.

Shapiro, J. A. (2011). *Evolution: A View from the 21st Century*. Upper Saddle River, NJ: Pearson Education.

Shapiro, L. (2011). *Embodied Cognition*. London: Routledge.

Shea, N. (2018). *Representation in Cognitive Science*. Oxford: Oxford University Press.

Shen, X. X., Hittinger, C. T., and Rokas, A. (2017). Contentious relationships in phylogenomic studies can be driven by a handful of genes. *Nature Ecology and Evolution* 1, 0126. doi:10.1038/s41559-017-0126.

Shepherd, G. M., and Erulkar, S. D. (1997). Centenary of the synapse: From Sherrington to the molecular biology of the synapse and beyond. *Trends in Neuroscience* 20, 385–392. doi:10.1016/S0166-2236(97)01059-X.

Sherrin, T., Blank, T., Hippel, C., Rayner, M., Davis, R. J., and Todorovic, C. (2010). Hippocampal c-Jun-N-terminal kinases serve as negative regulators of associative learning. *Journal of Neuroscience* 30, 13348–13361. doi:10.1523/JNEUROSCI.3492-10.2010.

Shettleworth, S. J. (1998). *Cognition, Evolution, and Behavior*. New York: Oxford University Press.

Shih, C.-T., Sporns, O., Yuan, S.-Y., Su, T.-S., Lin, Y.-J., Chuang, C.-C., Wang, T.-Y., Lo, C.-C., Greenspan, R. J., and Chiang, A.-S. (2015). Connectomics-based analysis of information flow in the *Drosophila* brain. *Current Biology* 25, 1249–1258.

Shizgal, P. (1999). On the neural computation of utility: Implications from studies of brain stimulation reward. In D. Kahneman, E. Diener, and N. Schwarz (Eds.), *Well-Being: The Foundations of Hedonic Psychology*, pp. 500–524. New York: Russell Sage.

Shomrat T., Graindorge, N., Bellanger, C., Fiorito, G., Loewenstein, Y., and Hochner, B. (2011). Alternative sites of synaptic plasticity in two homologous fan-out fan-in learning and memory networks. *Current Biology* 21, 1773–1782. doi:10.1016/j.cub.2011.09.011.

Shomrat, T., Turchetti-Maia, A. L., Stern-Mentch, N., Basil, J. A., and Hochner, B. (2015). The vertical lobe of cephalopods: An attractive brain structure for understanding the evolution of advanced learning and memory systems. *Journal of Comparative Physiology* 201, 947–956. doi:10.1007/s00359-015-1023-6.

Shu, D.-G., Conway Morris, S., Han, J., Li, Y., Zhang, X.-L., Hua, H., Zhang, Z.-F., Liu, J.-N., Guo, J.-F., Yao, Y., and Yasui, K. (2006). Lower Cambrian vendobionts from China and early diploblast evolution. *Science* 312, 731–734. doi:10.1126/science.1124565.

Shuai, Y., Hirokawa, A., Ai, Y., Zhang, M., Li, W., and Zhong, Y. (2015). Dissecting neural pathways for forgetting in *Drosophila* olfactory aversive memory. *Proceedings of the National Academy of Sciences USA* 112, E6663–E6672.

Shuai, Y., Hu, Y., Qin, H., Campbell R. A., and Zhong, Y. (2011). Distinct molecular underpinnings of *Drosophila* olfactory trace conditioning. *Proceedings of the National Academy of Sciences USA* 108, 20201–20206. doi:10.1073/pnas.1107489109.

Shuai, Y., Lu, B., Hu, Y., Wang, L., Sun, K., and Zhong, Y. (2010). Forgetting is regulated through Rac activity in *Drosophila*. *Cell* 140, 579–589.

Si, K., Choi, Y. B., White-Grindley, E., Majumdar, A., and Kandel, E. R. (2010). *Aplysia* CPEB can form prion-like multimers in sensory neurons that contribute to long-term facilitation. *Cell* 140, 421–435. doi:10.1016/j.cell.2010.01.008.

Siegel, J. M. (2005). Clues to the functions of mammalian sleep. *Nature* 437, 1264–1271.

Siegel, J. M. (2009). Sleep viewed as a state of adaptive inactivity. *Nature Reviews Neuroscience* 10, 747–753.

Simmons, D. K., Pang, K., and Martindale, M. Q. (2012). Lim homeobox genes in the Ctenophore *Mnemiopsis leidyi*: The evolution of neural cell type specification. *EvoDevo* 3, 2. doi:10.1186/2041-9139-3-2.

Simons, D., and Chabris, C. (1999). Gorilla in our midst: Sustained inattentional blindness for dynamic events. *Perception* 28, 1059–1074.

Singer, W. (2007). *Scholarpedia* 2(12). doi:10.4249/scholarpedia.1657 http://www.scholarpedia.org/article/Binding_by_synchrony.

Singer, W., and Gray, C. M. (1995). Visual feature integration and the temporal correlation hypothesis. *Annual Review of Neuroscience* 18, 555–586.

Skinner, B. F. (1981). Selection by consequences. *Science* 21, 501–504. doi:10.1126/science.7244649.

Slater M., and Sanchez-Vive, M. V. (2016). Enhancing our lives with immersive virtual reality. *Froniers in Robotics and AI* 3, 74. doi:10.3389/frobt.2016.00074.

Smith, M. P., and Harper, D. A. (2013). Causes of the Cambrian explosion. *Science* 31, 1355–1366. doi:10.1126/science.1239450.

Sneddon, L. U. (2011). Pain perception in fish: Evidence and implications for the use of fish. *Journal of Consciousness Studies* 18(9/10), 209–229.

Sneddon, L. U., Elwood, R. W., Adamo, S. A., and Leach, M. C. (2014). Defining and assessing animal pain. *Animal Behaviour* 97, 202–212. doi:10.1016/j.anbehav.2014.09.007.

Soares, J. J., and Öhman, A. (1993). Backward masking and skin conductance responses after conditioning to nonfeared but fear-relevant stimuli in fearful subjects. *Psychophysiology* 30, 460–466.

Soen, Y., Knafo, M., and Elgart, M. (2015). A principle of organization which facilitates broad Lamarckian-like adaptations by improvisation. *Biology Direct* 10, 68. doi:10.1186/s13062-015-0097-y.

Spafford, J. D., Spencer, A. N., and Gallin, W. J. (1998). A putative voltage-gated sodium channel α subunit (PpSCN1) from the hydrozoan jellyfish, *Polyorchis penicillatus*: Structural comparisons and evolutionary considerations. *Biochemical and Biophysical Research Communications* 244, 772–780.

Spencer, A. N. (1989). Chemical and electrical synaptic transmission in the Cnidaria. In P. A. V. Anderson (Ed.), *Evolution of the First Nervous Systems*, pp. 33–53. New York: Plenum.

Spencer, H. (1852). A theory of population, deduced from the general law of animal fertility. *Westminster Review* 57 [new series, vol. 1, no. 2], 468–501. http://www.victorianweb.org/science/science_texts/spencer2.html.

Spencer, H. (1855). *The Principles of Psychology.* 2 vols. 1st ed. London: Longman, Brown, Green and Longmans.

Spencer, H. (1857). Progress: Its law and cause. In *Essays: Scientific, Political, and Speculative. Library Edition, Containing Seven Essays Not Before Republished, and Various Other Additions*, vol. 1. London: Williams and Norgate.

Spencer, H. (1862). *First Principles*. 6th ed. London: Williams and Norgate.

Spencer, H. (1864/1867). *The Principles of Biology*. 2 vols. London: Williams and Norgate.

Spencer, H. (1870). *The Principles of Psychology*, vol 1. 2nd ed. London: Williams and Norgate.

Spencer, H. (1890). *The Principles of Psychology*. 2 vols. 3rd ed. London: William and Norgate.

Sperling, E. A., Frieder, C. A., Raman, A. V., Girguis, P. R., Levin, L. A., and Knoll, A. H. (2013). Oxygen, ecology, and the Cambrian radiation of animals. *Proceedings of the National Academy of Sciences USA* 110, 13446–13451.

Sperling, E. A., and Vinther, J. (2010). A placozoan affinity for Dickinsonia and the evolution of late Proterozoic metazoan feeding modes. *Evolution and Development* 12, 201–209.

Sperling, E. A., Wolock, C. J., Morgan, A. S., Gill, B. C., Kunzmann, M., Halverson, G. P., Macdonald, F. A., Knoll, A. H., and Johnston, D. T. (2015). Statistical analysis of iron geochemical data suggests limited late Proterozoic oxygenation. *Nature* 523, 451–454. doi:10.1038/nature14589.

Sperry, R. (1950). Neural basis of the spontaneous optokinetic response produced by visual inversion. *Journal of Comparative and Physiological Psychology* 43, 482–489.

Stanley, S. M. (1992). Can neurons explain the Cambrian explosion? *Geological Society of America Abstracts with Programs* 24, A45.

Staum, M. S. (1980). *Cabanis: Enlightenment and Medical Philosophy in the French Revolution*. Princeton, NJ: Princeton University Press.

Stefano, G. B., Cadet, P., Zhu, W., Rialas, C. M., Mantione, K., Benz, D., Fuentes, F., Casares, F., Fricchione, G. L., Fulop, Z., and Slingsby, B. (2002). The blueprint for stress can be found in invertebrates. *Neuroendocrinology Letters* 23, 85–93.

Stegner, M. E. J., Fritsch, M., and Richter, S. (2014). The central complex in Crustacea. In J. W. Wägele, T. Bartolomäus, B. Misof, and L. Vogt (Eds.), *Deep Metazoan Phylogeny: The Backbone of the Tree of Life*, pp. 361–384. Berlin: De Gruyter.

Stephenson, R., and Lewis, V. (2011). Behavioural evidence for a sleep-like quiescent state in a pulmonate mollusc, *Lymnaea stagnalis* (Linnaeus). *Journal of Experimental Biology* 214, 747–756.

Sterling, P., and Laughlin, S. (2015). How bigger brains are organized. In *Principles of Neural Design*, chap 4. Cambridge, MA: MIT Press.

Stern, S., Dror, T., Stolovicki, E., Brenner, N., and Braun, E. (2007). Genome-wide transcriptional plasticity underlies cellular adaptation to novel challenge. *Molecular Systems Biology* 3, 106. doi:10.1038/msb4100147.

Stern, S., Fridmann-Sirkis, Y., Braun, E., and Soen, Y. (2012). Epigenetically heritable alteration of fly development in response to toxic challenge. *Cell Reports* 5, 528–542. doi:10.1016/j.celrep.2012.03.012.

Stetler, R. A., Gan, Y., Zhang, W., Liou, A. K., Gao, Y., Cao, G., and Chen, J. (2010). Heat shock proteins: Cellular and molecular mechanisms in the CNS. *Progress in Neurobiology* 92, 184–211. doi:10.1016/j.pneurobio.2010.05.002.

Stöckl, A. L., Petie, R., and Nilsson, D.-E. (2011). Setting the pace: New insights into central pattern generator interactions in box jellyfish swimming. *PLoS One* 6, e27201. doi:10.1371/journal.pone.0027201.

Stolovicki, E., Dror, T., Brenner, N., and Braun, E. (2006). Synthetic gene recruitment reveals adaptive reprogramming of gene regulation in yeast. *Genetics* 173, 75e85. doi:10.1534/genetics.106.055442.

Stopfer, M. (2014). Central processing in the mushroom bodies. *Current Opinion in Insect Science* 6, 99–103. doi:10.1016/j.cois.2014.10.009.

Strausfeld, N. J., and Hirth, F. (2013). Deep homology of arthropod central complex and vertebrate basal ganglia. *Science* 340, 157–161. doi:10.1126/science.1231828.

Strauss, R. (2002). The central complex and the genetic dissection of locomotor behaviour. *Current Opinion in Neurobiology* 12, 633–638. doi:10.1016/S0959-4388(02)00385-9.

Suddendorf, T. (2013). Mental time travel: Continuities and discontinuities. *Trends in Cognitive Sciences* 17, 151–152.

Suddendorf, T., and Corballis, M. C. (2007). The evolution of foresight: What is mental time travel, and is it unique to humans? *Behavioral and Brain Sciences* 30, 299–351.

Sumbre, G., Gutfreund, Y., Fiorito, G., Flash, T., and Hochner, B. (2001). Control of octopus arm extension by a peripheral motor program. *Science* 293, 1845–1848.

Szabo Gendler, T., and Hawthorne, J. (Eds.). (2002). *Conceivability and Possibility*. Oxford: Oxford University Press.

Szathmáry, E. (2015a). Toward major evolutionary transitions theory 2.0. *Proceedings of the Natural Academy of Science USA* 112, 10104–10111.

Szathmáry, E. (Ed.). (2015b). Tibor Gánti (1933–2009): Towards the principles of life and systems chemistry. *Journal of Theoretical Biology* 381, 1–60.

Szathmáry, E., and Fernando, C. (2011). Concluding remarks. In B. Calcott and K. Sterelny (Eds.), *The Major Transitions in Evolution Revisited*, pp. 301–310. Cambridge, MA: MIT Press.

Tansey, E. M. (1997). Not committing barbarisms: Sherrington and the synapse, 1897. *Brain Research Bulletin* 44, 211–212. doi:10.1016/S0361-9230(97)00312-2.

Tatum, M. C., Ooi, F. K., Chikka, M. R., Chauve, L., Martinez-Velazquez, L. A., Steinbusch, H. W. M., Morimoto, R. I., and Prahlad, V. (2015). Neuronal serotonin release triggers the heat shock response in C. elegans in the absence of temperature increase. *Current Biology* 25, 163–174. doi:10.1016/j.cub.2014.11.040.

Tauber, A. I. (2017). *Immunity: The Evolution of an Idea*. Oxford: Oxford University Press.

Tavory, I., Ginsburg, S., and Jablonka, E. (2014). The reproduction of the social: A developmental system theory approach. In L. R. Caporael, J. Griesemer, and W. Wimsatt (Eds.), *Scaffolding in Evolution, Culture and Cognition: Vienna Series in Theoretical Biology*, pp. 307–327. Cambridge, MA: MIT Press.

Tavory, I., Jablonka, E., and Ginsburg, S. (2012). Culture and epigenesis: A Waddingtonian view. In J. Valsiner (Ed.), *The Oxford Handbook of Culture and Psychology*, pp. 662–676. New York: Oxford University Press.

Telford, M. J., Budd, G. E., and Philippe, H. (2015). Phylogenomic insights into animal evolution. *Current Biology* 25, R876–887. doi:10.1016/j.cub.2015.07.060.

Telford, M. J., Moroz, L. L., and Halanych, K. M. (2016). Evolution: A sisterly dispute. *Nature* 529, 286–287. doi:10.1038/529286a.

Templer, V. L., and Hampton, R. R. (2012). Rhesus monkeys (Macaca mulatta) show robust evidence for memory awareness across multiple generalization tests. *Animal Cognition* 15, 409–419. doi:10.1007/s10071-011-0468-4.

Templer, V. L., and Hampton, R. R. (2013). Episodic memory in nonhuman animals. *Current Biology* 23, R801–R806.

Teng, S., and Whitney, D. (2011). The acuity of echolocation: Spatial resolution in sighted persons compared to the performance of an expert who is blind. *Journal of Visual Impairment and Blindness* 105, 20–32.

Thaler, L., Arnott, S. R., and Goodale, M. A. (2011). Neural correlates of natural human echolocation in early and late blind echolocation experts. *PLoS One* 6, e20162. doi:10.1371/journal.pone.0020162.

Thompson, E. (2007). *Mind in Life: Biology, Phenomenology, and the Science of Mind*. Cambridge, MA: Harvard University Press.

Thompson, E. (2009). Life and mind. In B. Clarke and M. B. N. Hansen (Eds.), *Emergence and Embodiment—New Essays on Second-Order Systems Theory*, pp. 77–93. Durham, NC: Duke University Press.

Thompson, R., and McConnell, J. (1955). Classical conditioning in the planarian *Dugesia dorotocephala. Journal of Comparative and Physiological Psychology* 48, 65–68. doi:10.1037/h0041147.

Thorndike, E. (1898). Some experiments on animal intelligence. *Science* 7, 818–824. doi:10.1126/science.7.181.818.

Tinbergen, N. (1951). *The Study of Instinct.* Oxford: Clarendon Press.

Tinbergen, N. (1963). On aims and methods of ethology. *Zeitschrift fürTierpsychologie* 20, 410–433.

Todes, D. P. (2014). *Ivan Pavlov.* Oxford: Oxford University Press.

Tomasello, M. (2014). *A Natural History of Human Thinking.* Cambridge, MA: Harvard University Press.

Tomer, R., Denes, A. S., Tessmar-Raible, K., and Arendt, D. (2010). Profiling by image registration reveals common origin of annelid mushroom bodies and vertebrate pallium. *Cell* 142, 800–809. doi:10.1016/j.cell.2010.07.043.

Tomsic, D., and Romano, A. (2013). A multidisciplinary approach to learning and memory in the crab *Neohelice (Chasmagnathus) granulata.* In R. Menzel and P. R. Benjamin (Eds.), *Invertebrate Learning and Memory,* pp. 337–355. London: Academic Press.

Tonegawa, S., Liu, X., Ramirez, S., and Redondo, R. (2015). Memory engram cells have come of age. *Neuron* 8, 918–931. doi:org/10.1016/j.neuron.2015.08.002.

Tononi, G. (2012). *Phi: A Voyage from the Brain to the Soul.* New York: Pantheon Books.

Tononi, G. (2015). Integrated information theory. *Scholarpedia* 10, 4164. doi:10.4249/scholarpedia.4164.

Tononi, G., Boly, M., Massimini, M., and Koch, C. (2016). Integrated information theory: From consciousness to its physical substrate. *Nature Reviews Neuroscience* 17, 450–461. doi:10.1038/nrn.2016.44.

Tononi, G., and Edelman, G. M. (1998). Consciousness and complexity. *Science* 282, 1846–1851.

Tononi, G., and Koch, C. (2008). The neural correlates of consciousness: An update. *Annals of the New York Academy of Sciences USA* 1124, 239–261. doi:10.1196/annals.1440.004.

Tononi, G., and Koch, C. (2015). Consciousness: Here, there and everywhere? *Philosophical Transactions of the Royal Society of London B* 370, 20140167. doi:10.1098/rstb.2014.0167.

Torley, V. J. (2007). The anatomy of a minimal mind. PhD thesis, University of Melbourne.

Tort, L., and Teles, M. (2011). The endocrine response to stress—a comparative view. In F. Akin (Ed.), *Basic and Clinical Endocrinology Up-to-Date*, pp. 263–286. Rijeka, Croatia: InTech.

Tosches, M. A., and Arendt, D. (2013). The bilaterian forebrain: An evolutionary chimera? *Current Opinion in Neurobiology* 23, 1080–1089.

Tosches, M. A., Bucher, D., Vopalensky, P., and Arendt, D. (2014). Melatonin signaling controls circadian swimming behavior in marine zooplankton. *Cell* 159, 46–57.

Tremblay, M. È. (2011). The role of microglia at synapses in the healthy CNS: Novel insights from recent imaging studies. *Neuron Glia Biology* 7, 67–76. doi:10.1017/S1740925X12000038.

Trestman, M. (2013a). Which comes first in major transitions: The behavioral chicken, or the evolutionary egg? *Biological Theory* 7, 48–55.

Trestman, M. (2013b). The Cambrian explosion and the origins of embodied cognition. *Biological Theory* 8, 80–92.

Trifonov, E. N. (2011). Vocabulary of definitions of life suggests a definition. *Journal of Biomolecular Structure and Dynamics* 29, 259–266.

Tripodi, M., and Arber, S. (2012). Regulation of motor circuit assembly by spatial and temporal mechanisms. *Current Opinion in Neurobiology* 22, 615–623. doi:10.1016/j.conb.2012.02.011.

Trut, L., Oskina I., and Kharlamova, A. (2009). Animal evolution during domestication: The domesticated fox as a model. *BioEssays* 31, 349–360. doi:10.1002/bies.200800070.

Tsakiris, M., and Haggard, P. (2005). The rubber hand illusion revisited: Visuotactile integration and self-attribution. *Journal of Experimental Psychology: Human Perception and Performance* 31, 80–91.

Tse, P. U. (2013). *The Neural Basis of Free Will*. Cambridge, MA: MIT Press.

Tseng, A. S., and Levin, M. (2013). Cracking the bioelectric code: Probing endogenous ionic controls of pattern formation. *Communicative and Integrative Biology* 6, e22595. doi:10.4161/cib.22595.

Tulving, E. (1972). Episodic and semantic memory. In E. Tulving and W. Donaldson (Eds.), *Organization of Memory*, pp. 381–403. San Diego, CA: Academic Press.

Tulving, E. (2002). Episodic memory: From mind to brain. *Annual Review of Psychology* 53, 1–25.

Turner-Evans, D. B., and Jayaraman, V. (2016). The insect central complex. *Current Biology* 26, R445–R460.

Tuthill, J. C., and Azim, E. (2018). Proprioception. *Current Biology* 28, R194–R203. doi:10.1016/j.cub.2018.01.064.

Tye, M. (2017). *Tense Bees and Shell-Shocked Crabs: Are Animals Conscious?* New York: Oxford University Press.

Valentinĉiĉ, T. (1978). Learning in the starfish *Marthasterias glacialis*. In D. S. McLusky and A. J. Berry (Eds.), *Physiology and Behavior of Marine Organisms*, pp. 303–309. Oxford: Pergamon Press.

Valentinĉiĉ, T. (1983). Innate and learned responses to external stimuli in asteroids. In M. Jangoux and J. M. Lawrence (Eds.), *Echinoderm Studies*, vol. 1, pp. 111–137. Rotterdam, The Netherlands: A. A. Balkema.

Valentine, J. W. (2004). *On the Origin of Phyla*. Chicago: University of Chicago Press.

Valentine, J. W., Jablonski, D., and Erwin, D. H. (1999). Fossils, molecules and embryos: New perspectives on the Cambrian explosion. *Development* 126, 851–859.

Van Duijn, M., Keijzer, F., and Franken, D. (2006). Principles of minimal cognition: Casting cognition as sensorimotor coordination. *Adaptive Behavior* 14, 157–170.

van Gaal, S., and Lamme, V. A. F. (2012). Unconscious high-level information processing: Implication for neurobiological theories of consciousness. *Neuroscientist* 18, 287–301.

Varela, F. J. (1997). Patterns of life: Intertwining identity and cognition. *Brain and Cognition* 34, 72–87.

Varela, F. J. (1999). The specious present: A neurophenomenology of time consciousness. In J. Petiot, F. J. Varela, B. Pachoud, and J-M. Roy (Eds.), *Naturalizing Phenomenology: Issues in Contemporary Phenomenology and Cognitive Science*, pp. 266–314. Stanford, CA: Stanford University Press.

Varela, F. J., Thompson, E., and Rosch, E. (1991). *The Embodied Mind: Cognitive Science and Human Experience*. Cambridge, MA: MIT Press.

Vavouri, T., and Lehner, B. (2009). Conserved noncoding elements and the evolution of animal body plans. *BioEssays* 32, 727–735. doi:10.1002/bies.200900014.

Vecera, S., and Gilds, K. (1998). What processing is impaired in apperceptive agnosia? Evidence from normal subjects. *Journal of Cognitive Neuroscience* 10, 568–550. doi:10.1162/089892998562979.

Volkoff, H. (2012). Sleep and orexins in nonmammalian vertebrates. *Vitamins and Hormones* 89, 315–339.

von der Malsburg, C. (1981). The correlation theory of brain function; Internal Report 81–82, Dept. of Neurobiology, Max-Planck-Institute for Biophysical Chemistry,

Göttingen, Germany. In E. Domany, J. L. van Hemmen, and K. Schulten (Eds.), *Models of Neural Networks II*. Berlin: Springer.

von Holst, E., and Mittelstaedt, H. (1950/1973). *The Reafference Principle: Interaction between the Central Nervous System and the Periphery* (R. Martin, Trans.). In *Selected Papers of Erich von Holst: The Behavioural Physiology of Animals and Man*, pp. 39–73. London: Methuen.

Vorster, A. P., Krishnan, H. C., Cirelli, C., and Lyons, L. C. (2014). Characterization of sleep in *Aplysia californica*. *Sleep* 37, 1453–1463. doi:10.5665/sleep.3992.

Waddington, C. H. (1957). *The Strategy of the Genes*. London: George Allen and Unwin.

Wagner, M. J., Kim, T. H., Savallhou, J., Schnitzer, M. J., and Luo, L. (2017). Cerebellar granule cells encode the expectation of reward. *Nature* 544, 96–113. doi:10.1038/nature21726.

Walling, P. T., and Hicks, K. N. (2003). Dimensions of consciousness. *Baylor University Medical Center Proceedings* 16, 162–166.

Walling, P. T., and Hicks, K. N. (2009). *Consciousness: Anatomy of the Soul*. Bloomington, IN: AuthorHouse.

Walters, E. T., and Byrne, J. H. (1983). Associative conditioning of single sensory neurons suggests a cellular mechanism for learning. *Science* 219, 405–408.

Wang, J., Ren, K., Pérez, J., Silva, A. J., and Peña de Ortiz, S. (2003). The antimetabolite ara-CTP blocks long-term memory of conditioned taste aversion. *Learning and Memory* 10, 503–509. doi:10.1101/lm.63003.

Ward, L. M. (2011). The thalamic dynamic core theory of conscious experience. *Consciousness and Cognition* 20, 464–486. doi:10.1016/j.concog.2011.01.007.

Watanabe, S., Kirino, Y., and Gelperin, A. (2008). Neural and molecular mechanism of microcognition in *Limax*. *Learning and Memory* 15, 633–642. doi:10.1101/lm920908.

Watson, J. B. (1925). *Behaviorism*. New York: People's Institute.

Watts, D. J., and Strogatz, S. H. (1998). Collective dynamics of "small-world" networks. *Nature* 393, 440–442.

Weil, R. S., and Rees, G. (2010). Decoding the neural correlates of consciousness. *Current Opinion in Neurology* 23, 649–655. doi:10.1097/WCO.0b013e32834028c7.

Weismann, A. (1889). *Essays upon Heredity*. Oxford: Clarendon Press.

Wells, M. (1968). Sensitization and the evolution of associative learning. In J. Salánki (Ed.), *Neurobiology of Invertebrates*, pp. 391–411. New York: Plenum.

West-Eberhard, M. J. (1983). Sexual selection, social competition, and speciation. *Quarterly Review of Biology* 58, 155–183.

West-Eberhard, M. J. (2003). *Developmental Plasticity and Evolution.* Oxford: Oxford University Press.

West-Eberhard, M. J. (2005). Phenotypic accommodation: Adaptive innovation due to developmental plasticity. *Journal of Experimental Zoology Part B: Molecular and Developmental Evolution* 304, 610–618.

Williamson, D. I. (2006). Hybridization in the evolution of animal form and life-cycle. *Zoological Journal of the Linnean Society* 148, 585–602. doi:10.1111/j.1096-3642 .2006.00236.x.

Willows, A. O. D., and Corning, W. C. (1973). The echinoderms. In W. C. Corning, J. A. Dyal, and A. O. D. Willows (Eds.), *Invertebrate Learning*, vol. 3, pp. 103–135. New York: Plenum.

Wimber, M., Alink, A., Charest, I., Kriegeskorte, N., and Anderson, M. C. (2015). Retrieval induces adaptive forgetting of competing memories via cortical pattern suppression. *Nature Neuroscience* 18, 582–589. doi:10.1038/nn.3973.

Wittgenstein, L. (1922). *Tractatus Logico-Philosophicus* (C. K. Ogden, Trans.). London: Routledge and Kegan Paul.

Wittgenstein, L. (1988). *Wittgenstein's Lectures on Philosophical Psychology 1946–1947* (P. T. Geach, Ed.). Chicago: University of Chicago Press.

Woldemichael, B. T., Bohacek, J., Gapp, K., and Mansuy, I. M. (2014). Epigenetics of memory and plasticity. In Z. U. Khan and E. C. Muly (Eds.), *Progress in Molecular Biology and Translational Science* 122, pp. 305–340. Amsterdam, The Netherlands: Elsevier.

Wolff, G. H., and Strausfeld, N. J. (2015). Genealogical correspondence of mushroom bodies across invertebrate phyla. *Current Biology* 25, 38–44.

Wolff, G. H., and Strausfeld, N. J. (2016). Genealogical correspondence of a forebrain centre implies an executive brain in the protostome-deuterostome bilaterian ancestor. *Philosophical Transactions of the Royal Society B* 371, 20150055. doi:10.1098/ rstb.2015.0055.

Wood, M. A., Hawk, J. D., and Abel, T. (2006). Combinatorial chromatin modifications and memory storage: A code for memory? *Learning and Memory* 13, 241–244. doi:10.1101/lm.278206.

Woodruff, M. L. (2017). Consciousness in teleosts: There is something it feels like to be a fish. *Animal Sentience* 13(1), 1–21.

Wray, G. A. (2015). Molecular clocks and the early evolution of metazoan nervous systems. *Philosophical Transactions of the Royal Society B* 370, 20150046. doi:10.1098/rstb.2015.0046.

Wright, C. (1873). *Evolution of Self-Consciousness.* 1st ed. published in *North American Review.* Reprinted in C. E. Norton (Ed.), *Philosophical Discussions by Chauncey Wright,* pp. 199–266. New York: Lennox Hill, 1877.

Wright, L. (1973). Functions. *Philosophical Review* 82, 139–168.

Wyles, J. S., Kunkel, J. G., and Wilson, A. C. (1983). Birds, behavior, and anatomical evolution. *Proceedings of the National Academy of Sciences USA* 80, 4394–4397.

Xiao, S., and Laflamme, M. (2009). On the eve of animal radiation: Phylogeny, ecology and evolution of the Ediacara biota. *Trends in Ecology and Evolution* 24, 31–40. doi:10.1016/j.tree.2008.07.015.

Yang, Y., Yamada, T., Hill, K. K., Hemberg, M., Reddy, N. C., Cho, H. Y., Guthrie, A. N., Oldenborg, A., Heiney, S. A., S. Ohmae, S., Medina, J. F., Holy, T. E., and Bonni, A. (2016). Chromatin remodeling inactivates activity genes and regulates neural coding. *Science* 353, 300–305. doi:10.1126/science.aad4225.

Yehuda, R., Daskalakis, N. P., Lehrner, A., Desarnaud, F., Bader, H. N., Makotkine, I., Flory, J. D., Bierer, L. M., and Meaney, M. J. (2014). Influences of maternal and paternal PTSD on epigenetic regulation of the glucocorticoid receptor gene in Holocaust survivor offspring. *American Journal of Psychiatry* 171, 872–880. doi:10.1176/appi.ajp.2014.13121571.

Yin, Z., Zhua, M., Davidson, M. H., Bottjerc, D. J., Zhaoa, F., and Tafforeau, P. (2015). Sponge grade body fossil with cellular resolution dating 60 Myr before the Cambrian. *Proceedings of the National Academy of Sciences USA* 112, E1453–E1460. doi:10.1073/pnas.1414577112.

Yokawa, K., Kagenishi, T., Pavlovič, A., Gall, S., Weiland, M., Mancuso, S., and Baluška, F. (2018). Anaesthetics stop diverse plant organ movements, affect endocytic vesicle recycling and ROS homeostasis, and block action potentials in Venus flytraps. *Annals of Botany.* doi: 10.1093/aob/mcx155.

Young, J. M., Wessnitzer, J., Armstrong, J. D., and Webb, B. (2011). Elemental and non-elemental olfactory learning in *Drosophila. Neurobiology of Learning and Memory* 96, 339–352.

Young, J. Z. (1979). Learning as a process of selection and amplification. *Journal of the Royal Society of Medicine* 72, 801–814.

Young, R. M. (1968). Association of ideas. In P. P. Wiener (Ed.), *Dictionary of the History of Ideas,* vol. 1, pp. 111–118. New York: Scribner's.

Zahavi, D. (2003). Phenomenology of self. In T. Kircher and A. David (Eds.), *The Self in Neuroscience and Psychiatry*, pp. 56–75. Cambridge: Cambridge University Press.

Zeh, D. W., Zeh, J. A., and Ishida, Y. (2009). Transposable elements and an epigenetic basis for punctuated equilibria. *BioEssays* 31, 715–726. doi:10.1002/bies.200900026.

Zeki, S. (2003). The disunity of consciousness. *Trends in Cognitive Sciences* 7, 214–218. doi:10.1016/S1364-6613(03)00081-0.

Zettle, R. D., Hayes, S. C., Barnes-Holmes, D., and Biglan, A. (2016). *The Wiley Handbook of Contextual Behavioral Science*. Chichester: Wiley-Blackwell and Sons.

Zhang, X., Shu, D., Han, J., Zhang, Z., Liu, J., and Fu, D. (2014). Triggers for the Cambrian explosion: Hypotheses and problems. *Gondwana Research* 25, 896–909.

Zhou, J., Nagarkatti, P., Zhong, Y., Ginsberg, J. P., Singh, N. P., Zhang, J., and Nagarkatti, M. (2014). Dysregulation in microRNA expression is associated with alterations in immune functions in combat veterans with post-traumatic stress disorder. *PLoS One* 9, e94075. doi:10.1371/journal.pone.0094075.

Ziv, Y., and Schwartz, M. (2008). Immune-based regulation of adult neurogenesis: Implications for learning and memory. *Brain, Behavior, and Immunity* 22, 167–176. doi:10.1016/j.bbi.2007.08.006.

Zovkic, I. B., Guzman-Karlsson, M. C., and Sweatt, J. D. (2013). Epigenetic regulation of memory formation and maintenance. *Learning and Memory* 20, 61–74. doi:10.1101/lm.026575.112.

Index